REAL LINEAR ALGEBRA

MONOGRAPHS AND TEXTBOOKS IN PURE AND APPLIED MATHEMATICS

1. *K. Yano,* Integral Formulas in Riemannian Geometry (1970) *(out of print)*
2. *S. Kobayashi,* Hyperbolic Manifolds and Holomorphic Mappings (1970) *(out of print)*
3. *V. S. Vladimirov,* Equations of Mathematical Physics (A. Jeffrey, editor; A. Littlewood, translator) (1970) *(out of print)*
4. *B. N. Pshenichnyi,* Necessary Conditions for an Extremum (L. Neustadt, translation editor; K. Makowski, translator) (1971)
5. *L. Narici, E. Beckenstein, and G. Bachman,* Functional Analysis and Valuation Theory (1971)
6. *D. S. Passman,* Infinite Group Rings (1971)
7. *L. Dornhoff,* Group Representation Theory (in two parts). Part A: Ordinary Representation Theory. Part B: Modular Representation Theory (1971, 1972)
8. *W. Boothby and G. L. Weiss (eds.),* Symmetric Spaces: Short Courses Presented at Washington University (1972)
9. *Y. Matsushima,* Differentiable Manifolds (E. T. Kobayashi, translator) (1972)
10. *L. E. Ward, Jr.,* Topology: An Outline for a First Course (1972) *(out of print)*
11. *A. Babakhanian,* Cohomological Methods in Group Theory (1972)
12. *R. Gilmer,* Multiplicative Ideal Theory (1972)
13. *J. Yeh,* Stochastic Processes and the Wiener Integral (1973) *(out of print)*
14. *J. Barros-Neto,* Introduction to the Theory of Distributions (1973) *(out of print)*
15. *R. Larsen,* Functional Analysis: An Introduction (1973) *(out of print)*
16. *K. Yano and S. Ishihara,* Tangent and Cotangent Bundles: Differential Geometry (1973) *(out of print)*
17. *C. Procesi,* Rings with Polynomial Identities (1973)
18. *R. Hermann,* Geometry, Physics, and Systems (1973)
19. *N. R. Wallach,* Harmonic Analysis on Homogeneous Spaces (1973) *(out of print)*
20. *J. Dieudonné,* Introduction to the Theory of Formal Groups (1973)
21. *I. Vaisman,* Cohomology and Differential Forms (1973)
22. *B.-Y. Chen,* Geometry of Submanifolds (1973)
23. *M. Marcus,* Finite Dimensional Multilinear Algebra (in two parts) (1973, 1975)
24. *R. Larsen,* Banach Algebras: An Introduction (1973)
25. *R. O. Kujala and A. L. Vitter (eds.),* Value Distribution Theory: Part A; Part B: Deficit and Bezout Estimates by Wilhelm Stoll (1973)
26. *K. B. Stolarsky,* Algebraic Numbers and Diophantine Approximation (1974)
27. *A. R. Magid,* The Separable Galois Theory of Commutative Rings (1974)
28. *B. R. McDonald,* Finite Rings with Identity (1974)
29. *J. Satake,* Linear Algebra (S. Koh, T. A. Akiba, and S. Ihara, translators) (1975)

30. *J. S. Golan,* Localization of Noncommutative Rings (1975)
31. *G. Klambauer,* Mathematical Analysis (1975)
32. *M. K. Agoston,* Algebraic Topology: A First Course (1976)
33. *K. R. Goodearl,* Ring Theory: Nonsingular Rings and Modules (1976)
34. *L. E. Mansfield,* Linear Algebra with Geometric Applications: Selected Topics (1976)
35. *N. J. Pullman,* Matrix Theory and Its Applications (1976)
36. *B. R. McDonald,* Geometric Algebra Over Local Rings (1976)
37. *C. W. Groetsch,* Generalized Inverses of Linear Operators: Representation and Approximation (1977)
38. *J. E. Kuczkowski and J. L. Gersting,* Abstract Algebra: A First Look (1977)
39. *C. O. Christenson and W. L. Voxman,* Aspects of Topology (1977)
40. *M. Nagata,* Field Theory (1977)
41. *R. L. Long,* Algebraic Number Theory (1977)
42. *W. F. Pfeffer,* Integrals and Measures (1977)
43. *R. L. Wheeden and A. Zygmund,* Measure and Integral: An Introduction to Real Analysis (1977)
44. *J. H. Curtiss,* Introduction to Functions of a Complex Variable (1978)
45. *K. Hrbacek and T. Jech,* Introduction to Set Theory (1978)
46. *W. S. Massey,* Homology and Cohomology Theory (1978)
47. *M. Marcus,* Introduction to Modern Algebra (1978)
48. *E. C. Young,* Vector and Tensor Analysis (1978)
49. *S. B. Nadler, Jr.,* Hyperspaces of Sets (1978)
50. *S. K. Segal,* Topics in Group Rings (1978)
51. *A. C. M. van Rooij,* Non-Archimedean Functional Analysis (1978)
54. *L. Corwin and R. Szczarba,* Calculus in Vector Spaces (1979)
53. *C. Sadosky,* Interpolation of Operators and Singular Integrals: An Introduction to Harmonic Analysis (1979)
54. *J. Cronin,* Differential Equations: Introduction and Quantitative Theory (1980)
55. *C. W. Groetsch,* Elements of Applicable Functional Analysis (1980)
56. *I. Vaisman,* Foundations of Three-Dimensional Euclidean Geometry (1980)
57. *H. I. Freedman,* Deterministic Mathematical Models in Population Ecology (1980)
58. *S. B. Chae,* Lebesgue Integration (1980)
59. *C. S. Rees, S. M. Shah, and C. V. Stanojević,* Theory and Applications of Fourier Analysis (1981)
60. *L. Nachbin,* Introduction to Functional Analysis: Banach Spaces and Differential Calculus (R. M. Aron, translator) (1981)
61. *G. Orzech and M. Orzech,* Plane Algebraic Curves: An Introduction Via Valuations (1981)
62. *R. Johnsonbaugh and W. E. Pfaffenberger,* Foundations of Mathematical Analysis (1981)
63. *W. L. Voxman and R. H. Goetschel,* Advanced Calculus: An Introduction to Modern Analysis (1981)
64. *L. J. Corwin and R. H. Szcarba,* Multivariable Calculus (1982)
65. *V. I. Istrătescu,* Introduction to Linear Operator Theory (1981)
66. *R. D. Järvinen,* Finite and Infinite Dimensional Linear Spaces: A Comparative Study in Algebraic and Analytic Settings (1981)

67. *J. K. Beem and P. E. Ehrlich,* Global Lorentzian Geometry (1981)
68. *D. L. Armacost,* The Structure of Locally Compact Abelian Groups (1981)
69. *J. W. Brewer and M. K. Smith, eds.,* Emmy Noether: A Tribute to Her Life and Work (1981)
70. *K. H. Kim,* Boolean Matrix Theory and Applications (1982)
71. *T. W. Wieting,* The Mathematical Theory of Chromatic Plane Ornaments (1982)
72. *D. B. Gauld,* Differential Topology: An Introduction (1982)
73. *R. L. Faber,* Foundations of Euclidean and Non-Euclidean Geometry (1983)
74. *M. Carmeli,* Statistical Theory and Random Matrices (1983)
75. *J. H. Carruth, J. A. Hildebrant, and R. J. Koch,* The Theory of Topological Semigroups (1983)
76. *R. L. Faber,* Differential Geometry and Relativity Theory: An Introduction (1983)
77. *S. Barnett,* Polynomials and Linear Control Systems (1983)
78. *G. Karpilovsky,* Commutative Group Algebras (1983)
79. *F. Van Oystaeyen and A. Verschoren,* Relative Invariants of Rings: The Commutative Theory (1983)
80. *I. Vaisman,* A First Course in Differential Geometry (1984)
81. *G. W. Swan,* Applications of Optimal Control Theory in Biomedicine (1984)
82. *T. Petrie and J. D. Randall,* Transformation Groups on Manifolds (1984)
83. *K. Goebel and S. Reich,* Uniform Convexity, Hyperbolic Geometry, and Nonexpansive Mappings (1984)
84. *T. Albu and C. Năstăsescu,* Relative Finiteness in Module Theory (1984)
85. *K. Hrbacek and T. Jech,* Introduction to Set Theory, Second Edition, Revised and Expanded (1984)
86. *F. Van Oystaeyen and A. Verschoren,* Relative Invariants of Rings: The Noncommutative Theory (1984)
87. *B. R. McDonald,* Linear Algebra Over Commutative Rings (1984)
88. *M. Namba,* Geometry of Projective Algebraic Curves (1984)
89. *G. F. Webb,* Theory of Nonlinear Age-Dependent Population Dynamics (1985)
90. *M. R. Bremner, R. V. Moody, and J. Patera,* Tables of Dominant Weight Multiplicities for Representations of Simple Lie Algebras (1985)
91. *A. E. Fekete,* Real Linear Algebra (1985)

Other Volumes in Preparation

REAL LINEAR ALGEBRA

Antal E. Fekete
Memorial University of Newfoundland
St. John's, Newfoundland
Canada

MARCEL DEKKER, INC.   New York and Basel

Library of Congress Cataloging in Publication Data

Fekete, Antal E.
Real linear algebra.

(Monographs and textbooks in pure and applied mathematics; 91)
Bibliography: p.
Includes index.
1. Algebras, Linear. I. Title. II. Series: Monographs and textbooks in pure and applied mathematics; v. 91.
QA184.F45 1985 512′.5 84-23032
ISBN 0-8247-7238-5

MARCEL DEKKER, INC.
270 Madison Avenue, New York, New York 10016

Current printing (last digit):
10 9 8 7 6 5 4 3 2 1

PRINTED IN THE UNITED STATES OF AMERICA

Dedicated
to the memory of
Norman E. Steenrod

Norman E. Steenrod
(1910 – 1971)

Preface

In 1974 Professor D. C. Spencer of Princeton University invited me to spend a year at Princeton and he put at my disposal that part of the mathematical archives of his friend, the late Norman Earl Steenrod, which had to do with sophomore mathematics courses. I found a veritable treasure trove in those archives. Steenrod had written an axiomatic linear algebra text, published in 1959 as the first five chapters of the pioneering and influential *Advanced Calculus* by H. K. Nickerson, D. C. Spencer, and N. E. Steenrod. The archives revealed that Steenrod had started collecting material for a naive linear algebra textbook, shifting the emphasis from algebraic rigor to geometric intuition. His untimely death in 1971, at the age of 61, prevented him from carrying out his cherished plan. While at Princeton, I decided that I would complete his unfinished task.

Why this is an important task may best be told in Steenrod's own words, spoken in 1967 in Santa Barbara, California:

> Although geometry pervades all mathematics and is present at every stage of development, too often we fail to point this out to our students. We rely on analytic formulations since we realize that they are complete and we are in a hurry to get on. We do not take time to look at geometric formulations. We are too greatly impressed by the rigor of analysis. We seem to feel that geometry is not rigorous, or at least that the background needed for rigor is not available. We feel that it is better not to do anything that is not rigorous. When we do present geometry, it is too often the instructor who does the geometry while the student is merely a passive spectator. We present the geometry to him in order to explain the analysis, but then we require him to do only the analysis—no geometry. We tend to avoid geometric formulations of

questions in examinations. Questions are hard to formulate geometrically. Almost every time you try such a question, you find that a large group of students misinterprets it. Such questions are hard to grade because the answers are so varied. The absence of geometric questions on final exams tends to degrade the geometric content of the course, and leads to its neglect.

What has bothered me through the years is the control the exam seems to have of the course. Somehow the tail wags the dog. In the exam we are supposed to take a sample of what the student knows. This process of sampling has a feedback effect that is very serious. The most famous example is the College Board exam and its influence on the teaching of mathematics in secondary schools. The examiners, in order to be fair to students in all parts of the country, tended to take the intersection of the topics taught in various schools. In the 1920's and 1930's the exam had little effect on the teaching of mathematics, but by the early fifties the feedback effect became pronounced. A greater number of students was taking the exams, and schools were rated by the results. If a particular high school had a poor rating, they did something about it; they compared carefully what they were teaching with the kinds of questions asked on the exams; they altered their curriculum accordingly, and concentrated on topics of maximum frequency. The examiners, on their part, observed the shrinkage and narrowed the range of their questions accordingly. At one time it was projected that after forty years only one topic would survive this elimination process, and that would be the factoring of quadratics.

Some say that this cannot happen in college because the instructor is in charge of his course. Well, he is not, because in many colleges there are freshman courses with large enrollments and many sections. To avoid troubles with young instructors giving wide varieties of grades we insist on uniform exams and uniform grading. I have seen the feedback effect time and again while teaching a section of the freshman course. Along comes a bright fresh Ph.D. teaching his first class. Knowing that the concept of limit is central to the calculus, he settles down and does a good job of teaching limits for two months. His students do very well on that one question, but not so well on the other four of a more routine nature. The average for his students is ten points below the overall average, so he finds himself giving D's to students he thought were pretty good. Having learned his lesson, he runs a statistical analysis on the final exams for the last five years, and starts teaching his students how to turn the crank. By the end of the semester he normally brings their average up to where it should be.

I do not know how to defeat this, but I do have one suggestion to offer. Harness the feedback effect to upgrade geometry by putting more of the geometric questions into their final exam and then face the problem of grading them. If, in the earlier parts of the course, on the ten-minute quizzes and the homework, you have inflicted geometry on the students over and over again, then on the final exam you have some chance of getting a good reaction out of the geometric questions.

This is, of course, not the only, nor even the main reason why to bother with geometry at all. The main reason is that most problems are presented in geometric form in the first place. Reformulation and solution in analytic terms is merely a second step. To complete the process, there is an indispensable third, namely, the interpretation of the analytic solution in geometric terms. There is another reason, which is psychological. Two views of the same thing reinforce one another. Most of us are able to remember the multitudinous formulas of analysis mainly because we attach to each a geometric picture that keeps us from going astray. Even better than that, the geometric view of a problem helps us to focus on the invariants and to weed out the irrelevant details. A poor choice of coordinates may lead to a horrible mess in the analytic formulation, but with some geometric insight we may be able to choose the "best" coordinate system.

Now we come to the question of textbooks. The situation here is definitely sad. Books on linear algebra are written by algebraists for algebra students. Of course, this is not in itself a condemnation, but there are two features that show up from this. One is that geometry is treated in an offhand fashion, if it is treated at all. Secondly, most textbooks confuse the presentation of theory with the techniques of computation via matrices. One way to see that a book makes this mistake is to see that it has a chapter on determinants and matrices before linear transformations are defined. The linear transformation is an easy conceptual thing to talk about and give examples of without matrices. The matrix is a tool for computation, i.e., it is a set of coordinates subject to prior choice of a basis. The matrix, important as it may be for computation, is of *no* importance in the theoretical or conceptual part of the course, nor in the geometric pictures that come along. Presenting semi-theoretical material on matrices, and teaching the students to work with matrices before introducing linear transformations, is comparable to trying to teach someone to play the piano on a keyboard that isn't attached to any strings. There is no feedback, the student does not see the objective and finds no pleasure in what he is doing. True, historically matrices came first. For a long time a vector space was an $\mathbb{R}^n$ for some n and a linear transformation was a system of linear equations represented by the matrix of coefficients.

> Thus properties of linear transformations had to be formulated as properties of matrices. In this way an extensive theory of matrices arose. It is a cumbersome theory both in notation and conception. The conceptual point of view, that one could proceed on a different level and work without coordinates, developed during the twenties and thirties. It became clear that the matrix theory tended to obscure the geometric insight. With the new point of view the picture became quite easy and lovely, and the theory was disassociated from the mechanism of computation. Thus it is easy to see why the first books on linear algebra had to begin with matrices and determinants, but it seems to me that the conversion to the more recent and simpler view has been much too slow. It should be made clear to the student that matrices are not essential to understanding the theory and that the theory must not be confused with the computations which arise.
>
> Another inadequacy of many texts is that the structure theorems for linear operators are usually given only in the complex case. This case is algebraically easier and smoother because the characteristic equation splits into the product of linear factors. The details of the real case are omitted, in spite of the fact this is the case of interest because of the geometry, and it should be studied carefully before one proceeds to the complex case.

This book has adopted Steenrod's Santa Barbara Program without reservations. As the title *Real Linear Algebra* suggests, it takes an unhurried look at the vector spaces over the real numbers, in particular, at the three-dimensional real vector space $\mathbb{R}^3$ in which a major part of our lives is played out. It highlights the remarkable fact that the cross product converts $\mathbb{R}^3$ into a noncommutative Lie algebra that is compatible with the metric—an exclusive feature restricted to $n = 3$. Moreover, this Lie algebra is isomorphic to the Lie algebra of antisymmetric operators and, therefore, it is the Lie algebra of the group of rotations in three-dimensional space. Vector spaces over the field of complex numbers are not treated in this book. I have felt that this case was adequately covered in the existing literature.

The isomorphism between the algebra of linear operators and the algebra of matrices is put on the same footing as the isomorphism between the n-dimensional real vector space V and $\mathbb{R}^n$, the vector space of coordinates. Pseudoconcepts such as eigenvalues and eigenvectors of a *matrix*, and similarity or diagonalization of matrices are ignored altogether.

The geometric background to symmetric, antisymmetric, and orthogonal operators is provided. Of these, the antisymmetric case is omitted from most accounts. This oversight can hardly be condoned in view of the extraordinary importance of the Lie algebra of antisymmetric operators which gives rise to the orthogonal group under the exponential functor, and so it is the only reasonable candidate to generalize the cross product to higher dimensions.

I have furnished a large number of numerical exercises on orthogonal operators featuring matrices with *rational* coefficients. In this effort I was inspired by my own frustrations as an instructor in search for practice material. The perusal of all the linear algebra textbooks available in the university library turned up about half a dozen overlapping examples of orthogonal matrices, all of them (apart from trivial exceptions) having irrational coefficients and as such, useless for my purposes.

I have made an effort to segregate affine and metric ideas, such as parallelism and volume on the one hand, perpendicularity and length on the other. The organization of the material into chapters is also designed to serve this purpose. In teaching linear algebra from other introductory textbooks I have found it disturbing that it was impossible to disentangle ideas that depended on the dot product from those that did not. The abstract idea of orientation, and the concept of a normal subgroup, are given a geometric interpretation in the last chapter which I hope the instructor will find a challenge to teach.

While writing this book and trying to teach a course based on these ideas, I have found that the Santa Barbara Program is in many ways quite ambitious. Therefore, I must accept full responsibility for any inadequacy of this book. I am painfully aware of my own limitations and know that Steenrod, had he lived, would have put his brilliant program into effect far better than I ever could.

It remains for me to register my manifold indebtedness in connection with this project. Above all, my heartfelt thanks go to Professor Spencer, now in happy retirement in his beloved Rocky Mountains in Colorado, for his original suggestion that I might be the person to look at the Steenrod heritage concerning mathematical didactics. The Department of Mathematics at Princeton University was my gracious host in 1974-75, and while there I greatly benefited from a course by, and from many a private discussion with, Professor J. C. Moore, a collaborator of Steenrod. I have also been privileged by several private discussions with Dr. H. K. Nickerson of Rutgers University, another collaborator of Steenrod. She revealed many important historical details about the evolution of Steenrod's thinking on the teaching of sophomore courses in mathematics, which I could not otherwise have learned. I am also grateful to Acadia University in Wolfville, Nova Scotia, for having invited me in 1970-71 to teach an experimental course on linear algebra, from which many of the ideas presented here have originally sprung. Dr. Maurits Dekker, the Chairman of the Board, and Dr. Earl J. Taft, Executive Editor for the mathematics series of Marcel Dekker,

Inc., have taken an unusually friendly interest in this project, and my sincere thanks go to both of them. The editorial staff at Marcel Dekker, Inc., has made my job not only easier but in many ways outright pleasurable. The Mathematical Association of America, Inc., has kindly allowed me to quote extensively in this preface from an article entitled "The Geometric Content of Freshman and Sophomore Mathematics Courses" by N. E. Steenrod, of which the Association holds the copyright (see References and Selected Readings at the end of this book).

I dedicate this work to the memory of Norman Earl Steenrod, one of the greatest mathematicians this continent has given to the world. By courtesy of his widow, Mrs. Caroline Rosenblum, I am proud to be able to present a most impressive portrait of this great teacher for the inspiration of the new generation of young mathematicians. My hope is that his love of mathematics, and his love for those wanting to learn mathematics, will be conveyed through these pages. May my readers reach the love of mathematics through his example!

Antal E. Fekete

Contents

The Greek Alphabet

The Greek alphabet is of worldwide importance as a source of symbols in mathematics.

α	alpha	ν	nu
β	beta	ξ	xi
γ	gamma	o	omicron
δ	delta	π	pi
ε	epsilon	ρ	rho
ζ	zeta	σ	sigma
η	eta	τ	tau
θ	theta	υ	upsilon
ι	iota	ϕ	phi
κ	kappa	χ	chi
λ	lambda	ψ	psi
μ	mu	ω	omega

List of Applications

14 The Exponential Functor

Appendix 1: Numerical Methods of Linear Algebra: Determinants; Gaussian Elimination

REAL LINEAR ALGEBRA

0
Introductory Remarks Concerning Logic

THE NOTION OF A FUNCTION

In the present approach to linear algebra the notion of a set and that of a function will remain undefined. The descriptions that follow should not be treated as definitions as they do no more than substitute one word by a synonym.

A *set* X is a collection of elements. X is considered as given if we have a way of telling, for every object x, whether x belongs to X ($x \in X$) or not ($x \notin X$).

Examples

The set *Co* of all countries of the world
The set *Ci* of all cities of the world
The set *Pe* of all people of the world (presently living or deceased)

By contrast, the following are not sets:

The aggregate of people living in 1999
The aggregate of those sets X satisfying $X \notin X$

because we have no way, in principle or in practice, of establishing a membership list. The latter counterexample is quite famous and its discovery caused a veritable revolution in mathematics at the turn of the century. It is known as Russell's† paradox.

In mathematics, the most frequently occurring sets have standard symbols to denote them, such as:

†Bertrand Russell (1872-1970), English philosopher and mathematician, one of the greatest logicians of all time. More on Russell's paradox can be found, for example, in *Set Theory* by C. Pinter (1971, pp. 3, 11).

$\mathbb{N}$ is the set of nonnegative integers.
$\mathbb{Z}$ is the set of all integers.
$\mathbb{Q}$ is the set of rational numbers.
$\mathbb{R}$ is the set of real numbers.
$\mathbb{C}$ is the set of complex numbers.

Let X and Y be sets. A *function* f from X to Y, in symbols $f\colon X \to Y$, is a rule which to every element $x \in X$ makes a well-defined element, denoted $y = f(x) \in Y$, correspond. We shall also use the notation $x \rightsquigarrow f(x)$ to indicate this correspondence. The set X is called the *domain*, and Y the *codomain* of f; in symbols, $X = \operatorname{dom} f$, $Y = \operatorname{cod} f$.

Examples

capital: $Co \to Ci$, e.g., capital (U.S.) = Washington, D.C.
location: $Ci \to Co$, e.g., location (New York) = U.S., location (York) = U.K.

Indeed, every country has a well-defined capital city, i.e., the seat of its government. Also, every city is located in a well-defined country.

sin: $\mathbb{R} \to \mathbb{R}$, e.g., $\sin 0° = 0$, $\sin 30° = ½$, $\sin 90° = 1$

It is incorrect to refer to "the function $\sin x$." In fact, $\sin x$ is the *value* that the function sin assumes at the *place* x. If we find it awkward to talk about "the function sin" we may stay within the bounds of correct terminology by devising expressions such as "the function f such that $f(x) = \sin x$" or "the function $x \rightsquigarrow \sin x$." Similarly, in the case of the exponential function, instead of saying "the function e^x" we should say "the function f such that $f(x) = e^x$."

A function $f\colon X \to X$ (i.e., with $\operatorname{dom} f = \operatorname{cod} f$) is often called an *operator* on X.

Examples

mother: $Pe \to Pe$, e.g., mother (Mr. Smith, Jr.) = Mrs. Smith, Sr.
father: $Pe \to Pe$, e.g. father (Mr. Smith, Jr.) = Mr. Smith, Sr.

It is necessary to check any given rule, whether or not it qualifies as a function. For example, the relation "son" is not a function from the set of fathers to the set of people, because whereas some fathers have no sons, others may have several. Similarly, the rule $\sqrt{}\colon \mathbb{R} \to \mathbb{R}$ fails to qualify as a function, since negative numbers have no (real) square root, whereas positive numbers have two.

Intuitively, we may think of a function $f\colon X \to Y$ as a round of hits from X to Y, as if every element x in X were to throw one (and only one) ball into the target area Y, hitting the spot $y = f(x)$. Then the letter f, without any reference to x, is used to denote the complete scoreboard for this particular round of throws. This picture is helpful in allowing us to visualize correctly two patterns of deviant behavior from the standard of a one-to-one correspondence: namely, (i) it is possible that two elements $x_1 \neq x_2$ in X hit the same spot in Y: $f(x_1) = f(x_2)$; and (ii) it is possible that some spot in Y does not get hit at all: for some $y \in Y$ and for all $x \in X$ we may have $y \neq f(x)$.

Suppose that we have two functions $f: X \to Y$ and $g: Y \to Z$ (mark that $\text{cod} f = \text{dom}\, g$). Then we define the *product* of f and g; in symbols, $g \circ f: X \to Z$ by the formula $(g \circ f)(x) = g(f(x))$ for all $x \in X$. Indeed, $g \circ f$ makes a unique assignment in Z to every element in X, so it is a function from X to Z.

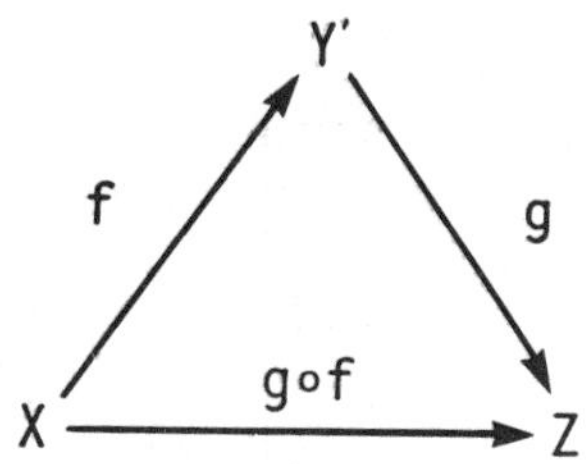

Examples

father ∘ mother = maternal grandfather
mother ∘ father = paternal grandmother

These examples also show that $g \circ f = f \circ g$ may fail; i.e., the product of functions (unlike the product of real numbers) is *not commutative*. However, the product of functions is *associative* (like the product of real numbers):

$$h \circ (g \circ f) = (h \circ g) \circ f$$

where $h: Z \to U$ is a third function. This is so because both functions $h \circ (g \circ f)$ and $(h \circ g) \circ f$ make the same assignment, namely $h(g(f(x)))$, to every $x \in X$.

Another example of the product function is

$$\text{location} \circ \text{capital}: Co \to Co$$

which to every country assigns the self-same country. This is so because the capital city of every country, without exception, is located in the same country. This function is denoted by the symbol 1_{Co}: $Co \to Co$, e.g., 1_{Co}(Canada) = Canada.

More generally, the function 1_X: $X \to X$ assigning to every element $x \in X$ the same element x, $1_X(x) = x$, is called the *identity operator* of the set

X. Thus every set X has its own identity operator 1_X; that of $\mathbb{R}$, $1_{\mathbb{R}}$, is the familiar linear function represented in the coordinate system by a straight line passing through the origin with a slope equal to 1. It is easy to see that $f \circ 1_X = f$ and $1_Y \circ f = f$ where f: $X \to Y$ is an arbitrary function.

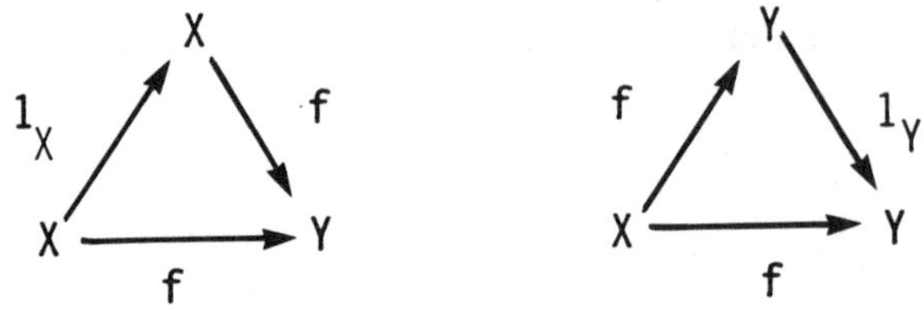

INJECTIONS AND SURJECTIONS

We shall say that a function f: $X \to Y$ is *injective* (*surjective*, *bijective*, respectively) or that f is an *injection* (*surjection*, *bijection*, respectively) if for every $y \in Y$ there exists *at most one* (*at least one*, *one and only one*, respectively) $x \in X$ such that $y = f(x)$. Intuitively, this means that under an injection f two different elements $x_1 \neq x_2$ in X always hit different spots: $f(x_1) \neq f(x_2)$ in Y [whereas a function g that fails to be injective would have a pair of elements $x_1 \neq x_2$ in its domain X which hit the same spot $g(x_1) = g(x_2)$ in the codomain Y]. Similarly, under a surjection f every spot $y \in Y$ gets hit; i.e., $y = f(x)$ for some $x \in X$ [whereas a function g that fails to be surjective would have a spot $y \in Y$ which does not get hit; i.e., $y \neq f(x)$ for any $x \in X$].

The terminology of injections, surjections, and bijections is borrowed from the French language. The original English terms were "one-to-one," "onto," and "one-to-one and onto," respectively.

Examples

(1) capital: $Co \to Ci$ is injective (but not surjective) because two different countries cannot have the same capital city (but not every city is the capital of a country).
(2) location: $Ci \to Co$ is surjective (but not injective) because every country has at least one city located within its boundaries—i.e., its capital city (but some countries have more than one city located within their boundaries).
(3) husband: $Wo \to Me$, where Wo is the set of married women and Me is the set of married men, is a bijective function. Note, however, that under Islamic law allowing a man to take several wives, "husband" would fail to be bijective because it would fail to be injective (although it would still be surjective).

(4) father: $Pe \to Pe$ is neither injective, because siblings have the same father, nor surjective, because not everybody is a father. Thus, for the stronger reason, "father" is not bijective either.
(5) 1_X: $X \to X$, the identity operator of X is bijective.
(6) $g \circ f$: $X \to Z$, the product function, is injective (surjective, bijective, respectively) provided that both f and g are injective (surjective, bijective, respectively).
(7) sin: $\mathbb{R} \to \mathbb{R}$ is neither injective nor surjective: $\sin 0° = \sin 180°$; $\sin x = 2$ has no solution for x in $\mathbb{R}$.
(8) f: $\mathbb{R} \to \mathbb{R}$ such that $f(x) = 2^x$ is injective but not surjective: $x_1 < x_2 \Rightarrow 2^{x_1} < 2^{x_2}$, but $2^x = -1$ has no solution for x in $\mathbb{R}$.
(9) f: $\mathbb{R} \to \mathbb{R}$ such that $f(x) = x^3 - x$ is surjective (as is every polynomial of *odd* degree). It is not injective, however, as $f(1) = 0 = f(-1)$.
(10) f: $\mathbb{R} \to \mathbb{R}$ such that $f(x) = x^3$ is bijective.

SUFFICIENT CONDITION. NECESSARY CONDITION

A *theorem*, at its simplest, is a combination of two propositions the first of which is called the *condition* and the second, the *desideratum*. A *proposition* is a sentence restricting a variable x. For example, "x is a red ball," "y is a bald man," and "$z \in \mathbb{Z}$" are propositions.

Theorems may be classified according as the condition they incorporate is sufficient, necessary, or, as indeed may happen in some lucky cases, both necessary and sufficient. We start with some nonmathematical "theorems" as a first example. In each case, the first proposition is the condition and the second the desideratum.

Examples

(1) x is a boy 18 years old $\Rightarrow$ x is a teenager. The condition here is sufficient but not necessary. It is not hard to weaken it to a necessary and sufficient condition:
(2) x is a boy or girl between 13 and 19 $\Leftrightarrow$ x is a teenager.
(3) Let us approach from the opposite direction: x is a woman living in France $\Leftarrow$ x is a Parisienne. The condition that x be a woman living in France is a necessary condition for the desideratum, of x be a Parisienne, to be true. However, that condition is far from sufficient; indeed, under that condition x could be a resident of Lyon.
(4) Let us strengthen the conditions in example 3: x is a female resident of France's capital $\Leftrightarrow$ x is a Parisienne. This condition is both necessary and sufficient.

Let us add some mathematical examples, already mentioned in another context.

$$f \text{ and } g \text{ injective} \Rightarrow g \circ f \text{ injective}$$
$$f \text{ and } g \text{ surjective} \Rightarrow g \circ f \text{ surjective}$$
$$f \text{ and } g \text{ bijective} \Rightarrow g \circ f \text{ bijective}$$

Proof: We prove the first two theorems.

$$\begin{aligned}(g \circ f)(x_1) = (g \circ f)(x_2) &\Rightarrow g(f(x_1)) = g(f(x_2)) \\ &\Rightarrow f(x_1) = f(x_2) \qquad \text{(because } g \text{ is injective)} \\ &\Rightarrow x_1 = x_2 \qquad \text{(because } f \text{ is injective)}\end{aligned}$$

Since $(g \circ f)(x_1) = (g \circ f)(x_2) \Rightarrow x_1 = x_2$, we conclude that $g \circ f$ is injective, proving the first theorem. Now we prove the second.

$$\begin{aligned}z \in Z &\Rightarrow z = g(y), \quad y \in Y \\ &\Rightarrow y = f(x), \quad x \in X \\ &\Rightarrow z = g(f(x)) = (g \circ f)(x)\end{aligned}$$

Since this is true for every $z \in Z = \text{cod}(g \circ f)$, we conclude that $g \circ f$ is surjective.

In the three propositions above, the desideratum is that $g \circ f$, the product of two functions, be injective (surjective, bijective, respectively). The condition that both f and g be injective (surjective, bijective, respectively) is sufficient. The example of the bijective function

$$\text{location} \circ \text{capital} = 1_{Co}$$

where "location" is not injective and "capital" is not surjective, shows that these conditions are far from being necessary.

In ordinary speech the particles "therefore," "hence," "if...then..." are used to join a sufficient condition to the desideratum, while the particles "because," "since," "only if...then..." are used to join a necessary condition to its desideratum. In the case of a necessary and sufficient condition the particle is "if, and only if." (Try each particle in the appropriate examples above!) In mathematics, where the formalism is all-pervasive, usage is standardized by the introduction of the symbol $\Rightarrow$ (read: "implies") to join a sufficient condition, the symbol $\Leftarrow$ (read: "is implied by") to join a necessary condition to the desideratum. The symbol $\Leftrightarrow$ (read: "is equivalent to") is reserved for the purposes of joining the two members of a necessary and sufficient condition.

Ideally, all theorems should express a condition that is both necessary and sufficient. Although this goal is not unattainable in everyday life situations, in mathematics the problem of finding a necessary and sufficient condition for a certain given desideratum could sometimes be very difficult, indeed impossible, without further refinement of the theory. It is precisely the solution of such problems in various areas which is the driving force behind mathematical research. What mathematicians do is this: They choose the desideratum. It is easy to come up with a sufficient condition that is much too strong, or with a necessary condition that is much too weak, for any particular desideratum. Then mathematicians try to weaken the sufficient condition and to strengthen the necessary condition, trusting to luck that they may strike it rich in between. If this is of no avail, mathematicians have to settle down to the laborious process of refining the theory to the point where, with the aid of new and more refined concepts, the desideratum can be captured. For example, let the desideratum be that the product $g \circ f$ of two functions f and g is injective. A sufficient condition, which is not necessary, is that f and g are both injective. A necessary condition, which is not sufficient, is that f (the function that applies first) be injective. It would take further development of the theory of sets and functions (which will be done in the exercises at the end of this introduction) if we wanted to close the gap between the two theorems, and state a necessary and sufficient condition for the product of two functions to be injective.

When it comes to proving theorems, we, in practice, always prove a sufficient condition. In fact, every necessary condition can be turned into a sufficient condition by the simple device of interchanging the condition and the desideratum and simultaneously replacing $\Leftarrow$ by $\Rightarrow$. For example, we want to prove that "f injective" is a necessary condition for the desideratum "$g \circ f$ injective," i.e.,

$$f \text{ injective} \Leftarrow g \circ f \text{ injective}$$

Instead of this necessary condition, we may prove the equivalent sufficient condition:

$$g \circ f \text{ injective} \Rightarrow f \text{ injective}$$

To prove that f is injective means to show that $f(x_1) = f(x_2) \Rightarrow x_1 = x_2$. This we can do as follows:

$$\begin{aligned} f(x_1) = f(x_2) &\Rightarrow g(f(x_1)) = g(f(x_2)) && \text{(because } g \text{ is a function)} \\ &\Rightarrow (g \circ f)(x_1) = (g \circ f)(x_2) && \text{(by the definition of } g \circ f\text{)} \\ &\Rightarrow x_1 = x_2 && \text{(because } g \circ f \text{ is injective)} \end{aligned}$$

This completes the proof that the injectivity of f is a necessary condition for the injectivity of $g \circ f$.

Similarly, in proving the necessary condition

$$g \text{ surjective} \Leftarrow g \circ f \text{ surjective}$$

we prove the equivalent sufficient condition

$$g \circ f \text{ surjective} \Rightarrow g \text{ surjective}$$

The surjectivity of g will be proved if we show that $z \in Z \Rightarrow z = g(y)$ for some $y \in Y$. This we can do as follows:

$$\begin{aligned} z \in Z \Rightarrow z &= (g \circ f)(x) \quad \text{for some } x \in X && \text{(because } g \circ f \text{ is surjective)} \\ &= g(f(x)) && \text{(by the definition of } g \circ f) \\ &= g(y) \quad \text{with } y = f(x) \in Y \end{aligned}$$

As a consequence of the two necessary conditions, we also have that

$$f \text{ injective and } g \text{ surjective} \Leftarrow g \circ f \text{ bijective}$$

i.e., the injectivity of f (the function that applies first) *and* the surjectivity of g (the function that applies last) is a necessary condition for the bijectivity of $g \circ f$. More than that cannot be said, since we have the example of a bijective function:

$$\text{location} \circ \text{capital} = 1_{Co}$$

where "location" is not injective and "capital" is not surjective.

This simple relationship between a necessary condition and its companion sufficient condition can be expressed in terms of a logical formula:

$$(A \Rightarrow B) \Leftrightarrow (B \Leftarrow A)$$

A, B signify propositions; $A \Rightarrow B$ is a theorem. The logical formula above asserts that the theorems $A \Rightarrow B$ and $B \Leftarrow A$ are equivalent. Theorems of the form $A \Rightarrow B$, called *implications*, are most common.

The implication $B \Rightarrow A$, obtained from the implication $A \Rightarrow B$ by interchanging A and B but leaving $\Rightarrow$ unchanged, is called the *converse* of the theorem $A \Rightarrow B$. Thus the two theorems $A \Rightarrow B$ and $B \Rightarrow A$ form a pair of converse theorems. Mark that the converse of a true theorem may or may not be true. For example,

$$x \text{ is a dog} \Rightarrow x \text{ has four legs}$$

is a true theorem with a false converse.

If, however, both the theorem and its converse are true, then the theorem expresses a necessary and sufficient condition. As a formula,

$$(A \Leftrightarrow B) \Leftrightarrow (A \Rightarrow B \text{ and } B \Rightarrow A)$$

i.e., a condition is necessary and sufficient if, and only if, both the sufficient condition and its converse are true. Moreover,

$$(A \Leftrightarrow B) \Leftrightarrow (B \Leftrightarrow A)$$

i.e., the converse of a theorem expressing a necessary and sufficient condition is the same theorem. In particular, the first of these two formulas tells us how to prove that a condition is both necessary and sufficient: We prove two implications.

THE PRINCIPLES OF NEGATION

The negation of a proposition A will be denoted $\bar{A}$, e.g., $\overline{f \text{ injective}} \Leftrightarrow f$ not injective. The following three principles are a priori; i.e., they come before any human experience and reasoning, and they are not capable of reduction to any simpler principles (although they may be expressed differently, via equivalent theorems).

The Principle of Double Negation

$$\bar{\bar{A}} \Leftrightarrow A$$

i.e., the negation of the negation of A is the same as A.

The Principle of the Excluded Middle

$$\overline{A \text{ and } \bar{A}}$$

i.e., "A and $\bar{A}$" is false.

The Principle of Negating an Implication

$$\overline{A \Rightarrow B} \Leftrightarrow (A \text{ and } \bar{B})$$

i.e., A does not imply B if, and only if, A is true and B is false.

The last of these three principles is especially important; it is thanks to the principle of negating an implication that we find it so much easier to disprove a false theorem than proving a true one. In *disproving* a theorem, all we need to do is to find a single counterexample. By contrast, in *proving* a theorem, examples are of no help. It does not matter how many affirmative examples we may have; as long as there are cases still unchecked, the theorem remains "not proven." In mathematics, unlike in the natural sciences, induction has no place, unless it is *total induction*, covering all conceivable cases. Mathematics is a *deductive science*, not an inductive one.

Example

$$ax = ay \Rightarrow x = y$$

is a false theorem. To see this, we look at its negation

$$ax = ay \qquad \text{and} \qquad x \neq y$$

which is true for $a = 0$, $x = 1$, $y = 2$. Therefore, by the principle of double negation, the original statement is false. Something can be saved, however. To find the correct form of the cancellation law, we must realize that $a = 0$ is the *only* exception:

$$a \neq 0 \Rightarrow (ax = ay \Rightarrow x = y)$$

We shall return later to the proof of this statement.

The Principle of Proof by Contradiction

$$(A \Rightarrow B) \Leftrightarrow (\bar{B} \Rightarrow \bar{A})$$

Proof: This is different from the previous principles in that it can be derived from them as follows:

$$\begin{aligned}
(A \Rightarrow B) &\Rightarrow \overline{\overline{A \Rightarrow B}} && \text{(by double negation)}\\
&\Rightarrow \overline{\overline{A \text{ and } \bar{B}}} && \text{(by negating the implication)}\\
&\Rightarrow \overline{\bar{B} \text{ and } A} && \\
&\Rightarrow \overline{\bar{B} \text{ and } \bar{\bar{A}}} && \text{(by double negation)}\\
&\Rightarrow \overline{\overline{\bar{B} \Rightarrow \bar{A}}} && \text{(by negating the implication)}\\
&\Rightarrow \bar{B} \Rightarrow \bar{A} && \text{(by double negation)}
\end{aligned}$$

We have shown that $(A \Rightarrow B) \Rightarrow (\bar{B} \Rightarrow \bar{A})$, which already implies its own converse if we apply it for the propositions $\bar{B}$, $\bar{A}$:

$$(\bar{B} \Rightarrow \bar{A}) \Rightarrow (\bar{\bar{A}} \Rightarrow \bar{\bar{B}}) \Leftrightarrow (A \Rightarrow B)$$

Hence $A \Rightarrow B$ is a necessary and sufficient condition for $\bar{B} \Rightarrow \bar{A}$. This completes the proof that the principle of proof by contradiction can be derived from other logical principles.

This principle earns its name by virtue of the fact that we are able, with its aid, to prove theorems in a most unexpected and unusual way, namely, by sacrificing the desideratum itself. This principle would not go down very well in the sports, where you can hardly *win* a game by resigning beforehand; you can only *lose* by default. To see how this works, suppose that we want to prove the theorem $A \Rightarrow B$. We proceed as follows. First we assume $\bar{B}$ (i.e., the negation of the desideratum is true, or what is the same, the desideratum is false) and thence try to conclude $\bar{A}$. If successful in this, we say that $A \Rightarrow B$ has been proved by contradiction (or, by the method of *reductio ad absurdum*).

Proof: We shall now illustrate this principle by proving the cancellation law,

$$a \neq 0 \Rightarrow (ax = ay \Rightarrow x = y)$$

We start by negating the desideratum:

$$\begin{aligned} \overline{ax = ay \Rightarrow x = y} &\Rightarrow ax = ay \text{ and } x \neq y \\ &\Rightarrow ax - ay = 0 \text{ and } x - y \neq 0 \\ &\Rightarrow a(x - y) = 0 \text{ and } x - y \neq 0 \\ &\Rightarrow a = 0 \end{aligned}$$

The last implication is justified by a theorem known as the absence of zero divisors for $\mathbb{R}$:

$$ab = 0 \Rightarrow (b \neq 0 \Rightarrow a = 0)$$

This is also proved by using the principle of proof by contradiction:

$$\begin{aligned} \overline{b \neq 0 \Rightarrow a = 0} &\Rightarrow b \neq 0 \text{ and } a \neq 0 \\ &\Rightarrow (a > 0 \text{ or } a < 0) \text{ and } (b > 0 \text{ or } b < 0) \\ &\Rightarrow ab > 0 \text{ or } ab < 0 \\ &\Rightarrow ab \neq 0 \end{aligned}$$

since the product of two nonzero real numbers is positive or negative according as the factors have the same or the opposite signature.

We have two versions of the cancellation law, which are equivalent:

$$(a \neq 0 \text{ and } ax = ay) \Rightarrow x = y$$
$$a \neq 0 \Rightarrow (ax = ay \Rightarrow x = y)$$

Similarly, we have two versions of the absence of zero divisors property, which are equivalent:

$$(ab = 0 \text{ and } b \neq 0) \Rightarrow a = 0$$
$$ab = 0 \Rightarrow (b \neq 0 \Rightarrow a = 0)$$

There examples illustrate the logical formula:

The Principle of Exportation

$$[(A \text{ and } B) \Rightarrow C] \Leftrightarrow [A \Rightarrow (B \Rightarrow C)]$$

This principle earns its name by telling us how to "export" part of the condition in an implication.

The principle of proof by contradiction can be formulated in several equivalent ways, e.g.,

The Principle of Duality

$$(A \Rightarrow B) \Leftrightarrow (\bar{A} \Leftarrow \bar{B})$$
$$(A \Leftarrow B) \Leftrightarrow (\bar{A} \Rightarrow \bar{B})$$

The theorem $\bar{A} \Leftarrow \bar{B}$ is called the *dual* of the theorem $A \Rightarrow B$, so that the two theorems $A \Rightarrow B$, $\bar{A} \Leftarrow \bar{B}$ are said to form a *dual pair of theorems*. To put it differently, we *dualize* a theorem by negating its members A, B and replacing $\Rightarrow$ by $\Leftarrow$, and vice versa. But whereas a theorem and its *converse* may or may not both be true simultaneously, it is different with a theorem and its *dual*. According to the principle of duality, a theorem and its dual stand or fall together.

Example

$$[f(x_1) = f(x_2) \Rightarrow x_1 = x_2] \Leftrightarrow [x_1 \neq x_2 \Rightarrow f(x_1) \neq f(x_2)]$$

and either of the two propositions may be used as the definition of injectivity of a function f.

Combining the two formulas of the principle of duality, we get

$$(A \Leftrightarrow B) \Leftrightarrow (\bar{A} \Leftrightarrow \bar{B})$$

i.e., the dual of a necessary and sufficient condition is also a necessary and sufficient condition; moreover,

$$(A \Leftrightarrow B) \Leftrightarrow (A \Rightarrow B \text{ and } \bar{A} \Rightarrow \bar{B})$$

i.e., a necessary and sufficient condition can be established by proving that the condition is sufficient for the desideratum, and that its negation is sufficient for the negation of the desideratum.

Here are the formulas that tell us how the particles "and," "or" fare under negation (also known as De Morgan's laws):

The Principle of Negating a Conjunction

$$\overline{A \text{ and } B} \Leftrightarrow \bar{A} \text{ or } \bar{B}$$

The Principle of Negating a Disjunction

$$\overline{A \text{ or } B} \Leftrightarrow \bar{A} \text{ and } \bar{B}$$

Examples

(1) The principle of the excluded middle can be restated as

$$A \text{ or } \bar{A}$$

i.e., "A or $\bar{A}$" is a true statement, on the strength of the principle of negating a conjunction: $(\overline{A \text{ and } \bar{A}}) \Leftrightarrow (\bar{A} \text{ or } A)$.

(2) From the theorem

$$f \text{ bijective} \Leftrightarrow f \text{ injective } \textit{and} \text{ surjective}$$

we may immediately conclude the theorem

$$f \text{ not bijective} \Leftrightarrow f \text{ not injective } \textit{or } f \text{ not surjective}$$

For functions, a weaker form of the cancellation law holds:

Left Cancellation Law

$$g \text{ injective} \Rightarrow (g \circ f = g \circ f' \Rightarrow f = f')$$

Right Cancellation Law

$$f \text{ surjective} \Rightarrow (g \circ f = g' \circ f \Rightarrow g = g')$$

Indeed,

$$\begin{aligned} g \circ f = g \circ f' &\Rightarrow (g \circ f)(x) = (g \circ f')(x) \quad \text{for all } x \in X \\ &\Rightarrow g(f(x)) = g(f'(x)) \Rightarrow f(x) = f'(x) \quad \text{for all } x \in X \\ &\Rightarrow f = f' \end{aligned}$$

proving the left cancellation law. On the other hand, for all $y \in Y$,

$$\begin{aligned} g(y) &= g(f(x)) \quad \text{for some } x \in X \\ &= (g \circ f)(x) = (g' \circ f)(x) = g'(f(x)) = g'(y) \end{aligned}$$

Therefore, $g = g'$, proving the right cancellation law.

THE AXIOM OF CHOICE

Let $g \circ f = 1_X$, where f: $X \to Y$ and g: $Y \to X$. Then we say that f is *left invertible* (g is *right invertible*) with g as a *left inverse* of f (f as a *right inverse* of g). For example, location $\circ$ capital $= 1_{Co}$, so that "capital" is left invertible with "location" as one of its left inverses, and "location" is right invertible with "capital" as one of its right inverses. We have the following necessary and sufficient conditions for left and right invertibility:

$$f \text{ left invertible} \Leftrightarrow f \text{ injective}$$

$$g \text{ right invertible} \Leftrightarrow g \text{ surjective}$$

Moreover, every left inverse of an injection is a surjection, and every right inverse of a surjection is an injection. These assertions, as well as the sufficiency of the conditions above, are immediate from

$$g \circ f \text{ bijective} \Rightarrow f \text{ injective and } g \text{ surjective}$$

To show that the conditions are also necessary, we must prove two existence theorems:

(1) Every injection has at least one left inverse.

(2) Every surjection has at least one right inverse.

There is a nice construction proof of theorem (1). Let f: $X \to Y$ be injective, and define a function g: $Y \to X$ by putting, for all $y \in Y$,

$$g(y) = \begin{cases} x & \text{with } y = f(x), \text{ whenever such an } x \in X \text{ exists} \\ x_0 & \text{otherwise, where } x_0 \in X \text{ is arbitrary but fixed} \end{cases}$$

The function g is well defined on the strength of the injectivity of f, and satisfies $g(f(x)) = x$ for all $x \in X$, i.e., $g \circ f = 1_X$.

The proof of theorem 2 is very different. Let $g: Y \to X$ be surjective; then for every $x \in X$ there exists at least one $y \in Y$ satisfying $g(y) = x$. Let us *choose* one of these values of y and denote it y'. Clearly, $g(y') = x$ for every $x \in X$. We now define $f: X \to Y$ by setting $f(x) = y'$ for every $x \in X$. Then

$$\begin{aligned} (g \circ f)(x) &= g(f(x)) \\ &= g(y') = x \end{aligned}$$

for all $x \in X$. Therefore $g \circ f = 1_X$ and g is right invertible, since it has at least one right inverse, namely, f.

This proof has not escaped challenge. The consensus of mathematicians today is that the proof is invalid in the case where X is an infinite set because, unlike the proof of theorem 1, it does not say explicitly how the *choice* of y' for every $x \in X$ is made. At the same time, no one could come up with a proof explicitly constructing a right inverse. In fact, theorem 2 has the aura of an *axiom*, and it has been called

The Axiom of Choice
Every surjection has at least one right inverse.

The axiom of choice earns its name by allowing us, in principle, to make infinitely many *simultaneous* choices. In linear algebra, it finds its most important application in the proof of the theorem (not carried out in this book) that every vector space has a basis (which may be infinite).

Injections are left invertible and surjections are right invertible, but their inverses may not be unique. However, for bijections (and only for them) the left and right inverses are uniquely determined and, in fact, they coincide. The uniqueness of the left and right inverses follow from the cancellation laws. To see that the unique left inverse f' is equal to the unique right inverse f'' of f, we write $f' = f' \circ 1_Y = f' \circ (f \circ f'') = (f' \circ f) \circ f'' = 1_X \circ f'' = f''$. Thus it will not be necessary to distinguish in notation between the left and right inverses of a bijection $f: X \to Y$, and for its unique inverse we shall write $f^{-1}: Y \to X$.

Formal Properties of the Inverse

$$f^{-1} \circ f = 1_X \qquad f \circ f^{-1} = 1_Y$$
$$(g \circ f)^{-1} = f^{-1} \circ g^{-1}$$
$$(f^{-1})^{-1} = f$$
$$1_X{}^{-1} = 1_X$$

For example, we prove that the inverse of the product of two bijections is the product of their inverses in the *reverse order*:

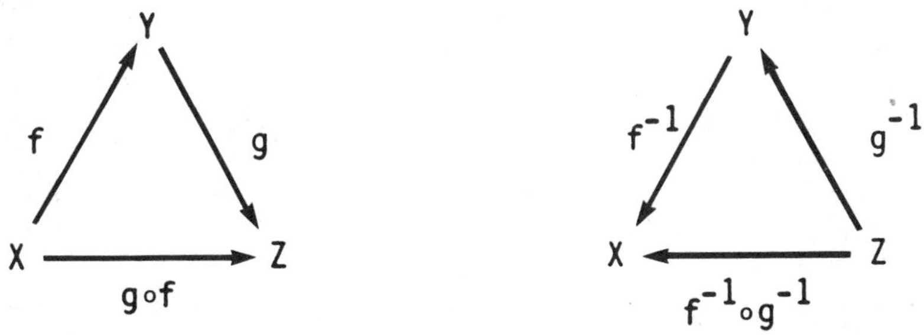

Indeed, $g \circ f$ is bijective and therefore has a unique inverse $(g \circ f)^{-1}$. The candidate for this job, $f^{-1} \circ g^{-1}$, does have the necessary qualifications:

$$(f^{-1} \circ g^{-1}) \circ (g \circ f) = f^{-1} \circ (g^{-1} \circ g) \circ f = f^{-1} \circ f = 1_X$$

and, similarly,

$$(g \circ f) \circ (f^{-1} \circ g^{-1}) = 1_Z$$

This concludes our heuristic introduction to logical principles.

EXERCISES

0.1 The condition that f be injective is necessary but not sufficient for $g \circ f$ to be injective. Strengthen it to a necessary and sufficient condition.

0.2 The condition that g be surjective is necessary but not sufficient for $g \circ f$ to be surjective. Strengthen it to a necessary and sufficient condition.

Define the *image* of a function $f\colon X \to Y$, in symbols im f, such that

$$\text{im } f = \{y \in Y \mid y = f(x),\ x \in X\}$$

Define the *restriction* of the function $f\colon X \to Y$ to a subset A of the domain of f, $f'\colon A \to Y$, by the formula

$$f'(x) = f(x) \qquad \text{for all } x \in A$$

In these terms,

0.3 State another necessary and sufficient condition for $g \circ f$ to be injective.

0.4 State another necessary and sufficient condition for $g \circ f$ to be surjective.

An operator p is called a *projection* if $p^2 = p$. An operator i is called an *involution* if $i^2 = 1$, the identify operator.

0.5 Show that capital $\circ$ location is a projection.

0.6 Show that, more generally, if g is a left inverse of f, then $f \circ g$ is a projection.

0.7 Define an operator called "spouse" on the set of all married people and show that it is an involution.

0.8 Show that an operator is an involution if, and only if, it is equal to its own inverse: $i^2 = 1 \Leftrightarrow i^{-1} = i$.

0.9 Show that the product of two involutions i, j is an involution if, and only if, they commute: $(i \circ j)^2 = 1 \Leftrightarrow i \circ j = j \circ i$.

0.10 Show that p is a projection if, and only if, $p = g \circ f$, where g is a right inverse of f (and f is a left inverse of g).

1

The Sum and the Scalar Multiple of Vectors

FORMAL RULES OF THE SUM

A vector is considered as given if its (a) length, (b) alignment, and (c) direction, are determined. Two vectors are said to have the same alignment if they are parallel to the same straight line. Two vectors may have the same length and the same alignment, yet be unequal. In this case they have opposite directions.

It is convenient to think of a vector as an oriented segment AB where A is the *initial point* and B is the *terminal point*. However, a vector is unlike an oriented segment in that it is not rigidly fixed in space but can be freely translated (i.e., shifted parallelly). Under such a translation the length, the alignment, and the orientation, and hence the vector itself, remain unchanged.

Let **u** and **v** be vectors. In order to define their *sum* $\mathbf{u}+\mathbf{v}$, we may assume that the initial point of **v** coincides with the terminal point of **u**: $\mathbf{u}=\overrightarrow{AB}$, $\mathbf{v}=\overrightarrow{BC}$; then $\mathbf{u}+\mathbf{v}=\overrightarrow{AB}+\overrightarrow{BC}=\overrightarrow{AC}$. This definition will be referred to as the *triangle rule*:

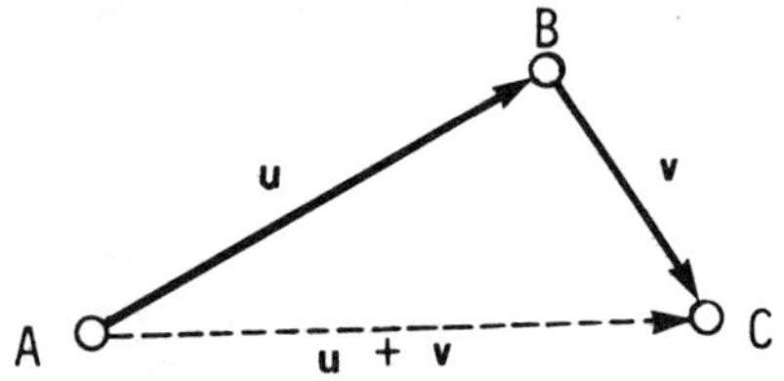

although the triangle may collapse:

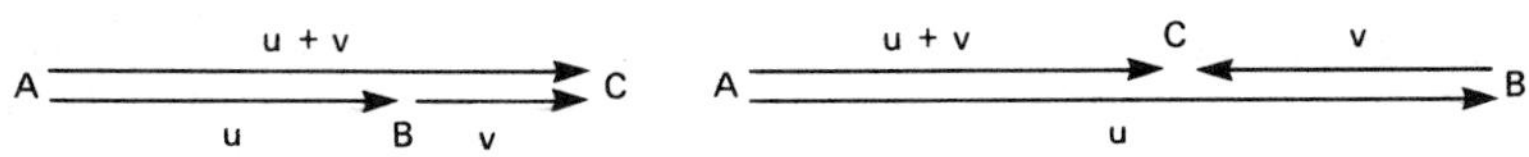

Let λ be a scalar (i.e., a real number; in symbols: $\lambda \in \mathbb{R}$) and let $\mathbf{v}$ be a vector. The scalar multiple $\lambda\mathbf{v}$ is defined as a vector whose (a) length is $|\lambda|$ times the length of $\mathbf{v}$ (the symbol $|\lambda|$ denotes the absolute value of λ, i.e., $|\lambda| = \lambda$ for $\lambda > 0$, $|\lambda| = -\lambda$ for $\lambda < 0$, and $|\lambda| = 0$ for $\lambda = 0$), (b) alignment is the same as that of $\mathbf{v}$; and (c) direction is the same as that of $\mathbf{v}$ if $\lambda > 0$, and it is the opposite if $\lambda < 0$. In short, to take the scalar multiple of vectors by the scalar λ is the same as applying a strain in ratio λ (with respect to some center).

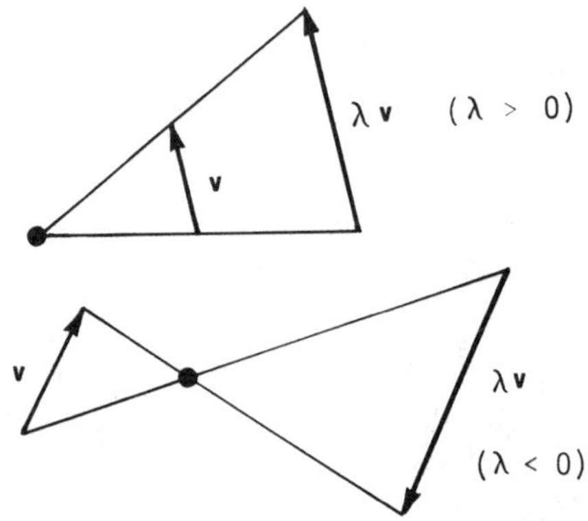

A *strain* (with respect to the origin) in ratio λ is a special case of the class of linear operators which are the main object of studies of linear algebra. It is defined as an operator sending a point P to P' such that

$$\overrightarrow{OP'} = \lambda \overrightarrow{OP}$$

(1) Commutative Law

$$\mathbf{u} + \mathbf{v} = \mathbf{v} + \mathbf{u}$$

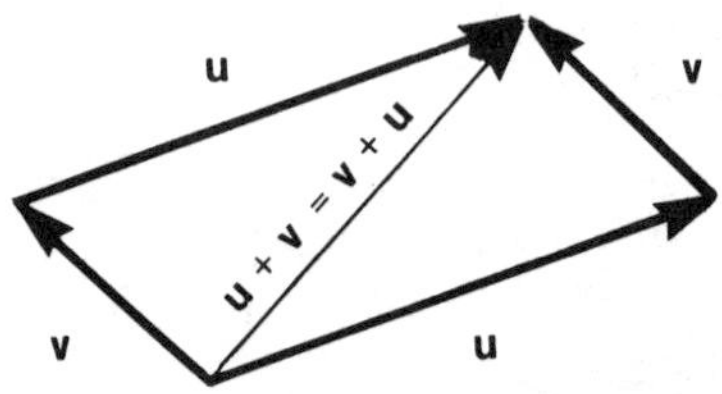

This figure shows the sum of two vectors as the diagonal of a parallelogram whose sides are the summands; for this reason (1) is also called the *parallelogram rule*.

(2) Associative Law

$$(\mathbf{u} + \mathbf{v}) + \mathbf{w} = \mathbf{u} + (\mathbf{v} + \mathbf{w})$$

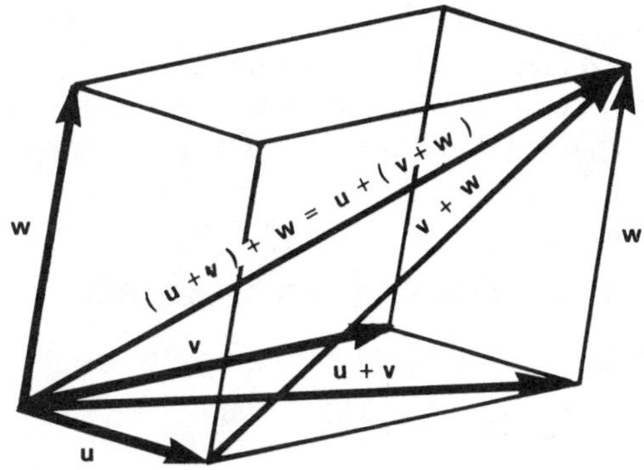

This figure shows the sum of three vectors as the diagonal of a parallelepiped whose sides are the summands; for this reason (2) is also called the *parallelepiped rule*.

(3) Neutrality of the Null Vector

$$\mathbf{v} + \mathbf{0} = \mathbf{v}$$

(4) The Negative of a Vector

$$\mathbf{v} + (-\mathbf{v}) = \mathbf{0}$$

(5) The Distributive Law

$$\lambda(\mathbf{u} + \mathbf{v}) = \lambda\mathbf{u} + \lambda\mathbf{v}$$

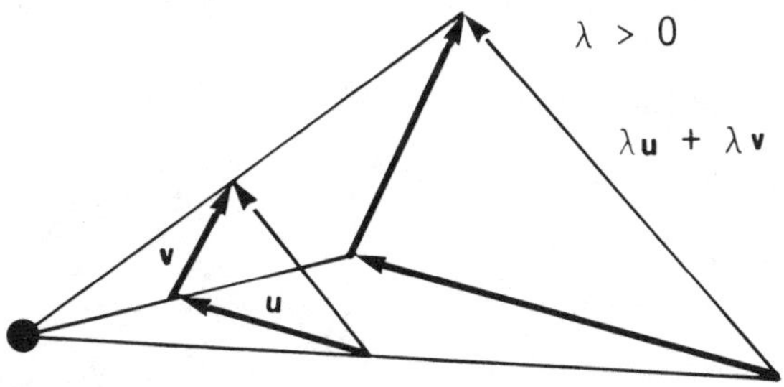

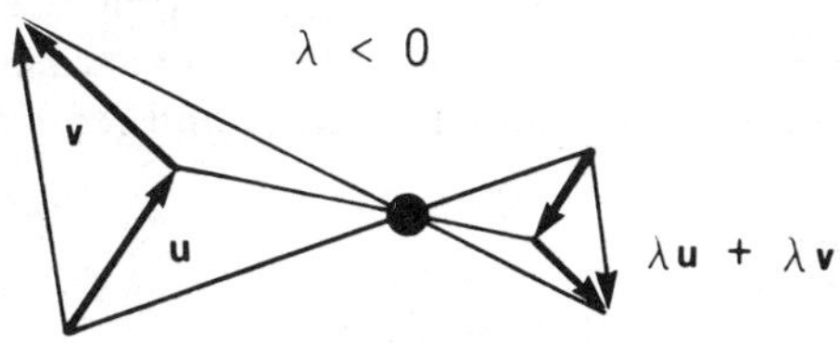

These figures show how a strain with ratio λ transforms a triangle into another triangle, in fact, into a *similar* triangle.

(5) is one of two distributive laws; see (5′) below. (5) can be paraphrased as: The scalar multiple of a vector is distributive over the sum of vectors.

FORMAL RULES OF THE SCALAR MULTIPLE

(2′) Scalar Associative Law

$$\lambda(\mu\mathbf{v}) = (\lambda\mu)\mathbf{v} \qquad (\lambda, \mu \in \mathbb{R})$$

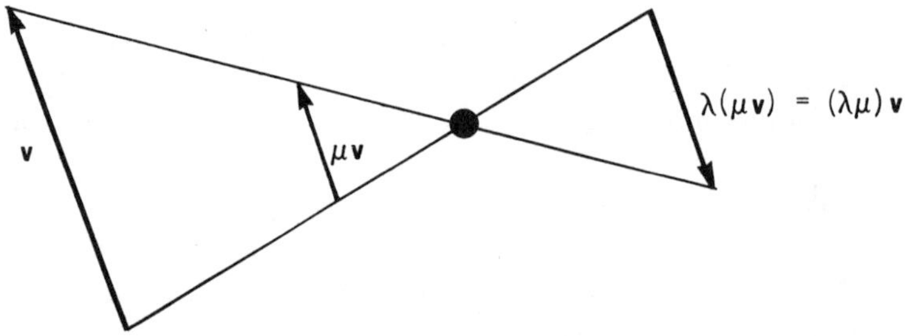

This figure shows that a strain in ratio μ followed by a strain in ratio λ is again a strain, namely, in ratio $\lambda\mu$ (in the case depicted, $\mu > 0$, $\lambda < 0$).

It is this scalar associative law that motivates the arithmetic rule, often a stumbling block in the elementary school, according to which $(-2)(-3) = 6$. The correct interpretation of this is as follows: A strain in ratio -3 followed by a strain in ratio -2 is again a strain, namely, in ratio 6 (and not -6, as there are two reflections involved, which cancel out).

(3′) Neutrality of the Scalar 1

$$1\mathbf{v} = \mathbf{v}$$

(5′) Distributive Law

$$(\lambda + \mu)\mathbf{v} = \lambda\mathbf{v} + \mu\mathbf{v}$$

(5′) is the second distributive law. (5′) can be paraphrased thus: The scalar multiple of a vector is distributive over the sum of scalars. The three laws (2′), (5), and (5′) are often lumped together and referred to as the

Law of Bilinearity

$$\lambda(\mu\mathbf{v}) = (\lambda\mu)\mathbf{v}$$

$$\lambda(\mathbf{u}+\mathbf{v})=\lambda\mathbf{u}+\lambda\mathbf{v} \qquad (\lambda,\ \mu \in \mathbb{R})$$
$$(\lambda+\mu)\mathbf{v}=\lambda\mathbf{v}+\mu\mathbf{v}$$

It is customary to include two more laws among the formal rules of the sum and the scalar multiple of vectors:

(0) Closure Under Addition

$$\mathbf{u},\ \mathbf{v} \text{ vectors} \Rightarrow \mathbf{u}+\mathbf{v} \text{ vector}$$

(0′) Closure Under Scalar Multiplication

$$\mathbf{v} \text{ vector},\ \lambda \text{ scalar} \Rightarrow \lambda\mathbf{v} \text{ vector}$$

These 10 formal rules are summarized in the following ready-reference table:

CLOSURE	0	0'	
COMMUTATIVITY	1		
ASSOCIATIVITY	2	2'	SCALAR ASSOCIATIVITY
NEUTRAL	3	3'	
NEGATIVE	4		
DISTRIBUTIVITY	5	5'	

The *difference* $\mathbf{u}-\mathbf{v}$ of two vectors (just as the difference of two numbers) is defined to be the vector such that the sum of the difference $\mathbf{u}-\mathbf{v}$ and the *subtrahend* $\mathbf{v}$ equals the *minuend* $\mathbf{u}$:

$$(\mathbf{u}-\mathbf{v})+\mathbf{v}=\mathbf{u}$$

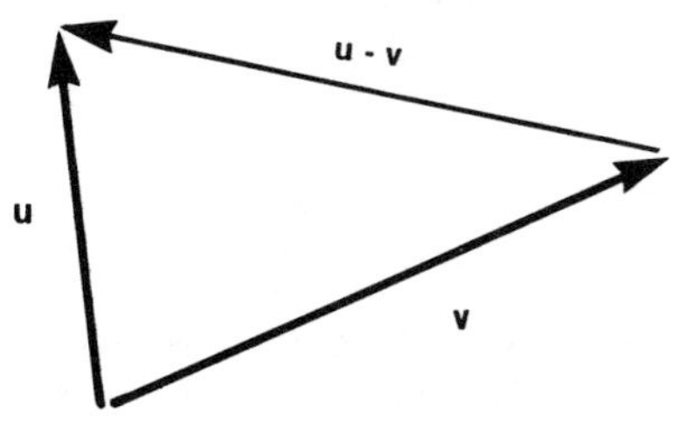

In constructing the difference vector $\mathbf{u}-\mathbf{v}$ special attention must be paid to its direction. The mnemotechnical key is: "*terminal minus initial*"; i.e., the terminal point of the difference coincides with that of the minuend (while its initial point coincides with the terminal point of the subtrahend).

The *null vector* is defined by the equation $\mathbf{0}=\mathbf{u}-\mathbf{u}$; i.e., it is the vector with coincident initial and terminal points.

The *negative* $-\mathbf{v}$ of a vector $\mathbf{v}$ is defined by the equation $-\mathbf{v}=\mathbf{0}-\mathbf{v}$. Thus $-\mathbf{v}$ is obtained from $\mathbf{v}$ by interchanging its initial and terminal points, i.e., by changing its direction.

FORMAL PROPERTIES OF THE NEGATIVE OF A VECTOR

Scalar Multiple by –1

$$(-1)\mathbf{v}=-\mathbf{v}$$

Negative of the Sum

$$-(\mathbf{u}+\mathbf{v})=(-\mathbf{u})+(-\mathbf{v})$$

Negative of the Null Vector

$$-\mathbf{0}=\mathbf{0}$$

Negative of the Negative

$$-(-\mathbf{v})=\mathbf{v}$$

The first of these properties asserts that in reflecting a vector in a point we get its negative:

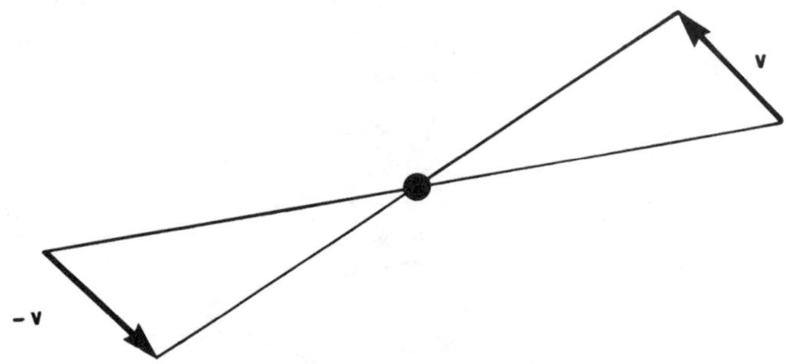

The other properties are an easy consequence of the foregoing; e.g., the negative of the sum is the sum of the negatives:

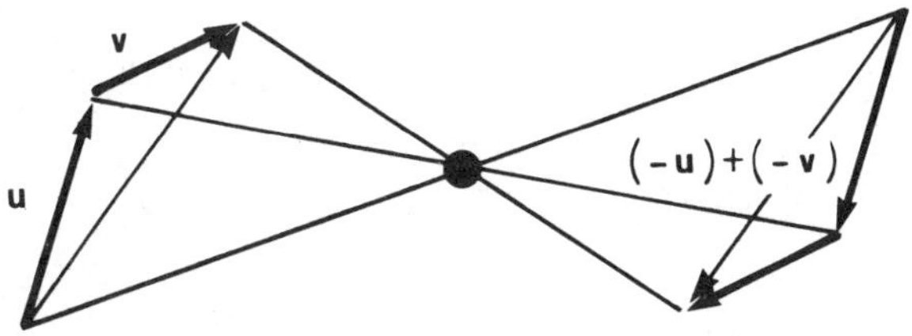

The last of these properties asserts that changing the direction of a vector twice will yield the original vector.

Scalar Multiple by Zero

$$0\mathbf{v} = \mathbf{0}$$

Scalar Multiple of the Null Vector

$$\lambda\mathbf{0} = \mathbf{0}$$

In fact, the converse statement is also true, so that a necessary and sufficient condition can be given for the scalar multiple to be equal to the null vector:

Absence of Zero Divisors

$$\lambda\mathbf{v} = \mathbf{0} \Leftrightarrow \lambda = 0 \text{ or } \mathbf{v} = \mathbf{0}$$

The formal properties of the null vector follow from the 10 formal rules of the vector sum and scalar multiple.

LOCATION VECTORS. COORDINATES. DIRECTION NUMBERS

A *location vector* is a vector whose initial point is anchored at the origin. If the terminal point of the location vector $\mathbf{v}$ is P, then $\mathbf{v} = \overrightarrow{OP}$. The location vector $\overrightarrow{OP}$ earns its name by fixing the location of the point P.

The coordinates of a location vector are the same as those of the point P that it locates. In order to find the coordinates of a free vector $\mathbf{v} = \overrightarrow{P_1P_2}$, we write it as a difference of location vectors:

$$\mathbf{v} = \overrightarrow{P_1P_2} = \overrightarrow{OP_2} - \overrightarrow{OP_1}$$

This suggests that the coordinates of a free vector are the differences of the corresponding coordinates of its terminal and initial points:

$$a = x_2 - x_1 \qquad b = y_2 - y_1 \qquad c = z_2 - z_1$$

and, once again, the mnemotechnic rule "terminal minus initial" applies.

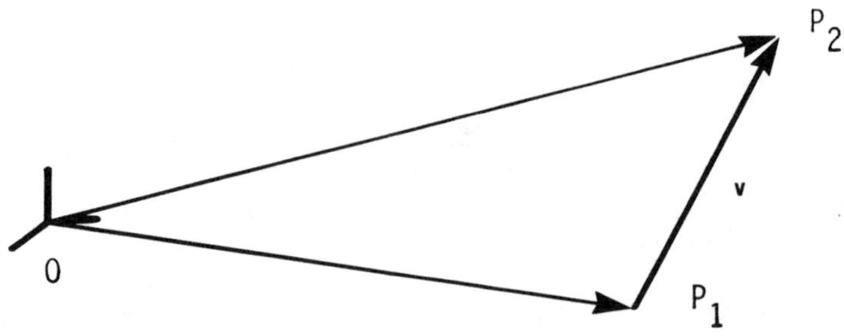

The coordinates of the vector $\mathbf{v} \neq \mathbf{0}$ are also called *direction numbers*. The direction numbers do not depend on the choice of the initial point, but depend on the vector **v** to which they belong.

In two-dimensional analytic geometry there are only two direction numbers: run $= a$, rise $= b$. In three-dimensional space, we may speak of east-west run $= a$, north-south run $= b$, rise $=$ c.

Complete the table (UNE, upper northeast; LNE, lower northeast; etc.).

Direction	Direction numbers	Direction	Direction numbers
East	(1,0,0)		(1,0,−1)
West			(0,1,1)
North			(0,−1,1)
South			(0,−1,−1)
Upper			(0,1,−1)
Lower		UNE	(1,1,1)
NE	(1,1,0)	UNW	
	(−1,1,0)	USW	
	(−1−1,0)	USE	
	(1,−1,0)	LNE	
	(1,0,1)	LNW	
	(−1,0,1)	LSW	
	(−1,0,−1)	LSE	

We shall be talking about the UNE equiangular line, etc. The UNE equiangular line earns its name by making equal angles with upper, north, and east.

Direction numbers	Vector parallel to:	Direction numbers	Vector parallel to:
(a,a,c)	NE & SW bisector plane	$(a,0,c)$	
$(a,-a,c)$		$(0,b,c)$	
(a,b,b)		$(a,a,0)$	(NE & SW bisector axis
$(a,b,-b)$			NW & SE bisector axis
(a,b,a)			UE & LW bisector axis
$(a,b,-a)$			UW & LE bisector axis
(a,a,a)	UNE & LSW equ. ang. line		UN & LS bisector axis
	UNW & LSE equ. ang. line		US & LN besector axis
	USW & LNE equ. ang. line	$(a.0,0)$	x axis
	USE & LNW equ. ang. line	$(0,b,0)$	
$(a,b,0)$	xy plane	$(0,0,c)$	

The notation **i**, **j**, **k** for the vectors described in the next table is standard.

Vector	Direction numbers	Direction	Vector	Direction numbers	Direction
i	(1,0,0)	E	−**i**	(−1,0,0)	W
j	(0,1,0)	N	−**j**	(0,−1,0)	S
k	(0,0,1)	U	−**k**	(0,0,−1)	L

We shall express the fact that the direction numbers of the vector **v** are a, b, c by the equation

$$\mathbf{v} = a\mathbf{i} + b\mathbf{j} + c\mathbf{k}$$

Complete the table.

Vector	Direction	Vector	Direction
$\mathbf{i}+\mathbf{j}$	NE		LS
	NW		LN
	SW	$\mathbf{i}+\mathbf{j}+\mathbf{k}$	UNE
	SE	$\mathbf{i}+\mathbf{j}-\mathbf{k}$	
	UE	$\mathbf{i}-\mathbf{j}+\mathbf{k}$	
	UW	$-\mathbf{i}+\mathbf{j}+\mathbf{k}$	
	LW	$\mathbf{i}-\mathbf{j}-\mathbf{k}$	
	LE	$-\mathbf{i}+\mathbf{j}-\mathbf{k}$	
	UN	$-\mathbf{i}-\mathbf{j}+\mathbf{k}$	
	US	$-\mathbf{i}-\mathbf{j}-\mathbf{k}$	
$a\mathbf{i}+b\mathbf{j}$	Horizontal	$c\mathbf{k}$	Vertical

LINEAR COMBINATIONS

A *linear combination* of two vectors **a**, **b** is the vector

$$\mathbf{v} = \alpha\mathbf{a} + \beta\mathbf{b}$$

where α, $\beta \in \mathbb{R}$ are scalars. Thus the notion of a linear combination generalizes those of the sum, difference, scalar multiple, and the negative of vectors: $\mathbf{a} + \mathbf{b} = 1\mathbf{a} + 1\mathbf{b}$, $\mathbf{a} - \mathbf{b} = 1\mathbf{a} + (-1)\mathbf{b}$, $-\mathbf{a} = (-1)\mathbf{a} + 0\mathbf{b}$, etc.

APPLICATIONS

1. Necessary and Sufficient Condition for Two Vectors to Be Parallel

$$\mathbf{a} \parallel \mathbf{b} \Leftrightarrow \alpha\mathbf{a} + \beta\mathbf{b} = \mathbf{0} \qquad \text{for some scalars } \alpha, \beta \text{ not both } 0$$

In words, two vectors are parallel if, and only if, there is a nontrivial linear combination of the vectors reproducing the null vector. A linear combination of three vectors **a**, **b**, **c** is the vector

$$\mathbf{v} = \alpha\mathbf{a} + \beta\mathbf{b} + \gamma\mathbf{c} \qquad \alpha, \beta, \gamma \in \mathbb{R}$$

2. Necessary and Sufficient Condition for Three Vectors to Be Coplanar

$$\mathbf{a}, \mathbf{b}, \mathbf{c} \text{ coplanar } \Leftrightarrow \alpha\mathbf{a} + \beta\mathbf{b} + \gamma\mathbf{c} = \mathbf{0} \qquad \text{for some } \alpha, \beta, \gamma \text{ not all } 0$$

In words, three vectors are coplanar (i.e., parallel to the same plane) if and only if there is a nontrivial linear combination of the vectors reproducing the null vector.

Parallel and coplanar vectors are examples of a *linearly dependent set* of vectors. A set of *four* vectors is always linearly dependent: we can easily find scalars α, β, γ, δ, not all zero, such that $\alpha\mathbf{a} + \beta\mathbf{b} + \gamma\mathbf{c} + \delta\mathbf{d} = \mathbf{0}$ (how?). On the other hand, suppose that **a**, **b**, **c** are such that $\alpha\mathbf{a} + \beta\mathbf{b} + \gamma\mathbf{c} = \mathbf{0} \Rightarrow \alpha = \beta = \gamma = 0$. Then we say that they form a *linearly independent set*. Thus a set of vectors is linearly dependent or independent according as the null vector can, or cannot, be obtained as a nontrivial linear combination of the vectors of the set. For example, $\{\mathbf{i}, \mathbf{j}, \mathbf{k}\}$ is a linearly independent set of vectors.

3. Point Dividing a Segment in a Given Ratio

Let A, B be distinct points; they will then determine a unique line. We want a necessary and sufficient condition for a point P to fall on the line of A, B in terms of their respective location vectors **p**, **a**, **b**.

$$P \text{ falls on line of } A, B \Leftrightarrow \mathbf{p} = \frac{\alpha\mathbf{a} + \beta\mathbf{b}}{\alpha + \beta}$$

for some $\alpha, \beta \in \mathbb{R}$. Indeed,

$$\begin{aligned} P \text{ falls on line of } A, B &\Leftrightarrow \mathbf{a} - \mathbf{p}, \mathbf{b} - \mathbf{p} \text{ are parallel} \\ &\Leftrightarrow \alpha(\mathbf{a} - \mathbf{p}) + \beta(\mathbf{b} - \mathbf{p}) = \mathbf{0} \\ &\Leftrightarrow \mathbf{p} = \frac{\alpha\mathbf{a} + \beta\mathbf{b}}{\alpha + \beta} \end{aligned}$$

because $\alpha + \beta \neq 0$ (otherwise, $A = B$). We may think of P as the center of gravity, or the barycenter, of the system of points weighted with weights α, β respectively. In fact, P divides the segment AB in the ratio β/α, i.e.,

$$\frac{\overline{AP}}{\overline{PB}} = \frac{\beta}{\alpha}$$

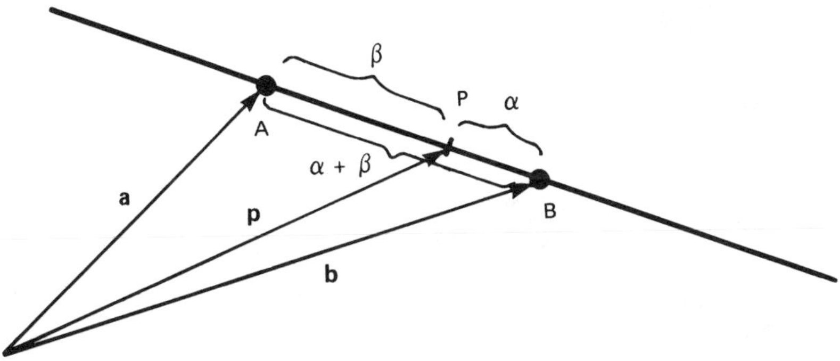

If $\alpha = \beta$, then the barycenter is the *midpoint* of the segment AB and its location vector is

$$\mathbf{m} = \frac{\mathbf{a} + \mathbf{b}}{2}$$

We may use negative weights as well, provided that $\alpha + \beta \neq 0$. If the weights α, β both have the same signature, the barycenter will fall inside the segment AB; otherwise, it will fall outside.

Every point P on the line of A, B is the barycenter, for some weights α, β, of the system of weighted points A, B. Of course, the weights are not uniquely determined by P because if we multiply both weights by the same number, the barycenter P is left unchanged.

We shall now describe a procedure on how to find the barycenter P if the weights α and β are integers (or, what is the same, how to find the point P that divides the segment AB in the given ratio β/α).

Case 1 $\alpha > 0$, $\beta > 0$. In this case the point P is between A and B. Divide the segment AB into $\alpha+\beta$ equal parts and call one part the unit of length. Measure β units from A toward B to get P.

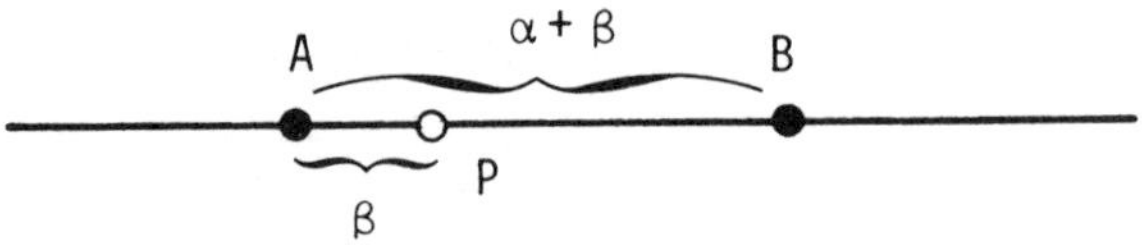

Case 2 The weights α, β have the opposite signature. In this case the point P will fall outside of the segment AB. Attach the negative signature to the weight that is smaller in absolute value. As in case 1, calculate the sum of weights (which is now a difference). Divide the segment AB into $\alpha+\beta$ equal parts and call one part the unit of length. Measure $|\beta|$ units from A toward B if β is positive, away from B if β is negative, to find P.

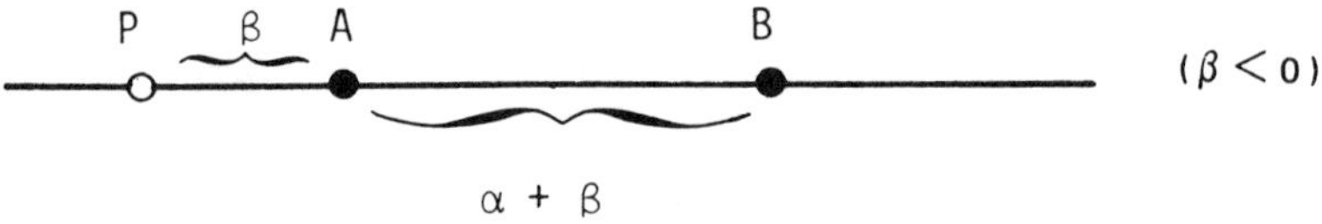

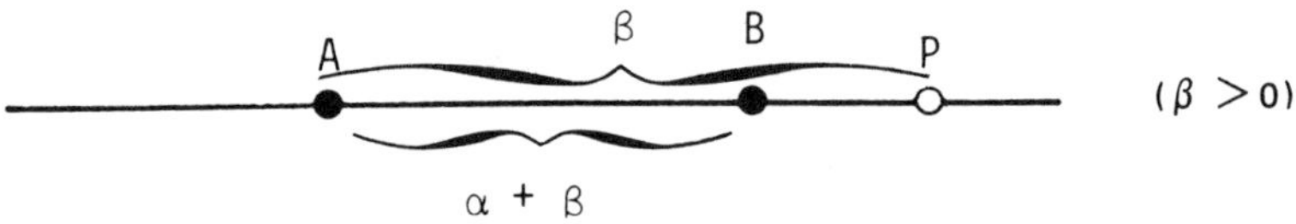

Complete the table:

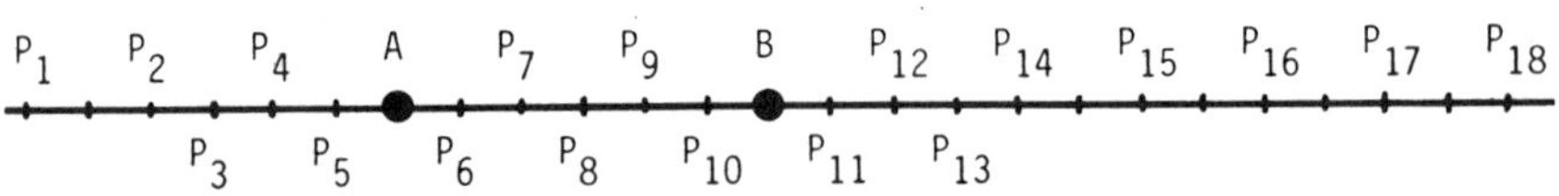

Point P	Ratio in which P divides AB	Weight at A	Weight at B	Location vector of point P
P	β/α	α	β	$(\alpha\mathbf{a}+\beta\mathbf{b})/(\alpha+\beta)$
A	0	1	0	$\mathbf{a}$
B	no ratio	0	1	$\mathbf{b}$
P_1	$-1/2$	2	-1	$2\mathbf{a}-\mathbf{b}$
P_2				
P_3				
P_4				
P_5				
P_6				

Point P	Ratio in which P divides AB	Weight at A	Weight at B	Location vector of point P
P_7				
P_8	1	1	1	$\frac{1}{2}(\mathbf{a}+\mathbf{b})$
	2			
	5			
	−7			
	−4			
		−1	3	
		−2	5	
		−1	2	
		−4	7	
		−5	8	
				$3\mathbf{b}-2\mathbf{a}$
no point	−1	−1	1	no barycenter

4. Barycentric Coordinates

We want to generalize the foregoing from a pair of weighted points to three and then to four weighted points.

Case of the plane $\mathbb{R}^2$ Consider the plane determined by the noncollinear points A, B, C whose respective location vectors are denoted by $\mathbf{a}$, $\mathbf{b}$, $\mathbf{c}$. Then $\mathbf{a}$, $\mathbf{b}$, $\mathbf{c}$ are not coplanar. We want a necessary and sufficient condition for a point P (with location vector $\mathbf{p}$) to lie in the plane of A, B, C.

$$P \text{ is in the plane of } A, B, C \Leftrightarrow \mathbf{p}=\frac{\alpha\mathbf{a}+\beta\mathbf{b}+\gamma\mathbf{c}}{\alpha+\beta+\gamma}$$

for some α, β, $\gamma \in \mathbb{R}$. Indeed,

$$\begin{aligned} A, B, C, P \text{ lie in the same plane} &\Leftrightarrow \mathbf{a}-\mathbf{p},\ \mathbf{b}-\mathbf{p},\ \mathbf{c}-\mathbf{p} \text{ coplanar}\\ &\Leftrightarrow \alpha(\mathbf{a}-\mathbf{p})+\beta(\mathbf{b}-\mathbf{p})+\gamma(\mathbf{c}-\mathbf{p})=\mathbf{0}\\ &\Leftrightarrow \mathbf{p}=\frac{\alpha\mathbf{a}+\beta\mathbf{b}+\gamma\mathbf{c}}{\alpha+\beta+\gamma}\end{aligned}$$

because $\alpha+\beta+\gamma\neq 0$ (otherwise, A, B, C would be collinear).

We shall say that P is the *barycenter* of the system with base points A, B, C with weights α, β, γ. The numbers α, β, γ are also called the *barycentric coordinates* of the point P (with respect to the base points A, B, C). The barycentric coordinates of P are not uniquely determined: multiplying each by the same nonzero number will leave the barycenter unchanged.

If $\alpha = \beta = \gamma$, the barycenter coincides with the *centroid* of the triangle ABC, and its location vector is

$$\mathbf{p} = \frac{\mathbf{a} + \mathbf{b} + \mathbf{c}}{3}$$

We may use negative weights as well, provided that $\alpha + \beta + \gamma \neq 0$. If all three weights have the same signature, the barycenter is inside the triangle ABC; otherwise, it is outside. If one of the weights is equal to zero, the barycenter is on a line containing one side of the triangle. If two of them are 0, the barycenter is a vertex.

Every point of the plane of the triangle ABC is the barycenter of the system for some appropriate weights α, β, γ.

Case of the Space $\mathbb{R}^3$ Let A, B, C, D be four points not in the same plane. They then determine a tetrahedron. Let $\mathbf{a}$, $\mathbf{b}$, $\mathbf{c}$, $\mathbf{d}$ denote their respective location vectors. If P is any point in the space and $\mathbf{p}$ is the location vector of P, then the four vectors $\mathbf{a} - \mathbf{p}$, $\mathbf{b} - \mathbf{p}$, $\mathbf{c} - \mathbf{p}$, $\mathbf{d} - \mathbf{p}$ are linearly dependent, so that

$$\alpha(\mathbf{a} - \mathbf{p}) + \beta(\mathbf{b} - \mathbf{p}) + \gamma(\mathbf{c} - \mathbf{p}) + \delta(\mathbf{d} - \mathbf{p}) = \mathbf{0}$$

or

$$(\alpha + \beta + \gamma + \delta)\mathbf{p} = \alpha\mathbf{a} + \beta\mathbf{b} + \gamma\mathbf{c} + \delta\mathbf{d}$$

We want to show that $\alpha + \beta + \gamma + \delta \neq 0$. Indeed, if $\alpha + \beta + \gamma + \delta = 0$, then $\delta = -(\alpha + \beta + \gamma)$. Here $\alpha + \beta + \gamma \neq 0$; otherwise, A, B, C would be collinear. Therefore, we have

$$\mathbf{d} = \frac{\alpha\mathbf{a} + \beta\mathbf{b} + \gamma\mathbf{c}}{\alpha + \beta + \gamma}$$

which is impossible, because then D would fall in the plane of A, B, C. We may therefore conclude that $\alpha + \beta + \gamma + \delta \neq 0$; hence

$$\mathbf{p} = \frac{\alpha\mathbf{a} + \beta\mathbf{b} + \gamma\mathbf{c} + \delta\mathbf{d}}{\alpha + \beta + \gamma + \delta}$$

We shall say that P is the *barycenter* of the system with base points A, B, C, D with weights α, β, γ, δ. These numbers α, β, γ, δ are also called the *barycentric coordinates* of the point P (with respect to the base points A, B, C, D). The barycentric coordinates of P are not uniquely determined: multiplying each by the same nonzero number will leave the barycenter unchanged. Every point P in the three-dimensional space is the barycenter of the system for some appropriate weights α, β, γ, δ.

If $\alpha = \beta = \gamma = \delta$, the barycenter coincides with the *centroid of the tetrahedron $ABCD$*, and its location vector is

$$\mathbf{p} = \frac{\mathbf{a} + \mathbf{b} + \mathbf{c} + \mathbf{d}}{4}$$

Some of the weights may be negative, but they are always subject to the condition $\alpha + \beta + \gamma + \delta \neq 0$. If all four weights are of the same sign, the barycenter is inside the tetrahedron $ABCD$; otherwise it is outside. If one of the weights is zero, the barycenter is on one of the faces, if two of the weights are zero, the barycenter is on one of the edges of the tetrahedron: and if three weights are zero, the barycenter coincides with one of the vertices.

COORDINATE FORMULAS

To summarize the rules of vector calculations in terms of the coordinates, let $\mathbf{v} = x\mathbf{i} + y\mathbf{j} + z\mathbf{k}$, $\mathbf{v}' = x'\mathbf{i} + y'\mathbf{j} + z'\mathbf{k}$. Then

$$\mathbf{v} + \mathbf{v}' = (x + x')\mathbf{i} + (y + y')\mathbf{j} + (z + z')\mathbf{k}$$

$$\lambda\mathbf{v} = (\lambda x)\mathbf{i} + (\lambda y)\mathbf{j} + (\lambda z)\mathbf{k}$$

$$-\mathbf{v} = -x\mathbf{i} - y\mathbf{j} - z\mathbf{k}$$

Three linearly independent vectors **a**, **b**, **c** furnish a *skew affine coordinate system*, whose axes may not be perpendicular to one another, and the scales on them may not have the same unit of length; moreover, the orientation of {**a**, **b**, **c**} may be negative (left-hand system). The *rectangular metric coordinate system* is just a special case of the skew affine coordinate systems, when $\mathbf{a} = \mathbf{i}$, $\mathbf{b} = \mathbf{j}$, $\mathbf{c} = \mathbf{k}$ are mutually perpendicular unit vectors forming a right-hand system. Every vector can be written

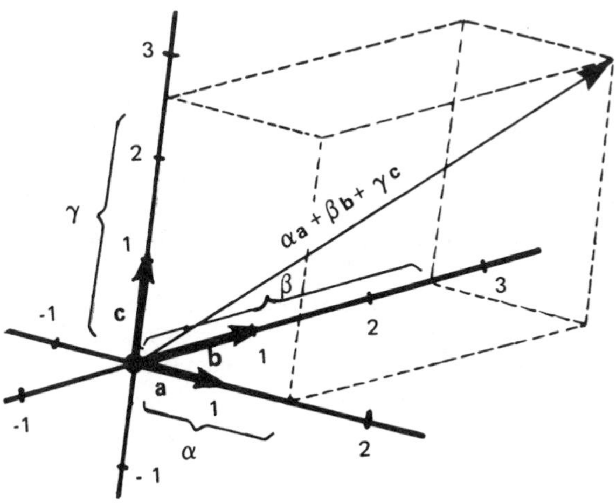

uniquely as a linear combination of the three vectors **a**, **b**, **c** if the latter are linearly independent†: $\mathbf{v} = \alpha\mathbf{a} + \beta\mathbf{b} + \gamma\mathbf{c}$.

EXERCISES

(*Note*: The calculational techniques needed for some of the exercises that follow include gaussian elimination. Readers wishing to review their knowledge of these techniques may find Appendix 1 at the end of this volume helpful.)

1.1 Show that the vectors $\mathbf{a} = \mathbf{i} + 3\mathbf{j} - \mathbf{k}$, $\mathbf{b} = 2\mathbf{i} - 3\mathbf{j} - 11\mathbf{k}$, $\mathbf{c} = -\mathbf{i} + 2\mathbf{j} + 6\mathbf{k}$ are linearly dependent.

1.2 Show that the vectors $\mathbf{i} - 2\mathbf{j} + 3\mathbf{k}$, $2\mathbf{i} - 2\mathbf{j}$, $\mathbf{j} + 7\mathbf{k}$ are linearly independent.

1.3 Are the vectors $\mathbf{i} + 5\mathbf{j} - \mathbf{k}$, $2\mathbf{i} - 4\mathbf{j} - 3\mathbf{k}$, $3\mathbf{i} + \mathbf{j} - 4\mathbf{k}$ linearly independent or dependent?

1.4 Are the vectors $\mathbf{i} + 2\mathbf{j} + 3\mathbf{k}$, $-\mathbf{i} + 2\mathbf{j} + 2\mathbf{k}$, $-\mathbf{i} - 2\mathbf{j} + 3\mathbf{k}$ linearly independent or dependent?

1.5 Can the vector $\mathbf{v} = 6\mathbf{i} + 3\mathbf{j} + 15\mathbf{k}$ be written as a linear combination of the vectors $\mathbf{a} = \mathbf{i} + 2\mathbf{j} + 4\mathbf{k}$, $\mathbf{b} = \mathbf{i} - \mathbf{j} + \mathbf{k}$, $\mathbf{c} = \mathbf{i} + \mathbf{j} + 3\mathbf{k}$, and if so, how?

1.6 Can the vector $\mathbf{i} + 2\mathbf{j}$ be written as a linear combination of the vectors $2\mathbf{i} + \mathbf{j} + 3\mathbf{k}$, $\mathbf{i} + \mathbf{j} + 2\mathbf{k}$, $6\mathbf{i} - \mathbf{j} + 5\mathbf{k}$, and if so, how?

1.7 Show that the vectors **a**, **b**, **c** form a linearly independent set, and express the vector **v** as a linear combination of **a**, **b**, **c** in each of the following cases.
(i) $\mathbf{a} = 3\mathbf{i} + 2\mathbf{j} + \mathbf{k}$, $\mathbf{b} = 2\mathbf{i} - 4\mathbf{j} + \mathbf{k}$, $\mathbf{c} = \mathbf{i} + 5\mathbf{j} - \mathbf{k}$; $\mathbf{v} = 6\mathbf{i} + 3\mathbf{j} + \mathbf{k}$
(ii) $\mathbf{a} = -3\mathbf{i} + 4\mathbf{j} + 2\mathbf{k}$, $\mathbf{b} = 7\mathbf{i} - \mathbf{j} + 3\mathbf{k}$, $\mathbf{c} = \mathbf{i} + 2\mathbf{j} + 8\mathbf{k}$; $\mathbf{v} = \mathbf{i}$
(iii) $\mathbf{a} = 2\mathbf{i} + 3\mathbf{j} + 3\mathbf{k}$, $\mathbf{b} = -\mathbf{i} + 4\mathbf{j} - 2\mathbf{k}$, $\mathbf{c} = -\mathbf{i} - 2\mathbf{j} + 4\mathbf{k}$; $\mathbf{v} = 4\mathbf{i} + 11\mathbf{j} + 11\mathbf{k}$
(iv) $\mathbf{a} = \mathbf{i} + \mathbf{j} + 2\mathbf{k}$, $\mathbf{b} = 2\mathbf{i} + 3\mathbf{j} + 5\mathbf{k}$, $\mathbf{c} = -3\mathbf{i} + \mathbf{j} - 4\mathbf{k}$; $\mathbf{v} = 4\mathbf{i} + 11\mathbf{j} + 13\mathbf{k}$

1.8 Find the coordinates of the point P which divides the segment AB in the ratio 2/3, where $A(1,-3,2)$ and $B(4,-2,-1)$.

1.9 With the same base points A, B as in Exercise 8, find the point P which divides AB in the ratio $-3/2$.

1.10 Find the centroid of the triangle with vertices $A(1,-2,3)$, $B(0,-1,-1)$, $C(4,-1,2)$.

1.11 Show that the three medians of a triangle meet at a point, namely, the centroid of the triangle. Show that, moreover, the centroid divides each median in the same ratio, 2. (A median is a line joining a vertex to the midpoint of the opposite side of the triangle.)

1.12 Show that the four medians of the tetrahedron meet at a point, namely, the centroid of the tetrahedron. Moreover, the centroid divides each median of the tetrahedron in the same ratio, 3. (A median is a line joining a vertex to the centroid of the opposite face of the tetrahedron.)

†The coefficients α, β, γ can be determined by Cramer's formula (see Chap. 4). There is more discussion on coordinate systems, bases, and orientation in Chap. 5.

1.13 The tetrahedron has three pairs of opposite edges; the join of the midpoints of a pair of opposite edges is called a *semimedian* of the tetrahedron. Show that the three semimedians meet at a point, namely, the centroid of the tetrahedron. Show that, moreover, the centroid divides each semimedian of the tetrahedron in the same ratio, 1.

1.14 Given two vertices $A(2,-3,-5)$, $B(-1,3,2)$ of a parallelogram $ABCD$ and the point $E(4,-1,7)$ of intersection of its diagonals, find the coordinates of the missing vertices.

1.15 Three vertices of a parallelogram $ABCD$ are given as $A(3,-4,7)$, $B(-5,3,-2)$, $C(1,2,-3)$. Find the fourth vertex, D, which is opposite B.

1.16 Find the coordinates of the points C, D, E, F which divide into five equal parts the line segment bounded by the points $A(-1,8,3)$, $B(9,-7,-2)$.

2

The Dot Product of Vectors

TRIGONOMETRIC FORMULA FOR THE DOT PRODUCT

The symbol $|\mathbf{a}|$ will be used to denote the *length* of vector **a**, and we agree to write, for simplicity's sake, $|\mathbf{a}|^2 = \mathbf{a}^2$.

The scalar

$$\mathbf{a} \cdot \mathbf{b} = \frac{1}{2}[(\mathbf{a} + \mathbf{b})^2 - \mathbf{a}^2 - \mathbf{b}^2]$$

is called the *dot product* of the vectors **a**, **b**. This definition was motivated by the desire to extend the validity of the important formulas

$$(\mathbf{a} + \mathbf{b})^2 = \mathbf{a}^2 + 2\mathbf{a} \cdot \mathbf{b} + \mathbf{b}^2$$
$$(\mathbf{a} - \mathbf{b})^2 = \mathbf{a}^2 - 2\mathbf{a} \cdot \mathbf{b} + \mathbf{b}^2$$

to vectors. Let us look at some significant special cases. If **a** and **b** are perpendicular vectors, in symbols, $\mathbf{a} \perp \mathbf{b}$, then by the theorem of Pythagoras, $(\mathbf{a} \pm \mathbf{b})^2 = \mathbf{a}^2 + \mathbf{b}^2$ and hence the dot product $\mathbf{a} \cdot \mathbf{b}$ must be zero. This rules out the cancellation law for the dot product. The converse is also true, so that

$$\mathbf{a} \cdot \mathbf{b} = 0 \Leftrightarrow \mathbf{a} \perp \mathbf{b}$$

The null vector is deemed to be perpendicular (and, by the same token, also parallel) to *every* vector, the only vector with this property.

$$\mathbf{a} \cdot \mathbf{b} > 0 \Leftrightarrow \mathbf{a}, \mathbf{b} \text{ make an acute angle}$$

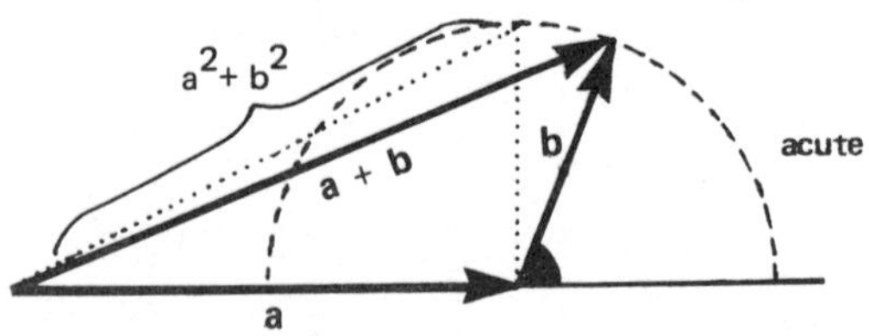

$$\mathbf{a}\cdot\mathbf{b} < 0 \Leftrightarrow \mathbf{a}, \mathbf{b} \text{ make an obtuse angle}$$

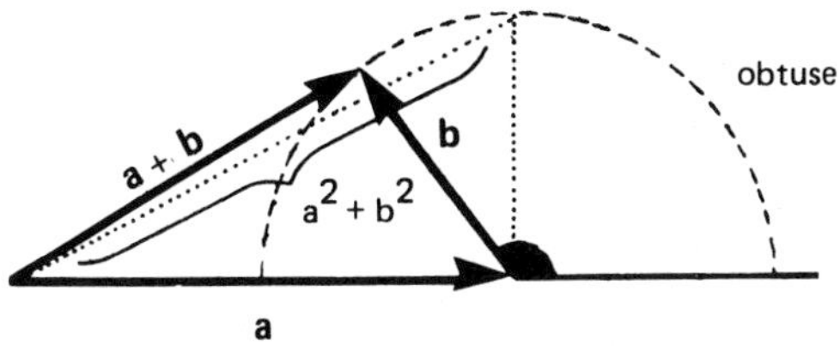

If the lengths of the vectors **a**, **b** are fixed, the maximum value of $\mathbf{a}\cdot\mathbf{b}$ is $|\mathbf{a}|\ |\mathbf{b}|$, whereas its minimum value is $-|\mathbf{a}|\ |\mathbf{b}|$:

$$-|\mathbf{a}|\ |\mathbf{b}| \leqq \mathbf{a}\cdot\mathbf{b} \leqq |\mathbf{a}|\ |\mathbf{b}|$$

Indeed, the maximum is assumed if **a** and **b** are parallel and have the same orientation, and the minimum is assumed if **a** and **b** are parallel and have the opposite orientation. In these cases the length of the sum is equal to the sum (difference) of the lengths of vectors and its square is calculated using the ordinary formula for squaring the sum (difference) of numbers.

We see that the dot product of **a**, **b** measures the deviation of the length of the third side $\mathbf{a}+\mathbf{b}$ of the triangle spanned by **a**, **b** from the standard set by the hypotenuse of the right triangle.

The foregoing definition of the dot product is a relationship between the squares of the sides of a triangle. But trigonometry has already furnished such a relationship, the law of cosines:

$$(\mathbf{a}-\mathbf{b})^2 = \mathbf{a}^2 + \mathbf{b}^2 - 2|\mathbf{a}|\ |\mathbf{b}|\cos\sphericalangle(\mathbf{a},\mathbf{b})$$

Comparing this with $(\mathbf{a}-\mathbf{b})^2 = \mathbf{a}^2 - 2\mathbf{a}\cdot\mathbf{b} + \mathbf{b}^2$, we have the

Trigonometric Formula for the Dot Product

$$\mathbf{a}\cdot\mathbf{b} = |\mathbf{a}|\ |\mathbf{b}|\cos\sphericalangle(\mathbf{a},\mathbf{b})$$

APPLICATION

1. The Theorem of Pythagoras

The *theorem of Pythagoras* is just a special case of the law of cosines: If $|\mathbf{a}| \neq 0$, $|\mathbf{b}| \neq 0$, then

$$\sphericalangle(\mathbf{a},\mathbf{b}) = 90 \Leftrightarrow |\mathbf{a}|^2 + |\mathbf{b}|^2 = |\mathbf{a}-\mathbf{b}|^2$$

It is remarkable that the condition is also necessary, not merely sufficient. That is, if in a triangle the sum of the squares of two sides equals

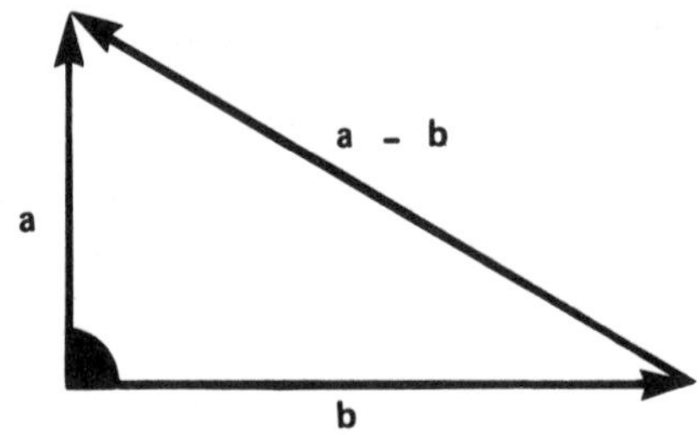

the square of the third side, then the angle opposite the third side must be a right angle—a fact that most accounts of this theorem fail to stress. Yet it was this fact that the Mesopotamians were using, some 4000 years ago, to construct a right triangle in practice. They observed that $3^2 + 4^2 = 9 + 16 = 25 = 5^2$ and concluded that the triangle of sides 3, 4, 5 has a right angle opposite to the side of length 5.

THE PERPENDICULAR PROJECTION THEOREM

A vector **u** is called a *unit vector* if it has length 1, i.e., $|\mathbf{u}| = 1$.

Theorem The dot product of a vector with a unit vector is just the signed length of the perpendicular projection of the vector onto any line parallel to the unit vector; the signature is positive or negative according as the vectors make an acute or an obtuse angle.

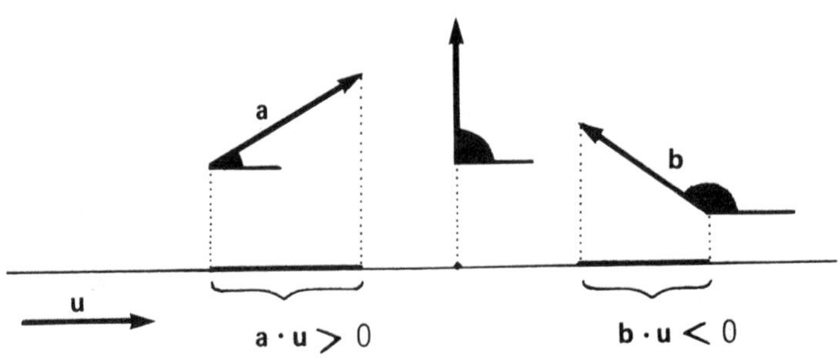

This theorem is an immediate consequence of the trigonometric formula for the dot product.

FORMAL RULES OF THE DOT PRODUCT

(0) Closure

$$\mathbf{a}, \mathbf{b} \text{ vectors} \Rightarrow \mathbf{a} \cdot \mathbf{b} \text{ scalar}$$

(1) Commutativity

$$\mathbf{a}\cdot\mathbf{b}=\mathbf{b}\cdot\mathbf{a}$$

(2) Scalar Associativity

$$\lambda(\mathbf{a}\cdot\mathbf{b})=(\lambda\mathbf{a})\cdot\mathbf{b}\qquad(\lambda\in\mathbb{R})$$

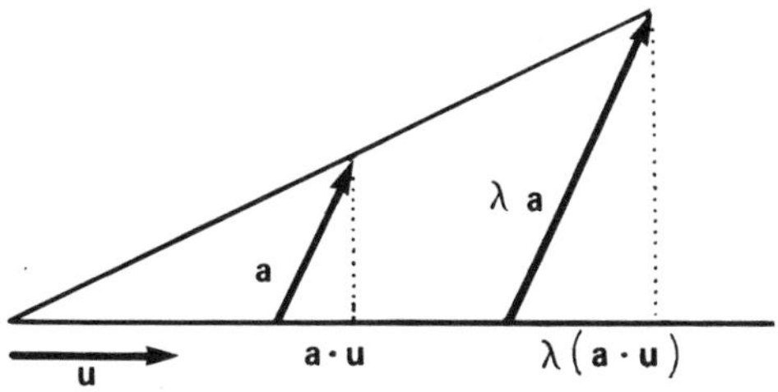

This figure shows that a strain in ratio λ transforms the right triangle with hypotenuse **a** into the right triangle with hypotenuse $\lambda\mathbf{a}$; in particular, the horizontal side of length $\mathbf{a}\cdot\mathbf{u}$ is transformed into the horizontal side of length $\lambda(\mathbf{a}\cdot\mathbf{u})=(\lambda\mathbf{a})\cdot\mathbf{u}$.

(4) Positive Definiteness

$$\mathbf{a}\cdot\mathbf{a}\geqq 0\qquad \mathbf{a}\cdot\mathbf{a}=0\Rightarrow\mathbf{a}=0$$

i.e., the dot product of a vector with itself is nonnegative, and it can be zero only for the null vector. This is a consequence of a special case of the dot product: The dot product of a vector with itself is just the square of its length:

$$\mathbf{a}\cdot\mathbf{a}=\mathbf{a}^2=|\mathbf{a}|^2$$

(5) Distributivity

$$(\mathbf{a}+\mathbf{b})\cdot\mathbf{c}=\mathbf{a}\cdot\mathbf{c}+\mathbf{b}\cdot\mathbf{c}$$

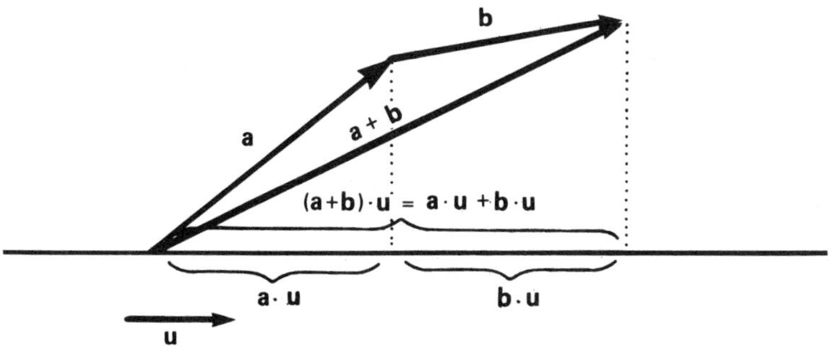

This figure shows that the perpendicular projection of a triangle onto a line is three collinear points and the three segments that they pairwise determine. In particular, the perpendicular projection of the side $\mathbf{a}+\mathbf{b}$ is the segment of length $(\mathbf{a}+\mathbf{b})\cdot\mathbf{u}=\mathbf{a}\cdot\mathbf{u}+\mathbf{b}\cdot\mathbf{u}$.

$(2^{\cdot})$ and $(5^{\cdot})$ are often lumped together and referred to as

Bilinearity

$$\lambda(\mathbf{a}\cdot\mathbf{b})=(\lambda\mathbf{a})\cdot\mathbf{b}$$

$$(\mathbf{a}+\mathbf{b})\cdot\mathbf{c}=\mathbf{a}\cdot\mathbf{c}+\mathbf{b}\cdot\mathbf{c}$$

The figures above prove bilinearity for the special case when $\mathbf{c}=\mathbf{u}$ is a unit vector. The general case follows from this special case if we apply a strain in ratio $\lambda=|\mathbf{c}|$.

The five formal rules of the dot product are combined with the 10 formal rules of vector sum and scalar multiple in the following table:

CLOSURE	0	0'	0·	
COMMUTATIVITY	1		1·	
ASSOCIATIVITY	2	2'	2·	SCALAR ASSOCIATIVITY
NEUTRAL	3	3'		
NEGATIVE	4		4·	POSITIVE DEFINITENESS
DISTRIBUTIVITY	5	5'	5·	

We say that the introduction of the dot product has converted our vector space into a *metric vector space*. Such concepts as the length of a vector, the angle made by two vectors, perpendicularity, the circle (and many others) depend, in an essential way, on the dot product.

APPLICATIONS

2. Diagonals of the Rhomb

Rhomb is the name of the equilateral parallelogram. We shall now prove that *the diagonals of a parallelogram are perpendicular if, and only if, it is a rhomb*.

Proof:

$$(\mathbf{a}+\mathbf{b})\cdot(\mathbf{a}-\mathbf{b}) = \mathbf{a}\cdot\mathbf{a}+\mathbf{a}\cdot\mathbf{b}-\mathbf{b}\cdot\mathbf{a}-\mathbf{b}\cdot\mathbf{b}$$
$$= \mathbf{a}\cdot\mathbf{a}-\mathbf{b}\cdot\mathbf{b} = |\mathbf{a}|^2-|\mathbf{b}|^2$$

Therefore, we may argue as follows:

$$\mathbf{a}+\mathbf{b} \text{ is perpendicular to } \mathbf{a}-\mathbf{b} \Leftrightarrow (\mathbf{a}+\mathbf{b})\cdot(\mathbf{a}-\mathbf{b})=0$$
$$\Leftrightarrow |\mathbf{a}| = |\mathbf{b}|$$
$$\Leftrightarrow \mathbf{a}, \mathbf{b} \text{ span a rhomb}$$

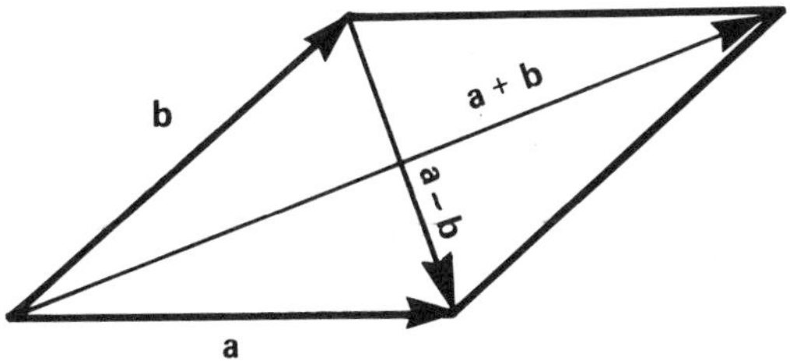

3. The Theorem of Thales

Let us consider the set of all right triangles with the segment between A and B as their common hypotenuse.

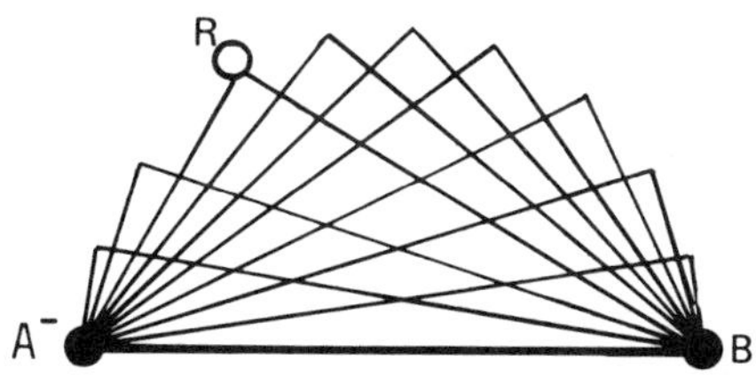

The theorem of Thales asserts that *a point R is the vertex of such a right triangle if, and only if, it falls on the circle a diameter of which is AB.*

Proof: Putting the origin at the midpoint of AB we have that

$$\mathbf{r}+\mathbf{a}\perp\mathbf{r}-\mathbf{a} \Leftrightarrow (\mathbf{r}+\mathbf{a})\cdot(\mathbf{r}-\mathbf{a})=0$$
$$\Leftrightarrow |\mathbf{r}|^2-|\mathbf{a}|^2=0$$
$$\Leftrightarrow |\mathbf{r}| = |\mathbf{a}| = \text{constant}$$

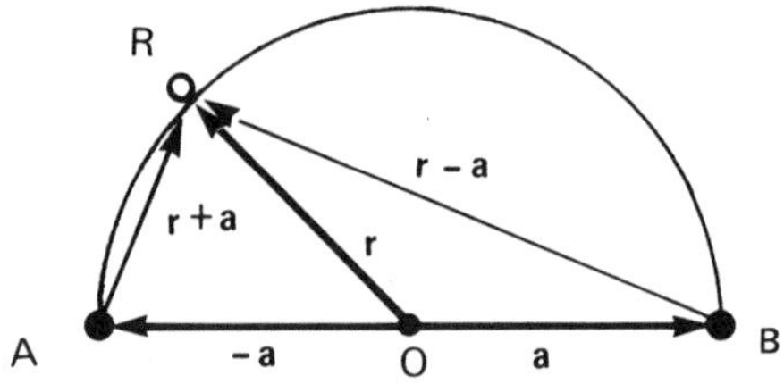

Thus R falls on the circle centered at 0 and having radius $|\mathbf{a}|$.

4. Orthocenter of the Triangle

We shall now prove that the three altitudes of a triangle meet in a point, called the *orthocenter* of the triangle.

Let $\mathbf{a}$, $\mathbf{b}$, $\mathbf{c}$ denote the location vectors of the vertices of the triangle, and let $\mathbf{h}$ be the location vector of the point of intersection between the two altitudes belonging to the vertices $\mathbf{a}$, $\mathbf{b}$. As a quick calculation (not reproduced here) shows, the equation

$$(\mathbf{a}-\mathbf{b})\cdot(\mathbf{h}-\mathbf{c})+(\mathbf{b}-\mathbf{c})\cdot(\mathbf{h}-\mathbf{a})+(\mathbf{c}-\mathbf{a})\cdot(\mathbf{h}-\mathbf{b})=0$$

is satisfied for every choice of the vectors $\mathbf{a}$, $\mathbf{b}$, $\mathbf{c}$, $\mathbf{h}$. So it must be satisfied for ours. However, by the definition of $\mathbf{h}$, we have that the last two terms vanish, thanks to the perpendicularity of the altitudes $\mathbf{h}-\mathbf{a}$, $\mathbf{h}-\mathbf{b}$ to the opposite sides $\mathbf{b}-\mathbf{c}$, $\mathbf{c}-\mathbf{a}$. Therefore, the first term must also vanish: $(\mathbf{a}-\mathbf{b})\cdot(\mathbf{h}-\mathbf{c})=0$. It follows that the line joining the terminal point of $\mathbf{h}$ to the vertex of the triangle at the terminal point of $\mathbf{c}$ is just the third altitude. This concludes the proof that the three altitudes of a triangle meet at the same point.

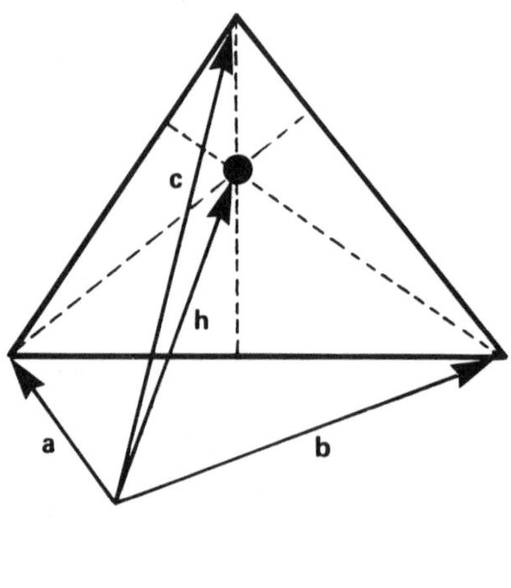

5. Circumcenter of the Triangle

We shall now prove that the three perpendicular bisectors of the sides of a triangle meet in a point, called the *circumcenter* of the triangle.

The proof is analogous to the previous one, and is based on the vector equation, satisfied for any vectors **a**, **b**, **c**, **m**:

$$(\mathbf{a}-\mathbf{b})\cdot\left(\mathbf{m}-\frac{\mathbf{a}+\mathbf{b}}{2}\right)+(\mathbf{b}-\mathbf{c})\cdot\left(\mathbf{m}-\frac{\mathbf{b}+\mathbf{c}}{2}\right)+(\mathbf{c}-\mathbf{a})\cdot\left(\mathbf{m}-\frac{\mathbf{c}+\mathbf{a}}{2}\right)=0$$

Let **m** be the location vector of the point of intersection between two perpendicular bisectors.

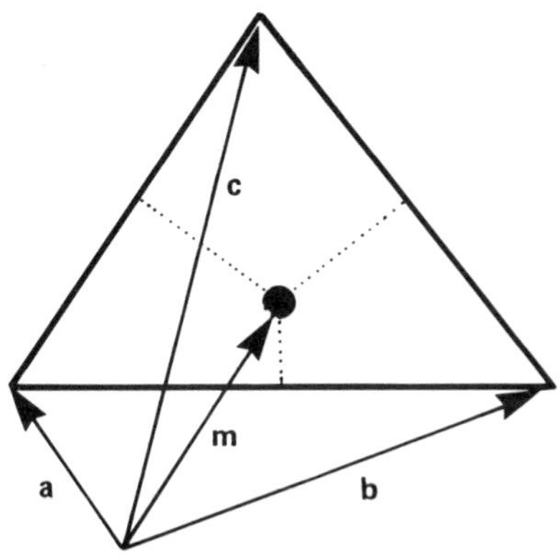

The circumcenter earns its name by serving as the center of the circumcircle. To see this, we shift the origin to the terminal point of **m**; then $\mathbf{m}=\mathbf{0}$ and we have

$$(\mathbf{a}-\mathbf{b})\cdot(\mathbf{a}+\mathbf{b})=0\Rightarrow \mathbf{a}^2-\mathbf{b}^2=0\Rightarrow |\mathbf{a}|=|\mathbf{b}|$$

and a similar calculation shows that $|\mathbf{b}|=|\mathbf{c}|$, $|\mathbf{c}|=|\mathbf{a}|$, so that $|\mathbf{a}|=|\mathbf{b}|=|\mathbf{c}|=\mathrm{r}$, which is the circumradius of the triangle.

6. Euler Line of the Triangle

As before, let the origin be at the circumcenter of the triangle ($\mathbf{m}=\mathbf{0}$). We have seen above that

$$(\mathbf{a}-\mathbf{b})\cdot(\mathbf{a}+\mathbf{b})=0$$

and

$$(\mathbf{a}-\mathbf{b})\cdot(\mathbf{h}-\mathbf{c})=0$$

By subtracting the latter from the former, we get

$$(\mathbf{a}-\mathbf{b})\cdot(\mathbf{a}+\mathbf{b}+\mathbf{c}-\mathbf{h})=0$$

and, similarly,

$$(\mathbf{b}-\mathbf{c})\cdot(\mathbf{a}+\mathbf{b}+\mathbf{c}-\mathbf{h})=0$$

$$(\mathbf{c}-\mathbf{a})\cdot(\mathbf{a}+\mathbf{b}+\mathbf{c}-\mathbf{h})=0$$

This means that the vector $\mathbf{a}+\mathbf{b}+\mathbf{c}-\mathbf{h}$ is perpendicular to all three sides of the triangle. But this is impossible, unless this vector is the null vector, or what is the same,

$$\mathbf{h}=\mathbf{a}+\mathbf{b}+\mathbf{c}=3\mathbf{p}$$

where

$$\mathbf{p}=\frac{\mathbf{a}+\mathbf{b}+\mathbf{c}}{3}$$

is the location vector of the centroid of the triangle. We see from the equation $\mathbf{h}=3\mathbf{p}$ that the circumcenter, the orthocenter, and the centroid of the triangle fall on the same straight line, called the Euler line of the triangle. Moreover, the centroid is between the circumcenter and the orthocenter; in fact, it divides the segment between the circumcenter and the orthocenter in the ratio ½. There is no Euler line if the circumcenter and the orthocenter of the triangle coincide, in which case these two points also coincide with the centroid of the triangle. It is clear from the foregoing that this happens if, and only if, the triangle is equilateral.

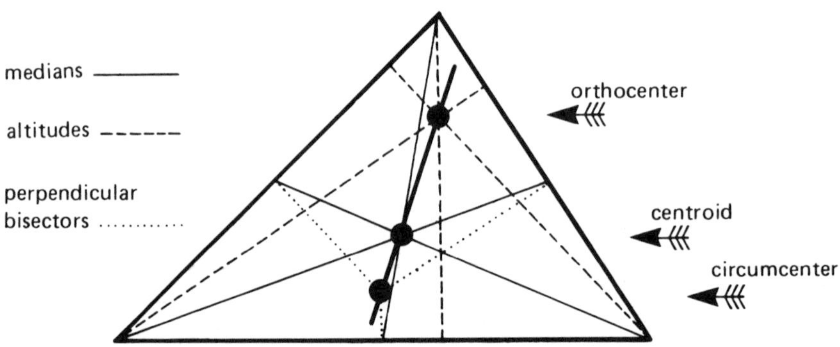

7. Nine-Point Circle of the Triangle

As before, let the origin be at the circumcenter of the triangle ($\mathbf{m}=\mathbf{0}$). Let $\mathbf{f}$ denote the location vector of the midpoint between the circumcenter and the orthocenter of the triangle, i.e., $\mathbf{f}=\frac{1}{2}\mathbf{h}$. The midpoints of the three sides of the triangle are located by the vectors

$$\frac{\mathbf{b}+\mathbf{c}}{2} \qquad \frac{\mathbf{c}+\mathbf{a}}{2} \qquad \frac{\mathbf{a}+\mathbf{b}}{2}$$

and the midpoints of the segments from the orthocenter to the three vertices of the triangle are located by the vectors

$$\frac{\mathbf{h}+\mathbf{a}}{2} \qquad \frac{\mathbf{h}+\mathbf{b}}{2} \qquad \frac{\mathbf{h}+\mathbf{c}}{2}$$

Thus we have six points; we wish to show that they are at the same distance from the point located by **f**; namely, they fall on a circle centered at the midpoint between the circumcenter and the orthocenter, with radius ½r, one-half of the circumradius. Indeed,

$$\mathbf{f}-\frac{\mathbf{b}+\mathbf{c}}{2}=\frac{\mathbf{a}+\mathbf{b}+\mathbf{c}}{2}-\frac{\mathbf{b}+\mathbf{c}}{2}=\frac{1}{2}\mathbf{a} \qquad \left|\mathbf{f}-\frac{\mathbf{b}+\mathbf{c}}{2}\right|=\left|\frac{1}{2}\mathbf{a}\right|=\frac{1}{2}\mathrm{r}$$

$$\mathbf{f}-\frac{\mathbf{h}+\mathbf{a}}{2}=\frac{1}{2}\mathbf{h}-\frac{1}{2}\mathbf{h}-\frac{1}{2}\mathbf{a}=-\frac{1}{2}\mathbf{a} \qquad \left|\mathbf{f}-\frac{\mathbf{h}+\mathbf{a}}{2}\right|=\left|-\frac{1}{2}\mathbf{a}\right|=\frac{1}{2}\mathrm{r}$$

etc.

This indicates that the segment joining the midpoint of a side to the midpoint between the opposite vertex and the orthocenter is a diameter of this circle. Furthermore, on the strength of the theorem of Thales, the feet of the three altitudes of the triangle also fall on this circle. We have thus identified nine points of this interesting circle associated with the triangle. This explains why it is known as the *nine-point circle* of the triangle.

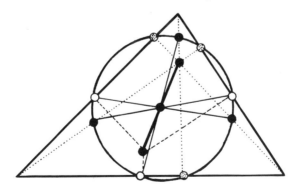

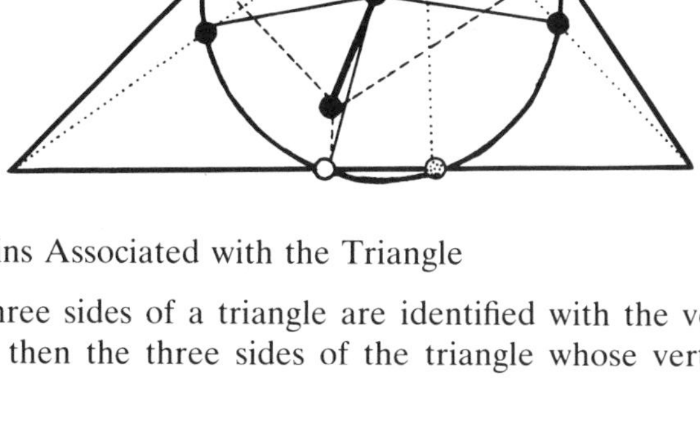

8. Two Strains Associated with the Triangle

If the three sides of a triangle are identified with the vectors $\mathbf{a}-\mathbf{b}$, $\mathbf{b}-\mathbf{c}$, $\mathbf{c}-\mathbf{a}$, then the three sides of the triangle whose vertices are the midpoints of the sides of the first triangle are identified with the vectors

$$\frac{\mathbf{b}+\mathbf{c}}{2}-\frac{\mathbf{c}+\mathbf{a}}{2}=\frac{\mathbf{b}-\mathbf{a}}{2}=-\frac{1}{2}(\mathbf{a}-\mathbf{b}) \qquad -\frac{1}{2}(\mathbf{b}-\mathbf{c}) \qquad -\frac{1}{2}(\mathbf{c}-\mathbf{a})$$

Therefore, the smaller triangle is obtained from the larger one through applying a strain in the ratio $-\frac{1}{2}$. The center of this strain is clearly the centroid of the large triangle. This strain transforms the circumcircle of the large triangle into the circumcircle of the small circle; but the latter circle is just the nine-point circle of the large triangle. It follows from this that the radius of the nine-point circle is exactly one-half of the circumradius. The strain leaves the Euler line of the large triangle unchanged:

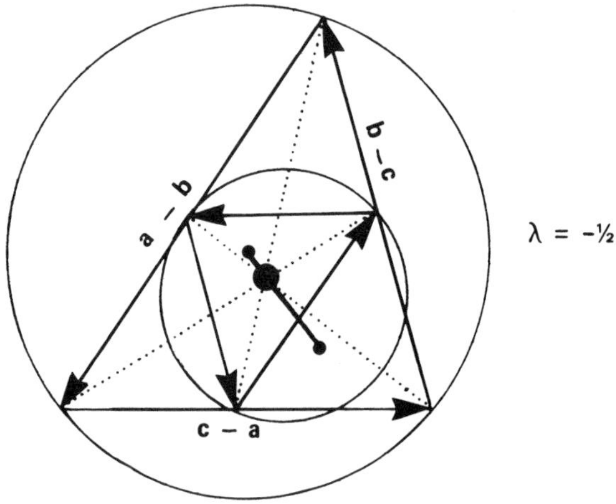

Again, we identify the three sides of the triangle with the vectors $\mathbf{a}-\mathbf{b}$, $\mathbf{b}-\mathbf{c}$, $\mathbf{c}-\mathbf{a}$. Another triangle is formed by the midpoints of the segments between the orthocenter and the three vertices of the large triangle; its sides can be identified with the vectors

$$\frac{\mathbf{h}+\mathbf{a}}{2}-\frac{\mathbf{h}+\mathbf{b}}{2}=\frac{1}{2}(\mathbf{a}-\mathbf{b}) \qquad \frac{1}{2}(\mathbf{b}-\mathbf{c}) \qquad \frac{1}{2}(\mathbf{c}-\mathbf{a})$$

Therefore, the small triangle is obtained from the large one through applying a strain in the ratio $\frac{1}{2}$. The center of this strain is clearly the orthocenter. This strain transforms the circumcircle of the large triangle into its nine-point circle, and transforms the Euler line into itself:

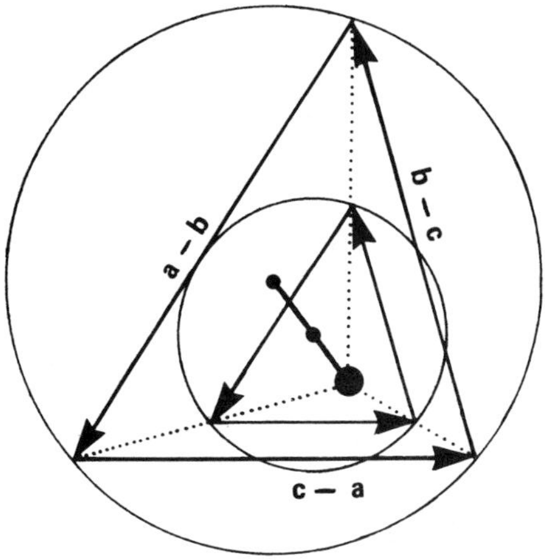

The dot product is indispensable in calculating distance, angle, in checking perpendicularity, in normalizing vectors, etc.

Length of a Vector

$$|\mathbf{a}| = \sqrt{\mathbf{a} \cdot \mathbf{a}}$$

Normalizing a Vector To normalize a vector means to find the unit vector (i.e., vector of length 1) having the same position and the same orientation as the given vector **a**. Denoting the normalized vector of **a** by the symbol $\mathbf{a}^0$, we have that

$$\mathbf{a}^0 = \frac{\mathbf{a}}{|\mathbf{a}|}$$

provided that $\mathbf{a} \neq 0$. The null vector **0** cannot be normalized.

Angle Made by Two Vectors **a**, **b**

$$\cos\sphericalangle(\mathbf{a},\mathbf{b}) = \frac{\mathbf{a} \cdot \mathbf{b}}{|\mathbf{a}|\,|\mathbf{b}|}$$

Equivalently, the cosine of $\sphericalangle(\mathbf{a},\mathbf{b})$ is the dot product of the normalized vectors $\mathbf{a}^0$, $\mathbf{b}^0$:

$$\cos\sphericalangle(\mathbf{a},\mathbf{b}) = \mathbf{a}^0 \cdot \mathbf{b}^0$$

COORDINATE FORMULAS

Since **i**, **j**, **k** are mutually perpendicular unit vectors, we have

$$\mathbf{i} \cdot \mathbf{i} = \mathbf{j} \cdot \mathbf{j} = \mathbf{k} \cdot \mathbf{k} = 1$$

and

$$\mathbf{i}\cdot\mathbf{j}=\mathbf{j}\cdot\mathbf{k}=\mathbf{k}\cdot\mathbf{i}=0$$

It follows from the distributive law that, for any vector $\mathbf{v}=x\mathbf{i}+y\mathbf{j}+z\mathbf{k}$,

$$x=\mathbf{i}\cdot\mathbf{v} \qquad y=\mathbf{j}\cdot\mathbf{v} \qquad z=\mathbf{k}\cdot\mathbf{v}$$

This gives the geometric interpretation of the coordinates of a vector:

x of $\mathbf{v}$ is the perpendicular projection of $\mathbf{v}$ onto the x axis.
y of $\mathbf{v}$ is the perpendicular projection of $\mathbf{v}$ onto the y axis.
z of $\mathbf{v}$ is the perpendicular projection of $\mathbf{v}$ onto the z axis.

Substituting these values of the coordinates in terms of the dot product, we get

$$\mathbf{v}=(\mathbf{i}\cdot\mathbf{v})\mathbf{i}+(\mathbf{j}\cdot\mathbf{v})\mathbf{j}+(\mathbf{k}\cdot\mathbf{v})\mathbf{k}$$

If $\mathbf{v}'=x'\mathbf{i}+y'\mathbf{j}+z'\mathbf{k}$ is another vector, and if we take the dot product of both sides of the expression above for $\mathbf{v}$ with $\mathbf{v}'$, we have

$$\mathbf{v}\cdot\mathbf{v}'=(\mathbf{i}\cdot\mathbf{v})(\mathbf{i}\cdot\mathbf{v}')+(\mathbf{j}\cdot\mathbf{v})(\mathbf{j}\cdot\mathbf{v}')+(\mathbf{k}\cdot\mathbf{v})(\mathbf{k}\cdot\mathbf{v}')$$

or, what is the same,

$$\mathbf{v}\cdot\mathbf{v}'=xx'+yy'+zz'$$

i.e., *the dot product of two vectors may be calculated by taking the sum of the products of their corresponding coordinates.*

The special case of this important formula for $\mathbf{v}'=\mathbf{v}$ yields the length of $\mathbf{v}$ in terms of its coordinates:

$$|\mathbf{v}|=\sqrt{x^2+y^2+z^2}$$

This also tells us how to normalize the vector $\mathbf{v}$ (provided that $\mathbf{v}\neq\mathbf{0}$):

$$\mathbf{v}^0=\frac{x\mathbf{i}+y\mathbf{j}+z\mathbf{k}}{\sqrt{x^2+y^2+z^2}}$$

DIRECTION COSINES

The coordinates of a vector are called direction numbers, and the coordinates of a *unit vector* are called *direction cosines*, for reasons to be made clear presently. Suppose that

$$\mathbf{u}=x\mathbf{i}+y\mathbf{j}+z\mathbf{k}$$

is a unit vector, i.e., $|\mathbf{u}|=1$; then clearly

$$x^2+y^2+z^2=1$$

Here $x = \mathbf{i}\cdot\mathbf{u} = \cos\alpha$, $y = \mathbf{j}\cdot\mathbf{u} = \cos\beta$, $z = \mathbf{k}\cdot\mathbf{u} = \cos\gamma$, where α, β, γ denote the angles made by **u** with **i**, **j**, **k**, respectively.

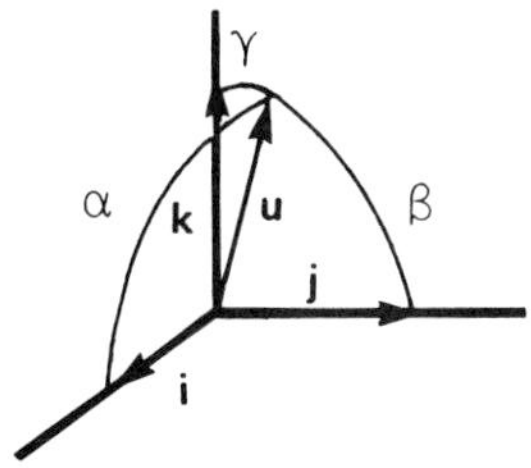

Thus every unit vector can be written in the form

$$\mathbf{u} = (\cos\alpha)\mathbf{i} + (\cos\beta)\mathbf{j} + (\cos\gamma)\mathbf{k}$$

where

$$\cos^2\alpha + \cos^2\beta + \cos^2\gamma = 1$$

Note that in the case of two-dimensional analytic geometry, when there are only two direction numbers, the same procedure yields

$$\mathbf{u} = (\cos\alpha)\mathbf{i} + (\sin\alpha)\mathbf{j}$$

where

$$\cos^2\alpha + \sin^2\alpha = 1$$

Here the direction cosines are $(\cos\alpha, \sin\alpha)$ because, in this simpler situation, $\alpha + \beta = 90°$ and therefore $\cos\beta = \cos(90° - \alpha) = \sin\alpha$.

Complete the table.

Direction	Direction cosines			Direction	Direction cosines
E	1	0	0	LS	
W				UE	
N				UW	
S				LW	
U				LE	
L				UNE	
NE				UNW	
NW				USW	
SW				USE	
SE				LNE	
UN				LNW	
US				LSW	
LN				LSE	

ORTHONORMAL BASES

We say that three vectors **u**, **v**, **w** form an *orthonormal basis* if they are mutually perpendicular unit vectors, i.e., $\mathbf{u}\cdot\mathbf{u}=\mathbf{v}\cdot\mathbf{v}=\mathbf{w}\cdot\mathbf{w}=1$, $\mathbf{u}\cdot\mathbf{v}=\mathbf{v}\cdot\mathbf{w}=\mathbf{w}\cdot\mathbf{u}=0$.

Example $\{\mathbf{i}, \mathbf{j}, \mathbf{k}\}$ is an orthonormal basis. Show that

$$\mathbf{u}=\frac{6}{31}\mathbf{i}+\frac{14}{31}\mathbf{j}+\frac{27}{31}\mathbf{k} \qquad \mathbf{v}=\frac{22}{31}\mathbf{i}-\frac{21}{31}\mathbf{j}+\frac{6}{31}\mathbf{k} \qquad \mathbf{w}=\frac{21}{31}\mathbf{i}+\frac{18}{31}\mathbf{j}-\frac{14}{31}\mathbf{k}$$

form an orthonormal basis.

Answer:

$$6^2+14^2+27^2=22^2+21^2+6^2=21^2+18^2+14^2=31^2$$

$$6(22)-14(21)+27(6)=22(21)-21(18)-6(14)=21(6)+18(14)-14(27)=0$$

APPLICATIONS

9. Power of a Point with Respect to a Circle. Radical Axis

By definition, a *circle* is the locus of those points in the plane whose distance from a fixed point, the *center*, is equal to a constant positive number, the *radius*. Let **c** be the location vector of the center, and $\rho>0$ the radius, of a circle C. We wish to find the equation of the circle C (in terms of metric rectangular coordinates of the plane $\mathbb{R}^2$). If **v** is the variable location vector of a point on the circle, then

$$\mathbf{v}\in C \Leftrightarrow |\mathbf{v}-\mathbf{c}|=\rho \Rightarrow (\mathbf{v}-\mathbf{c})^2=\rho^2$$

The last implication involves the squaring of the two sides of an equation. The converse of such an implication does not follow in general, without allowing points for which the signature of one side changes to the negative. In this case, however, there are no points such that $|\mathbf{v}-\mathbf{c}|$ is negative, i.e.,

$$\mathbf{v}\in C \Leftrightarrow (\mathbf{v}-\mathbf{c})^2-\rho^2=0$$

which gives us the vector form of the equation of the circle. Let now $\mathbf{v}=x\mathbf{i}+y\mathbf{j}$, $\mathbf{c}=\alpha\mathbf{i}+\beta\mathbf{j}$; then $(\mathbf{v}-\mathbf{c})^2=(x-\alpha)^2+(y-\beta)^2$ and we have the

Central Form of Circle

$$(x-\alpha)^2+(y-\beta)^2-\rho^2=0$$

in terms of rectangular metric coordinates. This can also be written as the

Normal Equation of Circle

$$x^2 + y^2 + Ax + By + C = 0$$

with $A = -2\alpha$, $B = -2\beta$, $C = \alpha^2 + \beta^2 - \rho^2$. The scalar-valued function of the vector variable $\mathbf{v}$,

$$C(\mathbf{v}) = (\mathbf{v} - \mathbf{c})^2 - \rho^2$$

or, what is the same, the scalar-valued function of two variables x, y,

$$C(x,y) = x^2 + y^2 + Ax + By + C$$

is called the *power* of the point $\mathbf{v} = x\mathbf{i} + y\mathbf{j}$ with respect to the circle C. Since $d = |\mathbf{v} - \mathbf{c}|$ is the distance from the point $\mathbf{v}$ to the center $\mathbf{c}$, we have that the power of the point is

$$C(\mathbf{v}) = d^2 - \rho^2$$

and hence

$C(\mathbf{v}) > 0 \Leftrightarrow$ the point is outside the circle.
$C(\mathbf{v}) < 0 \Leftrightarrow$ the point is inside the circle.
$C(\mathbf{v}) = 0 \Leftrightarrow$ the point is on the circle.

In the case $d > \rho$, $C(\mathbf{v}) > 0$ and therefore $C(\mathbf{v}) = \tau^2$ for some $\tau \in \mathbb{R}$ and

$$d^2 = \rho^2 + \tau^2$$

implies, on the strength of the theorem of Pythagoras, that there is a right triangle with sides ρ, τ and hypotenuse d. In fact, the side τ is the tangent that can be drawn to the circle C from the point $\mathbf{v}$. This gives us the geometric meaning of the power of the point with respect to the circle if the point is outside: *The power is the square of the length of the tangent from the point to the circle.*

We say that two circles, C_1 and C_2, intersect at a right angle, in symbols $C_1 \perp C_2$, if their tangents at the points of intersection are perpendicular to one another. It is clear that this happens if, and only if, the power of the center of one circle with respect to the other is equal to the square of the radius of the first circle. In fact, we have that *the following conditions on the circles C_1 and C_2 are equivalent*:

(1) $C_1 \perp C_2$

(2) $C_2(\mathbf{c}_1) = \rho_1^2$ [or $C_1(\mathbf{c}_2) = \rho_2^2$]

(3) $(\mathbf{c}_1 - \mathbf{c}_2)^2 = \rho_1^2 + \rho_2^2$

(4) $C_1(\mathbf{v}) + C_2(\mathbf{v}) = 2(\mathbf{v} - \mathbf{c}_1) \cdot (\mathbf{v} - \mathbf{c}_2)$ for all $\mathbf{v}$

Suppose now that the circles C_1 and C_2 are not concentric. We wish to find the locus of those points in the plane whose powers with respect to C_1 and C_2 are equal.

$$\begin{aligned}C_1(\mathbf{v}) = C_2(\mathbf{v}) &\Leftrightarrow (\mathbf{v}-\mathbf{c}_1)^2 - \rho_1^2 = (\mathbf{v}-\mathbf{c}_2)^2 - \rho_2^2\\ &\Leftrightarrow (2\mathbf{v} - (\mathbf{c}_1+\mathbf{c}_2))\cdot(\mathbf{c}_2-\mathbf{c}_1) = \rho_1^2 - \rho_2^2\\ &\Leftrightarrow 2\mathbf{v}\cdot(\mathbf{c}_2-\mathbf{c}_1) + \mathbf{c}_1^2 - \mathbf{c}_2^2 = \rho_1^2 - \rho_2^2\end{aligned}$$

Since the circles are not concentric, there is at least one point belonging to the locus and we may shift the origin to that point. Then

$$\begin{aligned}C_1(0) = C_2(0) &\Rightarrow \mathbf{c}_1^2 - \rho_1^2 = \mathbf{c}_2^2 - \rho_2^2\\ &\Rightarrow \rho_1^2 - \rho_2^2 = \mathbf{c}_1^2 - \mathbf{c}_2^2\end{aligned}$$

and the equation of the locus takes the form

$$\mathbf{v}\cdot(\mathbf{c}_2-\mathbf{c}_1) = 0$$

which is the equation of a straight line through the origin and perpendicular to the central $\mathbf{c}_2 - \mathbf{c}_1$ of the circles. (The *central* of a pair of nonconcentric circles is the line joining their centers.)

The straight line containing all the points having equal powers with respect to a pair of (nonconcentric) circles is called the *radical axis* of those circles. The equation of the radical axis can be obtained simply by subtracting the normal equation of one circle from that of the other: $C_1(\mathbf{v}) - C_2(\mathbf{v}) = 0$. If C_1, C_2 intersect in two distinct points, their radical axis is just the line joining the points of intersection. If C_1 and C_2 are tangent to one another (whether externally or internally) at a point, their radical axis is just their common tangent line at the point of contact.

10. Apollonius Circles

We shall solve the problem of Apollonius: Find the locus of those points in the plane whose distances from two fixed points in the plane give a constant positive ratio γ.

If the ratio is $\gamma = 1$, the locus is a straight line; in fact, it is the perpendicular bisector of the two fixed points. In what follows we shall assume that $\gamma \neq 1$. We may also take the fixed points as (1,0) and (−1,0) with location vectors $\mathbf{i}$, $-\mathbf{i}$, respectively. Then

$$\frac{|\mathbf{v}-\mathbf{i}|}{|\mathbf{v}+\mathbf{i}|} = \gamma \Rightarrow (\mathbf{v}-\mathbf{i})^2 = \gamma^2(\mathbf{v}+\mathbf{i})^2 \Rightarrow \mathbf{v}^2 + 1 + 2\left(\frac{\gamma^2+1}{\gamma^2-1}\right)\mathbf{v}\cdot\mathbf{i} = 0$$

$$\Rightarrow x^2 + y^2 + 1 + 2\left(\frac{\gamma^2+1}{\gamma^2-1}\right)x = 0$$

$$\Rightarrow \left(x + \frac{\gamma^2+1}{\gamma^2-1}\right)^2 + y^2 = \left(\frac{2\gamma}{\gamma^2-1}\right)^2$$

which is a circle centered at

$$\left(-\frac{\gamma^2+1}{\gamma^2-1}, 0\right) \quad \text{with radius} \quad \rho = \frac{2\gamma}{\gamma^2-1}$$

Every implication in the sequence can be reversed except the first one, where the equation is being squared. If we want to reverse this one, we have to admit negative values of γ. However, negative values of γ do not lead to new points that do not belong to the locus. Therefore, the points of the circle furnish the locus: the solution of the problem of Apollonius, for $\gamma \neq 1$, is a circle; for $\gamma = 1$, a straight line (namely, the perpendicular bisector). If we now allow γ to assume all positive values, we get a family of circles, known as the circles of Apollonius.

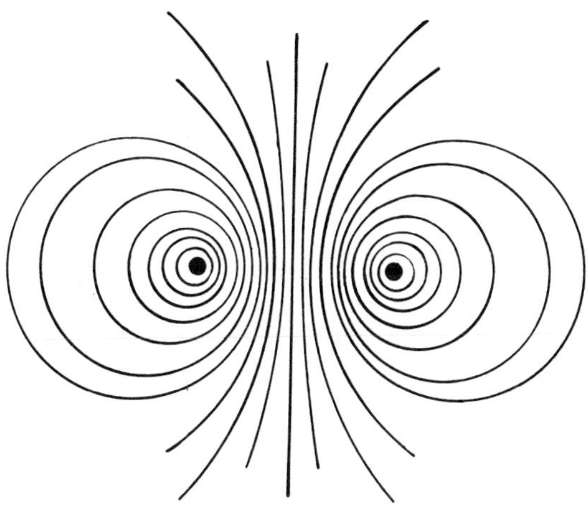

11. Thales Circles

We can paraphrase the theorem of Thales (Application 2 above) by saying that the locus of the points in the plane from which a fixed segment is seen at a right angle is the circle drawn around the segment as a diameter.

This leads to the following problem: Find the locus of those points in the plane from which a fixed segment can be seen at a constant angle α. If the angle is $\alpha = 180°$, the locus is the segment itself; if $\alpha = 0$, the locus is the complement of the segment in the straight line containing it. In what follows we shall assume that $\alpha \neq 0°$ or $180°$, and first we consider acute angles. We shall also fix the segment as the one with endpoints (1,0), (−1,0), having location vectors **i** and −**i**.

$$\frac{\mathbf{v}-\mathbf{i}}{|\mathbf{v}-\mathbf{i}|}\cdot\frac{\mathbf{v}+\mathbf{i}}{|\mathbf{v}+\mathbf{i}|}=\cos\alpha=\gamma\ \ (\gamma>0)\Rightarrow(\mathbf{v}^2-1)^2=\gamma^2(\mathbf{v}-\mathbf{i})^2(\mathbf{v}+\mathbf{i})^2$$

$$\begin{aligned}\Rightarrow(\mathbf{v}^2-1)^2&=\gamma^2(\mathbf{v}^2+1-2\mathbf{v}\cdot\mathbf{i})(\mathbf{v}^2+1+2\mathbf{v}\cdot\mathbf{i})\\&=\gamma^2((\mathbf{v}^2+1)^2-4(\mathbf{v}\cdot\mathbf{i})^2)\\&=\gamma^2((\mathbf{v}^2-1)^2+4(\mathbf{v}^2-(\mathbf{v}\cdot\mathbf{i})^2)\end{aligned}$$

$$\Rightarrow(\mathbf{v}^2-1)^2-4\delta^2(\mathbf{v}^2-(\mathbf{v}\cdot\mathbf{i})^2)=0\quad\text{with } \delta^2=\frac{\gamma^2}{1-\gamma^2}$$

$$\begin{aligned}&\Rightarrow(x^2+y^2-1)^2-4\delta^2y^2=0\\&\Rightarrow(x^2+(y-\delta)^2-(1+\delta^2))(x^2+(y+\delta)^2-(1+\delta^2))=0\end{aligned}$$

This is a pair of circles of the same radius $\rho=\sqrt{1+\delta^2}$, intersecting at the endpoints of the fixed segment. Each implication in the sequence can be reversed except the first one, which involves the squaring of the equation. If we insisted on reversing this one as well, we would have to admit points resulting in the value $-\gamma$. In terms of α this would mean that, along with the acute angle α, we would have to consider the obtuse angle $180°-\alpha$, since $\gamma=\cos\alpha\Leftrightarrow-\gamma=\cos(180°-\alpha)$. To summarize: The locus is the outer arcs of a pair of circles, of the same radius, intersecting at the endpoints of the segment, the inner arcs of these circles being the locus for the constant obtuse angle $180°-\alpha$. If we now let $\gamma=\cos\alpha$ assume all positive and negative values, we get the family of circles passing through the endpoints of the segment. This is called the family of Thales circles.

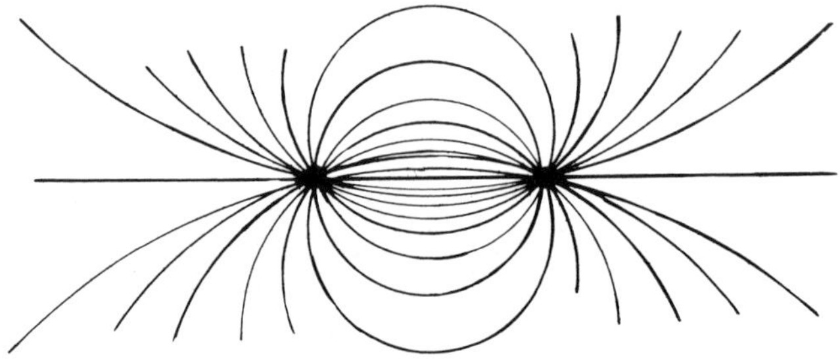

12. Parametric Forms of the Circle

The equation of the unit circle centered at the origin is $x^2+y^2=1$, where x, y are just the direction cosines of the unit vector $\overrightarrow{OP}$, the radius of the circle with $P(x,y)$. Therefore,

Parametric Form

$$x = \cos \tau \qquad (0 \leq \tau < 360°)$$
$$y = \sin \tau$$

where the parameter τ is just the angle made by the unit vectors **i** and $\overrightarrow{OP}$. This can be converted into the

Rational Parametric Form

$$x = \frac{1 - \omega^2}{1 + \omega^2} \qquad (\omega \in \mathbb{R})$$
$$y = \frac{2\omega}{1 + \omega^2}$$

by the substitution $\omega = \tan \tfrac{1}{2}\tau$ (the details of the calculation, involving trigonometric identities, are left to the reader). Note that the point $P_0(-1,0)$ cannot be obtained for any value of the parameter ω; it can be obtained only at the limit as $\omega \to \infty$ or $\omega \to -\infty$.

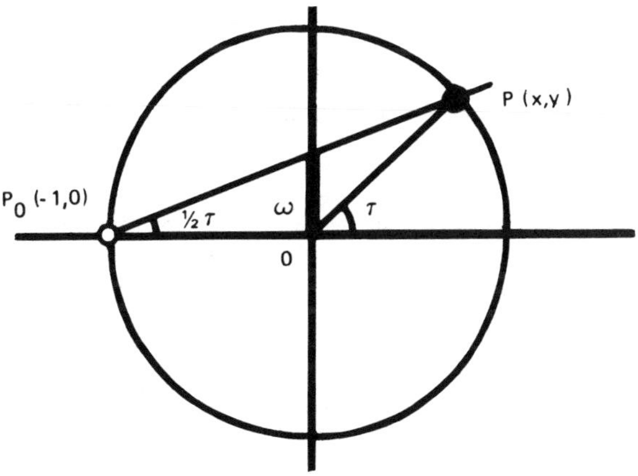

INEQUALITIES CONCERNING THE DOT PRODUCT

There are two famous inequalities connected with the dot product:

The Schwarz Inequality

$$|\mathbf{v} \cdot \mathbf{w}| \leqq |\mathbf{v}|\,|\mathbf{w}| \qquad \text{for all } \mathbf{v}, \mathbf{w}$$

wherein

$$|\mathbf{v} \cdot \mathbf{w}| = |\mathbf{v}|\,|\mathbf{w}| \Leftrightarrow \mathbf{w} = \lambda \mathbf{v} \qquad \text{for some } \lambda \in \mathbb{R}$$

The Triangle Inequality

$$|\mathbf{v} + \mathbf{w}| \leqq |\mathbf{v}| + |\mathbf{w}| \qquad \text{for all } \mathbf{v}, \mathbf{w}$$

wherein

$$|\mathbf{v} + \mathbf{w}| = |\mathbf{v}| + |\mathbf{w}| \Leftrightarrow \text{either } \mathbf{v} = 0 \text{ or } \mathbf{w} = \lambda \mathbf{v} \qquad \text{for some } \lambda \geqq 0$$

Proof: The Schwarz inequality follows from the trigonometric form of the dot product and the trigonometric inequality $-1 \leqq \cos\alpha \leqq 1$, $\alpha = \sphericalangle(\mathbf{v},\mathbf{w})$:

$$-|\mathbf{v}|\,|\mathbf{w}| \leqq \mathbf{v} \cdot \mathbf{w} \leqq |\mathbf{v}|\,|\mathbf{w}| \Leftrightarrow |\mathbf{v} \cdot \mathbf{w}| \leqq |\mathbf{v}|\,|\mathbf{w}|$$

and equality holds if, and only if, $\cos\alpha = 1$ or $-1 \Leftrightarrow \alpha = 0°$ or $180° \Leftrightarrow \mathbf{w} \,||\, \mathbf{v} \Leftrightarrow \mathbf{w} = \lambda \mathbf{v}$ for some $\lambda \in \mathbb{R}$. The proof of the triangle inequality is based on the Schwarz inequality:

$$|\mathbf{v} + \mathbf{w}|^2 = (\mathbf{v} + \mathbf{w})^2 = \mathbf{v}^2 + \mathbf{w}^2 + 2\mathbf{v} \cdot \mathbf{w} \leqq \mathbf{v}^2 + \mathbf{w}^2 + 2|\mathbf{v}|\,|\mathbf{w}| = (|\mathbf{v}| + |\mathbf{w}|)^2$$

Taking the nonnegative square root leaves the inequality unchanged, and the triangle inequality follows. Equality holds if, and only if, $\mathbf{v} \cdot \mathbf{w} = |\mathbf{v}|\,|\mathbf{w}| \Leftrightarrow \cos\alpha = 1 \Leftrightarrow \mathbf{v} \,||\, \mathbf{w}$ and have the same direction $\Leftrightarrow \mathbf{w} = \lambda \mathbf{v}$, $\lambda \geqq 0$.

The triangle inequality earns its name by its application to the triangle. Let $\mathbf{a}$, $\mathbf{b}$, $\mathbf{c}$ be the location vectors of the vertices of a triangle; then the lengths of the sides are $|\mathbf{a} - \mathbf{b}|$, $|\mathbf{b} - \mathbf{c}|$, and $|\mathbf{a} - \mathbf{c}|$. Since $(\mathbf{a} - \mathbf{b}) + (\mathbf{b} - \mathbf{c}) = \mathbf{a} - \mathbf{c}$, we have that $|\mathbf{a} - \mathbf{c}| \leqq |\mathbf{a} - \mathbf{b}| + |\mathbf{b} - \mathbf{c}|$; i.e., *the length of either side of the triangle is no greater than the sum of the lengths of the other two sides*. Equality holds if and only if the triangle collapses.

EXERCISES

2.1 Find the length of the diagonal of the unit cube.

2.2 Find the distance between the points $A(1,2,-3)$ and $B(2,-5,3)$.

2.3 Normalize the vector $\mathbf{a} = 5\mathbf{i} - 6\mathbf{j} + 30\mathbf{k}$.

2.4 Find the angle made by the vectors $\mathbf{a} = \mathbf{i} - 2\mathbf{j} + 3\mathbf{k}$ and $\mathbf{b} = 3\mathbf{i} + 4\mathbf{j} - 5\mathbf{k}$.

2.5 Prove that the triangle with vertices $A(3,-1,2)$, $B(0,-4,2)$, $C(-3,2,1)$ is isosceles.

2.6 Prove that the triangle with vertices $A(3,-2,1)$, $B(7,6,9)$, $C(9,1,-5)$ is right angled, and find the lengths of the sides.

2.7 Without calculating the angles, show that the triangle with vertices $A(4,-1,4)$, $B(0,7,-4)$, $C(3,1,-2)$ is obtuse.

2.8 Find the direction cosines determined by the vectors (i) $\mathbf{a} = 2\mathbf{i} - 3\mathbf{j} - 6\mathbf{k}$; (ii) $\mathbf{a} = 4\mathbf{i} + 5\mathbf{j} - 20\mathbf{k}$.

2.9 Show that each of the following sets of vectors is an orthonormal basis.

(i) $\mathbf{u} = \frac{2}{3}\mathbf{i} + \frac{1}{3}\mathbf{j} - \frac{2}{3}\mathbf{k}$, $\mathbf{v} = \frac{1}{3}\mathbf{i} + \frac{2}{3}\mathbf{j} + \frac{2}{3}\mathbf{k}$, $\mathbf{w} = \frac{2}{3}\mathbf{i} - \frac{2}{3}\mathbf{j} + \frac{1}{3}\mathbf{k}$

(ii) $\frac{2}{7}\mathbf{i} - \frac{3}{7}\mathbf{j} + \frac{6}{7}\mathbf{k}$, $\frac{3}{7}\mathbf{i} + \frac{6}{7}\mathbf{j} + \frac{2}{7}\mathbf{k}$, $-\frac{6}{7}\mathbf{i} + \frac{2}{7}\mathbf{j} + \frac{3}{7}\mathbf{k}$

(iii) $\frac{3}{13}\mathbf{i} + \frac{4}{13}\mathbf{j} + \frac{12}{13}\mathbf{k}$, $\frac{12}{13}\mathbf{i} + \frac{3}{13}\mathbf{j} - \frac{4}{13}\mathbf{k}$, $\frac{4}{13}\mathbf{i} - \frac{12}{13}\mathbf{j} + \frac{3}{13}\mathbf{k}$

(iv) $\frac{4}{9}\mathbf{i} - \frac{8}{9}\mathbf{j} + \frac{1}{9}\mathbf{k}$, $\frac{4}{9}\mathbf{i} + \frac{1}{9}\mathbf{j} - \frac{8}{9}\mathbf{k}$, $\frac{7}{9}\mathbf{i} + \frac{4}{9}\mathbf{j} + \frac{4}{9}\mathbf{k}$

(v) $-\frac{9}{11}\mathbf{i} + \frac{2}{11}\mathbf{j} + \frac{6}{11}\mathbf{k}$, $\frac{2}{11}\mathbf{i} - \frac{9}{11}\mathbf{j} + \frac{6}{11}\mathbf{k}$, $\frac{6}{11}\mathbf{i} + \frac{6}{11}\mathbf{j} + \frac{7}{11}\mathbf{k}$

(vi) $\frac{2}{15}\mathbf{i} + \frac{5}{15}\mathbf{j} + \frac{14}{15}\mathbf{k}$, $\frac{11}{15}\mathbf{i} - \frac{10}{15}\mathbf{j} + \frac{2}{15}\mathbf{k}$, $\frac{10}{15}\mathbf{i} + \frac{10}{15}\mathbf{j} - \frac{5}{15}\mathbf{k}$

2.10 Calculate the angle (approximately, within minutes) made by the diagonal of the cube with an adjacent edge.

2.11 Calculate the equal angles made by the equiangular line UNE with east, north, and upper.

2.12 Calculate the angles made by the equiangular line LSW with east, north, and upper.

2.13 Calculate the angles of each of the following triangles with given vertices.

(i) $A(1,3,2)$, $B(4,0,2)$, $C(4,3,-1)$

(ii) $A(3,-1,1)$, $B(4,7,5)$, $C(7,-5,8)$

(iii) $A(5,4,-1)$, $B(3,6,7)$, $C(4,2,6)$

(iv) $A(4,-8,1)$, $B(7,4,4)$, $C(-1,11,11)$

2.14 Find the acute angle made by the equiangular line UNE & LSW and the bisector axis NE & SW.

2.15 Find the acute angle made by the bisector axes UE & LW and NE & SW.

2.16 Find the acute angle made by the x axis and the NE & SW bisector axis.

2.17 Find the angle made by the equiangular line UNE & LSW and the equiangular line USE & LNW.

Exercises on the Circle

2.18 Determine, without plotting, whether the point $P(1,-2)$ is inside, on, or outside each of the following circles:

(i) $x^2 + y^2 = 1$

(ii) $x^2 + y^2 = 5$

(iii) $x^2 + y^2 = 9$

(iv) $x^2 + y^2 - 8x - 4y - 5 = 0$

(v) $x^2 + y^2 - 10x + 8y = 0$

2.19 Calculate the length of the tangent line from the point $P(1,-2)$ to the circle $x^2 + y^2 + 10x - 2y + 6 = 0$.

2.20 Find the condition under which the circles $(x - \alpha_1)^2 + (y - \beta_1)^2 = \rho_1^2$ and $(x - \alpha_2)^2 + (y - \beta_2)^2 = \rho_2^2$ will intersect at right angles.

2.21 Without calculating the coordinates of the points of intersection, write the equation of the straight line passing through the points of intersection of the pair of circles

$$x^2 + y^2 + 3x - y = 0 \qquad \text{and} \qquad 3x^2 + 3y^2 + 2x + y = 0$$

2.22 Without calculating the coordinates of the point of tangency, write the equation of the common tangent line to the pair of tangent circles

$$x^2 + y^2 + 2x + 4y - 20 = 0 \qquad \text{and} \qquad x^2 + y^2 - 5x + y - 4 = 0$$

2.23 Show that the point $P(5,5)$ is on the circle $x^2 + y^2 - 4x - 2y - 20 = 0$ and write the equation of the tangent to the circle at that point.

2.24 There is a unique point in the plane from which tangents of equal length can be drawn to the circles $x^2 + y^2 - 3x + 2y - 3 = 0$, $x^2 + y^2 - 8x + 2y + 12 = 0$, $x^2 + y^2 - 2 = 0$. Find the coordinates of that point. What is the equal length of the tangents?

2.25 Find the unique points on the x and y axes from which tangents of equal length can be drawn to the circles $x^2 + y^2 - 8x - 4y + 16 = 0$, $x^2 + y^2 + 4x = 0$. What are the lengths of the tangents?

3

The Cross Product and the Box Product of Vectors

TRIGONOMETRIC FORMULA FOR THE LENGTH OF THE CROSS PRODUCT

The cross product $\mathbf{a} \times \mathbf{b}$ of two vectors $\mathbf{a}$, $\mathbf{b}$ is defined as a vector such that its

(1) Length satisfies

$$|\mathbf{a} \times \mathbf{b}|^2 + (\mathbf{a} \cdot \mathbf{b})^2 = \mathbf{a}^2\mathbf{b}^2$$

(2) $\mathbf{a} \times \mathbf{b} \perp \mathbf{a}$ and $\mathbf{b}$.

(3) Direction is such that $\mathbf{a}$, $\mathbf{b}$, $\mathbf{a} \times \mathbf{b}$, in this order, form a right-hand system.

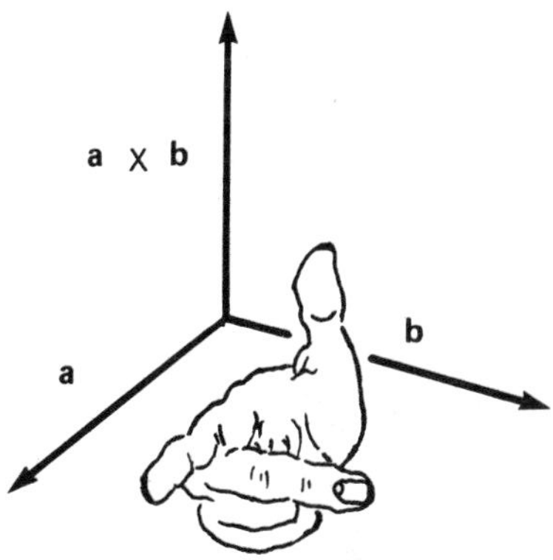

If one of the vectors $\mathbf{a}$, $\mathbf{b}$ is the null vector, or if $\mathbf{a}$ and $\mathbf{b}$ are parallel vectors, then the definition above does not determine the alignment and the direction of $\mathbf{a} \times \mathbf{b}$ unambiguously. This, however, is not disturbing, as in this case the length of $\mathbf{a} \times \mathbf{b}$ is zero and, therefore, $\mathbf{a} \times \mathbf{b} = \mathbf{0}$. Indeed,

$$\mathbf{a} \times \mathbf{b} = \mathbf{0} \Leftrightarrow \mathbf{a} \parallel \mathbf{b}$$

i.e., $\mathbf{a} \times \mathbf{b} = \mathbf{0}$ is a necessary and sufficient condition for two vectors **a**, **b** to be parallel. Recall that the null vector is deemed to be parallel to *every* vector. In particular, for all **a**, $\mathbf{a} \times \mathbf{a} = \mathbf{0}$. This rules out the possibility of a cancellation law for the cross product of vectors.

If we substitute the trigonometric expression for the dot product $\mathbf{a} \cdot \mathbf{b}$ into $|\mathbf{a} \times \mathbf{b}|^2 + (\mathbf{a} \cdot \mathbf{b})^2 = \mathbf{a}^2\mathbf{b}^2$ and solve for $|\mathbf{a} \times \mathbf{b}|$, we get the

Trigonometric Formula for the Length of the Cross Product

$$|\mathbf{a} \times \mathbf{b}| = |\mathbf{a}|\,|\mathbf{b}| \sin \sphericalangle(\mathbf{a},\mathbf{b})$$

This leads to the important geometric interpretation that $|\mathbf{a} \times \mathbf{b}|$ is *numerically* (but not dimensionally!) equal to the area of the parallelogram spanned by **a** and **b**.

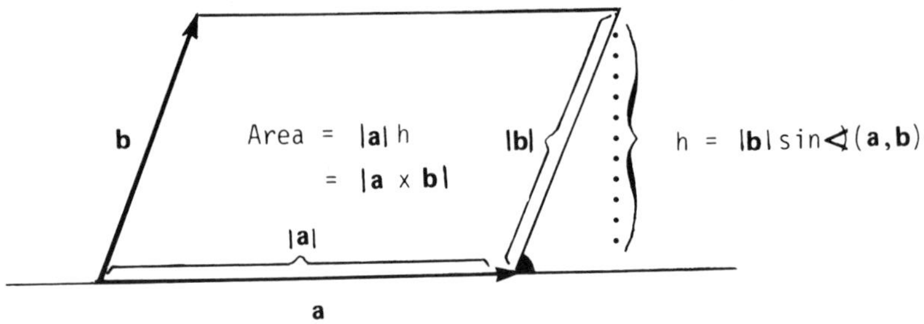

Examples

$\mathbf{i} \times \mathbf{i} = 0$, because **i** is parallel to itself.

$\mathbf{i} \times \mathbf{j} = \mathbf{k}$, because **i**, **j** span a square of area 1; moreover, **k** is perpendicular to both **i** and **j**; finally, **i**, **j**, **k** form a right-hand system.

$\mathbf{j} \times \mathbf{i} = -\mathbf{k}$, because **j**, **i**, **k** form a left-hand system.

Complete the table.

×	**i**	**j**	**k**
i			
j			
k			

APPLICATION

1. Heron's Formula

Heron's formula expresses the area of a triangle in terms of the length of its sides:

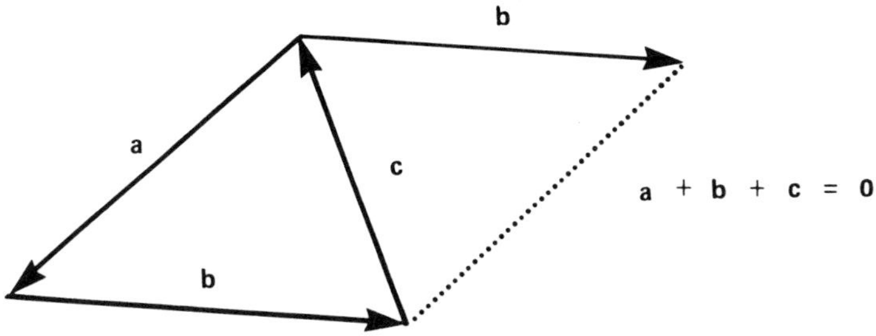

$$4A = \sqrt{(|\mathbf{a}| + |\mathbf{b}| + |\mathbf{c}|)(|\mathbf{b}| + |\mathbf{c}| - |\mathbf{a}|)(|\mathbf{c}| + |\mathbf{a}| - |\mathbf{b}|)(|\mathbf{a}| + |\mathbf{b}| - |\mathbf{c}|)}$$

Proof: It is clear that $A = \frac{1}{2}|\mathbf{a} \times \mathbf{b}|$, since A is one-half of the area of the parallelogram spanned by **a** and **b**. Then

$$\begin{aligned}
16A^2 &= 4(\mathbf{a} \times \mathbf{b})^2 \\
&= 4\mathbf{a}^2\mathbf{b}^2 - (2\mathbf{a} \cdot \mathbf{b})^2 = 4\mathbf{a}^2\mathbf{b}^2 - [\mathbf{a}^2 + \mathbf{b}^2 - (\mathbf{a} + \mathbf{b})^2]^2 \\
&= 4\mathbf{a}^2\mathbf{b}^2 - (\mathbf{a}^2 + \mathbf{b}^2 - \mathbf{c}^2)^2 \\
&= (2|\mathbf{a}|\,|\mathbf{b}| + \mathbf{a}^2 + \mathbf{b}^2 - \mathbf{c}^2)(2|\mathbf{a}|\,|\mathbf{b}| - \mathbf{a}^2 - \mathbf{b}^2 + \mathbf{c}^2) \\
&= ((|\mathbf{a}| + |\mathbf{b}|)^2 - |\mathbf{c}|^2)(|\mathbf{c}|^2 - (|\mathbf{a}| - |\mathbf{b}|)^2) \\
&= (|\mathbf{a}| + |\mathbf{b}| + |\mathbf{c}|)(|\mathbf{a}| + |\mathbf{b}| - |\mathbf{c}|)(|\mathbf{c}| + |\mathbf{a}| - |\mathbf{b}|)(|\mathbf{c}| - |\mathbf{a}| + |\mathbf{b}|)
\end{aligned}$$

THE CROSS PRODUCT AS A LINEAR OPERATOR

A *linear operator* is a vector-valued function of a vector variable preserving the sum and the scalar multiple of vectors. That is,

$$f(\mathbf{v} + \mathbf{v}') = f(\mathbf{v}) + f(\mathbf{v}')$$
$$f(\lambda\mathbf{v}) = \lambda f(\mathbf{v})$$

for all vectors **v**, $\mathbf{v}'$ and for every scalar λ.

Let **u** be a *fixed unit vector*. Then a linear operator t is defined by the formula

$$t(\mathbf{v}) = \mathbf{u} \times \mathbf{v}$$

where **v** is an arbitrary vector. To see this, we shall write t as a product

$t = r \circ p$ or, what is the same, $t(\mathbf{v}) = r(p(\mathbf{v}))$ for all $\mathbf{v}$, where p is the perpendicular projection onto the normal plane of the unit vector $\mathbf{u}$ (the normal plane of a vector is the plane perpendicular to that vector), and r is the rotation about the axis spanned by $\mathbf{u}$ through 90°. This follows from the facts that

$$|p(\mathbf{v})| = |\mathbf{v}| \sin \alpha = |r(p(\mathbf{v}))|, \; r(p(\mathbf{v})) \perp \mathbf{u}, \mathbf{v}, \text{ and that } \mathbf{u}, \mathbf{v}, r(p(\mathbf{v}))$$

form a right-hand system.

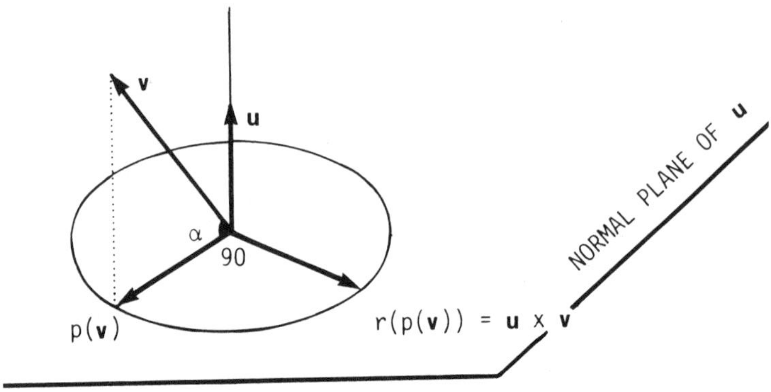

It is clear that the operators p and r preserve the sum and the scalar product of vectors; it follows that their product $t = r \circ p$ will do the same; i.e., t is also a linear operator.

The question arises of what will happen if we repeatedly apply the linear operator t to a vector $\mathbf{v}$. Since $t(\mathbf{v})$ is already in the normal plane of $\mathbf{u}$, a second application of t will leave it there, but will rotate it through another 90°, so that the third application will take $\mathbf{v}$ into the vector $-t(\mathbf{v})$. This means that $t^3(\mathbf{v}) = -t(\mathbf{v})$ for all $\mathbf{v}$, i.e., $t^3 = -t$. A linear operator with this property is called a *twist*. Twists play an important role in linear algebra, and in later chapters we look at them more closely.

FORMAL RULES OF THE CROSS PRODUCT

The cross product fails to satisfy the commutative law. As a partial compensation for this loss we shall have the anticommutative law (1^x). We also have the scalar associative law (2^x) and the distributive law (5^x).

(0^x) Closure

$$\mathbf{a}, \mathbf{b} \text{ vectors} \Rightarrow \mathbf{a} \times \mathbf{b} \text{ vector}$$

(1$^\times$) Anticommutativity

$$\mathbf{b} \times \mathbf{a} = -\mathbf{a} \times \mathbf{b}$$

Indeed, $\mathbf{a} \times \mathbf{b}$ and $\mathbf{b} \times \mathbf{a}$ have the same length and the same alignment, but the opposite orientation, as is seen from the definition.

2$^\times$) Scalar Associativity

$$\lambda(\mathbf{a} \times \mathbf{b}) = (\lambda \mathbf{a}) \times \mathbf{b} \qquad (\lambda \in \mathbb{R})$$

5$^\times$) Distributivity

$$\mathbf{a} \times (\mathbf{b} + \mathbf{c}) = \mathbf{a} \times \mathbf{b} + \mathbf{a} \times \mathbf{c}$$

(2$^\times$) and (5$^\times$) are often lumped together and referred to as

Bilinearity

$$\lambda(\mathbf{a} \times \mathbf{b}) = (\lambda \mathbf{a}) \times \mathbf{b}$$
$$\mathbf{a} \times (\mathbf{b} + \mathbf{c}) = \mathbf{a} \times \mathbf{b} + \mathbf{a} \times \mathbf{c}$$

Bilinearity is justified by the observation, already made, that $t(\mathbf{v}) = \mathbf{u} \times \mathbf{v}$ is a linear operator; i.e., it preserves the sum and the scalar multiple of vectors. The last of the five formal rules of the cross product is discussed in Chap. 4 (see the discussion of the Jacobi identity).

APPLICATION

2. The Law of Sines

The three sides of a triangle can be so oriented that they become vectors $\mathbf{a}$, $\mathbf{b}$, $\mathbf{c}$ satisfying the relation $\mathbf{a} + \mathbf{b} + \mathbf{c} = 0$. We shall denote the angles opposite to the sides $\mathbf{a}$, $\mathbf{b}$, $\mathbf{c}$ by α, β, γ, respectively.

The law of sines, which gives a relation between the lengths of the sides and the sines of the angles of a triangle, is familiar from trigonometry. Here we shall see a proof of it which is based on the cross product of vectors.

Proof:

$$\begin{aligned}
\mathbf{a} + \mathbf{b} + \mathbf{c} = \mathbf{0} &\Rightarrow \mathbf{a} \times \mathbf{a} + \mathbf{a} \times \mathbf{b} + \mathbf{a} \times \mathbf{c} = \mathbf{0} && \text{by } (5^x) \\
&\Rightarrow \mathbf{a} \times \mathbf{b} = \mathbf{c} \times \mathbf{a} && \text{by } (1^x) \\
&\Rightarrow |\mathbf{a}|\,|\mathbf{b}| \sin\sphericalangle(\mathbf{a},\mathbf{b}) = |\mathbf{c}|\,|\mathbf{a}| \sin\sphericalangle(\mathbf{c},\mathbf{a}) \\
&\Rightarrow |\mathbf{b}| \sin\gamma = |\mathbf{c}| \sin\beta \\
&\Rightarrow \frac{\sin\beta}{|\mathbf{b}|} = \frac{\sin\gamma}{|\mathbf{c}|} \left(= \frac{\sin\alpha}{|\mathbf{a}|} \right)
\end{aligned}$$

because

$$\gamma = 180° - \sphericalangle(\mathbf{a},\mathbf{b}) \qquad \text{and} \qquad \sin\gamma = \sin(180° - \sphericalangle(\mathbf{a},\mathbf{b})) = \sin\sphericalangle(\mathbf{a},\mathbf{b})$$

COORDINATE FORMULAS FOR THE CROSS PRODUCT

Let $\mathbf{v} = x\mathbf{i} + y\mathbf{j} + z\mathbf{k}$ and $\mathbf{v}' = x'\mathbf{i} + y'\mathbf{j} + z'\mathbf{k}$. Then

$$\mathbf{v} \times \mathbf{v}' = (yz' - zy')\mathbf{i} - (xz' - zx')\mathbf{j} + (xy' - yx')\mathbf{k}$$

This formula can be written in determinant form:

$$\mathbf{v} \times \mathbf{v}' = \begin{vmatrix} y & z \\ y' & z' \end{vmatrix}\mathbf{i} - \begin{vmatrix} x & z \\ x' & z' \end{vmatrix}\mathbf{j} + \begin{vmatrix} x & y \\ x' & y' \end{vmatrix}\mathbf{k}$$

We shall write it symbolically as a determinant of order 3, by virtue of the expansion property of determinants:

$$\mathbf{v} \times \mathbf{v}' = \begin{vmatrix} \mathbf{i} & \mathbf{j} & \mathbf{k} \\ x & y & z \\ x' & y' & z' \end{vmatrix}$$

The justification of these formulas involves the evaluation of $(x\mathbf{i} + y\mathbf{j} + z\mathbf{k}) \times (x'\mathbf{i} + y'\mathbf{j} + z'\mathbf{k})$ using the formal rules of the cross product.

It follows that the length of the vector $\mathbf{v} \times \mathbf{v}'$ can be evaluated in terms of coordinates as follows:

$$|\mathbf{v} \times \mathbf{v}'| = \sqrt{\begin{vmatrix} y & z \\ y' & z' \end{vmatrix}^2 + \begin{vmatrix} x & z \\ x' & z' \end{vmatrix}^2 + \begin{vmatrix} x & y \\ x' & y' \end{vmatrix}^2}$$

There is another method to calculate the length of the cross product, based on its definition: $|\mathbf{v} \times \mathbf{v}'|^2 = \mathbf{v}^2\mathbf{v}'^2 - (\mathbf{v} \cdot \mathbf{v}')^2$, i.e.,

$$|\mathbf{v} \times \mathbf{v}'| = \sqrt{(x^2 + y^2 + z^2)(x'^2 + y'^2 + z'^2) - (xx' + yy' + zz')^2}$$

APPLICATIONS

3. Area of the Triangle

If A, B, C are the vertices of a triangle Δ, then for the area of Δ we have

$$\text{Area}(\Delta) = \frac{1}{2}|\overrightarrow{CA} \times \overrightarrow{CB}|$$

Example The vertices of a triangle are $A(5,0,-6)$, $B(1,1,3)$, and $C(-1,-2,-3)$. Find the area.

Answer: $\overrightarrow{CA} = 6\mathbf{i} + 2\mathbf{j} - 3\mathbf{k} \qquad \overrightarrow{CB} = 2\mathbf{i} + 3\mathbf{j} + 6\mathbf{k}.$

$$\overrightarrow{CA} \times \overrightarrow{CB} = \begin{vmatrix} \mathbf{i} & \mathbf{j} & \mathbf{k} \\ 6 & 2 & -3 \\ 2 & 3 & 6 \end{vmatrix} = 21\mathbf{i} - 42\mathbf{j} + 14\mathbf{k} = 7(3\mathbf{i} - 6\mathbf{j} + 2\mathbf{k})$$

$$|\overrightarrow{CA} \times \overrightarrow{CB}| = 7|3\mathbf{i} - 6\mathbf{j} + 2\mathbf{k}| = 7\sqrt{9 + 36 + 4} = 7\sqrt{49} = 49 = 7^2$$

$$\text{Area}(\Delta) = \frac{1}{2}(49)$$

Notice that, in this particular case, $|\overrightarrow{CA}| = |\overrightarrow{CB}|$, $\overrightarrow{CA} \perp \overrightarrow{CB}$; hence the triangle is isosceles and right angled.

4. Perpendicular Distance from a Point to a Line

The perpendicular distance h between the point A and the line through the points B, C is

$$h = \frac{|\overrightarrow{CA} \times \overrightarrow{CB}|}{|\overrightarrow{CB}|}$$

Example Let $A(1,-2,3)$, $B(-1,0,1)$, $C(1,0,3)$ be given points. Find the perpendicular distance h from A to the line through B, C.

Answer: $\overrightarrow{CA} = -2\mathbf{j}$, $\overrightarrow{CB} = -2\mathbf{i} - 2\mathbf{k}$.

$$\overrightarrow{CA} \times \overrightarrow{CB} = \begin{vmatrix} \mathbf{i} & \mathbf{j} & \mathbf{k} \\ 0 & -2 & 0 \\ -2 & 0 & -2 \end{vmatrix} = 4 \begin{vmatrix} \mathbf{i} & \mathbf{j} & \mathbf{k} \\ 0 & 1 & 0 \\ 1 & 0 & 1 \end{vmatrix} = 4(\mathbf{i} - \mathbf{k})$$

$$|\overrightarrow{CA} \times \overrightarrow{CB}| = 4|\mathbf{i} - \mathbf{k}| = 4\sqrt{2}$$

$$|\overrightarrow{CB}| = |-2\mathbf{i} - 2\mathbf{k}| = 2|-\mathbf{i} - \mathbf{k}| = 2\sqrt{2}$$

$$h = \frac{4\sqrt{2}}{2\sqrt{2}} = 2$$

5. Parallel and Perpendicular Components of a Vector

We wish to solve the important problem of decomposing an arbitrary vector $\mathbf{v}$ into the sum

$$\mathbf{v} = \mathbf{v}_{\text{par}} + \mathbf{v}_{\text{per}}$$

where $\mathbf{v}_{\text{par}}$ is parallel and $\mathbf{v}_{\text{per}}$ is perpendicular to a certain unit vector $\mathbf{u}$.

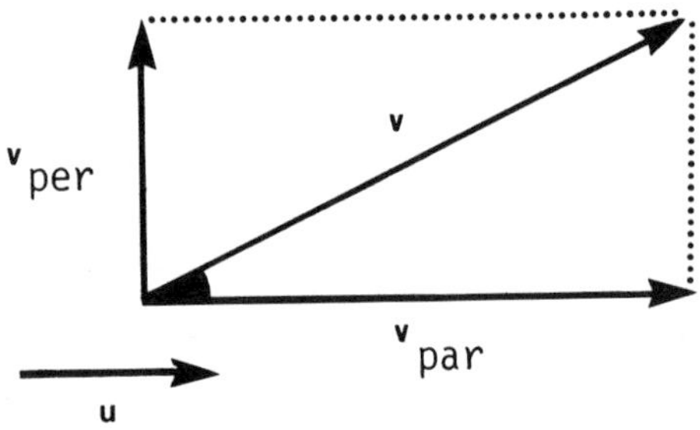

The problem has a unique solution, namely

$$\mathbf{v}_{\text{par}} = (\mathbf{u}\cdot\mathbf{v})\mathbf{u} \qquad \mathbf{v}_{\text{per}} = (\mathbf{u}\times\mathbf{v})\times\mathbf{u}$$

so that the decomposition is

$$\mathbf{v} = (\mathbf{v}\cdot\mathbf{v})\mathbf{u} + (\mathbf{u}\times\mathbf{v})\times\mathbf{u}$$

Indeed, it is immediate that $(\mathbf{u}\cdot\mathbf{v})\mathbf{u}$ and $(\mathbf{u}\times\mathbf{v})\times\mathbf{u}$ have the required alignment and orientation, and that $(\mathbf{u}\cdot\mathbf{v})\mathbf{u}$ also has the required length. Therefore, we have only to check that $(\mathbf{u}\times\mathbf{v})\times\mathbf{u}$ has the required length as well.

$$\begin{aligned}|(\mathbf{u}\times\mathbf{v})\times\mathbf{u}| &= |\mathbf{u}\times\mathbf{v}|\,|\mathbf{u}|\,\sin\sphericalangle(\mathbf{u}\times\mathbf{v},\mathbf{u}) = |\mathbf{u}\times\mathbf{v}|\,\sin 90^\circ\\ &= |\mathbf{u}\times\mathbf{v}| = |\mathbf{u}|\,|\mathbf{v}|\,\sin\sphericalangle(\mathbf{u},\mathbf{v}) = |\mathbf{v}|\,\sin\sphericalangle(\mathbf{u},\mathbf{v}) = |\mathbf{v}_{\text{per}}|\end{aligned}$$

Example Given the unit vector

$$\mathbf{u} = \frac{2}{7}\mathbf{i} + \frac{3}{7}\mathbf{j} + \frac{6}{7}\mathbf{k}$$

decompose the vector $\mathbf{v} = 8\mathbf{i} + 5\mathbf{j} + 3\mathbf{k}$ into parallel and perpendicular components.

Answer:

$$\mathbf{u}\cdot\mathbf{v} = \frac{16+15+18}{7} = \frac{49}{7} = 7 \qquad \text{and} \qquad \mathbf{v}_{\text{par}} = (\mathbf{u}\cdot\mathbf{v})\mathbf{u} = 2\mathbf{i} + 3\mathbf{j} + 6\mathbf{k}$$

There are two methods to calculate $\mathbf{v}_{\text{per}}$.
First method:

$$\begin{aligned}\mathbf{v}_{\text{per}} &= \mathbf{v} - \mathbf{v}_{\text{par}} = (8\mathbf{i} + 5\mathbf{j} + 3\mathbf{k}) - (2\mathbf{i} + 3\mathbf{j} + 6\mathbf{k})\\ &= 6\mathbf{i} + 2\mathbf{j} - 3\mathbf{k}\end{aligned}$$

Second method: $\mathbf{v}_{\text{per}} = (\mathbf{u}\times\mathbf{v})\times\mathbf{u}$.

$$\mathbf{u}\times\mathbf{v}=\begin{vmatrix}\mathbf{i} & \mathbf{j} & \mathbf{k}\\ 2/7 & 3/7 & 6/7\\ 8 & 5 & 3\end{vmatrix}=\frac{1}{7}\begin{vmatrix}\mathbf{i} & \mathbf{j} & \mathbf{k}\\ 2 & 3 & 6\\ 8 & 5 & 3\end{vmatrix}=\frac{1}{7}(-21\mathbf{i}+42\mathbf{j}-14\mathbf{k})$$

$$=-3\mathbf{i}+6\mathbf{j}-2\mathbf{k}$$

$$\mathbf{v}_{\text{per}}=(\mathbf{u}\times\mathbf{v})\times\mathbf{u}=\begin{vmatrix}\mathbf{i} & \mathbf{j} & \mathbf{k}\\ -3 & 6 & -2\\ 2/7 & 3/7 & 6/7\end{vmatrix}=\frac{1}{7}\begin{vmatrix}\mathbf{i} & \mathbf{j} & \mathbf{k}\\ -3 & 6 & -2\\ 2 & 3 & 6\end{vmatrix}$$

$$=\frac{1}{7}(42\mathbf{i}+14\mathbf{j}-21\mathbf{k})=6\mathbf{i}+2\mathbf{j}-3\mathbf{k}$$

Thus the decomposition is

$$8\mathbf{i}+5\mathbf{j}+3\mathbf{k}=(2\mathbf{i}+3\mathbf{j}+6\mathbf{k})+(6\mathbf{i}+2\mathbf{j}-3\mathbf{k})$$

and clearly, $2\mathbf{i}+3\mathbf{j}+6\mathbf{k}\perp 6\mathbf{i}+2\mathbf{j}-3\mathbf{k}$.

BOX PRODUCT OF THREE VECTORS

For any three vectors **a**, **b**, **c** we define their *box product* [**abc**] by the formula

$$[\mathbf{abc}]=(\mathbf{a}\times\mathbf{b})\cdot\mathbf{c}$$

There is no need to memorize where the cross goes the where the dot goes; we shall soon see that they can be interchanged, so that

$$[\mathbf{abc}]=(\mathbf{a}\times\mathbf{b})\cdot\mathbf{c}=\mathbf{a}\cdot(\mathbf{b}\times\mathbf{c})$$

(see the cross-dot formula below).

Examples

$$[\mathbf{ijk}]=(\mathbf{i}\times\mathbf{j})\cdot\mathbf{k}=\mathbf{k}\cdot\mathbf{k}=1$$
$$[\mathbf{ikj}]=(\mathbf{i}\times\mathbf{k})\cdot\mathbf{j}=(-\mathbf{j})\cdot\mathbf{j}=-1$$
$$[\mathbf{iij}]=(\mathbf{i}\times\mathbf{i})\cdot\mathbf{j}=0\cdot\mathbf{j}=0$$

Let us now look at some important special cases.

$$[\mathbf{abc}]=0\Leftrightarrow\mathbf{a},\ \mathbf{b},\ \mathbf{c}\text{ are coplanar}$$

Indeed,

$$\mathbf{a},\ \mathbf{b},\ \mathbf{c}\text{ coplanar}\Leftrightarrow\mathbf{a}\times\mathbf{b}\text{ perpendicular to }\mathbf{c}$$
$$\Leftrightarrow(\mathbf{a}\times\mathbf{b})\cdot\mathbf{c}=0\Leftrightarrow[\mathbf{abc}]=0$$

In other words, $[\mathbf{abc}] = 0$ *is a necessary and sufficient condition for the set of vectors* $\{\mathbf{a}, \mathbf{b}, \mathbf{c}\}$ *to be linearly dependent.*

Also,

$$[\mathbf{abc}] > 0 \Leftrightarrow \mathbf{a}, \mathbf{b}, \mathbf{c} \text{ form a right-hand system}$$

since in this case $\sphericalangle(\mathbf{a} \times \mathbf{b}, \mathbf{c})$ is acute. Similarly,

$$[\mathbf{abc}] < 0 \Leftrightarrow \mathbf{a}, \mathbf{b}, \mathbf{c} \text{ form a left-hand system}$$

The geometric significance of the box product is this:

The box product $[\mathbf{abc}]$ *is the signed volume of the parallelepiped spanned by the vectors* **a**, **b**, **c**; *the signature is positive if, and only if, the vectors form a right-hand system.*

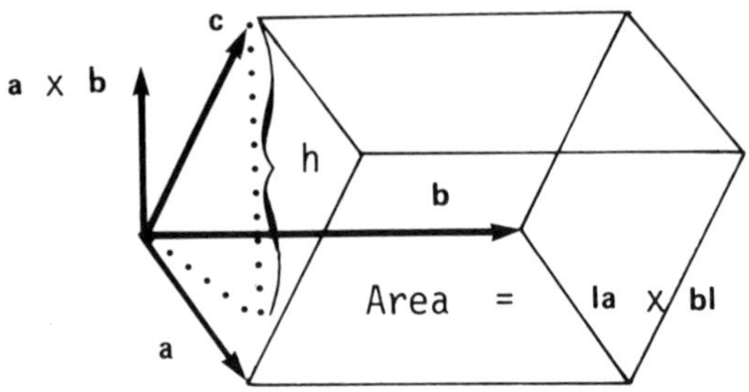

Proof: h = altitude of parallelepiped $= |\mathbf{c}| \cos\sphericalangle(\mathbf{a} \times \mathbf{b}, \mathbf{c})$

A = base area of parallelepiped $= |\mathbf{a} \times \mathbf{b}|$

Hence

$$\begin{aligned} V &= \text{volume of parallelepiped} \\ &= Ah \\ &= |\mathbf{a} \times \mathbf{b}|\,|\mathbf{c}| \cos\sphericalangle(\mathbf{a} \times \mathbf{b}, \mathbf{c}) \\ &= (\mathbf{a} \times \mathbf{b}) \cdot \mathbf{c} \\ &= [\mathbf{abc}] \end{aligned}$$

From this we can immediately derive the

Cross-Dot Formula

$$(\mathbf{a} \times \mathbf{b}) \cdot \mathbf{c} = \mathbf{a} \cdot (\mathbf{b} \times \mathbf{c})$$

asserting that the cross and the dot can be interchanged without affecting the value of the box product. This change has the effect of selecting another face, the face spanned by the vectors **b** and **c**, as the "base" of the parallelepiped. In fact, we can also write it in yet another way:

$$(\mathbf{a}\times\mathbf{b})\cdot\mathbf{c}=(\mathbf{c}\times\mathbf{a})\cdot\mathbf{b}=(\mathbf{b}\times\mathbf{c})\cdot\mathbf{a}$$

This means that $[\mathbf{abc}]=[\mathbf{bca}]=[\mathbf{cab}]$. In other words, an even permutation of the letters **a**, **b**, **c** leaves the box product unchanged. By contrast, an odd permutation will change the signature of the box product: $[\mathbf{abc}]=-[\mathbf{acb}]=-[\mathbf{cba}]=-[\mathbf{bac}]$.

COORDINATE FORMULA FOR THE BOX PRODUCT

$$[\mathbf{v}_1\mathbf{v}_2\mathbf{v}_3]=\begin{vmatrix} x_1 & y_1 & z_1\\ x_2 & y_2 & z_2\\ x_3 & y_3 & z_3\end{vmatrix}$$

where $\mathbf{v}_n=x_n\mathbf{i}+y_n\mathbf{j}+z_n\mathbf{k}$, $n=1, 2, 3$.

Proof:

$$[\mathbf{v}_1\mathbf{v}_2\mathbf{v}_3]=(\mathbf{v}_1\times\mathbf{v}_2)\cdot\mathbf{v}_3=\mathbf{v}_1\cdot(\mathbf{v}_2\times\mathbf{v}_3)$$

$$=x_1\begin{vmatrix} y_2 & z_2\\ y_3 & z_3\end{vmatrix}-y_1\begin{vmatrix} x_2 & z_2\\ x_3 & z_3\end{vmatrix}+z_1\begin{vmatrix} x_2 & y_2\\ x_3 & y_3\end{vmatrix}$$

$$=\begin{vmatrix} x_1 & y_1 & z_1\\ x_2 & y_2 & z_2\\ x_3 & y_3 & z_3\end{vmatrix}$$

where we have used the expansion property† of determinants. Note that the vanishing property† of determinants can be interpreted geometrically in terms of the cross and the box product of vectors, e.g.,

$$x_2\begin{vmatrix} y_2 & z_2\\ y_3 & z_3\end{vmatrix}-y_2\begin{vmatrix} x_2 & z_2\\ x_3 & z_3\end{vmatrix}+z_2\begin{vmatrix} x_2 & y_2\\ x_3 & y_3\end{vmatrix}=0$$

because $\mathbf{v}_2$ is perpendicular to $\mathbf{v}_2\times\mathbf{v}_3$, so that $\mathbf{v}_2\cdot(\mathbf{v}_2\times\mathbf{v}_3)=0$.

APPLICATIONS

6. Volume of the Tetrahedron

Let A, B, C, D be the vertices of a tetrahedron Δ. Then the volume of

†The expansion property and the vanishing property of determinants are discussed in Appendix 1.

Δ is

$$\mathrm{Vol}(\Delta) = \frac{1}{6}|[\overrightarrow{DA}, \overrightarrow{DB}, \overrightarrow{DC}]|$$

The volume of the tetrahedron spanned by the vectors $\overrightarrow{DA}$, $\overrightarrow{DB}$, $\overrightarrow{DC}$ is one-sixth of the volume of the parallelepiped spanned by the same three vectors. Indeed, the parallelepiped is cut into two rectangular prisms of equal volume by the diagonal plane,

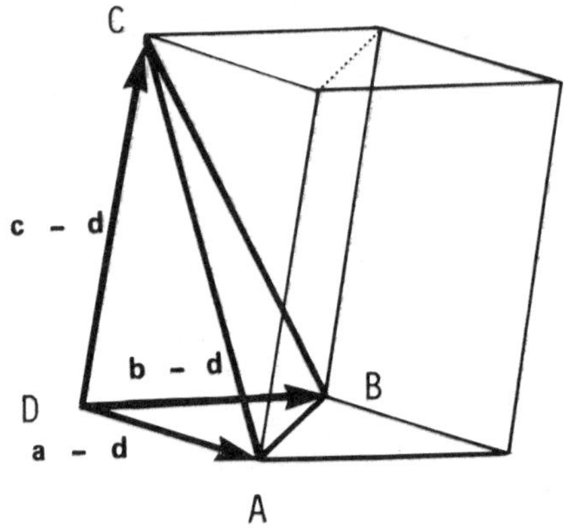

and the tetrahedron is just a triangular pyramid whose volume is one-third of the volume of the triangular prism, of the same base area and the same height.

7. Perpendicular Distance from a Point to a Plane

The perpendicular distance h from the point A to the plane determined by the three points B, C, D is

$$h = \frac{|[\overrightarrow{DA}, \overrightarrow{DB}, \overrightarrow{DC}]|}{|\overrightarrow{DB} \times \overrightarrow{DC}|}$$

Indeed, h is the altitude of the parallelepiped spanned by the vectors $\overrightarrow{DA}$, $\overrightarrow{DB}$, $\overrightarrow{DC}$ whose base area is $\overrightarrow{DB} \times \overrightarrow{DC}$.

8. Perpendicular Distance Between Two Lines (Length of the Orthogonal Transversal)

Any line joining a point on the first line to a point on the second is called a transversal. If the first and second lines are not in the same plane, then among the infinitely many transversals there is exactly one which intersects both at right angles: the *orthogonal transversal*. The distance between the two points of intersection on the orthogonal transversal furnishes the shortest distance between points on the first and points on the second line; the orthogonal transversal is the shortest transversal. The length of the orthogonal transversal is also called the perpendicular distance between the first and second lines. The perpendicular distance h between the lines through the points A, B and C, D is

$$h = \frac{|[\overrightarrow{CA}, \overrightarrow{AB}, \overrightarrow{CD}]|}{|\overrightarrow{AB} \times \overrightarrow{CD}|}$$

Indeed, h is the altitude of a parallelepiped spanned by $\overrightarrow{CA}$, $\overrightarrow{AB}$, $\overrightarrow{CD}$ whose base area is $\overrightarrow{AB} \times \overrightarrow{CD}$.

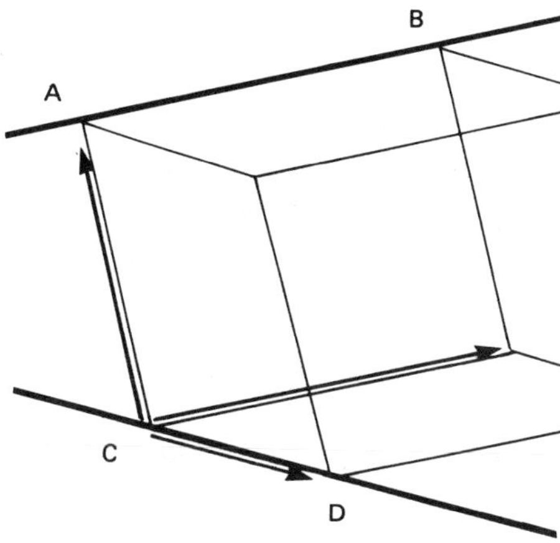

EXERCISES

(*Note*: The computational techniques needed for some of the exercises that follow include determinants of order 2 and 3. Readers wishing to review their knowledge of determinants may find Appendix 1 at the end of this volume helpful.)

3.1 Write a vector that is perpendicular to both vectors $\mathbf{a} = \mathbf{i} - 2\mathbf{j} + 3\mathbf{k}$ and $\mathbf{b} = 4\mathbf{j} - 5\mathbf{k}$; check your answer.

3.2 Given two perpendicular unit vectors $\mathbf{u} = \frac{1}{3}\mathbf{i} + \frac{2}{3}\mathbf{j} + \frac{2}{3}\mathbf{k}$ and $\mathbf{v} = \frac{2}{3}\mathbf{i} + \frac{1}{3}\mathbf{j} - \frac{2}{3}\mathbf{k}$, find a unit vector $\mathbf{w}$ such that $\mathbf{u}$, $\mathbf{v}$, $\mathbf{w}$ form an orthonormal basis. How many solutions are there?

3.3 Find the area of the parallelogram spanned by the vectors $\mathbf{a} = 3\mathbf{i} + 4\mathbf{j} + 12\mathbf{k}$ and $\mathbf{b} = 12\mathbf{i} + 3\mathbf{j} - 4\mathbf{k}$.

3.4 Calculate the area of the parallelogram spanned by $\mathbf{a} = \mathbf{i} + 2\mathbf{j} + 2\mathbf{k}$ and $\mathbf{b} = -2\mathbf{i} - 10\mathbf{j} + 11\mathbf{k}$ in two different ways: one through determinants, the other without determinants.

3.5 Given the vertices of a triangle ABC, find the area in each of the following cases.

(i) $A(5,0,-6)$, $B(1,1,3)$, $C(-1,-2,-3)$
(ii) $A(0,1,-1)$, $B(2,-1,-4)$, $C(4,1,5)$
(iii) $A(1,2,0)$, $B(3,0,-3)$, $C(5,2,6)$

3.6 Find the perpendicular distance from the point A to the line passing through B and C in each of the following cases.
(i) $A(1,-2,3)$, $B(-1,0,1)$, $C(1,0,3)$
(ii) $A(4,-8,1)$, $B(4,1,-8)$, $C(7,4,4)$

3.7 Given a vector $\mathbf{v}$ and a *unit* vector $\mathbf{u}$, decompose $\mathbf{v}$ as a sum of parallel and perpendicular components to $\mathbf{u}$ in each of the following cases.
(i) $\mathbf{v} = \mathbf{i} - 8\mathbf{j} - 7\mathbf{k}$, $\mathbf{u} = \frac{2}{3}\mathbf{i} + \frac{2}{3}\mathbf{j} + \frac{1}{3}\mathbf{k}$
(ii) $\mathbf{v} = 2\mathbf{i} - 3\mathbf{j} + 4\mathbf{k}$, $\mathbf{u} = \mathbf{k}$
(iii) $\mathbf{v} = 2\mathbf{i} + 2\mathbf{j} - \mathbf{k}$, $\mathbf{u} = \frac{2}{15}\mathbf{i} + \frac{5}{15}\mathbf{j} + \frac{14}{15}\mathbf{k}$

3.8 Using the box product, show that the vectors $\mathbf{a}$, $\mathbf{b}$, $\mathbf{c}$ are coplanar in each of the following cases.
(i) $\mathbf{a} = 3\mathbf{i} - 2\mathbf{j} + \mathbf{k}$, $\mathbf{b} = 5\mathbf{i} + 4\mathbf{j} - 3\mathbf{k}$, $\mathbf{c} = 11\mathbf{i} - \mathbf{k}$
(ii) $\mathbf{a} = \mathbf{i} + 2\mathbf{j} + 3\mathbf{k}$, $\mathbf{b} = 4\mathbf{i} + 5\mathbf{j} + 6\mathbf{k}$, $\mathbf{c} = 7\mathbf{i} + 8\mathbf{j} + 9\mathbf{k}$
(iii) $\mathbf{a} = \mathbf{i} - 2\mathbf{j} + 3\mathbf{k}$, $\mathbf{b} = 2\mathbf{i} - 2\mathbf{j}$, $\mathbf{c} = \mathbf{j} + 7\mathbf{k}$

3.9 Find the volume of the parallelepiped spanned by the vectors $\mathbf{a}$, $\mathbf{b}$, $\mathbf{c}$ in each of the following cases.
(i) $\mathbf{a} = -8\mathbf{i} + \mathbf{j} + 4\mathbf{k}$, $\mathbf{b} = \mathbf{i} - 8\mathbf{j} + 4\mathbf{k}$, $\mathbf{c} = 4\mathbf{i} + 4\mathbf{j} + 7\mathbf{k}$
(ii) $\mathbf{a} = 3\mathbf{i} + 2\mathbf{j} + \mathbf{k}$, $\mathbf{b} = 2\mathbf{i} - 4\mathbf{j} + \mathbf{k}$, $\mathbf{c} = \mathbf{i} + 5\mathbf{j} - \mathbf{k}$
(iii) $\mathbf{a} = \mathbf{i}$, $\mathbf{b} = \mathbf{i} + \mathbf{j}$, $\mathbf{c} = \mathbf{i} + \mathbf{j} + \mathbf{k}$

3.10 Show that the four points A, B, C, D lie in the same plane in each of the following cases.
(i) $A(1,2,-1)$, $B(0,1,5)$, $C(-1,2,1)$, $D(2,1,3)$
(ii) $A(1,2,-3)$, $B(3,5,-1)$, $C(0,-2,7)$, $D(2,1,3)$

3.11 Find the volume of the tetrahedron with vertices A, B, C, D in each of the following cases.
(i) $A(-1,0,0)$, $B(-2,1,4)$, $C(-1,-3,3)$, $D(-3,-1,2)$
(ii) $A(2,-1,1)$, $B(5,5,4)$, $C(3,2,-1)$, $D(4,1,3)$
(iii) $A(1,0,-2)$, $B(-1,1,0)$, $C(2,-1,1)$, $D(0,3,1)$

3.12 Find the perpendicular distance from the point A to the plane determined by the points B, C, D in each of the following cases.
(i) $A(5,5,7)$, $B(9,4,-2)$, $C(6,-4,3)$, $D(3,2,1)$
(ii) $A(1,0,-2)$, $B(-1,1,0)$, $C(2,-1,1)$, $D(0,3,1)$

3.13 Show that the line determined by A, B intersects the line determined by C, D in each of the following cases.
(i) $A(1,-2,3)$, $B(2,0,2)$, $C(-1,-6,1)$, $D(1,-2,7)$
(ii) $A(4,6,8)$, $B(7,10,13)$, $C(5,7,9)$, $D(9,12,15)$

3.14 Find the perpendicular distance (the length of the orthogonal transversal) between the line determined by A, B and the line determined by C, D in each of the following cases.
(i) $A(2,4,5)$, $B(4,5,3)$, $C(1,2,3)$, $D(3,0,4)$
(ii) $A(-7,-4,-3)$, $B(-4,0,-5)$, $C(21,-5,2)$, $D(27,-9,1)$

(iii) $A(-2,-3,4)$, $B(2,3,0)$, $C(-2,3,2)$, $D(2,0,1)$
(iv) $A(-2,4,3)$, $B(2,-8,0)$, $C(1,-3,5)$, $D(4,1,-7)$
(v) $A(2,3,1)$, $B(0,-1,2)$, $C(1,2,5)$, $D(-3,1,0)$

3.15 In each of the following cases, show that the vectors are perpendicular unit vectors, and find the unique third vector such that the triple forms a positively oriented orthonormal basis.

(i) $-\frac{9}{17}\mathbf{i} + \frac{8}{17}\mathbf{j} + \frac{12}{17}\mathbf{k}$, $\frac{8}{17}\mathbf{i} - \frac{9}{17}\mathbf{j} + \frac{12}{17}\mathbf{k}$
(ii) $\frac{1}{19}\mathbf{i} + \frac{6}{19}\mathbf{j} + \frac{18}{19}\mathbf{k}$, $\frac{18}{19}\mathbf{i} - \frac{6}{19}\mathbf{j} + \frac{1}{19}\mathbf{k}$
(iii) $\frac{4}{21}\mathbf{i} + \frac{16}{21}\mathbf{j} + \frac{13}{21}\mathbf{k}$, $\frac{8}{21}\mathbf{i} + \frac{11}{21}\mathbf{j} - \frac{16}{21}\mathbf{k}$
(iv) $\frac{5}{21}\mathbf{i} + \frac{20}{21}\mathbf{j} + \frac{4}{21}\mathbf{k}$, $\frac{4}{21}\mathbf{i} - \frac{5}{21}\mathbf{j} + \frac{20}{21}\mathbf{k}$
(v) $\frac{3}{23}\mathbf{i} + \frac{18}{23}\mathbf{j} - \frac{14}{23}\mathbf{k}$, $\frac{6}{23}\mathbf{i} + \frac{13}{23}\mathbf{j} + \frac{18}{23}\mathbf{k}$
(vi) $\frac{10}{27}\mathbf{i} + \frac{2}{27}\mathbf{j} - \frac{25}{27}\mathbf{k}$, $-\frac{10}{27}\mathbf{i} + \frac{25}{27}\mathbf{j} - \frac{2}{27}\mathbf{k}$
(vii) $-\frac{21}{29}\mathbf{i} + \frac{12}{29}\mathbf{j} + \frac{16}{29}\mathbf{k}$, $\frac{12}{29}\mathbf{i} - \frac{11}{29}\mathbf{j} + \frac{24}{29}\mathbf{k}$
(viii) $\frac{30}{31}\mathbf{i} + \frac{5}{31}\mathbf{j} + \frac{6}{31}\mathbf{k}$, $-\frac{6}{31}\mathbf{i} + \frac{30}{31}\mathbf{j} + \frac{5}{31}\mathbf{k}$
(ix) $\frac{6}{31}\mathbf{i} + \frac{22}{31}\mathbf{j} + \frac{21}{31}\mathbf{k}$, $\frac{14}{31}\mathbf{i} - \frac{21}{31}\mathbf{j} + \frac{18}{31}\mathbf{k}$
(x) $\frac{8}{33}\mathbf{i} + \frac{1}{33}\mathbf{j} + \frac{32}{33}\mathbf{k}$, $\frac{8}{33}\mathbf{i} - \frac{32}{33}\mathbf{j} - \frac{1}{33}\mathbf{k}$
(xi) $-\frac{17}{33}\mathbf{i} + \frac{28}{33}\mathbf{j} + \frac{4}{33}\mathbf{k}$, $\frac{28}{33}\mathbf{i} + \frac{16}{33}\mathbf{j} + \frac{7}{33}\mathbf{k}$
(xii) $-\frac{17}{33}\mathbf{i} - \frac{20}{33}\mathbf{j} + \frac{20}{33}\mathbf{k}$, $-\frac{20}{33}\mathbf{i} - \frac{8}{33}\mathbf{j} - \frac{25}{33}\mathbf{k}$
(xiii) $\frac{30}{35}\mathbf{i} + \frac{15}{35}\mathbf{j} - \frac{10}{35}\mathbf{k}$, $-\frac{17}{35}\mathbf{i} + \frac{30}{35}\mathbf{j} - \frac{6}{35}\mathbf{k}$
(xiv) $\frac{3}{37}\mathbf{i} + \frac{8}{37}\mathbf{j} - \frac{36}{37}\mathbf{k}$, $\frac{24}{37}\mathbf{i} + \frac{27}{37}\mathbf{j} + \frac{8}{37}\mathbf{k}$
(xv) $\frac{2}{39}\mathbf{i} + \frac{19}{39}\mathbf{j} + \frac{34}{39}\mathbf{k}$, $\frac{29}{39}\mathbf{i} + \frac{22}{39}\mathbf{j} - \frac{14}{39}\mathbf{k}$
(xvi) $-\frac{39}{41}\mathbf{i} + \frac{4}{41}\mathbf{j} + \frac{12}{41}\mathbf{k}$, $\frac{12}{41}\mathbf{i} + \frac{24}{41}\mathbf{j} + \frac{31}{41}\mathbf{k}$
(xvii) $\frac{7}{43}\mathbf{i} + \frac{6}{43}\mathbf{j} + \frac{42}{43}\mathbf{k}$, $-\frac{42}{43}\mathbf{i} + \frac{7}{43}\mathbf{j} + \frac{6}{43}\mathbf{k}$
(xviii) $-\frac{7}{43}\mathbf{i} - \frac{30}{43}\mathbf{j} - \frac{30}{43}\mathbf{k}$, $-\frac{30}{43}\mathbf{i} - \frac{18}{43}\mathbf{j} + \frac{25}{43}\mathbf{k}$
(xix) $\frac{4}{49}\mathbf{i} + \frac{48}{49}\mathbf{j} - \frac{9}{49}\mathbf{k}$, $-\frac{36}{49}\mathbf{i} + \frac{9}{49}\mathbf{j} + \frac{32}{49}\mathbf{k}$
(xx) $\frac{1}{51}\mathbf{i} + \frac{50}{51}\mathbf{j} + \frac{10}{51}\mathbf{k}$, $\frac{10}{51}\mathbf{i} - \frac{10}{51}\mathbf{j} + \frac{49}{51}\mathbf{k}$
(xxi) $\frac{1}{51}\mathbf{i} + \frac{38}{51}\mathbf{j} - \frac{34}{51}\mathbf{k}$, $-\frac{22}{51}\mathbf{i} + \frac{31}{51}\mathbf{j} + \frac{34}{51}\mathbf{k}$
(xxii) $\frac{28}{53}\mathbf{i} - \frac{45}{53}\mathbf{j}$, $\frac{45}{53}\mathbf{i} + \frac{28}{53}\mathbf{j}$
(xxiii) $-\frac{7}{57}\mathbf{i} - \frac{40}{57}\mathbf{j} - \frac{40}{57}\mathbf{k}$, $-\frac{40}{57}\mathbf{i} + \frac{25}{57}\mathbf{j} + \frac{32}{57}\mathbf{k}$
(xxiv) $\frac{9}{59}\mathbf{i} + \frac{50}{59}\mathbf{j} + \frac{30}{59}\mathbf{k}$, $\frac{30}{59}\mathbf{i} - \frac{30}{59}\mathbf{j} + \frac{41}{59}\mathbf{k}$
(xxv) $\frac{39}{65}\mathbf{i} + \frac{20}{65}\mathbf{j} - \frac{48}{65}\mathbf{k}$, $\frac{60}{65}\mathbf{j} + \frac{25}{65}\mathbf{k}$
(xxvi) $-\frac{21}{79}\mathbf{i} - \frac{70}{79}\mathbf{j} - \frac{30}{79}\mathbf{k}$, $-\frac{30}{79}\mathbf{i} - \frac{21}{79}\mathbf{j} + \frac{70}{79}\mathbf{k}$
(xxvii) $\frac{1}{81}\mathbf{i} + \frac{76}{81}\mathbf{j} - \frac{28}{81}\mathbf{k}$, $\frac{44}{81}\mathbf{i} + \frac{23}{81}\mathbf{j} + \frac{64}{81}\mathbf{k}$
(xxviii) $\frac{16}{81}\mathbf{i} + \frac{79}{81}\mathbf{j} - \frac{8}{81}\mathbf{k}$, $-\frac{47}{81}\mathbf{i} + \frac{16}{81}\mathbf{j} + \frac{64}{81}\mathbf{k}$
(xxix) $-\frac{1}{99}\mathbf{i} + \frac{70}{99}\mathbf{j} + \frac{70}{99}\mathbf{k}$, $-\frac{70}{99}\mathbf{i} + \frac{49}{99}\mathbf{j} - \frac{50}{99}\mathbf{k}$
(xxx) $-\frac{10}{111}\mathbf{i} - \frac{11}{111}\mathbf{j} - \frac{110}{111}\mathbf{k}$, $-\frac{11}{111}\mathbf{i} + \frac{110}{111}\mathbf{j} - \frac{10}{111}\mathbf{k}$
(xxxi) $0.8\mathbf{j} + 0.6\mathbf{k}$, $0.6\mathbf{i} + 0.48\mathbf{j} - 0.64\mathbf{k}$
(xxxii) $0.6\mathbf{i} + 0.64\mathbf{j} + 0.48\mathbf{k}$, $0.48\mathbf{i} - 0.768\mathbf{j} + 0.424\mathbf{k}$
(xxxiii) $0.744\mathbf{i} - 0.64\mathbf{j} + 0.192\mathbf{k}$, $0.64\mathbf{i} + 0.6\mathbf{j} - 0.48\mathbf{k}$
(xxxiv) $-0.352\mathbf{i} - 0.36\mathbf{j} - 0.864\mathbf{k}$, $-0.36\mathbf{i} - 0.8\mathbf{j} + 0.48\mathbf{k}$

4

The Lie Algebra of Vectors in Three-Dimensional Space

THE JACOBI IDENTITY. THE CONCEPT OF A LIE ALGEBRA

We have encountered several products so far which failed to satisfy one or another of the customary properties of a product such as the commutative law or the cancellation law. However, all the products have obeyed the associative law. Indeed, the associative law may appear as a truly universal law, since the product of functions is associative.

Therefore, it is with considerable interest that we observe the fact, having no precedent in our previous experience, that the cross product of vectors is a nonassociative product, i.e., in general,

$$(\mathbf{a} \times \mathbf{b}) \times \mathbf{c} \neq \mathbf{a} \times (\mathbf{b} \times \mathbf{c})$$

To see this one may, for example, check that

$$\mathbf{0} = (\mathbf{i} \times \mathbf{i}) \times \mathbf{j} \neq \mathbf{i} \times (\mathbf{i} \times \mathbf{j}) = \mathbf{i} \times \mathbf{k} = -\mathbf{j}$$

The cross product of vectors is "nicely" nonassociative, however. We have the "next best thing" to the associative law, which is the Jacobi† identity:

($3^{\times}$) Jacobi Identity

$$\mathbf{a} \times (\mathbf{b} \times \mathbf{c}) + \mathbf{b} \times (\mathbf{c} \times \mathbf{a}) + \mathbf{c} \times (\mathbf{a} \times \mathbf{b}) = \mathbf{0}$$

The Jacobi identity completes the list of the formal rules for the cross product, which are combined with the 10 formal rules of the vector sum and the scalar multiple in the following table:

†Carl Gustav Jacobi (1804 – 1851), German mathematician.

CLOSURE	0	0'	$0^{\times}$	
COMMUTATIVITY	1		$1^{\times}$	ANTICOMMUTATIVITY
ASSOCIATIVITY	2	2'	$2^{\times}$	SCALAR ASSOCIATIVITY
NEUTRAL	3	3'	$3^{\times}$	JACOBI IDENTITY
NEGATIVE	4			
DISTRIBUTIVITY	5	5'	$5^{\times}$	

We say that the introduction of the cross product has converted our vector space into a *Lie algebra*.† A Lie algebra is called *commutative* if the product of any two of its elements is zero. Thus every vector space can be converted into a Lie algebra in a trivial way. What we have here in $\mathbb{R}^3$ is the simplest example of a noncommutative Lie algebra.

Moreover, the Lie algebra structure of $\mathbb{R}^3$ is compatible with the metric. By this we mean that

$$(\mathbf{a} \times \mathbf{b})^2 + (\mathbf{a} \cdot \mathbf{b})^2 = \mathbf{a}^2\mathbf{b}^2$$

It is a remarkable fact (one that we shall not prove in this book) that $n = 3$ is the only case in which it is possible to convert $\mathbb{R}^n$ into a noncommutative Lie algebra in a way that is compatible with the metric. There is one other case, $n = 1$, but there the Lie algebra is commutative.

Lie algebras are indispensable in modern mathematics: linear algebra, advanced calculus, and differential geometry. Later in this chapter, as well as in Chap. 13, we shall have a glimpse of why this is so.

GEOMETRIC INTERPRETATION OF THE JACOBI IDENTITY

The Jacobi identity is one of the nontrivial, nonetheless highly important, formulas of linear algebra. Therefore, it will be useful to find its geometric interpretation.

A three-dimensional analog of the triangle is the *trihedron*, i.e., the figure formed by three noncoplanar vectors **a**, **b**, **c**. These vectors correspond to the vertices of the triangle; what will correspond to its sides? To the sides of the triangle there correspond the faces of the trihedron. The faces of the trihedron are planes for which we may substitute their normal

†After Sophus Lie, Norwegian mathematician (1842–1899). Lie is pronounced "lee."

vectors, i.e., the vectors perpendicular to them, $\mathbf{b} \times \mathbf{c}$, $\mathbf{c} \times \mathbf{a}$, $\mathbf{a} \times \mathbf{b}$. Using the same correspondence between planes and vectors, we see that the vectors $\mathbf{a} \times (\mathbf{b} \times \mathbf{c})$, $\mathbf{b} \times (\mathbf{c} \times \mathbf{a})$, $\mathbf{a} \times (\mathbf{b} \times \mathbf{c})$ correspond to the altitudes of the trihedron, i.e., the planes containing an edge and perpendicular to the opposite face.

If the sum of three vectors is the null vector, the three vectors must be coplanar. The normal vectors of three planes having a point in common are coplanar if, and only if, the planes also have a line in common. Hence the geometric interpretation of the Jacobi identity: The altitudes of the trihedron are three planes having a line in common. This is a generalization of the familiar theorem from plane geometry asserting that the altitudes of the triangle are three lines having a point (the orthocenter) in common.

THE DOUBLE CROSS FORMULA

The proof of the Jacobi identity 3^x is based on the important

Double Cross Formula

$$(\mathbf{a} \times \mathbf{b}) \times \mathbf{c} = (\mathbf{a} \cdot \mathbf{c})\mathbf{b} - (\mathbf{b} \cdot \mathbf{c})\mathbf{a}$$

valid for any three vectors **a**, **b**, **c**. A special case of this we have already proved when we decomposed the vector **v** into the sum of parallel and perpendicular components to the unit vector **u**: $\mathbf{v}_{\text{per}} = \mathbf{v} - \mathbf{v}_{\text{par}}$, or

$$(\mathbf{u} \times \mathbf{v}) \times \mathbf{u} = \mathbf{v} - (\mathbf{u} \cdot \mathbf{v})\mathbf{u} = (\mathbf{u} \cdot \mathbf{u})\mathbf{v} - (\mathbf{u} \cdot \mathbf{v})\mathbf{u}$$

because here $\mathbf{u} \cdot \mathbf{u} = 1$. The proof of the double cross formula that follows is based on this special case.

Proof: If **a** is parallel to **b**, then both sides yield **0**. We may, therefore, assume in the sequel that **a** and **b** are nonparallel vectors. Then $\mathbf{a} \times \mathbf{b} \neq \mathbf{0}$ is a normal vector of the plane spanned by **a** and **b** (a normal vector of the plane is just any vector, other than the null vector, that is perpendicular to the plane). For this reason $(\mathbf{a} \times \mathbf{b}) \times \mathbf{c}$ is also in the plane spanned by **a** and **b**, as this vector must also be perpendicular to $\mathbf{a} \times \mathbf{b}$. Therefore,

$$\alpha\mathbf{a} + \beta\mathbf{b} = (\mathbf{a} \times \mathbf{b}) \times \mathbf{c} \qquad (*)$$

for suitable scalars α, β. The proof of the formula will be complete if we show that $\alpha = -\mathbf{b} \cdot \mathbf{c}$ and $\beta = \mathbf{a} \cdot \mathbf{c}$. We do this by finding two linear equations for α, β and solving.

We begin by assuming, for the time being, that **a** and **b** are both unit vectors. Taking the dot product of both sides of (*) with the vector **a**, we get

$$\begin{aligned}
\mathbf{a}\cdot(\alpha\mathbf{a}+\beta\mathbf{b}) &= \mathbf{a}\cdot((\mathbf{a}\times\mathbf{b})\times\mathbf{c}) \\
&= (\mathbf{a}\times(\mathbf{a}\times\mathbf{b}))\cdot\mathbf{c} && \text{by the cross-dot formula} \\
&= -((\mathbf{a}\times\mathbf{b})\times\mathbf{a})\cdot\mathbf{c} && \text{by } (1^{\times}) \\
&= -(\mathbf{b}-(\mathbf{a}\cdot\mathbf{b})\mathbf{a})\cdot\mathbf{c} && \text{by the special case already proved} \\
&= -\mathbf{b}\cdot\mathbf{c}+(\mathbf{a}\cdot\mathbf{b})(\mathbf{a}\cdot\mathbf{c}) && \text{by } (5^{\cdot}) \\
&= (\mathbf{a}\cdot\mathbf{b})(\mathbf{a}\cdot\mathbf{c})-\mathbf{b}\cdot\mathbf{c}
\end{aligned}$$

or, what is the same by virtue of $\mathbf{a}\cdot\mathbf{a}=1$,

$$\alpha+(\mathbf{a}\cdot\mathbf{b})\beta=(\mathbf{a}\cdot\mathbf{b})(\mathbf{a}\cdot\mathbf{c})-\mathbf{b}\cdot\mathbf{c} \qquad (^{**})$$

If we now take the dot product of both sides of (*) with **b**, a completely analogous computation yields

$$(\mathbf{a}\cdot\mathbf{b})\,\alpha+\beta=\mathbf{a}\cdot\mathbf{c}-(\mathbf{a}\cdot\mathbf{b})(\mathbf{b}\cdot\mathbf{c}) \qquad (^{***})$$

The equations (**) and (***) form a linear system with a unique solution for the unknowns α and β as the determinant

$$\begin{vmatrix} 1 & \mathbf{a}\cdot\mathbf{b} \\ \mathbf{a}\cdot\mathbf{b} & 1 \end{vmatrix} = 1-(\mathbf{a}\cdot\mathbf{b})^2 \neq 0$$

[Otherwise, $\mathbf{a}\cdot\mathbf{b}=\pm 1 \Rightarrow \cos\sphericalangle(\mathbf{a},\mathbf{b})=\pm 1 \Rightarrow \mathbf{a}\|\mathbf{b}$, which is impossible.] The quickest way to solve this system of linear equations is by adding and subtracting (**) and (***), which yields

$$\begin{aligned}
(1+\mathbf{a}\cdot\mathbf{b})\alpha+(1+\mathbf{a}\cdot\mathbf{b})\beta &= (1+\mathbf{a}\cdot\mathbf{b})(\mathbf{a}\cdot\mathbf{c})-(1+\mathbf{a}\cdot\mathbf{b})(\mathbf{b}\cdot\mathbf{c}) \\
(1-\mathbf{a}\cdot\mathbf{b})\alpha-(1-\mathbf{a}\cdot\mathbf{b})\beta &= -(1-\mathbf{a}\cdot\mathbf{b})(\mathbf{a}\cdot\mathbf{c})-(1-\mathbf{a}\cdot\mathbf{b})(\mathbf{b}\cdot\mathbf{c})
\end{aligned}$$

or, what is the same,

$$\begin{aligned}
\alpha+\beta &= \mathbf{a}\cdot\mathbf{c}-\mathbf{b}\cdot\mathbf{c} \\
\alpha-\beta &= -\mathbf{a}\cdot\mathbf{c}-\mathbf{b}\cdot\mathbf{c}
\end{aligned}$$

implying that $\alpha=-\mathbf{b}\cdot\mathbf{c}$ and $\beta=\mathbf{a}\cdot\mathbf{c}$. This proves the double cross formula in the case when **a**, **b** are unit vectors. If **a**, **b** are not unit vectors, then $\mathbf{a}=|\mathbf{a}|\;\mathbf{a}^0$, $\mathbf{b}=|\mathbf{b}|\;\mathbf{b}^0$, and

$$\begin{aligned}
(\mathbf{a}\times\mathbf{b})\times\mathbf{c} &= (|\mathbf{a}|\;\mathbf{a}^0\times|\mathbf{b}|\;\mathbf{b}^0)\times\mathbf{c} \\
&= |\mathbf{a}|\,|\mathbf{b}|\;(\mathbf{a}^0\times\mathbf{b}^0)\times\mathbf{c} \\
&= |\mathbf{a}|\,|\mathbf{b}|((\mathbf{a}^0\cdot\mathbf{c})\mathbf{b}^0-(\mathbf{b}^0\cdot\mathbf{c})\mathbf{a}^0) \\
&= (|\mathbf{a}|\;\mathbf{a}^0\cdot\mathbf{c})(|\mathbf{b}|\;\mathbf{b}^0)-(|\mathbf{b}|\;\mathbf{b}^0\cdot\mathbf{c})(|\mathbf{a}|\;\mathbf{a}^0) \\
&= (\mathbf{a}\cdot\mathbf{c})\mathbf{b}-(\mathbf{b}\cdot\mathbf{c})\mathbf{a}
\end{aligned}$$

The proof of the double cross formula is now complete.

The Jacobi identity (3^x) is a simple consequence:

$$\begin{aligned}&\mathbf{a}\times(\mathbf{b}\times\mathbf{c})+\mathbf{b}\times(\mathbf{c}\times\mathbf{a})+\mathbf{c}\times(\mathbf{a}\times\mathbf{b})\\&\quad=(\mathbf{a}\cdot\mathbf{c})\mathbf{b}-(\mathbf{a}\cdot\mathbf{b})\mathbf{c}+(\mathbf{b}\cdot\mathbf{a})\mathbf{c}-(\mathbf{b}\cdot\mathbf{c})\mathbf{a}+(\mathbf{c}\cdot\mathbf{b})\mathbf{a}-(\mathbf{c}\cdot\mathbf{a})\mathbf{b}=\mathbf{0}\end{aligned}$$

In view of the great importance of the double cross formula we shall give a second proof, which is also based on the special case

$$(\mathbf{u}\times\mathbf{v})\times\mathbf{u}=(\mathbf{u}\cdot\mathbf{u})\mathbf{v}-(\mathbf{u}\cdot\mathbf{v})\mathbf{u}$$

Second Proof: It is easy to show that this is valid not only for unit vectors **u**, but for an arbitrary pair of vectors **u**, **v**. The details are left to the reader, who will find that it is a consequence of the scalar associative law (2^x). The general case of the double cross formula can now be proved as follows. If **a** and **b** are parallel vectors, then both sides of the double cross formula are zero and there is nothing to prove. We may therefore assume that **a**, **b** are not parallel. In this case **a**, **b**, $\mathbf{a}\times\mathbf{b}$ are not coplanar (i.e., they are linearly independent) vectors, and hence every vector can be written as a linear combination of the three, such as $\mathbf{c}=\alpha\mathbf{a}+\beta\mathbf{b}+\gamma(\mathbf{a}\times\mathbf{b})$. Then

$$\begin{aligned}(\mathbf{a}\times\mathbf{b})\times\mathbf{c}&=\alpha(\mathbf{a}\times\mathbf{b})\times\mathbf{a}+\beta(\mathbf{a}\times\mathbf{b})\times\mathbf{b}+\gamma(\mathbf{a}\times\mathbf{b})\times(\mathbf{a}\times\mathbf{b})\\&=\alpha(\mathbf{a}\times\mathbf{b})\times\mathbf{a}-\beta(\mathbf{b}\times\mathbf{a})\times\mathbf{b}\\&=\alpha((\mathbf{a}\cdot\mathbf{a})\mathbf{b}-(\mathbf{b}\cdot\mathbf{a})\mathbf{a})-\beta((\mathbf{b}\cdot\mathbf{b})\mathbf{a}-(\mathbf{a}\cdot\mathbf{b})\mathbf{b})\\&=(\alpha(\mathbf{a}\cdot\mathbf{a})+\beta(\mathbf{a}\cdot\mathbf{b}))\mathbf{b}-(\alpha(\mathbf{b}\cdot\mathbf{a})+\beta(\mathbf{b}\cdot\mathbf{b}))\mathbf{a}\\&=(\mathbf{a}\cdot(\alpha\mathbf{a}+\beta\mathbf{b}))\mathbf{b}-(\mathbf{b}\cdot(\alpha\mathbf{a}+\beta\mathbf{b}))\mathbf{a}\\&=(\mathbf{a}\cdot(\alpha\mathbf{a}+\beta\mathbf{b}+\gamma(\mathbf{a}\times\mathbf{b})))\mathbf{b}-(\mathbf{b}\cdot(\alpha\mathbf{a}+\beta\mathbf{b}+\gamma(\mathbf{a}\times\mathbf{b})))\mathbf{a}\\&=(\mathbf{a}\cdot\mathbf{c})\mathbf{b}-(\mathbf{b}\cdot\mathbf{c})\mathbf{a}\end{aligned}$$

and the second proof is complete.

There are, in fact, two double cross formulas:

$$\begin{aligned}(\mathbf{a}\times\mathbf{b})\times\mathbf{c}&=(\mathbf{a}\cdot\mathbf{c})\mathbf{b}-(\mathbf{b}\cdot\mathbf{c})\mathbf{a}\\\mathbf{a}\times(\mathbf{b}\times\mathbf{c})&=(\mathbf{a}\cdot\mathbf{c})\mathbf{b}-(\mathbf{a}\cdot\mathbf{b})\mathbf{c}\end{aligned}$$

Either formula follows from the other, which can be seen by applying the anticommutative law 1^x.

Three remarks will facilitate the retention of both formulas. The vector product on the left-hand side is a linear combination of those two vectors which are inside the parentheses, since it is in the plane spanned by them. [For example, $(\mathbf{a}\times\mathbf{b})\times\mathbf{c}$ is in the plane spanned by **a** and **b**, because it is perpendicular to $\mathbf{a}\times\mathbf{b}$, the normal vector of that plane.] This then is the first remark: The two vectors that fall *inside* the parentheses on the left-hand side of the cross dot formula will fall *outside* the parentheses on the right-hand side. Next, the coefficients of the linear combination on the right-hand side are easy to remember; we get them if we take the dot products of the other two vectors. The third remark helps us remember the signatures on the right-hand side of the double cross formula: The positive signature is taken by the vector that is between the two crosses on the left-hand side; the other vector takes the negative signature.

APPLICATIONS

1. The Triple Cross Formula. Cramer's Formula

The question arises naturally how to evaluate other multiple products of vectors. In the absence of the associative law, the answer is not trivial and has to be worked out in each individual case.

The following is a list of formulas for various multiple products of vectors. Each formula is an easy consequence of the double cross formula and the cross-dot formula.

Cross-Dot Formula

$$(\mathbf{a} \times \mathbf{b}) \cdot \mathbf{c} = \mathbf{a} \cdot (\mathbf{b} \times \mathbf{c})$$

Double Cross Formulas

$$(\mathbf{a} \times \mathbf{b}) \times \mathbf{c} = (\mathbf{a} \cdot \mathbf{c})\mathbf{b} - (\mathbf{b} \cdot \mathbf{c})\mathbf{a}$$
$$\mathbf{a} \times (\mathbf{b} \times \mathbf{c}) = (\mathbf{a} \cdot \mathbf{c})\mathbf{b} - (\mathbf{a} \cdot \mathbf{b})\mathbf{c}$$

Triple Cross Formulas

$$(\mathbf{a} \times \mathbf{b}) \times (\mathbf{u} \times \mathbf{v}) = [\mathbf{auv}]\mathbf{b} - [\mathbf{buv}]\mathbf{a}$$
$$(\mathbf{a} \times \mathbf{b}) \times (\mathbf{u} \times \mathbf{v}) = [\mathbf{abv}]\mathbf{u} - [\mathbf{abu}]\mathbf{v}$$

We get two expressions for $(\mathbf{a} \times \mathbf{b}) \times (\mathbf{u} \times \mathbf{v})$ because we may treat either $\mathbf{a} \times \mathbf{b}$ or $\mathbf{u} \times \mathbf{v}$ as a single vector and apply the double cross formula. Let us equate the right-hand sides of the triple cross formulas. We get (after putting $\mathbf{u} = \mathbf{c}$)

Cramer's† Formula

$$[\mathbf{abc}]\mathbf{v} = [\mathbf{vbc}]\mathbf{a} + [\mathbf{avc}]\mathbf{b} + [\mathbf{abv}]\mathbf{c}$$

†Gabriel Cramer (1704–1752), Swiss mathematician, one of the discoverers of determinants. Cramer's rule (see Appendix 1) is a generalization of this formula to $\mathbb{R}^n$.

Cramer's formula furnishes the numerical solution to a problem to which the geometric solution was given in Chap. 1. Let **a**, **b**, **c** be fixed vectors, and let **v** be an arbitrary vector. Then **v** can be uniquely expressed as a linear combination of **a**, **b**, **c**:

$$\mathbf{v} = \alpha\mathbf{a} + \beta\mathbf{b} + \gamma\mathbf{c} \qquad (\alpha, \beta, \gamma \in \mathbb{R})$$

if, and only if, the vectors **a**, **b**, **c** are not parallel to the same plane or, what is the same, $[\mathbf{abc}] \neq 0$. Cramer's formula furnishes the unique values of the coefficients:

$$\alpha = \frac{[\mathbf{vbc}]}{[\mathbf{abc}]} \qquad \beta = \frac{[\mathbf{avc}]}{[\mathbf{abc}]} \qquad \gamma = \frac{[\mathbf{abv}]}{[\mathbf{abc}]}$$

2. The Cross-Dot-Cross Formula. The Double Box Formula

Cross-Dot-Cross Formula

$$(\mathbf{a} \times \mathbf{b}) \cdot (\mathbf{u} \times \mathbf{v}) = \begin{vmatrix} \mathbf{a} \cdot \mathbf{u} & \mathbf{b} \cdot \mathbf{u} \\ \mathbf{a} \cdot \mathbf{v} & \mathbf{b} \cdot \mathbf{v} \end{vmatrix}$$

In order to prove the cross-dot-cross formula we first apply the cross-dot formula and then the double cross formula to the left-hand side.

Double Box Formula

$$[\mathbf{abc}][\mathbf{uvw}] = \begin{vmatrix} \mathbf{a} \cdot \mathbf{u} & \mathbf{b} \cdot \mathbf{u} & \mathbf{c} \cdot \mathbf{u} \\ \mathbf{a} \cdot \mathbf{v} & \mathbf{b} \cdot \mathbf{v} & \mathbf{c} \cdot \mathbf{v} \\ \mathbf{a} \cdot \mathbf{w} & \mathbf{b} \cdot \mathbf{w} & \mathbf{c} \cdot \mathbf{w} \end{vmatrix}$$

Proof:

$$\begin{aligned}
[\mathbf{abc}][\mathbf{uvw}] &= [\mathbf{abc}](\mathbf{u} \times \mathbf{v}) \cdot \mathbf{w} \\
&= ([\mathbf{u} \times \mathbf{v},\mathbf{b},\mathbf{c}]\mathbf{a} + [\mathbf{a},\mathbf{u} \times \mathbf{v},\mathbf{c}]\mathbf{b} + [\mathbf{a},\mathbf{b},\mathbf{u} \times \mathbf{v}]\mathbf{c}) \cdot \mathbf{w} \\
&= (\mathbf{b} \times \mathbf{c}) \cdot (\mathbf{u} \times \mathbf{v})(\mathbf{a} \cdot \mathbf{w}) + (\mathbf{c} \times \mathbf{a}) \cdot (\mathbf{u} \times \mathbf{v})(\mathbf{b} \cdot \mathbf{w}) + (\mathbf{a} \times \mathbf{b}) \cdot (\mathbf{u} \times \mathbf{v})(\mathbf{c} \cdot \mathbf{w}) \\
&= \begin{vmatrix} \mathbf{b} \cdot \mathbf{u} & \mathbf{c} \cdot \mathbf{u} \\ \mathbf{b} \cdot \mathbf{v} & \mathbf{c} \cdot \mathbf{v} \end{vmatrix}(\mathbf{a} \cdot \mathbf{w}) - \begin{vmatrix} \mathbf{a} \cdot \mathbf{u} & \mathbf{c} \cdot \mathbf{u} \\ \mathbf{a} \cdot \mathbf{v} & \mathbf{c} \cdot \mathbf{v} \end{vmatrix}(\mathbf{b} \cdot \mathbf{w}) + \begin{vmatrix} \mathbf{a} \cdot \mathbf{u} & \mathbf{b} \cdot \mathbf{u} \\ \mathbf{a} \cdot \mathbf{v} & \mathbf{b} \cdot \mathbf{v} \end{vmatrix}(\mathbf{c} \cdot \mathbf{w}) \\
&= \begin{vmatrix} \mathbf{a} \cdot \mathbf{u} & \mathbf{b} \cdot \mathbf{u} & \mathbf{c} \cdot \mathbf{u} \\ \mathbf{a} \cdot \mathbf{v} & \mathbf{b} \cdot \mathbf{v} & \mathbf{c} \cdot \mathbf{v} \\ \mathbf{a} \cdot \mathbf{w} & \mathbf{b} \cdot \mathbf{w} & \mathbf{c} \cdot \mathbf{w} \end{vmatrix}
\end{aligned}$$

wherein we have applied Cramer's formula for $[\mathbf{abc}](\mathbf{u} \times \mathbf{v})$, then the cross-dot-cross formula three times, and finally, have used the expansion property of determinants (with respect to the third row).

3. Twists

There is no question of defining higher powers of a vector via the cross product because $\mathbf{v} \times \mathbf{v} = \mathbf{0}$ for all $\mathbf{v}$. However, we can define a linear operator via the cross product and consider its higher powers defined via composition. This idea is of great importance in linear algebra. First we shall consider once more the linear operator t such that $t(\mathbf{v}) = \mathbf{u} \times \mathbf{v}$ for all $\mathbf{v}$, where $\mathbf{u}$ is a fixed unit vector. Then

$$t^2(\mathbf{v}) = t(t(\mathbf{v})) = \mathbf{u} \times t(\mathbf{v}) = \mathbf{u} \times (\mathbf{u} \times \mathbf{v}) = (\mathbf{u} \cdot \mathbf{v})\mathbf{u} - \mathbf{v}$$

by the double cross formula, and

$$t^3(\mathbf{v}) = t(t^2(\mathbf{v})) = \mathbf{u} \times ((\mathbf{u} \cdot \mathbf{v})\mathbf{u} - \mathbf{v}) = -\mathbf{u} \times \mathbf{v} = -t(\mathbf{v}) \qquad \text{for all } \mathbf{v}$$

i.e., $t^3 = -t$. A linear operator satisfying this property is called a *twist*. All the higher powers of a twist can be expressed in terms of t and t^2:

$$\begin{aligned}
t &= t^5 = t^9 = t^{13} = \cdots = t^{4n+1} = t \\
t^2 &= t^6 = t^{10} = t^{14} = \cdots = t^{4n+2} = t^2 \\
t^3 &= t^7 = t^{11} = t^{15} = \cdots = t^{4n+3} = -t \\
t^4 &= t^8 = t^{12} = t^{16} = \cdots = t^{4n} \quad = -t^2
\end{aligned}$$

It should be noted that not every twist has the form $t(\mathbf{v}) = \mathbf{u} \times \mathbf{v}$ ($|\mathbf{u}| = 1$) Those that do are called *antisymmetric*. It is left as an exercise to show that the linear operator t defined by the formula

$$t(\mathbf{v}) = (\mathbf{e}' \times \mathbf{v}) \times \mathbf{n}' - (\mathbf{e} \times \mathbf{v}) \times \mathbf{n}$$

where $\mathbf{e} \cdot \mathbf{n} = 0 = \mathbf{e}' \cdot \mathbf{n}'$, $\mathbf{e} \cdot \mathbf{n}' = 1 = \mathbf{e}' \cdot \mathbf{n}$ also satisfies $t^3 = -t$ (i.e., is a twist) and that t is antisymmetric if, and only if, $\mathbf{e} = \mathbf{n}'$, $\mathbf{e}' = \mathbf{n}$. In Chap. 10 we shall see that every twist of $\mathbb{R}^3$ can be written in this form.

4. Projections

Next we define a linear operator by the formula $p(\mathbf{v}) = (\mathbf{u} \times \mathbf{v}) \times \mathbf{u}$ for all $\mathbf{v}$, where $\mathbf{u}$ is a fixed unit vector. Then

$$\begin{aligned}
p^2(\mathbf{v}) &= p(p(\mathbf{v})) = (\mathbf{u} \times p(\mathbf{v})) \times \mathbf{u} = p(\mathbf{v}) - (\mathbf{u} \cdot p(\mathbf{v}))\mathbf{u} \\
&= p(\mathbf{v}) - [\mathbf{u}, \mathbf{u} \times \mathbf{v}, \mathbf{u}]\mathbf{u} = p(\mathbf{v}) \qquad \text{for all } \mathbf{v}
\end{aligned}$$

i.e., $p^2 = p$. A linear operator satisfying this property is called a *projection*. All the higher powers of a projection are equal: $p = p^2 = p^3 = p^4 = \cdots$. We recognize that $p(\mathbf{v})$ is the perpendicular component of $\mathbf{v}$ with respect to $\mathbf{u}$, so that p is the *perpendicular projection* onto the normal plane of $\mathbf{u}$. Not every projection onto a plane is perpendicular, however. It is left as an exercise to show that the linear operator defined by the formula

$$p(\mathbf{v}) = (\mathbf{e} \times \mathbf{v}) \times \mathbf{n}$$

where $\mathbf{e} \cdot \mathbf{n} = 1$ also satisfies $p^2 = p$ and, in fact, is a projection onto the normal plane of $\mathbf{n}$ along the line spanned by $\mathbf{e}$, and that p is a perpendicular projection if, and only if, $\mathbf{e} = \mathbf{n} = \mathbf{u}$ is a unit vector. In Chap. 10 we shall see that every projection onto a plane can be written in this form.

Example Show that the perpendicular projection p onto the plane spanned by the pair of perpendicular unit vectors $\mathbf{u}_1$, $\mathbf{u}_2$ is given by the formula

$$p(\mathbf{v}) = (\mathbf{u}_1 \cdot \mathbf{v})\mathbf{u}_1 + (\mathbf{u}_2 \cdot \mathbf{v})\mathbf{u}_2$$

Answer: Take the perpendicular component of $\mathbf{v}$ with respect to the unit vector $\mathbf{u} = \mathbf{u}_1 \times \mathbf{u}_2$ and apply the double cross formula repeatedly to get

$$\begin{aligned} p(\mathbf{v}) = \mathbf{v}_{\text{per}} &= (\mathbf{u} \times \mathbf{v}) \times \mathbf{u} = ((\mathbf{u}_1 \times \mathbf{u}_2) \times \mathbf{v}) \times (\mathbf{u}_1 \times \mathbf{u}_2) \\ &= ((\mathbf{u}_1 \cdot \mathbf{v})\mathbf{u}_2 - (\mathbf{u}_2 \cdot \mathbf{v})\mathbf{u}_1) \times (\mathbf{u}_1 \times \mathbf{u}_2) = (\mathbf{u}_1 \cdot \mathbf{v})\mathbf{u}_1 + (\mathbf{u}_2 \cdot \mathbf{v})\mathbf{u}_2 \end{aligned}$$

5. Lifts

Let us now define a linear operator by the formula

$$\ell(\mathbf{v}) = (\mathbf{e} \times \mathbf{v}) \times \mathbf{n} \qquad \text{for all } \mathbf{v}$$

where $\mathbf{e} \perp \mathbf{n}$ are fixed vectors. Then

$$\ell^2(\mathbf{v}) = (\mathbf{e} \times \ell(\mathbf{v})) \times \mathbf{n} = -\mathbf{n} \cdot \ell(\mathbf{v}))\mathbf{e} = 0$$

because $\ell(\mathbf{v}) = (\mathbf{e} \times \mathbf{v}) \times \mathbf{n} = -(\mathbf{n} \cdot \mathbf{v})\mathbf{e} \perp \mathbf{n}$. A linear operator that satisfies

$$\ell^2 = 0$$

is called a *lift*. All the higher powers of a lift are trivial:

$$\ell^2 = \ell^3 = \ell^4 = \ell^5 = \cdots = 0$$

In Chap. 10 we shall see that every lift of $\mathbb{R}^3$ can be written in this form.

6. Spherical Trigonometry

Let A, B, C denote the vertices of a spherical triangle on the sphere of unit radius, and let $\mathbf{a}$, $\mathbf{b}$, $\mathbf{c}$ denote the corresponding location vectors. We shall further assume that the notation is so chosen that $\mathbf{a}$, $\mathbf{b}$, $\mathbf{c}$ form a right-hand system, i.e., $[\mathbf{abc}] > 0$. Then the *sides* of the spherical triangle are

$$\alpha = \measuredangle(\mathbf{b},\mathbf{c}) \qquad \beta = \measuredangle(\mathbf{c},\mathbf{a}) \qquad \gamma = \measuredangle(\mathbf{a},\mathbf{b})$$

If A, B, C denote (by abuse of notation) the *interior angles* of the spherical triangle, then

$$A = 180° - \sphericalangle(\mathbf{c} \times \mathbf{a}, \mathbf{a} \times \mathbf{b}) \qquad B = 180° - \sphericalangle(\mathbf{a} \times \mathbf{b}, \mathbf{b} \times \mathbf{c})$$
$$C = 180° - \sphericalangle(\mathbf{b} \times \mathbf{c}, \mathbf{c} \times \mathbf{a})$$

because $\mathbf{c} \times \mathbf{a}$ is the normal vector of the plane spanned by $\mathbf{c}$, $\mathbf{a}$ (i.e., $\mathbf{c} \times \mathbf{a}$ is a vector perpendicular to the plane of $\mathbf{c}$, $\mathbf{a}$) and therefore $\sphericalangle(\mathbf{c} \times \mathbf{a}, \mathbf{a} \times \mathbf{b})$ is the *exterior angle* of the spherical triangle at the vertex A.

Of course, the vectors $\mathbf{a} \times \mathbf{b}$, $\mathbf{b} \times \mathbf{c}$, $\mathbf{c} \times \mathbf{a}$ may not be unit vectors, but we can certainly normalize them:

$$(\mathbf{a} \times \mathbf{b})^0 = \frac{\mathbf{a} \times \mathbf{b}}{|\mathbf{a} \times \mathbf{b}|} = \mathbf{c}' \qquad (\mathbf{b} \times \mathbf{c})^0 = \mathbf{a}' \qquad (\mathbf{c} \times \mathbf{a})^0 = \mathbf{b}'$$

since none of them is $\mathbf{0}$. Then $\mathbf{a}'$, $\mathbf{b}'$, $\mathbf{c}'$ are the location vectors of the points A', B', C' on the unit sphere, determining another spherical triangle $\Delta A'B'C'$, called the *polar triangle* of ΔABC. There is a complete reciprocity between the spherical triangles ΔABC and $\Delta A'B'C'$; i.e., the polar triangle of $\Delta A'B'C'$ is ΔABC. The polarity relationship between spherical triangles can be characterized by the condition that the *sides* of ΔABC and the *angles* of the polar triangle $\Delta A'B'C'$ are supplementary angles, and vice versa, i.e.,

$$\alpha + A' = \beta + B' = \gamma + C' = 180°$$

and

$$A + \alpha' = B + \beta' = C + \gamma' = 180°$$

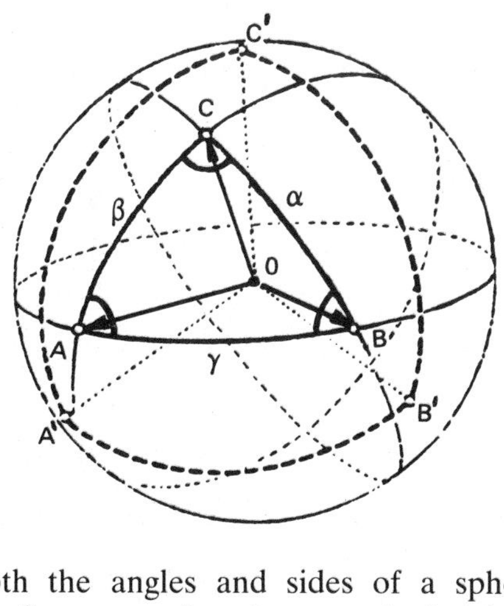

Note that both the angles and sides of a spherical triangle are measured as angles (i.e., as arc length on a unit circle). It is clear that the sum of angles of a spherical triangle is strictly *greater than* 180°. In fact, it is easy to visualize a spherical triangle with three right angles (or one with two right angles and an obtuse angle).

Spherical trigonometry is based on the spherical law of sines and cosines. Of the latter, there are two kinds, the spherical law of cosines relating to the sides, and the spherical law of cosines relating to the angles. All these are valid for any spherical triangle; a special case of the spherical law of cosines relating to the sides is the theorem of Pythagoras, which is valid for spherical right triangles only.

Spherical Law of Sines

$$\frac{\sin A}{\sin \alpha} = \frac{\sin B}{\sin \beta} = \frac{\sin C}{\sin \gamma}$$

Proof:

$$\begin{aligned}(\mathbf{a} \times \mathbf{b}) \times (\mathbf{b} \times \mathbf{c}) &= (\mathbf{a} \cdot (\mathbf{b} \times \mathbf{c}))\mathbf{b} - (\mathbf{b} \cdot (\mathbf{b} \times \mathbf{c}))\mathbf{a} \quad &\text{(by the double cross formula)}\\ &= [\mathbf{abc}]\mathbf{b} - 0 \quad &\text{(because } [\mathbf{bbc}] = 0)\end{aligned}$$

Since $[\mathbf{abc}] > 0$ and $|\mathbf{b}| = 1$, we have

$$|(\mathbf{a} \times \mathbf{b}) \times (\mathbf{b} \times \mathbf{c})| = [\mathbf{abc}]$$

On the other hand, by the trigonometric formula for the length of the cross product,

$$\begin{aligned}|(\mathbf{a} \times \mathbf{b}) \times (\mathbf{b} \times \mathbf{c})| &= |\mathbf{a} \times \mathbf{b}|\,|\mathbf{b} \times \mathbf{c}| \sin(180° - B)\\ &= \sin \gamma \sin \alpha \sin B = [\mathbf{abc}]\end{aligned}$$

by equating the right-hand sides. By the same token, we also have

$$\sin \alpha \sin \beta \sin C = [\mathbf{abc}]$$

and

$$\sin \beta \sin \gamma \sin A = [\mathbf{abc}]$$

The spherical law of sines follows by equating the left-hand sides.

Spherical Law of Cosines Relating to the Sides

$$\cos \gamma = \cos \alpha \cos \beta + \sin \alpha \sin \beta \cos C$$

and two more formulas, expressing the cosines of the other two sides α, β.

Proof:

$$\begin{aligned}(\mathbf{a} \times \mathbf{b}) \cdot (\mathbf{b} \times \mathbf{c}) &= ((\mathbf{a} \times \mathbf{b}) \times \mathbf{b}) \cdot \mathbf{c} \quad &\text{(by the cross-dot formula)}\\ &= ((\mathbf{a} \cdot \mathbf{b})\mathbf{b} - \mathbf{a}) \cdot \mathbf{c} \quad &\text{(by the double-cross formula)}\end{aligned}$$

$$= (\mathbf{a}\cdot\mathbf{b})(\mathbf{b}\cdot\mathbf{c}) - \mathbf{a}\cdot\mathbf{c}$$
$$= \cos\gamma\cos\alpha - \cos\beta$$

On the other hand, by the trigonometric formulas for the dot product and for the length of the cross product,

$$(\mathbf{a}\times\mathbf{b})\cdot(\mathbf{b}\times\mathbf{c}) = |\mathbf{a}\times\mathbf{b}|\,|\mathbf{b}\times\mathbf{c}|\,\cos(180^\circ - B) = -\sin\gamma\sin\alpha\cos B$$

and, by equating the right-hand sides, the spherical law of cosines relating to the side β,

$$\cos\beta = \cos\gamma\cos\alpha + \sin\gamma\sin\alpha\cos B$$

follows.

The other two formulas can be obtained in an analogous way.

Spherical Pythagorean Theorem

$$C = 90^\circ \Leftrightarrow \cos\gamma = \cos\alpha\cos\beta$$

which is obviously a special case.

Spherical Law of Cosines Relating to the Angles

$$\cos C = -\cos A\cos B + \sin A\sin B\cos\gamma$$

and two more formulas, expressing the cosines of the other two angles A, B. They follow from the spherical law of cosines relating to the sides, as applied to the polar triangle. The details are left to the reader.

THE LIE ALGEBRA OF ANTISYMMETRIC OPERATORS

It is customary to talk about the paradox of linear algebra: while every vector space can be converted into a metric vector space by endowing it with a dot product, regardless of its dimension, only the three-dimensional vector space can be converted into a Lie algebra by the introduction of the cross product. The cross product seems to lack a higher-dimensional generalization.

This paradox is the result of a vicious formulation of the problem, as we shall now see. Let **a** be a fixed vector, the linear operator f defined by the formula $f(\mathbf{v}) = \mathbf{a}\times\mathbf{v}$ for all $\mathbf{v}$ is called an *antisymmetric* operator. An alternative definition of this concept given in Chap. 11 can readily be generalized to higher-dimensional vector spaces. Let g be another antisymmetric operator: $g(\mathbf{v}) = \mathbf{b}\times\mathbf{v}$; then we can talk about the sum $f+g$ and the scalar multiple λf of antisymmetric operators, which are seen to be antisymmetric: $(f+g)(\mathbf{v}) = (\mathbf{a}+\mathbf{b})\times\mathbf{v}$, $(\lambda f)(\mathbf{v}) = (\lambda\mathbf{a})\times\mathbf{v}$. However, the product $g\circ f$ of antisymmetric operators, in general, fails to be antisymmetric. This failure led Sophus Lie to introduce what we today call the Lie product of linear operators:

$$[f,g]=f\circ g-g\circ f$$

For a pair of antisymmetric operators we then have

$$\begin{aligned}[f,g](\mathbf{v}) &= (f\circ g-g\circ f)(\mathbf{v})=f(g(\mathbf{v}))-g(f(\mathbf{v}))\\ &= \mathbf{a}\times(\mathbf{b}\times\mathbf{v})-\mathbf{b}\times(\mathbf{a}\times\mathbf{v})=(\mathbf{a}\times\mathbf{b})\times\mathbf{v}\end{aligned}$$

by the Jacobi identity 3^x and the anticommutative law 1^x. This means that the Lie product of two antisymmetric operators is antisymmetric (and in $\mathbb{R}^3$, the notion of the Lie product of antisymmetric operators coincides with the notion of the cross product of vectors). In fact, the antisymmetric operators form a Lie algebra, in particular, the Lie product satisfies the Jacobi identity.

This result is true for all finite-dimensional vector spaces with a dot product. The Lie algebra of antisymmetric operators plays an important role in these vector spaces. If $\dim V=n$, then the Lie algebra of antisymmetric operators has dimension

$$\frac{n(n-1)}{2}$$

as we shall see in Chap. 14. Therefore, if $n\neq 0$, then

$$\frac{n(n-1)}{2}=n\Leftrightarrow n=3$$

It appears that the special nature of $\mathbb{R}^3$ consists in having a natural isomorphism between vectors and antisymmetric operators; no such isomorphism exists for any other dimension n. Moreover, $\mathbb{R}^3$ is the simplest example of a (noncommutative) Lie algebra.

The full importance of the cross product (and, more generally, of antisymmetric operators) will be brought out in Chaps. 13 and 14, where the "infinitesimal structure" of linear operators is discussed. We shall see that antisymmetric operators are distinguished by the property that they induce (via the exponential functor) all the one-parameter groups of rotations in the metric vector space. In these terms, antisymmetric operators serve as the "axis" and "angular velocity" of rotations.

It is a different problem to convert the metric vector space $\mathbb{R}^n$ into a Lie algebra over $\mathbb{R}$ so that the Lie product is compatible with the dot product, i.e.,

$$[\mathbf{a},\mathbf{b}]^2+(\mathbf{a}\cdot\mathbf{b})^2=\mathbf{a}^2\mathbf{b}^2$$

One of the great classical problems of algebra was to find all the n for which the metric vector space $\mathbb{R}^n$ could be so converted. The answer is $n = 1, 3$: $\mathbb{R}$ with the trivial product ($[\mathbf{a},\mathbf{b}] = 0$ for all $\mathbf{a}, \mathbf{b} \in \mathbb{R}$), a commutative Lie algebra, and $\mathbb{R}^3$ with the cross product. If we drop the Jacobi identity, then we also have $n = 7$. However, no metric vector space $\mathbb{R}^n$, other than $\mathbb{R}$, $\mathbb{R}^3$, and $\mathbb{R}^7$, can be given a bilinear vector product. These questions are rather deep and must remain outside the scope of this book.

EXERCISES

4.1 Show that a necessary and sufficient condition that four points A, B, C, P be coplanar is that their location vectors $\mathbf{a}$, $\mathbf{b}$, $\mathbf{c}$, $\mathbf{p}$ satisfy

$$[\mathbf{pbc}] + [\mathbf{apc}] + [\mathbf{abp}] - [\mathbf{abc}] = 0$$

Prove the following vector identities:

4.2 $[\mathbf{a} \times \mathbf{b}, \mathbf{b} \times \mathbf{c}, \mathbf{c} \times \mathbf{a}] = [\mathbf{abc}]^2$

4.3 $(\mathbf{b} \times \mathbf{c}) \cdot (\mathbf{a} \times \mathbf{d}) + (\mathbf{c} \times \mathbf{a}) \cdot (\mathbf{b} \times \mathbf{d}) + (\mathbf{a} \times \mathbf{b}) \cdot (\mathbf{c} \times \mathbf{d}) = 0$

4.4 $(\mathbf{b} \times \mathbf{c}) \times (\mathbf{a} \times \mathbf{d}) + (\mathbf{c} \times \mathbf{a}) \times (\mathbf{b} \times \mathbf{d}) + (\mathbf{a} \times \mathbf{b}) \times (\mathbf{c} \times \mathbf{d}) = -2[\mathbf{abc}]\mathbf{d}$

4.5 $(\mathbf{a} - \mathbf{d}) \cdot (\mathbf{b} - \mathbf{c}) + (\mathbf{b} - \mathbf{d}) \cdot (\mathbf{c} - \mathbf{a}) + (\mathbf{c} - \mathbf{d}) \cdot (\mathbf{a} - \mathbf{b}) = 0$

4.6 $(\mathbf{a} - \mathbf{d}) \times (\mathbf{b} - \mathbf{c}) + (\mathbf{b} - \mathbf{d}) \times (\mathbf{c} - \mathbf{a}) + (\mathbf{c} - \mathbf{d}) \times (\mathbf{a} - \mathbf{b})$
$= 2(\mathbf{a} \times \mathbf{b} + \mathbf{b} \times \mathbf{c} + \mathbf{c} \times \mathbf{a})$

4.7 The *power* of a point P with respect to a sphere of center C and radius ρ is defined (by analogy with the circle) as $(CP)^2 - \rho^2$. Prove that the power of P with respect to a sphere having AB as diameter is $\overline{PA} \cdot \overline{PB}$.

4.8 Prove that the locus of points having the same power with respect to two (nonconcentric) spheres is a plane, called the *radical plane* of the two spheres.

4.9 Prove that the radical planes of three spheres (the centers of which are not collinear) meet in a line, called the *radical axis* of the three spheres. Moreover, if the location vectors of the centers of the spheres are $\mathbf{a}$, $\mathbf{b}$, $\mathbf{c}$, then the direction vector of the radical axis is $\mathbf{a} \times \mathbf{b} + \mathbf{b} \times \mathbf{c} + \mathbf{c} \times \mathbf{a}$.

4.10 Prove the law of cosines relating to the angles of the spherical triangle.

4.11 Show that the linear operator t defined by the formula

$$t(\mathbf{v}) = (\mathbf{e}' \times \mathbf{v}) \times \mathbf{n}' - (\mathbf{e} \times \mathbf{v}) \times \mathbf{n}$$

where $\mathbf{e} \cdot \mathbf{n} = 0 = \mathbf{e}' \cdot \mathbf{n}'$, $\mathbf{e} \cdot \mathbf{n}' = 1 = \mathbf{e}' \cdot \mathbf{n}$ is a twist (i.e., $t^3 = -t$).

4.12 Show that the twist t in Exercise 11 is antisymmetric if, and only if, $\mathbf{e} = \mathbf{n}'$, $\mathbf{e}' = \mathbf{n}$. (See Chap. 10, Exercise 9, and Chap. 11, Exercise 25.)

4.13 Show that the linear operator p defined by the formula

$$p(\mathbf{v}) = (\mathbf{e} \times \mathbf{v}) \times \mathbf{n}$$

where $\mathbf{e} \cdot \mathbf{n} = 1$ is a projection, i.e., $p^2 = p$. Show that, in fact, p is a projection onto the normal plane of **n** (but not necessarily a perpendicular projection). (See Chap. 10, Exercise 2, and Chap. 11, Exercises 26 and 27.)

4.14 Show that the projection p in Exercise 13 is a perpendicular projection if, and only if, $\mathbf{e} = \mathbf{n}$. (See Chap. 10, Exercise 2, and Chap. 11, Exercises 26 and 27.)

5
Lines and Planes†

COORDINATES IN THREE-DIMENSIONAL SPACE

An orthonormal basis was defined in Chap. 2. An orthonormal basis gives rise to a *rectangular metric coordinate system*. Rectangular means that the coordinate axes are mutually perpendicular; metric means that the three coordinates are just the signed distances from the point to the appropriate coordinate plane. As a consequence, the scales on the three coordinate axes will be the same. By contrast, there are other types of coordinate systems: skew metric (axes not pairwise perpendicular), rectangular affine (axes pairwise perpendicular, but with different scales on them), and skew affine (axes not pairwise perpendicular, with different scales on them). A *skew affine coordinate system* arises when an arbitrary basis is chosen. We shall be using only rectangular metric or skew affine coordinates. It is important to know what kind of coordinates are specified, because certain formulas are valid in one types of coordinates but not in the other, e.g., the coordinate formulas for the dot, cross, and box products are valid in rectangular metric coordinates only. However, it is true that the equation of lines and planes will be linear whether in rectangular metric or in skew affine coordinates. From now on, at the beginning of each chapter we shall specify what kind of basis (i.e., what type of coordinates) is meant in that chapter.

Positive orientation in the three-dimensional space $\mathbb{R}^3$ means that the preferred order of the three axes corresponds to the order of the first finger, second finger, and the thumb of the right hand. For this reason, a positively oriented basis (coordinate system) is also called a *right-hand system*.

†*Note*: In this chapter all bases are orthonormal and all coordinate systems are rectangular metric.

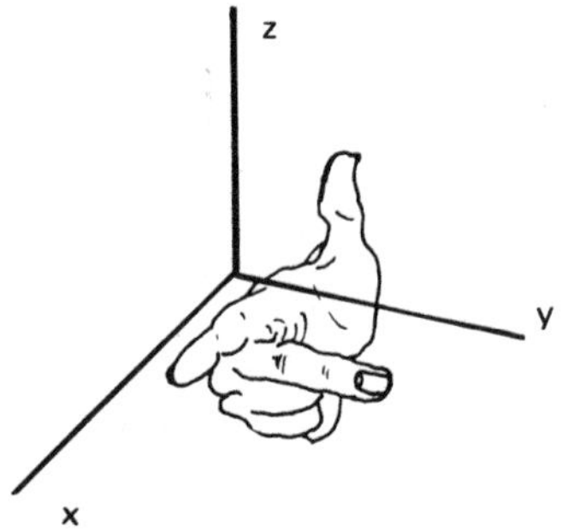

In plane analytic geometry orientation is described in terms of the distinction between counterclockwise and clockwise rotation. There positive orientation means that the preferred order of the two axes correspond to a *counterclockwise* rotation of the axes. Thus negative orientation corresponds to clockwise movement, an anomaly for which the historical choice of the clockmakers can be blamed. The distinction between counterclockwise and clockwise movement can also be used to characterize orientation in the three-dimensional coordinate system. Thus positive orientation means that, from the vantage point of an observer in the positive half of the z axis, a counterclockwise rotation through 90° will move the x axis into the y axis.

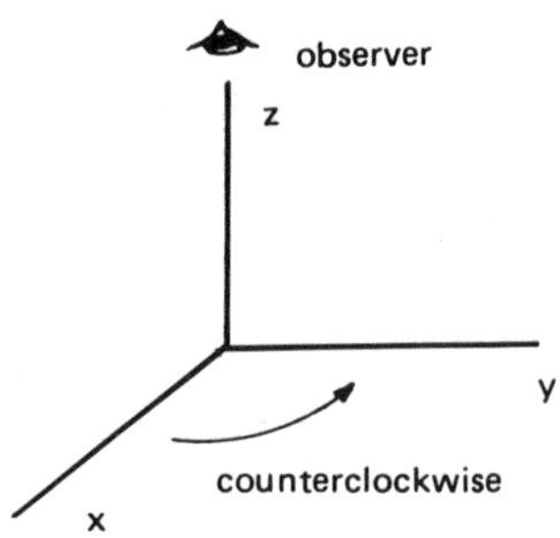

A third way of describing orientation is that if the x axis points eastward and the y axis points northward, then the z axis will point upward for the positive (downward for the negative) orientation. For this reason we may use the letters E and W to indicate the direction of the positive and negative x axis; the letters N and S the positive and negative y axis; the letters U and L (for upper and lower) the positive and negative z axis.

Then, if a cube is fitted to the coordinate system so that the midpoints of its six faces correspond to E, W, N, S, U, L, the eight vertices of the cube will correspond to the eight octants UNE (upper northeast), UNW,

USW, USE, LNE, LNW, LSW, LSE into which the *xy* plane, the *yz* plane, and the *zx* plane divide the space.

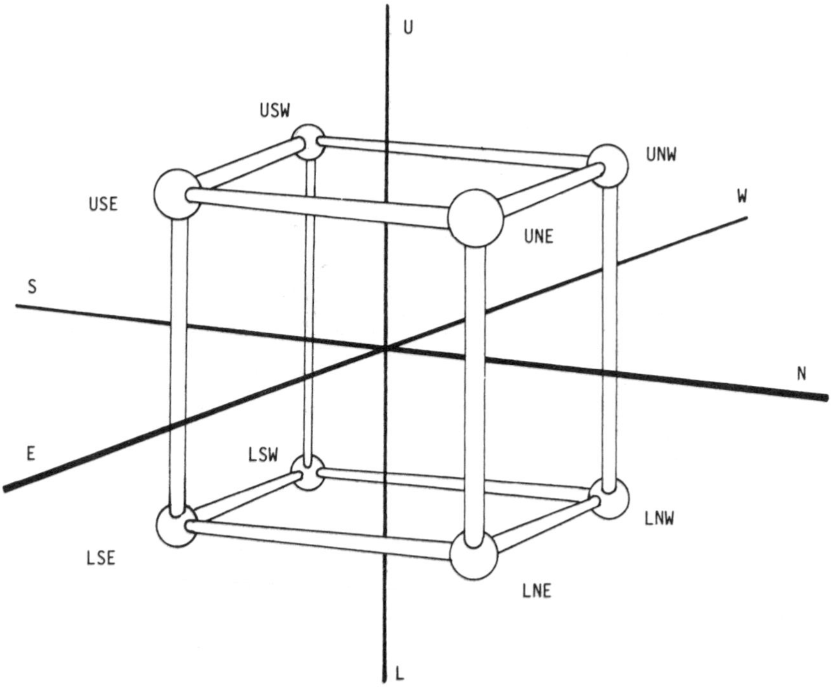

Complete the table.

Coordinates of the point	Location of the point
(+,+,+)	UNE octant
(+,+,−)	
(+,−,+)	
(+,−,−)	
(−,+,+)	
(−,+,−)	
(−,−,+)	
(−,−,−)	
(+,+,0)	NE quadrant of the *xy* plane
(+,−,0)	
(−,+,0)	

(continued)

Coordinates of the point	Location of the point
(–,–,0)	
(+,0,+)	
(+,0,–)	
(–,0,+)	
(–,0,–)	
(0,+,+)	
(0,+;–)	
(0,–,+)	
(0,–,–)	
(+,0,0)	E half of the x axis
(–,0,0)	
(0,+,0)	
(0,–,0)	
(0,0,+)	
(0,0,–)	

Bisector Planes First we wish to clarify what is meant by an equation of a geometric figure such as a plane or a line. An equation is a necessary and sufficient condition which the coordinates of a point must satisfy in order for the point to belong to the figure. This, then, means two things: (a) the coordinates of every point of the figure do satisfy the equation, and (b) the coordinates of a point that does not belong to the figure do not satisfy the equation.

The equations of the coordinate planes are $z = 0$ (xy plane), $x = 0$ (yz plane), and $y = 0$ (zx plane). Since $x^2 + y^2 = 0 \Leftrightarrow (x = 0 \text{ and } y = 0) \Leftrightarrow$ the point falls on the z axis, we have the equations of the coordinate axes: $x^2 + y^2 = 0$ (z axis), $y^2 + z^2 = 0$ (x axis), $z^2 + x^2 = 0$ (y axis). We may, however, avoid quadratic equations in writing equations of lines, as follows.

Let $F_1(x,y,z) = 0$ and $F_2(x,y,z) = 0$ be the equations of two geometric figures. Then the *intersection* of the two figures can be characterized by the system of equations

$$F_1(x,y,z) = 0$$
$$F_2(x,y,z) = 0$$

For example, the x axis is the intersection of the xy plane and the zx plane, so we have

$$\left.\begin{array}{l} y=0 \\ z=0 \end{array}\right\} (x \text{ axis}) \qquad \left.\begin{array}{l} z=0 \\ x=0 \end{array}\right\} (y \text{ axis}) \qquad \left.\begin{array}{l} x=0 \\ y=0 \end{array}\right\} (z \text{ axis})$$

Similarly, the *union* of the two figures $F_1(x,y,z) = 0$ and $F_2(x,y,z) = 0$ has an equation $F_1(x,y,z)\ F_2(x,y,z) = 0$ since a product is zero if, and only if, one of the factors is equal to zero. For example,

$$xyz = 0$$

is an equation of the figure obtained as the union of the three coordinate planes; and

$$(x^2 + y^2)(y^2 + z^2)(z^2 + x^2) = 0$$

is an equation of the figure obtained as the union of the three coordinate axes.

The equation

$$x = y$$

characterizes those points which are at the same distance from the yz and zx planes in the NE & SW direction. These lie in a plane containing the z axis and bisecting the interior angle made by the yz and zx planes. We shall call it the NE & SW *bisector plane*. Similarly, the equation

$$x = -y$$

characterizes those points which are at the same distance from the yz and zx planes in the NW & SE direction. These points lie in a plane containing the z axis and bisecting the exterior angle made by the yz and zx planes. This plane is called the NW & SE bisector plane. There are six bisector planes.

Complete the table.

Equation of plane	Description of plane	Contains coordinate axis
$x = y$	NE & SW bisector plane	z axis
$x = -y$		
$y = z$		
$y = -z$		
$z = x$		
$z = -x$		

Equiangular Lines The intersection of the NE & SW and the UN & LS bisector planes is a line characterized by

$$x = y$$
$$y = z$$

It is possible to "telescope" the two equations into one:

$$x = y = z$$

Clearly, this telescoped equation characterizes those points which are at the same distance from the three coordinate planes and make equal angles with the coordinate axes in the U, N, E directions. For this reason, we shall call it the UNE & LSW *equiangular line*. There are four such equiangular lines, corresponding to the four diagonals of the cube (see the figure on page 298).

Complete the table:

Equation of line	Description of line
$x = y = z$	UNE & LSW equiangular line
$x = y = -z$	
$x = -y = z$	
$-x = y = z$	

Bisector Axes There are six bisector axes, joining the midpoints of opposite edges of the cube (see the figure on page 298).

Complete the table.

Equation of line	Description of line
$x = y,\ z = 0$	NE & SW bisector axis
$x = -y,\ z = 0$	
$y = z,\ x = 0$	
$y = -z,\ x = 0$	
$z = x,\ y = 0$	
$z = -x,\ y = 0$	

GENERAL EQUATION OF THE PLANE

A vector $\mathbf{n} \neq \mathbf{0}$ perpendicular to a plane is called a *normal vector* of that plane. Two planes are parallel (perpendicular) if, and only if, their normal vectors are parallel (perpendicular). More generally, we define the angle made by two planes as the angle made by their normal vectors.

If $\mathbf{v}$ denotes the location vector of the generic point and $\mathbf{v}_0$ denotes the location vector of a particular point of the plane, then a necessary and sufficient condition for the point to belong to the plane is that $\mathbf{v}-\mathbf{v}_0$ be perpendicular to the normal vector $\mathbf{n}$, i.e.,

$$\mathbf{n}\cdot(\mathbf{v}-\mathbf{v}_0)=0$$

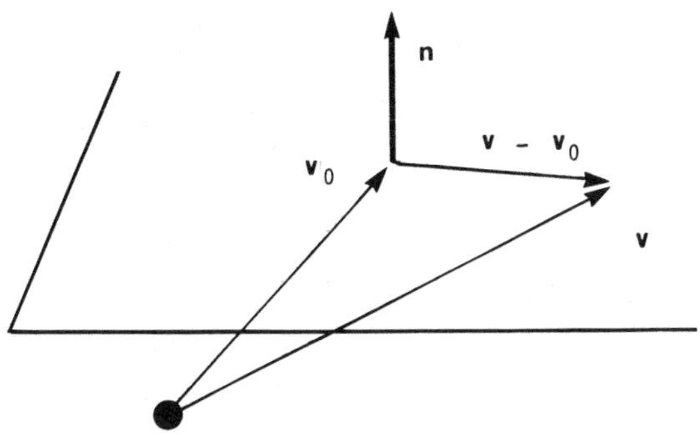

This vector equation can be translated into a scalar equation by writing these vectors in coordinate form: $\mathbf{n}=A\mathbf{i}+B\mathbf{j}+C\mathbf{k}$, $\mathbf{v}_0=x_0\mathbf{i}+y_0\mathbf{j}+z_0\mathbf{k}$; then $\mathbf{n}\cdot(\mathbf{v}-\mathbf{v}_0)=0$, i.e.,

$$A(x-x_0)+B(y-y_0)+C(z-z_0)=0$$

If we collect the constant terms and call them

$$D=-Ax_0-By_0-Cz_0$$

we get the

General Equation of the Plane

$$Ax+By+Cz+D=0$$

It should be noted that the coefficients A, B, C cannot all be zero, since they are the coordinates of the normal vector of the plane $\mathbf{n}\neq\mathbf{0}$. Also note that

$$D=0 \Leftrightarrow \text{the plane passes through the origin}$$

Examples

The normal plane of the equiangular line UNE & LSW passing through the origin is $x+y+z=0$.

$x+y=0$ is the equation of the NW & SE bisector plane, since its normal vector is $\mathbf{n}=\mathbf{i}+\mathbf{j}$.

The equation of the xy plane is $z = 0$ because its normal vector is $\mathbf{n} = \mathbf{k}$.

Complete the table.

A	B	C	Description of plane
0	B	C	Parallel to the x axis
A	0	C	
A	B	0	
0	0	C	Parallel to the xy plane
0	B	0	
A	0	0	
A	A	C	Parallel to the NW & SE bisector axis
A	B	A	
A	B	B	
A	A	A	Perpendicular to the UNE & LSW equiangular line
A	A	0	
A	0	A	
0	B	B	
A	$-A$	C	
A	B	$-A$	
A	B	$-B$	
A	$-A$	0	
A	0	$-A$	
0	B	$-B$	
A	A	$-A$	
A	$-A$	A	
$-A$	A	A	
A	B	0	A vertical plane
0	0	C	A horizontal plane

In various applications, we may find the following equation useful:

Intercept Form of the Plane

$$\frac{x}{a} + \frac{y}{b} + \frac{z}{c} = 1$$

where a, b, c are the x, y, z intercepts, respectively. Note that a plane passing through the origin, or a plane which is parallel to one of the coordinate axes, has no intercept form. The intercept form of the plane is especially handy if we want to make a sketch of the plane's position relative to the coordinate system.

Examples

1. Sketch the plane $x + 2y + 3z - 6 = 0$.

   Answer: The intercept form is

$$\frac{x}{6} + \frac{y}{3} + \frac{z}{2} = 1$$

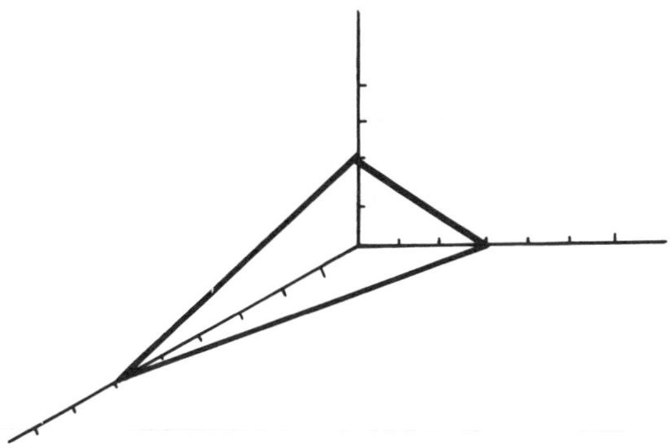

2. Find the vertices of the tetrahedron that the plane $3x - 4y + 5z + 6 = 0$ forms together with the coordinate planes.

   Answer: The intercept form is

$$\frac{x}{-2} + \frac{y}{3/2} + \frac{z}{-6/5} = 1$$

The requested points are the origin, $A(-2,0,0)$, $B(0,3/2,0)$, $C(0,0,-6/5)$.

NORMAL EQUATION OF THE PLANE

A plane has infintely many equations depending on the length and orientation of its normal vector. Indeed, multiplying an equation by a nonzero scalar will not change the plane, only replace one normal vector by another. Two of those equations stand out, namely the ones whose normal vectors are unit vectors, i.e., $\mathbf{n}^0$ and $-\mathbf{n}^0$. We can get these if we divide both sides of the equation by $|\mathbf{n}|$; this is called "normalizing the equation," and the result is the

Normal Equation

$$\frac{Ax + By + Cz + D}{\sqrt{A^2 + B^2 + C^2}} = 0$$

Note that

$$\frac{A}{\sqrt{A^2 + B^2 + C^2}} = \cos \alpha \quad \frac{B}{\sqrt{A^2 + B^2 + C^2}} = \cos \beta \quad \frac{C}{\sqrt{A^2 + B^2 + C^2}} = \cos \gamma$$

are just the direction cosines of the normal vector of the plane, so that the normal equation can also be written in the form

$$(\cos \alpha)x + (\cos \beta)y + (\cos \gamma)z + d = 0$$

where

$$d = \frac{D}{\sqrt{A^2 + B^2 + C^2}}$$

is the signed distance between the origin and the plane; it is positive if, and only if, the origin falls on the side of the plane pointed out by the normal vector.

The advantage of the normal equation of the plane is that it provides the machinery whereby the perpendicular distance from a point to the plane can be obtained immediately. Indeed, if we substitute the coordinates of any point not in the plane, into the normal equation of the plane, we do not get zero, of course, but what we get is precisely the signed distance from the point to the plane. This follows from the projection theorem and the fact that the distance sought is just the perpendicular projection of the vector $\mathbf{v} - \mathbf{v}_0$ onto the direction of the normal vector, i.e., $\mathbf{n}^0 \cdot (\mathbf{v} - \mathbf{v}_0)$.

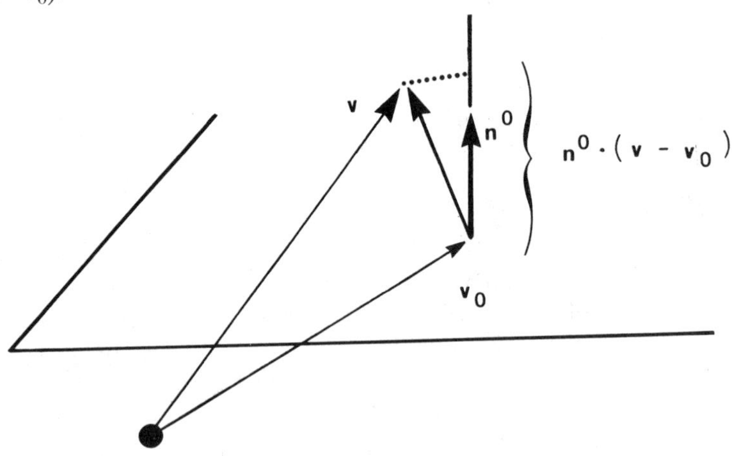

The normal equation is also useful if we wish to find the angle made by two planes, or by a plane and a line.

Examples

(1) Find the perpendicular distance from the point $P(1,-1,2)$ to the plane $2x-3y+6z+5=0$.

Answer: The normal vector is $\mathbf{n} = 2\mathbf{i} - 3\mathbf{j} + 6\mathbf{k}$; $|\mathbf{n}| = \sqrt{4+9+36} = \sqrt{49} = 7$. The normal equation of the plane is

$$\frac{2}{7}x - \frac{3}{7}y + \frac{6}{7}z + \frac{5}{7} = 0$$

The distance is 22/7.

(2) Find the angles that the plane $3x-4y-12z+10=0$ makes with the coordinate axes.

Answer:

$$\mathbf{n} = 3\mathbf{i} - 4\mathbf{j} - 12\mathbf{k} \qquad \mathbf{n}^0 = \frac{\mathbf{n}}{|\mathbf{n}|} = \frac{3}{13}\mathbf{i} - \frac{4}{13}\mathbf{j} - \frac{12}{13}\mathbf{k}$$

is a unit vector, so its coordinates are direction cosines: $\cos\alpha = 3/13$, $\cos\beta = -4/13$, $\cos\gamma = -12/13$. The required angles are just the complementary angles of α, β, γ.

(3) Find the angle made by the planes $2x+3y-6z+7=0$ and $3x-4y+12z+13=0$.

Answer: The normal equations are

$$\frac{2}{7}x + \frac{3}{7}y - \frac{6}{7}z + 1 = 0 \qquad \text{and} \qquad \frac{3}{13}x - \frac{4}{13}y + \frac{12}{13}z + 1 = 0$$

The angle made by the planes is the same as the angle made by their normal vectors, i.e., δ where

$$\cos\delta = \mathbf{n}_1 \cdot \mathbf{n}_2 = \frac{2(3) + 3(-4) + (-6)12}{91} = \frac{78}{91}$$

APPLICATIONS

1. Plane Through Three Points

Let $\mathbf{v}_0$, $\mathbf{v}_1$, $\mathbf{v}_2$ be the location vectors of the three given points, then the normal vector is $\mathbf{n} = (\mathbf{v}_1 - \mathbf{v}_0) \times (\mathbf{v}_2 - \mathbf{v}_0)$. The required vector equation is

$$(\mathbf{v}-\mathbf{v}_0)\cdot(\mathbf{v}_1-\mathbf{v}_0)\times(\mathbf{v}_2-\mathbf{v}_0)=0 \qquad \text{or} \qquad [\mathbf{v}-\mathbf{v}_0,\ \mathbf{v}_1-\mathbf{v}_0,\ \mathbf{v}_2-\mathbf{v}_0]=0$$

It translates into the scalar equation

$$\begin{vmatrix} x-x_0 & y-y_0 & z-z_0 \\ x_1-x_0 & y_1-y_0 & z_1-z_0 \\ x_2-x_0 & y_2-y_0 & z_2-z_0 \end{vmatrix} = 0$$

2. Plane Through Two Points, Perpendicular to Another Plane

If the location vectors of the points are $\mathbf{v}_0$ and $\mathbf{v}_1$, and the normal vector of the given plane is $\mathbf{n}$, the normal vector of the required plane is $\mathbf{n}' = (\mathbf{v}_1 - \mathbf{v}_0) \times \mathbf{n}$. The vector equation is

$$(\mathbf{v}-\mathbf{v}_0)\cdot(\mathbf{v}_1-\mathbf{v}_0)\times\mathbf{n}=0, \qquad \text{or} \qquad [\mathbf{v}-\mathbf{v}_0,\ \mathbf{v}_1-\mathbf{v}_0,\ \mathbf{n}]=0$$

This translates into

$$\begin{vmatrix} x-x_0 & y-y_0 & z-z_0 \\ x_1-x_0 & y_1-y_0 & z_1-z_0 \\ A & B & C \end{vmatrix} = 0$$

3. Plane Through a Point, Perpendicular to Two Other Planes

If the location vector of the point is $\mathbf{v}_0$, and the normal vectors of the given planes are $\mathbf{n}_1$ and $\mathbf{n}_2$, then the normal vector of the required plane is $\mathbf{n} = \mathbf{n}_1 \times \mathbf{n}_2$. The vector equation is

$$(\mathbf{v}-\mathbf{v}_0)\cdot(\mathbf{n}_1\times\mathbf{n}_2)=0 \qquad \text{or} \qquad [\mathbf{v}-\mathbf{v}_0,\ \mathbf{n}_1,\ \mathbf{n}_2]=0$$

The scalar version is

$$\begin{vmatrix} x-x_0 & y-y_0 & z-z_0 \\ A_1 & B_1 & C_1 \\ A_2 & B_2 & C_2 \end{vmatrix} = 0$$

PARAMETRIC AND SYMMETRIC FORMS OF A LINE

A vector $\mathbf{d} \neq \mathbf{0}$ which is parallel to a line will be called a *direction vector* of the line. Thus a line is parallel to a plane if, and only if, its direction vector is perpendicular to the plane's normal vector, i.e., $\mathbf{d}\cdot\mathbf{n} = 0$. In the sequel a line contained in a plane will be regarded as parallel to it. Similarly, a line is perpendicular to a plane if, and only if, its direction vector is parallel to the plane's normal vector, i.e., $\mathbf{d} \times \mathbf{n} = \mathbf{0}$.

If $\mathbf{v}$ is the location vector of the generic point, and $\mathbf{v}_0$ is the location vector of a particular point of the line, then a necessary and sufficient condition for the point to belong to the line is that $\mathbf{v} - \mathbf{v}_0$ be parallel to the direction vector $\mathbf{d}$, i.e.,

$$\mathbf{v} = \mathbf{v}_0 + t\mathbf{d}$$

where $t \in \mathbb{R}$ is a scalar parameter (denoted by t, as it is often interpreted as "time").

This vector equation can be translated into three scalar equations, if we introduce coordinates: $\mathbf{d} = a\mathbf{i} + b\mathbf{j} + c\mathbf{k}$:

Parametric Form

$$x = x_0 + at$$
$$y = y_0 + bt$$
$$z = z_0 + ct$$

Assuming that neither of the direction numbers a, b, c vanishes, we can solve each equation for t and "telescope" them into the so-called

Symmetric Form

$$\frac{x - x_0}{a} = \frac{y - y_0}{b} = \frac{z - z_0}{c}$$

APPLICATIONS

4. Line Through Two Points

Let $\mathbf{v}_0$, $\mathbf{v}_1$ be the location vectors of the given points. Then the direction vector of the required line is $\mathbf{d} = \mathbf{v}_1 - \mathbf{v}_0$ and its vector equation: $\mathbf{v} = \mathbf{v}_0 + t(\mathbf{v}_1 - \mathbf{v}_0)$. The symmetric form:

$$\frac{x - x_0}{x_1 - x_0} = \frac{y - y_0}{y_1 - y_0} = \frac{z - z_0}{z_1 - z_0}$$

5. Line Through a Point, Parallel to Two Planes

Let $\mathbf{v}_0$ be the location vector of the point, and $\mathbf{n}_1$, $\mathbf{n}_2$ the normal vectors of the planes. The direction vector of the required line is $\mathbf{d} = \mathbf{n}_1 \times \mathbf{n}_2$; its vector equation: $\mathbf{v} = \mathbf{v}_0 + t(\mathbf{n}_1 \times \mathbf{n}_2)$. The symmetric form:

$$\frac{x-x_0}{\begin{vmatrix} B_1 & C_1 \\ B_2 & C_2 \end{vmatrix}} = \frac{y-y_0}{-\begin{vmatrix} A_1 & C_1 \\ A_2 & C_2 \end{vmatrix}} = \frac{z-z_0}{\begin{vmatrix} A_1 & B_1 \\ A_2 & B_2 \end{vmatrix}}$$

6. Line Through a Point, Perpendicular to Two Lines

If the location vector of the given point is $\mathbf{v}_0$, and the direction vectors of the given lines are $\mathbf{d}_1$, $\mathbf{d}_2$, then the vector equation of the required line is $\mathbf{v} = \mathbf{v}_0 + t(\mathbf{d}_1 \times \mathbf{d}_2)$. The symmetric form:

$$\frac{x-x_0}{\begin{vmatrix} b_1 & c_1 \\ b_2 & c_2 \end{vmatrix}} = \frac{y-y_0}{-\begin{vmatrix} a_1 & c_1 \\ a_2 & c_2 \end{vmatrix}} = \frac{z-z_0}{\begin{vmatrix} a_1 & b_1 \\ a_2 & b_2 \end{vmatrix}}$$

7. Line Through a Point, Parallel to a Plane, Perpendicular to a Line

If the location vector of the point is $\mathbf{v}_0$, the normal vector of the given plane is $\mathbf{n}$, and the direction vector of the given line is $\mathbf{d}$, then the direction vector of the required line is $\mathbf{n} \times \mathbf{d}$. Its vector equation: $\mathbf{v} = \mathbf{v}_0 + t(\mathbf{n} \times \mathbf{d})$. The symmetric form:

$$\frac{x-x_0}{\begin{vmatrix} B & C \\ b & c \end{vmatrix}} = \frac{y-y_0}{-\begin{vmatrix} A & C \\ a & c \end{vmatrix}} = \frac{z-z_0}{\begin{vmatrix} A & B \\ a & b \end{vmatrix}}$$

8. Point of Intersection Between Line and Plane

Example Given a line as

$$\frac{x-1}{2} = \frac{y-1}{3} = \frac{z-1}{4}$$

and a plane as $2x + y - 3z + 4 = 0$, find their point of intersection.

Answer: Put the equation of the line into parametric form: $x = 2t + 1$, $y = 3t + 1$, $z = 4t + 1$ and substitute these expressions for x, y, z into the equation of the plane:

$$(4t + 2) + (3t + 1) - (12t + 3) + 4 = 0 \qquad \text{or} \qquad -5t = -4$$

Thus $t = 4/5$; substituting this value of t back into the parametric form we get $x = 13/5$, $y = 17/5$, $z = 21/5$; the point required is $P(13/5, 17/5, 21/5)$.

9. Line of Intersection Between Two Planes

The direction vector of the line of intersection is $\mathbf{d} = \mathbf{n}_1 \times \mathbf{n}_2$, where $\mathbf{n}_1$

and $\mathbf{n}_2$ are the normal vectors of the given planes. We also need the coordinates of a (fixed but arbitrary) point in the line of intersection. Such a point could, of course, be one on a coordinate plane.

Example Find the parametric form of the line of intersection between the planes $5x + y + z = 0$, $2x + y - 4z = 0$.

Answer: A common point is, obviously, the origin, so that we set $P_0 = O$. The direction vector is

$$\mathbf{d} = \mathbf{n}_1 \times \mathbf{n}_2 = \begin{vmatrix} \mathbf{i} & \mathbf{j} & \mathbf{k} \\ 5 & 1 & 1 \\ 2 & 1 & -4 \end{vmatrix} = -5\mathbf{i} + 22\mathbf{j} + 3\mathbf{k}$$

The required line in parametric form is

$$x = -5t$$

$$y = 22t$$

$$z = 3t$$

Second method, by gaussian elimination:

$$\left.\begin{matrix} 5x + y + \ \ z = 0 \\ 2x + y - 4z = 0 \end{matrix}\right\} \Rightarrow \begin{pmatrix} 5 & 1 & 1 \\ 2 & 1 & -4 \end{pmatrix} \Rightarrow \begin{pmatrix} 1 & -1 & 9 \\ 2 & 1 & -4 \end{pmatrix} \Rightarrow \begin{pmatrix} 1 & -1 & 9 \\ 0 & 3 & -22 \end{pmatrix}$$

$$\Rightarrow \begin{pmatrix} 1 & 0 & 5/3 \\ 0 & 1 & -22/3 \end{pmatrix} \Rightarrow \begin{cases} x = -5t \\ y = 22t \\ z = 3t \end{cases}$$

10. Point of Intersection of Three Planes

In the typical case three planes have exactly one point in common; this will happen if, and only if, the system of linear equations formed of the equations of the planes,

$$A_1x + B_1y + C_1z = -D_1$$

$$A_2x + B_2y + C_2z = -D_2$$

$$A_3x + B_3y + C_3z = -D_3$$

has a nonvanishing determinant:

$$\begin{vmatrix} A_1 & B_1 & C_1 \\ A_2 & B_2 & C_2 \\ A_3 & B_3 & C_3 \end{vmatrix} \neq 0$$

Geometrically, this means that the three normal vectors satisfy

$$[\mathbf{n}_1,\ \mathbf{n}_2,\ \mathbf{n}_3] \neq 0$$

i.e., they are *not* parallel to the same plane. Then the coordinates of the unique point of intersection can be found by Cramer's rule. (Gaussian elimination can also be used.)

If, on the other hand, $[\mathbf{n}_1,\ \mathbf{n}_2,\ \mathbf{n}_3] = 0$, the three planes have normal vectors which are parallel to a plane or, what is the same, they are perpendicular to the same plane. This fourth plane must also be perpendicular to the line of intersection, if any, of pairs of the given planes. There are several cases. (a) The planes have a line in common. Then any point on that line is a solution. We say that the system is redundant. (b) The three planes pairwise intersect in three parallel lines. No solution exists. (c) There is a pair of parallel planes while the third plane intersects them in a pair of parallel lines. No solution exists. (d) The three planes are pairwise parallel. No solution exists. In cases (b) through (d) we say that the system is inconsistent.

11. Plane Through a Point, Parallel to Two Lines

If the location vector of the point is $\mathbf{v}_0$, and the direction vectors of the given lines are $\mathbf{d}_1$, $\mathbf{d}_2$, the normal vector of the plane required is $\mathbf{n} = \mathbf{d}_1 \times \mathbf{d}_2$. Its vector equation is

$$(\mathbf{v}-\mathbf{v}_0)\cdot(\mathbf{d}_1 \times \mathbf{d}_2) = 0 \qquad \text{or} \qquad [\mathbf{v}-\mathbf{v}_0,\ \mathbf{d}_1,\ \mathbf{d}_2] = 0$$

and the scalar equation is

$$\begin{vmatrix} x-x_0 & y-y_0 & z-z_0 \\ a_1 & b_1 & c_1 \\ a_2 & b_2 & c_2 \end{vmatrix} = 0$$

A necessary and sufficient condition for two lines $\mathbf{v} = \mathbf{v}_1 + t\mathbf{d}_1$ and $\mathbf{v} = \mathbf{v}_2 + t\mathbf{d}_2$ to lie in the same plane is that

$$[\mathbf{v}_1-\mathbf{v}_2,\ \mathbf{d}_1,\ \mathbf{d}_2] = 0$$

or, what is the same,

$$\begin{vmatrix} x_1-x_2 & y_1-y_2 & z_1-z_2 \\ a_1 & b_1 & c_1 \\ a_2 & b_2 & c_2 \end{vmatrix} = 0$$

Example Show that the lines

$$x = -3 + 2t \qquad\qquad x = 5 + t$$
$$y = -2 + 3t \qquad \text{and} \qquad y = -1 - 4t$$
$$z = 6 - 4t \qquad\qquad z = -4 + t$$

are in the same plane, and find the coordinates of the point of intersection.

Answer

$$[\mathbf{v}_1 - \mathbf{v}_2, \mathbf{d}_1, \mathbf{d}_2] = \begin{vmatrix} -8 & -1 & 10 \\ 2 & 3 & -4 \\ 1 & -4 & 1 \end{vmatrix} = 0$$

Hence the lines lie in the same plane. Equating the x, y, z coordinates (and paying attention to the fact that the parameter t has a different meaning on each line), we have a system of linear equations that can be solved by gaussian elimination:

$$\left.\begin{aligned} -3 + 2t_1 &= 5 + t_2 \\ -2 + 3t_1 &= -1 - 4t_2 \\ 6 - 4t_1 &= -4 + t_2 \end{aligned}\right\} \Rightarrow \left.\begin{aligned} 2t_1 - t_2 &= 8 \\ 3t_1 + 4t_2 &= 1 \\ -4t_1 - t_2 &= -10 \end{aligned}\right\} \Rightarrow \left(\begin{array}{cc|c} 2 & -1 & 8 \\ 3 & 4 & 1 \\ -4 & -1 & -10 \end{array}\right) \Rightarrow \left(\begin{array}{cc|c} 2 & -1 & 8 \\ 3 & 4 & 1 \\ 0 & -3 & 6 \end{array}\right)$$

$$\Rightarrow \left(\begin{array}{cc|c} 1 & 5 & -7 \\ 3 & 4 & 1 \\ 0 & -3 & 6 \end{array}\right) \Rightarrow \left(\begin{array}{cc|c} 1 & 5 & -7 \\ 0 & 1 & -2 \\ 0 & 0 & 0 \end{array}\right) \Rightarrow \left(\begin{array}{cc|c} 1 & 0 & 3 \\ 0 & 1 & -2 \\ 0 & 0 & 0 \end{array}\right) \Rightarrow \begin{cases} t_1 = 3 \\ t_2 = -2 \end{cases}$$

Substituting these parameter values into the equations, we get the coordinates of the point of intersection as $P(3,7,-6)$.

12. Plane Through Two Points, Parallel to a Line

If the location vectors of the points are $\mathbf{v}_0$ and $\mathbf{v}_1$, and the direction vector of the line is $\mathbf{d}$, the normal vector of the required plane is $\mathbf{n} = (\mathbf{v}_1 - \mathbf{v}_2) \times \mathbf{d}$. Its vector equation is

$$(\mathbf{v} - \mathbf{v}_0) \cdot (\mathbf{v}_1 - \mathbf{v}_0) \times \mathbf{d} = 0 \qquad \text{or} \qquad [\mathbf{v} - \mathbf{v}_0, \mathbf{v}_1 - \mathbf{v}_0, \mathbf{d}] = 0$$

The scalar equation becomes

$$\begin{vmatrix} x - x_0 & y - y_0 & z - z_0 \\ x_1 - x_0 & y_1 - y_0 & z_1 - z_0 \\ a & b & c \end{vmatrix} = 0$$

13. Plane Through a Point, Parallel to a Line, Perpendicular to a Plane

If the location vector of the point is $\mathbf{v}_0$, and the direction vector of the line is $\mathbf{d}$, and the normal vector of the given plane is $\mathbf{n}$, then the normal vector of the required plane is $\mathbf{n}' = \mathbf{d} \times \mathbf{n}$ and its vector equation is

$$(\mathbf{v}-\mathbf{v}_0)\cdot\mathbf{d}\times\mathbf{n}=0 \qquad \text{or} \qquad [\mathbf{v}-\mathbf{v}_0,\ \mathbf{d},\ \mathbf{n}]=0$$

and the scalar equation is

$$\begin{vmatrix} x-x_0 & y-y_0 & z-z_0 \\ a & b & c \\ A & B & C \end{vmatrix} = 0$$

TRANSVERSALS: APPLICATIONS

14. Line Through a Point, Intersecting Two Lines

Given a point with location vector $\mathbf{v}_0$ and two lines: $\mathbf{v} = \mathbf{v}_1 + t\mathbf{d}_1$ and $\mathbf{v} = \mathbf{v}_2 + t\mathbf{d}_2$, the plane $[\mathbf{v}-\mathbf{v}_0,\ \mathbf{v}_1-\mathbf{v}_0,\ \mathbf{d}_1]=0$ contains the point and the first line, and the plane $[\mathbf{v}-\mathbf{v}_0,\ \mathbf{v}_2-\mathbf{v}_0,\ \mathbf{d}_2]=0$ contains the point and the second line. Therefore, the required line passing through the point and intersecting both lines (called tranversal through the given point) is the line of intersection between the planes.

Example

$$P_0(-4,-5,3);\qquad \frac{x+1}{3}=\frac{y+3}{-2}=\frac{z-2}{-1},\qquad \frac{x-2}{2}=\frac{y+1}{3}=\frac{z-1}{-5}$$

$$\begin{vmatrix} x+4 & y+5 & z-3 \\ 3 & 2 & -1 \\ 3 & -2 & -1 \end{vmatrix} = 0 \Rightarrow x+3z-5=0$$

$$\begin{vmatrix} x+4 & y+5 & z-3 \\ 6 & 4 & -2 \\ 2 & 3 & -5 \end{vmatrix} = 0 \Rightarrow 7x-13y-5z-22=0$$

$$\left.\begin{aligned} 7x-13y-5z-22&=0 \\ x \qquad\quad +3z-\ 5&=0 \end{aligned}\right\} \Rightarrow \begin{cases} x=5-3t \\ y=1-2t \\ z=t \end{cases}$$

In symmetric form:

$$\frac{x-5}{-3}=\frac{y-1}{-2}=z$$

15. Orthogonal Transversal of Two Lines

Given two lines $\mathbf{v} = \mathbf{v}_1 + t\mathbf{d}_1$, $\mathbf{v} = \mathbf{v}_2 + t\mathbf{d}_2$, the direction vector of their orthogonal transversal, as we have seen earlier, is $\mathbf{d}_3 = \mathbf{d}_1 \times \mathbf{d}_2$. The plane

$$(\mathbf{v}-\mathbf{v}_1)\cdot(\mathbf{d}_1 \times \mathbf{d}_3) = [\mathbf{v}-\mathbf{v}_1,\ \mathbf{d}_1,\ \mathbf{d}_1 \times \mathbf{d}_2] = 0$$

contains the first line plus the orthogonal transversal, and the plane

$$(\mathbf{v}-\mathbf{v}_2)\cdot(\mathbf{d}_2 \times \mathbf{d}_3) = [\mathbf{v}-\mathbf{v}_2,\ \mathbf{d}_2,\ \mathbf{d}_1 \times \mathbf{d}_2] = 0$$

contains the second line plus the orthogonal transversal. Therefore, the orthogonal transversal is just the line of intersection between these two planes.

Example Find the orthogonal transversal of the lines

$$\frac{x-1}{1} = \frac{y-3}{2} = \frac{z}{2} \qquad \text{and} \qquad \frac{x-2}{1} = \frac{y+1}{3} = \frac{z}{1}$$

Answer: $\mathbf{d}_1 = \mathbf{i} + 2\mathbf{j} + 2\mathbf{k}$, $\mathbf{d}_2 = \mathbf{i} + 3\mathbf{j} + \mathbf{k}$. Then

$$\mathbf{d}_3 = \mathbf{d}_1 \times \mathbf{d}_2 = \begin{vmatrix} \mathbf{i} & \mathbf{j} & \mathbf{k} \\ 1 & 2 & 2 \\ 1 & 3 & 1 \end{vmatrix} = -4\mathbf{i} + \mathbf{j}\mathbf{k}$$

$$[\mathbf{v}-\mathbf{v}_1,\ \mathbf{d}_1,\ \mathbf{d}_1 \times \mathbf{d}_2] = \begin{vmatrix} x-1 & y-3 & z \\ 1 & 2 & 2 \\ -4 & 1 & 1 \end{vmatrix} = 0 \Rightarrow y - z - 3 = 0$$

$$[\mathbf{v}-\mathbf{v}_2,\ \mathbf{d}_2,\ \mathbf{d}_1 \times \mathbf{d}_2] = \begin{vmatrix} x-2 & y+1 & z \\ 1 & 3 & 1 \\ -4 & 1 & 1 \end{vmatrix} = 0 \Rightarrow 2x - 5y + 13z - 9 = 0$$

The orthogonal transversal is just the intersection between the planes above:

$$\left.\begin{aligned} 2x - 5y + 13z - 9 &= 0 \\ y - \quad z - 3 &= 0 \end{aligned}\right\} \Rightarrow \begin{cases} x = 12 - 4t \\ y = \ 3 + t \\ z = t \end{cases}$$

In symmetric form, the orthogonal transversal is

$$\frac{x-12}{-4} = \frac{y-3}{1} = \frac{z}{1}$$

16. Line Through a Point, Parallel to a Plane, Intersecting a Line

Given a point with location vector $\mathbf{v}_0$, a plane $\mathbf{n}\cdot(\mathbf{v}-\mathbf{v}_1) = 0$, and a line $\mathbf{v} = \mathbf{v}_2 + t\mathbf{d}$, the plane $\mathbf{n}\cdot(\mathbf{v}-\mathbf{v}_0) = 0$ contains the point and is parallel to the given plane, and the plane $[\mathbf{v}-\mathbf{v}_0,\ \mathbf{v}_2-\mathbf{v}_0,\ \mathbf{d}] = 0$ contains the point and the given line. The required line is the line of intersection between these two planes.

Example

$$P_0(3,-2,-4) \qquad 3x-2y-3z-7=0 \qquad \frac{x-2}{3}=\frac{y+4}{-2}=\frac{z-1}{2}$$

The plane through P_0 and parallel to the given plane is $3x-2y-3z-25=0$. The plane containing P_0 and the given line

$$\begin{vmatrix} x-3 & y+2 & z+4 \\ -1 & -2 & 5 \\ 3 & -2 & 2 \end{vmatrix} = 0 \Rightarrow 6x+17y+8z+48=0$$

$$\left.\begin{aligned} 3x - 2y - 3z &= 25 \\ 6x + 17y + 8z &= -48 \end{aligned}\right\} \Rightarrow \left(\begin{array}{ccc|c} 3 & -2 & -3 & 25 \\ 6 & 17 & 8 & -48 \end{array}\right) \Rightarrow \left(\begin{array}{ccc|c} 3 & -2 & -3 & 25 \\ 0 & 3 & 2 & -14 \end{array}\right)$$

$$\Rightarrow \left(\begin{array}{ccc|c} 3 & -2 & -3 & 25 \\ 0 & 1 & 2/3 & -14/3 \end{array}\right) \Rightarrow \left(\begin{array}{ccc|c} 3 & 0 & -5/3 & 47/3 \\ 0 & 1 & 2/3 & -14/3 \end{array}\right) \Rightarrow \left(\begin{array}{ccc|c} 1 & 0 & -5/9 & 47/9 \\ 0 & 1 & 2/3 & -14/3 \end{array}\right)$$

$$\Rightarrow \begin{cases} x = 5t + 47/9 \\ y = -6t - 14/3 \\ z = 9t \end{cases}$$

PARAMETRIC FORM OF THE PLANE

Let $\mathbf{v}_0$ be the location vector of a point in the plane, and let $\mathbf{d}_1$ and $\mathbf{d}_2$ be a pair of vectors, not parallel to one another, but both parallel to the plane. Then the vector equation

$$\mathbf{v} = \mathbf{v}_0 + s\mathbf{d}_1 + t\mathbf{d}_2$$

where $s, t \in \mathbb{R}$ are free parameters (i.e., variables free to assume any scalar value) describes the plane through the point and parallel to the given directions. This vector equation translates into the scalar equations called the

Parametric Form of the Plane

$$\begin{aligned} x &= x_0 + a_1 s + a_2 t \\ y &= y_0 + b_1 s + b_2 t \\ z &= z_0 + c_1 s + c_2 t \end{aligned}$$

Examples

(1) Write the parametric form of the plane $3x + 2y - z + 4 = 0$.

Answer: We may put $x = s$, $y = t$; then $z = 4 + 3x + 2y = 4 + 3s + t$, and the parametric form is

$$x = s$$
$$y = t$$
$$z = 4 + 3s + 2t$$

(2) A plane is given in parametric form

$$\begin{aligned} x &= 1 - 2s + 3t \\ y &= -3 + 5s - t \\ z &= 2 + s \end{aligned}$$

Write the general equation of the plane.

Answer: We solve the system of linear equations for s and t (as if x, y, z were known) by using gaussian elimination:

$$\left.\begin{aligned} -2s + 3t &= x - 1 \\ 5s - t &= y + 3 \\ s &= z - 2 \end{aligned}\right\} \Rightarrow \left(\begin{array}{rr|l} -2 & 3 & x-1 \\ 5 & -1 & y+3 \\ 1 & 0 & z-2 \end{array}\right) \Rightarrow \left(\begin{array}{rr|l} 1 & 0 & z-2 \\ 0 & 3 & x+2z-5 \\ 0 & -1 & y-5z+13 \end{array}\right)$$

$$\Rightarrow \left(\begin{array}{rr|l} 1 & 0 & z-2 \\ 0 & 1 & -y+5z+13 \\ 0 & 0 & x+3y-13z+34 \end{array}\right)$$

Thus the system has a solution if, and only if, $x + 3y - 13z + 34 = 0$. So, if there is a plane, then this is its general equation.

EXERCISES

5.1 Given the point $P_0(1,-2,3)$ and the planes $2x - 3y - z - 1 = 0$, $x - 4y + 5z + 2 = 0$, write the symmetric form of the line through P_0 and parallel to the given planes.

5.2 Given the points $P_0(0,-1,3)$, $P_1(3,-2,1)$ write the symmetric form of the line joining P_0 to P_1.

5.3 Given the point $P_0(1,0,-1)$ and a pair of lines $x/3 = y/5 = z/-2$, $x - 1 = y + 2 = -z + 3$, write the symmetric form of the line through P_0 and parallel to the orthogonal transversal of the given pair of lines.

5.4 Find the symmetric form of the line through $P_0(1,1,-3)$ which lies in the plane $2x - 3y + z + 4 = 0$ and is parallel to the plane $3x + y - z = 0$.

5.5 Find the line of intersection of the planes $2x - 3y + 4z - 5 = 0$ and $x + y + z + 1 = 0$.

5.6 Given a point $P_0(1,-2,3)$, a plane $4x-3y+2z-1=0$, and a line $x-1=y+2=z-3$, find the symmetric form of the line passing through P_0, parallel to the given plane and perpendicular to the given line.

5.7 Given a plane $5x-2y+3z=0$ and a line $x=y=-z$, find the symmetric form of the line which lies in the given plane, passes through the origin, and is perpendicular to the given line.

5.8 Find the line of intersection between the NE & SW bisector plane and the normal plane of the UNE & LSW equiangular line.

5.9 Find the line of intersection between the NE & SW and the UN & LE bisector planes.

5.10 Find the symmetric equation of the line of intersection between the planes $x-2y+3z-4=0$, $3x+2y-5z-4=0$.

5.11 Check whether a unique point of intersection for the three planes $7x+2y+3z-15=0$, $5x-3y+2z-15=0$, $10x-11y+5z-36=0$ exists. If so, find that unique point.

5.12 Given the point $P_0(3,4,-5)$ and the lines $x/3=y-1=-z$, $x+1=\frac{1}{2}(-y-1)=z+1$; find the plane through P_0 which is parallel to the given lines.

5.13 Show that the lines $(x-1)/2=(y+2)/-3=(z-5)/5$; $x=3t+7$, $y=2t+2$, $z=-2t+1$ lie in the same plane. Find that plane.

5.14 Given the points $P_0(2,-1,3)$, $P_1(3,1,2)$ and the line $(x-1)/3=(y+1)/-1=z/-4$, find the plane through P_0, P_1 which is parallel to the given line.

5.15 Given $P_0(3,-2,1)$ and a line $(x+1)/2=(y-1)/3=z/4$, find a plane that contains them both.

5.16 Find a plane containing both of the parallel lines $(x-2)/3=(y+1)/2=(z-3)/-2$, $(x-1)/3=(y-2)/2=(z+3)/-2$.

5.17 Find the plane passing through the line $(x-1)/2=(y+2)/-3=(z-2)/2$ and is perpendicular to the plane $3x+2y-z-5=0$.

5.18 Find the orthogonal transversal of $(x-4)/7=(y+8)/4=(z-1)/4$ and $(x-4)4=(y-1)/-8=z+8$.

5.19 Find the plane passing through the points $P_1(1,-2,3)$, $P_2(3,4,-5)$, $P_3(1,0,-1)$.

5.20 Find the plane that passes through the origin and is perpendicular to both planes $2x-y+3z-1=0$ and $x+2y+z=0$.

5.21 Find the plane passing through the points $P_0(1,-1,-2)$, $P_1(3,1,1)$ and perpendicular to the plane $x-2y+3z-5=0$.

5.22 Find the perpendicular distance from the origin to the plane $x-y+4=0$.

5.23 Find the vertices of the tetrahedron that the plane $15x+10y+6z-30=0$ forms with the coordinate planes.

5.24 Write the equation of the plane passing through $P_0(3,4,-5)$ and parallel to both vectors $3\mathbf{i}+\mathbf{j}-\mathbf{k}$, $\mathbf{i}-2\mathbf{j}+\mathbf{k}$.

5.25 Find the angle made by the planes $6x+3y-2z=0$ and $x+2y+6z-12=0$.

5.26 Calculate the perpendicular distance between the parallel planes $3x-4y+12z-13=0$ and $3x-4y+12z+26=0$.

5.27 Write the equation of the straight line, both in parametric and in symmetric form, which passes through $P_0(2,0,-3)$ and is parallel to the vector $2\mathbf{i}-3\mathbf{j}+5\mathbf{k}$.

5.28 Show that the lines represented by the parametric equations

$$\begin{aligned} x &= -3+2t \\ y &= -2+3t \\ z &= 6-4t \end{aligned} \qquad \text{and} \qquad \begin{aligned} x &= 5+t \\ y &= -1-4t \\ z &= -4+t \end{aligned}$$

intersect.

5.29 Find the perpendicular distance between the parallel lines $x=y=z$ and $x-1=y-2=z-3$.

5.30 Find the volume of the tetrahedron that is bounded by the coordinate planes and the plane $x+2y+3z-6=0$.

5.31 Find the perpendicular distance between the lines

$$\begin{aligned} x &= 2t-4 \\ y &= -t+4 \\ z &= -2t-1 \end{aligned} \qquad \text{and} \qquad \begin{aligned} x &= 4t-5 \\ y &= -3t+5 \\ z &= -5t+5 \end{aligned}$$

5.32 Find the perpendicular distance from the point $P_0(1,-1,-2)$ to the line $(x+3)/3=(y+2)/2=(z-8)/-2$.

5.33 Given $A(1,-2,-1)$, $B(4,0,-3)$; $C(1,2,-1)$, $D(2,-4,-5)$, find the points P, Q on the lines AB, CD, respectively, such that PQ is the orthogonal transversal of AB, CD. What is the length of the orthogonal transversal?

6

Linear Operators[†]

SOME SIMPLE EXAMPLES OF LINEAR OPERATORS

1. *Reflection in the origin:* If we change the signatures of all three coordinates of a point $P(x,y,z)$, i.e.,

$$(x,y,z) \rightsquigarrow (-x,-y,-z)$$

then the point P is sent to its reflection in the origin. We shall be using the symbol $P \rightsquigarrow P'$, or $(x,y,y) \rightsquigarrow (x',y',z')$ to mean that P is sent to (moved to) P' by an operator (i.e., a function from the space to itself). In our instance this means that

$$x' = -x$$
$$y' = -y$$
$$z' = -z$$

Suppose that we look at this as a system of three linear equations with three variables x, y, z; then the nine coefficients on the right-hand side yield a matrix, called the *matrix of the operator*,

$$\begin{pmatrix} -1 & 0 & 0 \\ 0 & -1 & 0 \\ 0 & 0 & -1 \end{pmatrix}$$

2. *Reflections in the coordinate planes:* If we change the signature of the first coordinate of P while leaving the other coordinates unchanged:

[†]In this chapter all bases are orthonormal and all coordinate systems are rectangular metric, except where otherwise stipulated.

$$(x,y,z) \rightsquigarrow (-x,y,z)$$

then the point P is sent to its reflection in the yz plane. The matrix of this reflection is

$$\begin{pmatrix} -1 & 0 & 0 \\ 0 & 1 & 0 \\ 0 & 0 & 1 \end{pmatrix}$$

3. *Reflections in the coordinate axes:* If we change the signature of the first two coordinates of a point P while leaving the third coordinate unchanged:

$$(x,y,z) \rightsquigarrow (-x,-y,z)$$

then P is sent to its reflection in the z axis. The matrix of this reflection is

$$\begin{pmatrix} -1 & 0 & 0 \\ 0 & -1 & 0 \\ 0 & 0 & 1 \end{pmatrix}$$

A reflection in a line is the same as a rotation through 180° about that line.

Complete the table.

Operator	How it changes coordinates	Matrix
Reflection in origin	$(x,y,z) \rightsquigarrow (-x,-y,-z)$	$\begin{pmatrix} -1 & 0 & 0 \\ 0 & -1 & 0 \\ 0 & 0 & -1 \end{pmatrix}$
	$(x,y,z) \rightsquigarrow (-x,y,z)$	
	$(x,y,z) \rightsquigarrow (x,-y,z)$	
	$(x,y,z) \rightsquigarrow (x,y,-z)$	
	$(x,y,z) \rightsquigarrow (-x,-y,z)$	
	$(x,y,z) \rightsquigarrow (-x,y,-z)$	
	$(x,y,z) \rightsquigarrow (x,-y,-z)$	

4. *Reflections in the bisector planes:* If we interchange the first two coordinates of P while leaving the third coordinate unchanged:

$$(x,y,z) \rightsquigarrow (y,x,z)$$

then P is sent to its reflection in the NE & SW bisector plane. This means that

$$x' = y$$
$$y' = x$$
$$z' = z$$

From this we may read off the matrix of the reflection:

$$\begin{pmatrix} 0 & 1 & 0 \\ 1 & 0 & 0 \\ 0 & 0 & 1 \end{pmatrix}$$

Complete the table.

Operator	How it changes coordinates	Matrix
Reflection in the NE & SW bisector plane	$(x,y,z) \rightsquigarrow (y,x,z)$	$\begin{pmatrix} 0 & 1 & 0 \\ 1 & 0 & 0 \\ 0 & 0 & 1 \end{pmatrix}$
	$(x,y,z) \rightsquigarrow (x,z,y)$	
	$(x,y,z) \rightsquigarrow (z,y,x)$	
Reflection in the NW & SE bisector plane	$(x,y,z) \rightsquigarrow (-y,-x,z)$	
	$(x,y,z) \rightsquigarrow (x,-z,-y)$	
	$(x,y,z) \rightsquigarrow (-z,y,-x)$	

5. *Rotations about the equiangular lines:* $(x,y,z) \rightsquigarrow (z,x,y)$ is a rotation through 120° (or, what is the same, through −240°) about the UNE & LSW equiangular line. Its matrix is

$$\begin{pmatrix} 0 & 0 & 1 \\ 1 & 0 & 0 \\ 0 & 1 & 0 \end{pmatrix}$$

Complete the table.

Operator	How it changes coordinates	Matrix
Rotation through 120° about UNE & LSW	$(x,y,z) \rightsquigarrow (z,x,y)$	$\begin{pmatrix} 0 & 0 & 1 \\ 1 & 0 & 0 \\ 0 & 1 & 0 \end{pmatrix}$
Rotation through 240° about UNE & LSW	$(x,y,z) \rightsquigarrow (y,z,x)$	$\begin{pmatrix} 0 & 1 & 0 \\ 0 & 0 & 1 \\ 1 & 0 & 0 \end{pmatrix}$
Rotation through 120° about UNW & LSE		
Rotation through 240° about UNW & LSE		$\begin{pmatrix} 0 & 0 & -1 \\ -1 & 0 & 0 \\ 0 & 1 & 0 \end{pmatrix}$
		$\begin{pmatrix} 0 & 1 & 0 \\ 0 & 0 & -1 \\ -1 & 0 & 0 \end{pmatrix}$
Rotation through 240° about USW & LNE		$\begin{pmatrix} 0 & 0 & -1 \\ 1 & 0 & 0 \\ 0 & -1 & 0 \end{pmatrix}$
		$\begin{pmatrix} 0 & -1 & 0 \\ 0 & 0 & -1 \\ 1 & 0 & 0 \end{pmatrix}$
Rotation through 240° about USE & LNW		$\begin{pmatrix} 0 & 0 & 1 \\ -1 & 0 & 0 \\ 0 & -1 & 0 \end{pmatrix}$

Given the axis and the angle, in general there are two rotations satisfying those conditions (and they are inverses of one another in the sense of

Chap. 8). However, if we attribute a direction to the axis, and a signature to the angle, the rotation will be uniquely determined. Our convention in this book is that all axes have a direction attributed to them, e.g., on the UNE & LSW equiangular line the first-mentioned octant UNE determines the direction; on the coordinate axes, the positive coordinates determine the direction.

6. *Rotation around the coordinate axes:* $(x,y,z) \rightsquigarrow (x,-z,y)$ is a rotation around the x axis through 90° (or, what is the same, through –270°). Its matrix is

$$\begin{pmatrix} 1 & 0 & 0 \\ 0 & 0 & -1 \\ 0 & 1 & 0 \end{pmatrix}$$

Complete the table.

Operator	How it changes coordinates	Matrix
Rotation through 90° about the x axis	$(x,y,z) \rightsquigarrow (x,-z,y)$	$\begin{pmatrix} 1 & 0 & 0 \\ 0 & 0 & -1 \\ 0 & 1 & 0 \end{pmatrix}$
Rotation through 270° about the x axis	$(x,y,z) \rightsquigarrow (x,z,-y)$	$\begin{pmatrix} 1 & 0 & 0 \\ 0 & 0 & 1 \\ 0 & -1 & 0 \end{pmatrix}$
Rotation through 90° about the y axis		
Rotation through 270° about the y axis		
Rotation through 90° about the z axis		
Rotation through 270° about the z axis		

7. *Reflection in the bisector axes:* $(x,y,z) \rightsquigarrow (y,x,-z)$ is a reflection in the NE & SW bisector axis (or, what is the same, a rotation through 180° about the NE & SW bisector axis). Its matrix is

$$\begin{pmatrix} 0 & 1 & 0 \\ 1 & 0 & 0 \\ 0 & 0 & -1 \end{pmatrix}$$

Complete the table.

Operator	How it changes coordinates	Matrix
Reflection in the NE & SW bisector axis	$(x,y,z) \rightsquigarrow (y,x,-z)$	$\begin{pmatrix} 0 & 1 & 0 \\ 1 & 0 & 0 \\ 0 & 0 & -1 \end{pmatrix}$
Reflection in the NW & SE bisector axis		$\begin{pmatrix} 0 & -1 & 0 \\ -1 & 0 & 0 \\ 0 & 0 & -1 \end{pmatrix}$
Reflection in the UE&LW bisector axis		
Reflection in the UW & LE bisector axis		
Reflection in the UN & LS bisector axis		
Reflection in the US & LN bisector axis		

8. *Projections onto the coordinate planes and axes:* $(x,y,z) \rightsquigarrow (x,y,0)$ is a perpendicular projection onto the xy plane. Its matrix is

$$\begin{pmatrix} 1 & 0 & 0 \\ 0 & 1 & 0 \\ 0 & 0 & 0 \end{pmatrix}$$

Complete the table.

Operator	How it changes coordinates	Matrix
Projection onto the *xy* plane	$(x,y,z) \rightsquigarrow (x,y,0)$	$\begin{pmatrix} 1 & 0 & 0 \\ 0 & 1 & 0 \\ 0 & 0 & 0 \end{pmatrix}$
Projection onto the *yz* plane		
Projection onto the *zx* plane		
Projection onto the *x* axis	$(x,y,z) \rightsquigarrow (x,0,0)$	$\begin{pmatrix} 1 & 0 & 0 \\ 0 & 0 & 0 \\ 0 & 0 & 0 \end{pmatrix}$
		$\begin{pmatrix} 0 & 0 & 0 \\ 0 & 1 & 0 \\ 0 & 0 & 0 \end{pmatrix}$
		$\begin{pmatrix} 0 & 0 & 0 \\ 0 & 0 & 0 \\ 0 & 0 & 1 \end{pmatrix}$
Projection into the origin (*trivial or zero operator*)	$(x,y,z) \rightsquigarrow (0,0,0)$	$\begin{pmatrix} 0 & 0 & 0 \\ 0 & 0 & 0 \\ 0 & 0 & 0 \end{pmatrix}$
The *identity operator*	$(x,y,z) \rightsquigarrow (x,y,z)$	$\begin{pmatrix} 1 & 0 & 0 \\ 0 & 1 & 0 \\ 0 & 0 & 1 \end{pmatrix}$

The trivial operator and the identity operator will play an important role, comparable to the roles of the numbers 0 and 1.

CONCEPT OF A LINEAR OPERATOR

We want to capture the basic idea that is in common to these simple examples of operators. We see that they all preserve the sum and the scalar multiple of vectors, because the diagonal of a parallelogram is taken into the diagonal of the image parallelogram. This motivates the following definition.

A vector-valued function f of a vector variable $\mathbf{v}$ is called a *linear operator* if

$$f(\mathbf{v}+\mathbf{v}') = f(\mathbf{v}) + f(\mathbf{v}')$$
$$f(\lambda\mathbf{v}) = \lambda f(\mathbf{v}) \qquad (\lambda \in \mathbb{R})$$

It follows that f must also preserve the null vector: $f(\mathbf{0}) = \mathbf{0}$, the negative: $f(-\mathbf{v}) = -f(\mathbf{v})$, and the difference: $f(\mathbf{v}-\mathbf{v}') = f(\mathbf{v}) - f(\mathbf{v}')$. Moreover, f preserves linear combinations:

$$f(\alpha\mathbf{a}+\beta\mathbf{b}+\gamma\mathbf{c}) = \alpha f(\mathbf{a}) + \beta f(\mathbf{b}) + \gamma f(\mathbf{c}) \qquad (\alpha, \beta, \gamma \in \mathbb{R})$$

Therefore, a linear operator f takes a line into a line (or into a point), and it takes a plane into a plane (or into a line, or into a point, according as two or more of the vectors $f(\mathbf{a})$, $f(\mathbf{b})$, $f(\mathbf{c})$ are parallel).

We also have

$$f\left(\frac{\alpha\mathbf{a}+\beta\mathbf{b}}{\alpha+\beta}\right) = \frac{\alpha f(\mathbf{a}) + \beta f(\mathbf{b})}{\alpha+\beta}$$

Therefore, a linear operator f preserves the ratio in which a point divides a segment. In more detail, f takes three collinear points P, Q, R into three collinear points $f(P)$, $f(Q)$, $f(R)$ such that

$$(PQR) = (f(P), f(Q), f(R))$$

assuming that $f(P)$, $f(Q)$, $f(R)$ are distinct.

Examples

(1) $f(\mathbf{v}) = (\mathbf{k}\cdot\mathbf{v})\mathbf{k}$ is perpendicular projection onto the vertical axis. More generally, $f(\mathbf{v}) = (\mathbf{u}\cdot\mathbf{v})\mathbf{u}$ is *perpendicular projection onto the line spanned by the unit vector* $\mathbf{u}$.

(2) $f(\mathbf{v}) = (\mathbf{k}\times\mathbf{v})\times\mathbf{k} = \mathbf{v} - (\mathbf{k}\cdot\mathbf{v})\mathbf{k}$ is perpendicular projection onto the horizontal plane. More generally, $f(\mathbf{v}) = (\mathbf{u}\times\mathbf{v})\times\mathbf{u} = \mathbf{v} - (\mathbf{u}\cdot\mathbf{v})\mathbf{u}$ is *perpendicular projection onto the normal plane of the unit vector* $\mathbf{u}$ (see parallel and perpendicular components, Chap. 3, Application 5).

(3) $f(\mathbf{v}) = (\mathbf{k}\cdot\mathbf{v})\mathbf{k} + \mathbf{k}\times\mathbf{v}$ is rotation through 90° around the vertical axis.

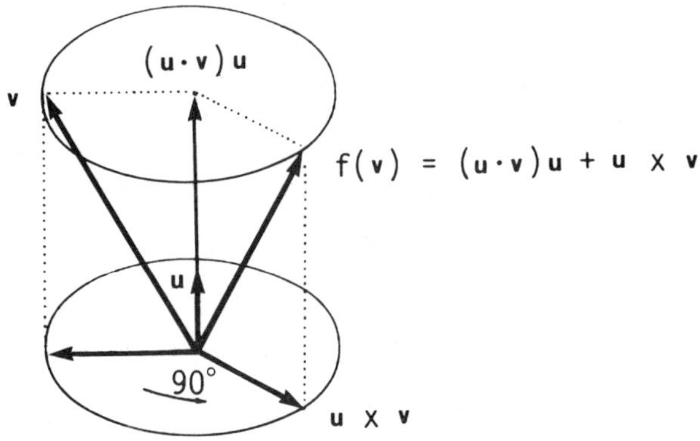

More generally, $f(\mathbf{v}) = (\mathbf{u} \cdot \mathbf{v})\mathbf{u} + \mathbf{u} \times \mathbf{v}$ is *rotation through* 90° *around the axis spanned by the unit vector* $\mathbf{u}$.

(4) $f(\mathbf{v}) = 2(\mathbf{k} \cdot \mathbf{v})\mathbf{k} - \mathbf{v}$ is reflection in the vertical axis (or, what is the same, rotation through 180° around the vertical axis). More generally, $f(\mathbf{v}) = 2(\mathbf{u} \cdot \mathbf{v})\mathbf{u} - \mathbf{v}$ is *reflection in the line spanned by the unit vector* $\mathbf{u}$ (or, what is the same, *rotation through* 180° *around the line spanned by the unit vector* $\mathbf{u}$).

(5) $f(\mathbf{v}) = \mathbf{v} - 2(\mathbf{k} \cdot \mathbf{v})\mathbf{k}$ is reflection in the horizontal plane. More generally, $f(\mathbf{v}) = \mathbf{v} - 2(\mathbf{u} \cdot \mathbf{v})\mathbf{u}$ is *reflection in the normal plane of the unit vector* $\mathbf{u}$.

(6) $f(\mathbf{v}) = -\mathbf{v}$ is *reflection in the origin*.

(7) $f(\mathbf{v}) = \lambda\mathbf{v}$, where $\lambda \in \mathbb{R}$ is called a *strain* in ratio λ. In more detail, it is a *stretching* if $\lambda > 1$ or a contraction is $0 < \lambda < 1$. If $\lambda = 1$, it is the *identity operator* leaving every vector unchanged; if $\lambda = -1$, it is reflection in the origin. If λ is a negative scalar, f can be described as a stretching or contraction followed by a reflection in the origin, according as $|\lambda| > 1$ or $|\lambda| < 1$. We may confuse the strain f in ratio λ with the scalar λ in notation, i.e., write $\lambda(\mathbf{v}) = \lambda\mathbf{v}$. Thus 2 is a linear operator, viz., $2(\mathbf{v}) = 2\mathbf{v}$ (strain in the ratio 2).

(8) We have seen in Chap. 3 that the cross product by a fixed vector $\mathbf{a}$, $f(\mathbf{v}) = \mathbf{a} \times \mathbf{v}$, is a linear operator. The same is, of course, true of $g(\mathbf{v}) = \mathbf{v} \times \mathbf{b}$, where $\mathbf{b}$ is a fixed vector.

MATRIX OF A LINEAR OPERATOR†

Let f be a linear transformation, and let $\mathbf{v} = x\mathbf{i} + y\mathbf{j} + z\mathbf{k}$ be a vector. Then

$$\begin{aligned} f(\mathbf{v}) &= f(x\mathbf{i} + y\mathbf{j} + z\mathbf{k}) \\ &= xf(\mathbf{i}) + yf(\mathbf{j}) + zf(\mathbf{k}) \end{aligned}$$

†*Note*: In this section the basis may be arbitrary.

where we put

Column Formulas

$$f(\mathbf{i}) = a_{11}\mathbf{i} + a_{21}\mathbf{j} + a_{31}\mathbf{k}$$
$$f(\mathbf{j}) = a_{12}\mathbf{i} + a_{22}\mathbf{j} + a_{32}\mathbf{k}$$
$$f(\mathbf{k}) = a_{13}\mathbf{i} + a_{23}\mathbf{j} + a_{33}\mathbf{k}$$

Then

$$\begin{aligned} f(\mathbf{v}) = &\quad x(a_{11}\mathbf{i} + a_{21}\mathbf{j} + a_{31}\mathbf{k}) \\ &+ y(a_{12}\mathbf{i} + a_{22}\mathbf{j} + a_{32}\mathbf{k}) \\ &+ z(a_{13}\mathbf{i} + a_{23}\mathbf{j} + a_{33}\mathbf{k}) \\ = &\quad (a_{11}x + a_{12}y + a_{13}z)\mathbf{i} \\ &+ (a_{21}x + a_{22}y + a_{23}z)\mathbf{j} \\ &+ (a_{31}x + a_{32}y + a_{33}z)\mathbf{k} \end{aligned}$$

This means that under the impact of f, the coordinates of a point (vector) will change as follows:

$$(x,y,z) \stackrel{f}{\rightsquigarrow} (a_{11}x + a_{12}y + a_{13}z,\ a_{21}x + a_{22}y + a_{23}z,\ a_{31}x + a_{32}y + a_{33}z)$$

or, what is the same,

Row Formulas

$$x' = a_{11}x + a_{12}y + a_{13}z$$
$$y' = a_{21}x + a_{22}y + a_{23}z$$
$$z' = a_{31}x + a_{32}y + a_{33}z$$

where $\mathbf{v}' = x'\mathbf{i} + y'\mathbf{j} + z'\mathbf{k} = f(\mathbf{v})$. This shows that the matrix of the linear transformation f is

$$f = \begin{pmatrix} a_{11} & a_{12} & a_{13} \\ a_{21} & a_{22} & a_{23} \\ a_{31} & a_{32} & a_{33} \end{pmatrix}$$

By the column formulas, the column vectors of the matrix f have an immediate geometric meaning: they are just the images of the vectors **i**, **j**, **k** under f: $f(\mathbf{i})$, $f(\mathbf{j})$, $f(\mathbf{k})$. By contrast, the row vectors of the matrix have no immediate geometric meaning; their significance is algebraic; by virtue of the row formulas, they are used in finding the impact of f on the coordinates of a point (or its location vector). The geometric meaning of the column vectors of the matrix, spelled out by the column formulas, helps us find the matrix of a linear operator described geometrically and, conversely, find the geometric description of a given matrix.

Examples

(1) Find the matrix of the identity operator 1.

Answer: $1(\mathbf{i}) = \mathbf{i}$, $1(\mathbf{j}) = \mathbf{j}$, $1(\mathbf{k}) = \mathbf{k}$. Hence the column vectors of the matrix 1 will be **i**, **j**, **k**:

$$1 = \begin{pmatrix} 1 & 0 & 0 \\ 0 & 1 & 0 \\ 0 & 0 & 1 \end{pmatrix}$$

(2) Find the geometric description of the linear operator with matrix

$$\begin{pmatrix} -1 & 0 & 0 \\ 0 & -1 & 0 \\ 0 & 0 & -1 \end{pmatrix}$$

Answer: $f(\mathbf{i}) = -\mathbf{i}$, $f(\mathbf{j}) = -\mathbf{j}$, $f(\mathbf{k}) = -\mathbf{k}$. Therefore, f is the reflection in the origin, $f = -1$.

(3) Find the matrix of the rotation through 180° around the UW & LE bisector axis.

Answer: $f(\mathbf{i}) = -\mathbf{k}$, $f(\mathbf{j}) = -\mathbf{j}$, $f(\mathbf{k}) = -\mathbf{i}$; hence the matrix is

$$f = \begin{pmatrix} 0 & 0 & -1 \\ 0 & -1 & 0 \\ -1 & 0 & 0 \end{pmatrix}$$

(4) Find the geometric description of the linear transformation

$$f = \begin{pmatrix} 0 & -1 & 0 \\ 0 & 0 & 1 \\ -1 & 0 & 0 \end{pmatrix}$$

Answer: $f(\mathbf{i}) = -\mathbf{k}$, $f(\mathbf{j}) = -\mathbf{i}$, $f(\mathbf{k}) = \mathbf{j}$. Thus f sends the vectors **i**, $-\mathbf{k}$, $-\mathbf{j}$ into one another cyclically. Therefore, f is the rotation through 240° around the LSE & UNW equiangular line.

(5) Find the matrix of the rotation around the z axis through 45°.

Answer: $f(\mathbf{i}) = \frac{1}{\sqrt{2}}\mathbf{i} + \frac{1}{\sqrt{2}}\mathbf{j}$ $\quad f(\mathbf{j}) = -\frac{1}{\sqrt{2}}\mathbf{i} + \frac{1}{\sqrt{2}}\mathbf{j}$ $\quad f(\mathbf{k}) = \mathbf{k}$

$$f = \begin{pmatrix} 1/\sqrt{2} & -1/\sqrt{2} & 0 \\ 1/\sqrt{2} & 1/\sqrt{2} & 0 \\ 0 & 0 & 1 \end{pmatrix}$$

In the literature matrices are accorded a much greater prominence than their usefulness deserves. The reason for this lies in the history of the development of the idea of a linear transformation. In the beginning, linear operators were treated through their matrices and their properties were described as properties of matrices. In this way an extensive theory of matrices arose. Later it became clear that the matrix theory tends to obscure the geometric insight. The abstract point of view, that one could proceed on a different level and work without coordinates and matrices developed during the 1920s and 1930s. With this new point of view the picture became quite easy and lovely and the theory was disassociated from the mechanism of computation. As a result, *matrices are no longer used in the development of the theory*. Matrices are still used, and are useful, in applications and in computations. But they are not essential to understanding the theory, and the theory should not be confused with the computations that arise.

IMPACT OF A LINEAR OPERATOR ON A VECTOR†

As our introductory examples showed, one of the most important applications of matrices is to calculate the impact of a linear operator f on a vector $\mathbf{v}$, i.e., to help find $f(\mathbf{v})$ in terms of its coordinates. The general rule was given above in the row formulas describing how the coordinates of $f(\mathbf{v})$ may be computed by using the rows of the matrix f.

Examples

(1) Find where the rotation through 90° around the y axis takes the vector $\mathbf{v} = 2\mathbf{i} - 3\mathbf{j} + 4\mathbf{k}$.

Answer: In order to find the matrix of the rotation we find its column vectors: $f(\mathbf{i}) = -\mathbf{k}$, $f(\mathbf{j}) = \mathbf{j}$, $f(\mathbf{k}) = \mathbf{i}$;

$$f = \begin{pmatrix} 0 & 0 & 1 \\ 0 & 1 & 0 \\ -1 & 0 & 0 \end{pmatrix}$$

Then $f(\mathbf{v}) = x'\mathbf{i} + y'\mathbf{j} + z'\mathbf{k}$, where $x' = 0(2) + 0(-3) + 1(4) = 4$, $y' = -3$, $z' = -2$; i.e., $f(\mathbf{v}) = 4\mathbf{i} - 3\mathbf{j} - 2\mathbf{k}$.

† *Note*: In this section the basis may be arbitrary.

(2) Find where the perpendicular projection onto the UNE & LSW equiangular line takes the point $P(1,-2,3)$.

Answer: The matrix of the projection onto the UNE & LSW equiangular line is

$$f = \begin{pmatrix} 1/3 & 1/3 & 1/3 \\ 1/3 & 1/3 & 1/3 \\ 1/3 & 1/3 & 1/3 \end{pmatrix}$$

Thus $(1,-2,3) \rightsquigarrow (2/3,2/3,2/3)$.

The matrix is a property of the linear operator but is subject to prior choice of coordinates. Changing the coordinate system will change the matrix of most linear operators. There are exceptions, though: the matrix of a strain in ratio λ, in any coordinate system, is

$$\lambda = \begin{pmatrix} \lambda & 0 & 0 \\ 0 & \lambda & 0 \\ 0 & 0 & \lambda \end{pmatrix}$$

In particular, the matrix of the identity operator, in any coordinate system, is

$$1 = \begin{pmatrix} 1 & 0 & 0 \\ 0 & 1 & 0 \\ 0 & 0 & 1 \end{pmatrix}$$

and the matrix of the reflection in the origin, in any coordinate system is

$$-1 = \begin{pmatrix} -1 & 0 & 0 \\ 0 & -1 & 0 \\ 0 & 0 & -1 \end{pmatrix}$$

In the applications that follow we shall gain valuable geometric information from the sum of the diagonal entries of the matrix of f. Therefore, we define a new concept, the *trace of a linear operator*, by the formula

$$\operatorname{trace} f = a_{11} + a_{22} + a_{33}$$

The formal treatment of trace f is given in Chap. 13.

APPLICATIONS

1. Perpendicular Projection onto a Line

Given the unit vector

$$\mathbf{u} = (\cos\alpha)\mathbf{i} + (\cos\beta)\mathbf{j} + (\cos\gamma)\mathbf{k}$$

the perpendicular projection p onto the line spanned by $\mathbf{u}$ is

$$p(\mathbf{v}) = (\mathbf{u} \cdot \mathbf{v})\mathbf{u}$$

Then

$$\begin{aligned} p(\mathbf{i}) &= (\mathbf{u} \cdot \mathbf{i})\mathbf{u} \\ &= (\cos^2\alpha)\mathbf{i} + (\cos\alpha\cos\beta)\mathbf{j} + (\cos\alpha\cos\gamma)\mathbf{k} \end{aligned}$$

and a similar calculation of $p(\mathbf{j})$, $p(\mathbf{k})$ yields the matrix

$$p = \begin{pmatrix} \cos^2\alpha & \cos\beta\cos\alpha & \cos\gamma\cos\alpha \\ \cos\alpha\cos\beta & \cos^2\beta & \cos\gamma\cos\beta \\ \cos\alpha\cos\gamma & \cos\beta\cos\gamma & \cos^2\gamma \end{pmatrix}$$

The remarkable fact that this matrix is symmetric† is explained in Chap. 11. Note that the sum of the entries in the main diagonal, i.e., trace p, is 1, and that the direction cosines of the unit vector $\mathbf{u}$ spanning the line onto which p projects the space can be obtained by taking the square roots of the diagonal entries. (Assign the appropriate signature!)

Examples

(1) Given the unit vector

$$\frac{1}{\sqrt{14}}\mathbf{i} - \frac{2}{\sqrt{14}}\mathbf{j} + \frac{3}{\sqrt{14}}\mathbf{k}$$

find the perpendicular projection p onto the line spanned by $\mathbf{u}$.

Answer:

$$p = \begin{pmatrix} 1/14 & -2/14 & 3/14 \\ -2/14 & 4/14 & -6/14 \\ 3/14 & -6/14 & 9/14 \end{pmatrix}$$

(2) Show that

$$p = \begin{pmatrix} 1/9 & 2/9 & 2/9 \\ 2/9 & 4/9 & 4/9 \\ 2/9 & 4/9 & 4/9 \end{pmatrix}$$

†A matrix is said to be symmetric if it remains unchanged under reflection in its main diagonal, i.e., if its rows and columns can be interchanged (see Chap. 11).

is a perpendicular projection onto a line and find the unit vector spanning that line.

Answer: The sum of the diagonal elements is 1, hence they are the squares of the direction cosines of a line spanned by the unit vector

$$\mathbf{u} = \frac{1}{3}\mathbf{i} + \frac{2}{3}\mathbf{j} + \frac{2}{3}\mathbf{k}$$

2. Perpendicular Projection onto a Plane

Given the unit vector $\mathbf{u} = (\cos\alpha)\mathbf{i} + (\cos\beta)\mathbf{j} + (\cos\gamma)\mathbf{k}$, the perpendicular projection q onto the normal plane of $\mathbf{u}$ is

$$q(\mathbf{v}) = (\mathbf{u} \times \mathbf{v}) \times \mathbf{u}$$

or, what is the same (see Chap. 3, Application 5; or by the double cross formula),

$$q(\mathbf{v}) = \mathbf{v} - (\mathbf{u} \cdot \mathbf{v})\mathbf{u}$$

Calculation of $q(\mathbf{i})$, $q(\mathbf{j})$, $q(\mathbf{k})$ yields the matrix

$$q = \begin{pmatrix} 1-\cos^2\alpha & -\cos\beta\cos\alpha & -\cos\gamma\cos\alpha \\ -\cos\alpha\cos\beta & 1-\cos^2\beta & -\cos\gamma\cos\beta \\ -\cos\alpha\cos\gamma & -\cos\beta\cos\gamma & 1-\cos^2\gamma \end{pmatrix}$$

The remarkable fact that this matrix is symmetric is explained in Chap. 11. Note that the sum of the entries in the main diagonal of the matrix, trace q, is 2. The trace of a projection is therefore 1 or 2, according as it is a projection onto a line or a plane. Furthermore, the direction cosines of the normal vector of the plane onto which q projects the space can be obtained as follows: $\cos^2\alpha = 1 - a_{11}$, $\cos^2\beta = 1 - a_{22}$, $\cos^2\gamma = 1 - a_{33}$. In calculating these direction cosines it will be necessary to inspect other entries of the matrix in order to find the correct signature. Recall (Chap. 4, Application 4) that perpendicular projections belong to a larger class of linear operators, called *projections*, satisfying $p^2 = p$.

SUM AND SCALAR MULTIPLE OF LINEAR OPERATORS

Let f and g be linear operators, and let $\lambda \in \mathbb{R}$. We shall define $f+g$ and λf by the formulas

$$(f+g)(\mathbf{v}) = f(\mathbf{v}) + g(\mathbf{v})$$

$$(\lambda f)(\mathbf{v}) = \lambda(f(\mathbf{v}))$$

It is clear that the sum and the scalar multiple of linear operators are also linear operators, as they both preserve the sum and the scalar multiple of vectors. We can also speak of the negative and the difference of linear operators: $-f=(-1)f$, $f-g=f+(-1)g$. These definitions also tell us how to find the sum and the scalar multiple of matrices: we must perform these operations "elementwise."

The formal rules of the sum and scalar multiple of linear operators:

Commutative Law

$$f+g=g+f$$

Associative Law

$$(f+g)+h=f+(g+h)$$

Scalar Associativity

$$(\lambda\mu)f=\lambda(\mu f) \qquad (\lambda,\mu\in\mathbb{R})$$

Distributive Laws

$$(\lambda+\mu)f=\lambda f+\mu f$$
$$\lambda(f+g)=\lambda f+\lambda g$$

The *trivial operator* (also called the zero operator), in symbols 0, is such that $0(\mathbf{v})=\mathbf{0}$, the null vector for all $\mathbf{v}$. It satisfies

$$f+0=0 \qquad f+(-f)=0$$

We also have

$$1f=f \qquad 0f=0$$

APPLICATIONS

3. Complementary Projections

Let p be the perpendicular projection onto the vertical line and let q be the perpendicular projection onto the horizontal plane. Then

$$p+q=1$$

the identity operator, because $p(\mathbf{v})+q(\mathbf{v})=(\mathbf{u}\cdot\mathbf{v})\mathbf{u}+(\mathbf{u}\times\mathbf{v})\times\mathbf{u}=\mathbf{v}$ for all $\mathbf{v}$, by the double cross formula.

If p is any perpendicular projection, the linear operator $q=1-p$

(where 1 denotes the identity operator) is also a perpendicular projection, called the *complementary projection* of p. If p is a projection onto a line, then $q = 1 - p$ is the projection onto the normal plane of that line, and vice versa.

Example Show that

$$p = \begin{pmatrix} 1/3 & 1/3 & 1/3 \\ 1/3 & 1/3 & 1/3 \\ 1/3 & 1/3 & 1/3 \end{pmatrix}$$

is a perpendicular projection onto a line. Find that line, and find the complementary projection of p and describe it geometrically.

Answer: trace $p = 1/3 + 1/3 + 1/3 = 1$; if p is a projection, it is a projection onto a line spanned by the unit vector

$$\mathbf{u} = \frac{1}{\sqrt{3}}\mathbf{i} + \frac{1}{\sqrt{3}}\mathbf{j} + \frac{1}{\sqrt{3}}\mathbf{k}$$

which is the UNE & LSW equiangular line. The complementary projection is

$$1 - p = \begin{pmatrix} 1 & 0 & 0 \\ 0 & 1 & 0 \\ 0 & 0 & 1 \end{pmatrix} - \begin{pmatrix} 1/3 & 1/3 & 1/3 \\ 1/3 & 1/3 & 1/3 \\ 1/3 & 1/3 & 1/3 \end{pmatrix} = \begin{pmatrix} 2/3 & -1/3 & -1/3 \\ -1/3 & 2/3 & -1/3 \\ -1/3 & -1/3 & 2/3 \end{pmatrix}$$

perpendicular projection onto the plane $x + y + z = 0$.

4. Reflections in a Line (Rotations Through 180°)

Given a unit vector $\mathbf{u} = \cos\alpha\mathbf{i} + \cos\beta\mathbf{j} + \cos\gamma\mathbf{k}$; if p is the perpendicular projection onto the line spanned by $\mathbf{u}$, then

$$h = 2p - 1$$

is the reflection in the line spanned by $\mathbf{u}$ or, what is the same, the rotation through 180° around the axis spanned by $\mathbf{u}$. Since we know the matrix of p (see Application 1 above), we can find the matrix of h by scalar multiplication and subtraction:

$$h = \begin{pmatrix} 2\cos^2\alpha - 1 & 2\cos\beta\cos\alpha & 2\cos\gamma\cos\alpha \\ 2\cos\alpha\cos\beta & 2\cos^2\beta - 1 & 2\cos\gamma\cos\beta \\ 2\cos\alpha\cos\gamma & 2\cos\beta\cos\gamma & 2\cos^2\gamma - 1 \end{pmatrix}$$

The remarkable fact that this matrix is symmetric is explained in Chap. 11.

Note that the sum of the diagonal elements, trace h, is –1. Moreover, the direction cosines of the vector $\mathbf{u}$ satisfy $\cos^2\alpha = \frac{1}{2}(a_{11}+1)$, $\cos^2\beta = \frac{1}{2}(a_{22}+1)$, $\cos^2\gamma = \frac{1}{2}(a_{33}+1)$. Therefore, if we know that h is a reflection in a line, then from the matrix h we can determine the direction cosines of the axis of reflection. Furthermore, the perpendicular projection onto the axis of reflection (i.e., onto the axis of rotation through 180°) is

$$p = \tfrac{1}{2}(1+h)$$

This is the simplest special case of the famous ergodic theorem that can be used to determine the axis of an arbitrary rotation (see Chap. 7, Application 10; Chap. 12, Application 3).

5. Reflections in a Plane

Let the normal vector of the plane of reflection be $\mathbf{u} = \cos\alpha\mathbf{i} + \cos\beta\mathbf{j} + \cos\gamma\mathbf{k}$, a unit vector. Then the reflection h' in the normal plane of $\mathbf{u}$ is

$$h' = 1 - 2p = -h$$

where h is the reflection in the line spanned by $\mathbf{u}$. Therefore, the matrix of h' can be obtained simply by taking the negative of h:

$$h' = \begin{pmatrix} 1-2\cos^2\alpha & -2\cos\beta\cos\alpha & -2\cos\gamma\cos\alpha \\ -2\cos\alpha\cos\beta & 1-2\cos^2\beta & -2\cos\gamma\cos\beta \\ -2\cos\alpha\cos\gamma & -2\cos\beta\cos\gamma & 1-2\cos^2\gamma \end{pmatrix}$$

The remarkable fact that the matrix of a reflection in a plane is always symmetric will be explained later (see Chap. 11). Note that the sum of the diagonal elements is trace $h' = 1$ (it was –1 for a reflection in a line). Furthermore, the direction cosines of the normal vector of the plane of reflection satisfy $\cos^2\alpha = \frac{1}{2}(1-a_{11})$, $\cos^2\beta = \frac{1}{2}(1-a_{22})$, $\cos^2\gamma = \frac{1}{2}(1-a_{33})$. Therefore, if we know that h' is a reflection, then trace $h' = a_{11} + a_{22} + a_{33}$ will tell us whether it is a reflection in a plane or in a line; if it is a reflection in a plane, the formulas above can help us find the equation of the plane of reflection.

Later we shall see that reflections belong to a larger class of linear operators called *involutions*, satisfying $h^2 = 1$, the identity operator.

Examples

(1) Given the unit vector

$$\mathbf{u} = \frac{1}{\sqrt{14}}\mathbf{i} - \frac{2}{\sqrt{14}}\mathbf{j} + \frac{3}{\sqrt{14}}\mathbf{k}$$

find the reflections h and h' in the line spanned by **u** and in the normal plane of **u**, respectively:

Answer:

$$h = \begin{pmatrix} -6/7 & -2/7 & 3/7 \\ -2/7 & -3/7 & -6/7 \\ 3/7 & -6/7 & 2/7 \end{pmatrix} \qquad h' = \begin{pmatrix} 6/7 & 2/7 & -3/7 \\ 2/7 & 3/7 & 6/7 \\ -3/7 & 6/7 & -2/7 \end{pmatrix}$$

(2) Show that

$$h = \begin{pmatrix} 1/3 & -2/3 & -2/3 \\ -2/3 & 1/3 & -2/3 \\ -2/3 & -2/3 & 1/3 \end{pmatrix}$$

is a reflection in a plane and find the equation of that plane.

Answer: trace $h = 1$; therefore, if h is a reflection, it must be a reflection in a plane. The direction cosines of the normal vector **u** of the plane would satisfy $\cos^2\alpha = ½(1-1/3) = 1/3 = \cos^2\beta = \cos^2\gamma$. In this case all three direction cosines are positive, as the inspection of the other entries of the matrix shows. Therefore, $\mathbf{u} = (1/\sqrt{3})(\mathbf{i}+\mathbf{j}+\mathbf{k})$ which spans the UNE & LSW equiangular line. The equation of the plane of reflection is $x + y + z = 0$.

(3) Show that

$$h = \begin{pmatrix} -8/9 & 1/9 & 4/9 \\ 1/9 & -8/9 & 4/9 \\ 4/9 & 4/9 & 7/9 \end{pmatrix}$$

is a reflection in a line. Find the symmetric form of that line.

Answer: trace $h = -1$; therefore, if h is a reflection, it must be a reflection in a line spanned by the unit vector $\mathbf{u} = \cos\alpha\mathbf{i} + \cos\beta\mathbf{j} + \cos\gamma\mathbf{k}$, where $\cos^2\alpha = 1/18 = \cos^2\beta$, $\cos^2\gamma = 8/9 = 16/18$. Since $\cos\alpha\cos\beta$, $\cos\beta\cos\gamma$, $\cos\gamma\cos\alpha$ are all positive, we conclude that $\cos\alpha$, $\cos\beta$, $\cos\gamma$ have the same signature (say, positive). Therefore, $\mathbf{u} = (1/\sqrt{18})(\mathbf{i}+\mathbf{j}+4\mathbf{k})$. The symmetric form of the line is $x = y = z/4$.

6. Antisymmetric Twists

We have seen in Chapter 4 that $t(\mathbf{v}) = \mathbf{u}\times\mathbf{v}$, where $\mathbf{u} = (\cos\alpha)\mathbf{i} + (\cos\beta)\mathbf{j} + (\cos\gamma)\mathbf{k}$ is a unit vector, is a twist. We now find its matrix by calculating $t(\mathbf{i}) = \mathbf{u}\times\mathbf{i} = (\cos\gamma)\mathbf{j} - (\cos\beta)\mathbf{k}$, etc.:

$$t = \begin{pmatrix} 0 & -\cos\gamma & \cos\beta \\ \cos\gamma & 0 & -\cos\alpha \\ -\cos\beta & \cos\alpha & 0 \end{pmatrix}$$

Since its matrix is antisymmetric, t is called an *antisymmetric twist*. In Chap. 7 we shall see that there are other twists as well.

7. Rotations Through 90°

The matrix of the rotation r through 90° can be derived from the formula

$$r(\mathbf{v}) = (\mathbf{u}\cdot\mathbf{v})\mathbf{u} + \mathbf{u}\times\mathbf{v}$$

(already mentioned earlier in this chapter) as the sum of a symmetric and an antisymmetric operator

$$r = \begin{pmatrix} \cos^2\alpha & \cos\alpha\cos\beta & \cos\alpha\cos\gamma \\ \cos\alpha\cos\beta & \cos^2\beta & \cos\beta\cos\gamma \\ \cos\alpha\cos\gamma & \cos\beta\cos\gamma & \cos^2\gamma \end{pmatrix} + \begin{pmatrix} 0 & -\cos\gamma & \cos\beta \\ \cos\gamma & 0 & -\cos\alpha \\ -\cos\beta & \cos\alpha & 0 \end{pmatrix}$$

$$= \begin{pmatrix} \cos^2\alpha & \cos\alpha\cos\beta - \cos\gamma & \cos\alpha\cos\gamma + \cos\beta \\ \cos\alpha\cos\beta + \cos\gamma & \cos^2\beta & \cos\beta\cos\gamma - \cos\alpha \\ \cos\alpha\cos\gamma - \cos\beta & \cos\beta\cos\gamma + \cos\alpha & \cos^2\gamma \end{pmatrix}$$

Since in the main diagonal we find the squares of the direction cosines of the axis of rotation, we have trace $r = 1$. In Chap. 13 we shall see a simple formula showing how to calculate the angle of any rotation r from trace r. Note also that the axis of the rotation r through 90° can be readily determined as the square roots of the diagonal elements are just the direction cosines.

Later we shall see that rotations through 90° or –90° belong to a larger class of linear operators called *invections* satisfying $r^4 = 1$ ($r^2 \neq 1$).

Examples

(1) Find the matrix of the rotation through 90° if the axis is spanned by the unit vector

$$\mathbf{u} = \frac{1}{3}\mathbf{i} + \frac{2}{3}\mathbf{j} + \frac{2}{3}\mathbf{k}$$

Answer:

$$r = \begin{pmatrix} 1/9 & -4/9 & 8/9 \\ 8/9 & 4/9 & 1/9 \\ -4/9 & 7/9 & 4/9 \end{pmatrix}$$

(2) Given that

$$r = \begin{pmatrix} 9/25 & 12/25 & 4/5 \\ 12/25 & 16/25 & -3/5 \\ -4/5 & 3/5 & 0 \end{pmatrix}$$

is a rotation through 90°, find the axis of rotation.

Answer: Taking the square roots of the diagonal elements, we get $\cos\alpha = \pm 3/5$, $\cos\beta = \pm 4/5$, $\cos\gamma = 0$; the axis is horizontal. In order to determine the signatures of the direction cosines of the axis, we note that $a_{23} = -3/5 = -\cos\alpha$ and $a_{31} = -\cos\beta = -4/5$. We conclude that the axis of the rotation is spanned by the unit vector

$$\mathbf{u} = \frac{3}{5}\mathbf{i} + \frac{4}{5}\mathbf{j}$$

8. Lifts

The linear operator $\ell(\mathbf{v}) = (\mathbf{i}\cdot\mathbf{v})\mathbf{k}$ whose matrix is

$$\ell = \begin{pmatrix} 0 & 0 & 0 \\ 0 & 0 & 0 \\ 1 & 0 & 0 \end{pmatrix}$$

is a *lift*, in more detail, an UL lift of E. The terminology is borrowed from aerodynamics. We may visualize an aircraft flying against a prevailing easterly wind. If the pilot increases the velocity of the aircraft by $\mathbf{v}$, then the uplift acting upon the wings will increase by $\ell(\mathbf{v})$, the length of which is proportional to the perpendicular projection of $\mathbf{v}$ onto $\mathbf{i}$, the direction of the flying path.

More generally, let us consider a pair of perpendicular vectors $\mathbf{e} = a\mathbf{i} + b\mathbf{j} + c\mathbf{k}$, $\mathbf{n} = A\mathbf{i} + B\mathbf{j} + C\mathbf{k}$; $\mathbf{n}\cdot\mathbf{e} = Aa + Bb + Cc = 0$. The linear operator

$$\ell(\mathbf{v}) = (\mathbf{n}\cdot\mathbf{v})\mathbf{e} \qquad (\mathbf{n}\cdot\mathbf{e} = 0)$$

is a $\pm\mathbf{e}$ lift of $\mathbf{n}$, and its matrix is

$$\ell = \begin{pmatrix} Aa & Ba & Ca \\ Ab & Bb & Cb \\ Ac & Bc & Cc \end{pmatrix}$$

Note that if ℓ is a lift, trace $\ell = Aa + Bb + Cc = 0$. On the strength of the double cross formula, the lift ℓ can also be written as

$$\ell(\mathbf{v}) = \mathbf{n} \times (\mathbf{e} \times \mathbf{v}) \qquad (\mathbf{n}\cdot\mathbf{e} = 0)$$

Later we shall see that all lifts satisfy the interesting property that $\ell \neq 0$ but $\ell^2 = 0$ (also, see Chap. 4, Application 5).

Example Find the matrix of the UNE & LSW lift of NW.

Answer: $\mathbf{n} = \mathbf{i} + \mathbf{j} + \mathbf{k}$, $\mathbf{e} = -\mathbf{i} + \mathbf{j}$; $(\mathbf{n} \cdot \mathbf{e} = 0)$. Therefore,

$$\ell = \begin{pmatrix} -1 & -1 & -1 \\ 1 & 1 & 1 \\ 0 & 0 & 0 \end{pmatrix}$$

9. Shears

Let ℓ be a lift. Then the linear operator $f = 1 + \ell$ is called a *shear*.

Example A lift ℓ is given by the formula $\ell(\mathbf{v}) = (\mathbf{k} \cdot \mathbf{v})\mathbf{i}$. Then $f = 1 + \ell$ is a horizontal shear toward E and its matrix is

$$f = 1 + \ell = \begin{pmatrix} 1 & 0 & 0 \\ 0 & 1 & 0 \\ 0 & 0 & 1 \end{pmatrix} + \begin{pmatrix} 0 & 0 & 1 \\ 0 & 0 & 0 \\ 0 & 0 & 0 \end{pmatrix} = \begin{pmatrix} 1 & 0 & 1 \\ 0 & 1 & 0 \\ 0 & 0 & 1 \end{pmatrix}$$

A shear can be demonstrated with a deck of cards. The name "shear" is borrowed from physics, where it is used to describe a type of deformation in which parallel planes in a body remain parallel but are relatively displaced in a direction parallel to themselves, as if by sliding the individual cards in the deck.

In the general case, the shear parallel to the plane $Ax + By + Cz = 0$ along the line $x/a = y/b = z/c$ (where $Aa + Bb + Cc = 0$, i.e., the line is parallel to the plane) is given by the formula

$$f(\mathbf{v}) = \mathbf{v} + (\mathbf{n} \cdot \mathbf{v})\mathbf{e} \qquad (\mathbf{n} \cdot \mathbf{e} = 0)$$

with $\mathbf{n} = A\mathbf{i} + B\mathbf{j} + C\mathbf{k}$, $\mathbf{e} = a\mathbf{i} + b\mathbf{j} + c\mathbf{k}$ and the matrix of the shear is

$$f = \begin{pmatrix} 1 + Aa & Ba & Ca \\ Ab & 1 + Bb & Cb \\ Ac & Bc & 1 + Cc \end{pmatrix}$$

Note that trace $f = 3$ for every shear f.

The shear f has the effect of deforming the unit cube into a parallelepiped of the same volume.

Example

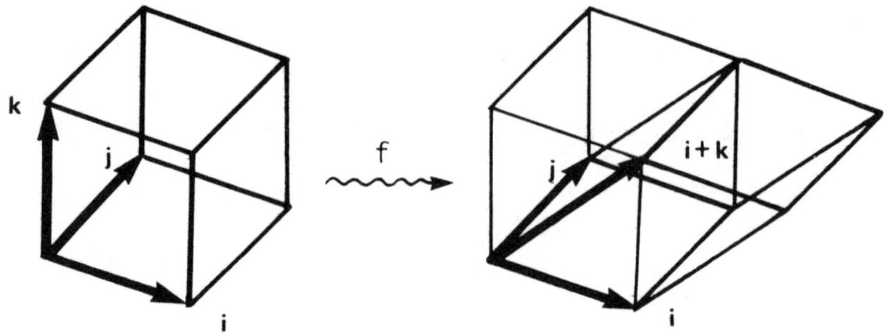

Find the matrix of the deformation depicted by the figure above. Name the deformation.

Answer: $f(\mathbf{i}) = \mathbf{i}$, $f(\mathbf{j}) = \mathbf{j}$, $f(\mathbf{k}) = \mathbf{i} + \mathbf{k}$. Hence

$$f = \begin{pmatrix} 1 & 0 & 1 \\ 0 & 1 & 0 \\ 0 & 0 & 1 \end{pmatrix}$$

the horizontal shear to E.

EXERCISES

6.1 Find the matrix of rotation through ϕ around the z axis. Find the matrix of rotation through ϕ around the y axis.

6.2 Find the impact of the rotation through 180° around $3\mathbf{i} - 4\mathbf{j}$ on the point $P(7,-4,3)$.

6.3 Given the unit vector

$$\mathbf{u} = \frac{1}{\sqrt{14}}\mathbf{i} - \frac{2}{\sqrt{14}}\mathbf{j} + \frac{3}{\sqrt{14}}\mathbf{k}$$

find the matrix of the perpendicular projection onto the plane with normal vector $\mathbf{u}$.

6.4 Show that

$$q = \begin{pmatrix} 8/9 & -2/9 & -2/9 \\ -2/9 & 5/9 & -4/9 \\ -2/9 & -4/9 & 5/9 \end{pmatrix}$$

is a perpendicular projection onto a plane. Find the normal vector of that plane.

6.5 Find the matrix of rotation through 90° if the axis is spanned by the vector $3\mathbf{i} + 4\mathbf{j}$.

6.6 Show that

$$h = \begin{pmatrix} 9/11 & 2/11 & 6/11 \\ 2/11 & 9/11 & -6/11 \\ 6/11 & -6/11 & -7/11 \end{pmatrix}$$

is a reflection in a plane. Find the equation of that plane.

6.7 Find the matrix of rotation through –90° around the axis with direction cosines $\cos\alpha$, $\cos\beta$, $\cos\gamma$.

6.8 Find the matrix of perpendicular projection onto the NE & SW bisector plane.

6.9 Find the geometrical meaning of the matrix

$$\begin{pmatrix} 1/2 & 1/2 & 0 \\ 1/2 & 1/2 & 0 \\ 0 & 0 & 0 \end{pmatrix}$$

6.10 Find the matrix of the perpendicular projection onto the UNE & LSW equiangular line.

6.11 Show that

$$\begin{pmatrix} -8/9 & 1/9 & 4/9 \\ 1/9 & -8/9 & 4/9 \\ 4/9 & 4/9 & 7/9 \end{pmatrix}$$

is a reflection. Is it a reflection in a line or in a plane? Determine the equation of that line or plane.

6.12 Show that

$$\begin{pmatrix} 5/6 & 1/6 & 2/6 \\ 1/6 & 5/6 & -2/6 \\ 2/6 & -2/6 & 2/6 \end{pmatrix}$$

is a perpendicular projection. Is it a projection onto a line or onto a plane? Determine the equation of that line or plane.

6.13 Write the matrix of the rotation around the UNE & LSW equiangular line through 180°.

6.14 Describe the matrix

$$\begin{pmatrix} 1/3 & -2/3 & -2/3 \\ -2/3 & 1/3 & -2/3 \\ -2/3 & -2/3 & 1/3 \end{pmatrix}$$

geometrically.

6.15 Given that the matrix

$$\begin{pmatrix} 1/2 & 1/2 & \sqrt{2}/2 \\ 1/2 & 1/2 & -\sqrt{2}/2 \\ -\sqrt{2}/2 & \sqrt{2}/2 & 0 \end{pmatrix}$$

is a rotation, find its axis and the angle.

6.16 Show that the matrix of reflection of the plane in the line $y = mx$ is

$$\begin{pmatrix} \dfrac{1-m^2}{1+m^2} & \dfrac{2m}{1+m^2} \\ \dfrac{2m}{1+m^2} & -\dfrac{1-m^2}{1+m^2} \end{pmatrix}$$

6.17 Show that

$$\ell = \begin{pmatrix} Aa & Ba \\ Ab & Bb \end{pmatrix}$$

is a lift of $\mathbb{R}^2$, provided that $Aa + Bb = 0$. Describe ℓ geometrically.

6.18 Show that

$$f = \begin{pmatrix} 1+Aa & Ba \\ Ab & 1+Bb \end{pmatrix}$$

is a shear of $\mathbb{R}^2$, provided that $Aa + Bb = 0$. Describe f geometrically.

7

The Product of Linear Operators†

FORMAL RULES OF THE PRODUCT OF LINEAR OPERATORS

The *product* of two linear operators is defined to be their product as functions. The product of functions was defined in the Introduction. Thus, if f and g are linear operators, their product $g \circ f$ is defined by the formula

$$(g \circ f)(\mathbf{v}) = g(f(\mathbf{v}))$$

The matrix of $g \circ f$ (in terms of the same coordinate system) is called the *product of the matrices* of f and g.

Since both f and g preserve the sum and the scalar multiple of vectors, $g \circ f$ will also do so and is therefore a linear operator.

Examples

1. $f \circ 1 = f = 1 \circ f$: The product of any linear operator with the identity is the same linear operator; therefore

$$\begin{pmatrix} 1 & 0 & 0 \\ 0 & 1 & 0 \\ 0 & 0 & 1 \end{pmatrix} \circ \begin{pmatrix} a_{11} & a_{12} & a_{13} \\ a_{21} & a_{22} & a_{23} \\ a_{31} & a_{32} & a_{33} \end{pmatrix} = \begin{pmatrix} a_{11} & a_{12} & a_{13} \\ a_{21} & a_{22} & a_{23} \\ a_{31} & a_{32} & a_{33} \end{pmatrix}$$

2. $f \circ 0 = 0 = 0 \circ f$: The product of any linear operator with the trivial operator is trivial; therefore,

$$\begin{pmatrix} 0 & 0 & 0 \\ 0 & 0 & 0 \\ 0 & 0 & 0 \end{pmatrix} \circ \begin{pmatrix} a_{11} & a_{12} & a_{13} \\ a_{21} & a_{22} & a_{23} \\ a_{31} & a_{32} & a_{33} \end{pmatrix} = \begin{pmatrix} 0 & 0 & 0 \\ 0 & 0 & 0 \\ 0 & 0 & 0 \end{pmatrix}$$

†*Note*: In this chapter all bases are orthonormal and all coordinate systems are rectangular metric, except where stipulated otherwise.

3. Let p be the perpendicular projection onto the vertical axis, and let q be the perpendicular projection onto the horizontal plane. Then $p \circ q = 0 = q \circ p$, because their product projects all vectors into the origin. In matrix language this means that

$$\begin{pmatrix} 0 & 0 & 0 \\ 0 & 0 & 0 \\ 0 & 0 & 1 \end{pmatrix} \circ \begin{pmatrix} 1 & 0 & 0 \\ 0 & 1 & 0 \\ 0 & 0 & 0 \end{pmatrix} = \begin{pmatrix} 0 & 0 & 0 \\ 0 & 0 & 0 \\ 0 & 0 & 1 \end{pmatrix}$$

Note that $p \neq 0$, $q \neq 0$, yet $p \circ q = 0$. This phenomenon is quite unknown among the scalars, and we shall say that among the linear operators, unlike among the scalars, there exist *zero divisors*. This fact rules out any possibility of a cancellation law for the product of linear operators.

4. Let p and q be as above. Then $p^2 = p \circ p = p$ and $q^2 = q \circ q = q$ because a vector remains unchanged under a projection if it is already in the "target," such as $p(\mathbf{v})$ is in the target of p, so that $p(p(\mathbf{v})) = p(\mathbf{v})$. Therefore,

$$\begin{pmatrix} 1 & 0 & 0 \\ 0 & 1 & 0 \\ 0 & 0 & 0 \end{pmatrix}^2 = \begin{pmatrix} 1 & 0 & 0 \\ 0 & 1 & 0 \\ 0 & 0 & 0 \end{pmatrix}$$

5. Let -1 denote the reflection in the origin; then $(-1) \circ (-1) = 1$ is the identity operator, because the second reflection sends every vector back, so

$$\begin{pmatrix} -1 & 0 & 0 \\ 0 & -1 & 0 \\ 0 & 0 & -1 \end{pmatrix}^2 = \begin{pmatrix} 1 & 0 & 0 \\ 0 & 1 & 0 \\ 0 & 0 & 1 \end{pmatrix}$$

6. Let h denote the reflection in the vertical axis or, what is the same, the rotation through 180° around the vertical axis. Then $h^2 = h \circ h = 1$; therefore,

$$\begin{pmatrix} -1 & 0 & 0 \\ 0 & -1 & 0 \\ 0 & 0 & 1 \end{pmatrix}^2 = \begin{pmatrix} 1 & 0 & 0 \\ 0 & 1 & 0 \\ 0 & 0 & 1 \end{pmatrix}$$

The same is true for $-h$, the reflection in the horizontal plane: $(-h)^2 = 1$; therefore,

$$\begin{pmatrix} 1 & 0 & 0 \\ 0 & 1 & 0 \\ 0 & 0 & -1 \end{pmatrix}^2 = \begin{pmatrix} 1 & 0 & 0 \\ 0 & 1 & 0 \\ 0 & 0 & 1 \end{pmatrix}$$

7. Let ℓ be a lift, e.g., an EW lift of U: $\ell(\mathbf{v}) = (\mathbf{k} \cdot \mathbf{v})\mathbf{i}$. Then $\ell^2(\mathbf{v}) = \ell(\ell(\mathbf{v})) = (\mathbf{k} \cdot \ell(\mathbf{v}))\mathbf{i} = (\mathbf{k} \cdot (\mathbf{k} \cdot \mathbf{v})\mathbf{i})\mathbf{i} = (\mathbf{k} \cdot \mathbf{v})(\mathbf{k} \cdot \mathbf{i}) = (\mathbf{k} \cdot \mathbf{v})0 = 0$. Hence $\ell^2 = 0$. Again, this is a phenomenon quite unknown among

scalars. In matrix notation:

$$\begin{pmatrix} 0 & 0 & 1 \\ 0 & 0 & 0 \\ 0 & 0 & 0 \end{pmatrix}^2 = \begin{pmatrix} 0 & 0 & 0 \\ 0 & 0 & 0 \\ 0 & 0 & 0 \end{pmatrix}$$

The product of linear operators is, in general, not commutative. However, it has the following formal properties:

Associativity

$$h \circ (g \circ f) = (h \circ g) \circ f$$

Scalar Associativity

$$(\lambda g) \circ f = \lambda (g \circ f) = g \circ (\lambda f) \qquad (\lambda \in \mathbb{R})$$

Distributivity

$$g \circ (f + f') = g \circ f + g \circ f'$$
$$(g + g') \circ f = g \circ f + g' \circ f$$

Each of these can be established by a simple calculation involving the appropriate definitions. For example, in the case of the second distributive law, we write

$$\begin{aligned} ((g + g') \circ f)(\mathbf{v}) &= (g + g')(f(\mathbf{v})) = g(f(\mathbf{v})) + g'(f(\mathbf{v})) \\ &= (g \circ f)(\mathbf{v}) + (g' \circ f)(\mathbf{v}) = (g \circ f + g' \circ f)(\mathbf{v}) \end{aligned}$$

APPLICATIONS

1. Projections

Projections were introduced in Chapter 4, Application 4. In Chap. 6, Applications 2 and 3, we encountered a particular type of this species, namely, the *perpendicular* projections. Later we shall see that these are precisely the projections whose matrix is symmetric. There are infinitely many perpendicular projections, as for every choice of the unit vector **u** we have two of them, the perpendicular projection onto the line spanned by **u** and the perpendicular projection onto the normal plane of **u**. There is an even greater abundance of nonsymmetric projections.

Example The projection onto the UNE & LSW equiangular line, along the xy plane, is such that $(x,y,z) \rightsquigarrow (z,z,z)$. The matrix of this projection is

$$p = \begin{pmatrix} 0 & 0 & 1 \\ 0 & 0 & 1 \\ 0 & 0 & 1 \end{pmatrix}$$

and we conclude that

$$\begin{pmatrix} 0 & 0 & 1 \\ 0 & 0 & 1 \\ 0 & 0 & 1 \end{pmatrix}^2 = \begin{pmatrix} 0 & 0 & 1 \\ 0 & 0 & 1 \\ 0 & 0 & 1 \end{pmatrix}$$

In nature, projections occur in pairs:

$$p \text{ projection} \Leftrightarrow 1-p \text{ projection}$$

Indeed,

$$\begin{aligned} p^2 = p \Rightarrow (1-p)^2 = (1-p)\circ(1-p) &= 1-p-p+p\circ p \\ &= 1-p-p+p = 1-p \end{aligned}$$

and since $1-(1-p) = p$, the converse follows at once. We shall call the projections p and $1-p$ *complementary*.

The product of complementary projections is 0:

$$p\circ(1-p) = p-p^2 = p-p = 0 = (1-p)\circ p$$

We may now survey all the projections as follows. Let $Ax + By + Cz = 0$ be a plane and let $x/a = y/b = z/c$ be a nonparallel line. Then $Aa + Bb + Cc \neq 0$; in fact, we shall assume that the lengths of the vectors $\mathbf{n} = A\mathbf{i} + B\mathbf{j} + C\mathbf{k}$ and $\mathbf{d} = a\mathbf{i} + b\mathbf{j} + c\mathbf{k}$ have been so fixed that $\mathbf{n}\cdot\mathbf{d} = Aa + Bb + Cc = 1$. Then the projection p onto the line with direction vector $\mathbf{d}$, along the plane with normal vector $\mathbf{n}$, is given by the formula

$$p(\mathbf{v}) = (\mathbf{n}\cdot\mathbf{v})\mathbf{d} \qquad \mathbf{n}\cdot\mathbf{d} = 1$$

and its matrix is

$$p = \begin{pmatrix} Aa & Ba & Ca \\ Ab & Bb & Cb \\ Ac & Bc & Cc \end{pmatrix}$$

Notice that $\operatorname{trace} p = Aa + Bb + Cc = \mathbf{n}\cdot\mathbf{d} = 1$. A projection p with $\operatorname{trace} p = 1$ is called a *principal projection* (i.e., projection onto a line). A perpendicular projection onto the line spanned by the unit vector $\mathbf{u}$ is a special case of principal projections, namely, $\mathbf{n} = \mathbf{d} = \mathbf{u}$ in which case $A = a$, $B = b$, $C = c$ are just the direction cosines of the unit vector $\mathbf{u}$.

The projection $q = 1 - p$ is not a principal projection. q projects vectors onto the plane with normal vector $\mathbf{n}$, along the line with direction vector $\mathbf{d}$. Its matrix is

$$q = \begin{pmatrix} 1-Aa & -Ba & -Ca \\ -Ab & 1-Bb & -Cb \\ -Ac & -Bc & 1-Cc \end{pmatrix}$$

Notice that trace $q = 3 - (Aa + Bb + Cc) = 2$. A perpendicular projection onto a plane with normal vector $\mathbf{u}$ is a special case of this, namely, $\mathbf{n} = \mathbf{d} = \mathbf{u}$, which must be a unit vector and $A = a$, $B = b$, $C = c$, the direction cosines of the normal vector. As

$$\begin{aligned} q(\mathbf{v}) &= \mathbf{v} - (\mathbf{n} \cdot \mathbf{v})\mathbf{d} = (\mathbf{n} \cdot \mathbf{d})\mathbf{v} - (\mathbf{n} \cdot \mathbf{v})\mathbf{d} \\ &= \mathbf{n} \times (\mathbf{v} \times \mathbf{d}) \qquad \text{for all } \mathbf{v} \end{aligned}$$

by the double cross formula, we have that

$$q(\mathbf{v}) = \mathbf{n} \times (\mathbf{v} \times \mathbf{d}) \qquad \mathbf{n} \cdot \mathbf{d} = 1$$

We shall see (in Chap. 10) that these two types of projections, together with the identity operator 1 and the trivial operator 0, exhaust the set of all projections.

Examples

1. Write the matrix of the projection onto the line $x/4 = -y = y/-7$ along the plane $8x - y + 5z = 0$.

Answer:

$$p = \begin{pmatrix} 8(4) & (-4)4 & 5(4) \\ 8(-1) & (-4)(-1) & 5(-1) \\ 8(-7) & (-4)(-7) & 5(-7) \end{pmatrix} = \begin{pmatrix} 32 & -16 & 20 \\ -8 & 4 & -5 \\ -56 & 28 & -35 \end{pmatrix}$$

because $Aa + Bb + Cc = 32 + 4 - 35 = 1$.

2. Write the matrix of the projection onto the plane $5x - 3y + 2z = 0$ along the line $x/2 = y/-3 = z/-7$

Answer: $Aa + Bb + Cc = 10 + 9 - 14 = 5$; hence q, the complementary projection of the principal projection p, is

$$q = 1 - p = \begin{pmatrix} 1 & 0 & 0 \\ 0 & 1 & 0 \\ 0 & 0 & 1 \end{pmatrix} - \begin{pmatrix} 10/5 & -6/5 & 4/5 \\ -15/5 & 9/5 & -6/5 \\ -35/5 & 21/5 & -14/5 \end{pmatrix} = \begin{pmatrix} -1 & 6/5 & -4/5 \\ 3 & -4/5 & 6/5 \\ 7 & -21/5 & 19/5 \end{pmatrix}$$

2. Involutions

A linear operator h satisfying $h^2 = 1$ is called an *involution*. So far we have encountered a particular type only of this species, namely, the reflections. Later we shall see that these are precisely those involutions with a symmetric matrix. There are, of course, infinitely many symmetric involutions; every choice of a unit vector $\mathbf{u}$ yields two of them: the reflection in the line spanned by $\mathbf{u}$ and the reflection in the plane with normal vector $\mathbf{u}$. There is an even greater abundance of (nonsymmetric) involutions.

Example The involution in the NE & SW bisector plane along the y axis is

$$h = \begin{pmatrix} 1 & 0 & 0 \\ 2 & -1 & 0 \\ 0 & 0 & 1 \end{pmatrix}$$

It follows that

$$\begin{pmatrix} 1 & 0 & 0 \\ 2 & -1 & 0 \\ 0 & 0 & 1 \end{pmatrix} \circ \begin{pmatrix} 1 & 0 & 0 \\ 2 & -1 & 0 \\ 0 & 0 & 1 \end{pmatrix} = \begin{pmatrix} 1 & 0 & 0 \\ 0 & 1 & 0 \\ 0 & 0 & 1 \end{pmatrix}$$

The higher powers of involutions are easy to get:

$$h = h^3 = h^5 = h^7 = \cdots = h^{2n+1} \qquad 1 = h^2 = h^4 = h^6 = \cdots = h^{2n}$$

i.e., the odd powers of an involution h are equal to h and the even powers to 1, the identity operator.

In nature, involutions occur in pairs:

$$h \text{ involution} \Leftrightarrow -h \text{ involution}$$

because $h^2 = 1 \Rightarrow (-h)^2 = h^2 = 1$ and the converse also follows from this since $-(-h) = h$. The involutions h and $-h$ are called *complementary*.

Furthermore, there is a close relationship between involutions and projections:

$$h \text{ involution} \Rightarrow p = \tfrac{1}{2}(1 + h) \text{ projection}$$

and

$$p \text{ projection} \Rightarrow h = 2p - 1 \text{ involution}$$

Indeed,

$$h^2 = 1 \Rightarrow p^2 = [\tfrac{1}{2}(1 + h)]^2 = \tfrac{1}{4}(1 + 2h + h^2) = \tfrac{1}{2}(1 + h) = p$$

and

$$p^2 = p \Rightarrow h^2 = (2p - 1)^2 = 4p^2 - 2p - 2p + 1 = 1$$

It is clear that $h \rightsquigarrow p = \tfrac{1}{2}(1 + h)$ is a bijective correspondence between involutions and projections, and we shall call the involution h and the projection $p = \frac{1}{2}(1 + h)$ *associated*.

We may now survey the involutions as follows. Let $Ax + By + Cz = 0$ be a plane with normal vector $\mathbf{n} = A\mathbf{i} + B\mathbf{j} + C\mathbf{k}$, and let $x/a = y/b = z/c$ be a nonparallel line with direction vector $\mathbf{d} = a\mathbf{i} + b\mathbf{j} + c\mathbf{k}$. Then $\mathbf{n} \cdot \mathbf{d} = Aa + Bb + Cc \neq 0$; in fact, we may assume that the lengths of the vectors $\mathbf{n}$ and $\mathbf{d}$ have been so fixed that $\mathbf{n} \cdot \mathbf{d} = Aa + Bb + Cc = 1$.

The involution h in the line with direction vector $\mathbf{d}$, along the plane with normal vector $\mathbf{n}$ is given by the formula

$$h(\mathbf{v}) = 2(\mathbf{n} \cdot \mathbf{v})\mathbf{d} - \mathbf{v} \qquad \mathbf{n} \cdot \mathbf{d} = 1$$

and its matrix is

$$h = \begin{pmatrix} 2Aa-1 & 2Ba & 2Ca \\ 2Ab & 2Bb-1 & 2Cb \\ 2Ac & 2Bc & 2Cc-1 \end{pmatrix}$$

By the double cross formula, the involution h can also be put in the form

$$h(\mathbf{v}) = (\mathbf{n} \cdot \mathbf{v})\mathbf{d} - \mathbf{n} \times (\mathbf{v} \times \mathbf{d}) \qquad \mathbf{n} \cdot \mathbf{d} = 1$$

To summarize, *h is an involution if, and only if, it is the difference of a pair of complementary projections*:

$$h^2 = 1 \Leftrightarrow h = p - q, \quad p^2 = p, \quad q^2 = q, \quad p + q = 1$$

Indeed, $(p-q)^2 = p^2 + q^2 = p + q = 1$ because $p \circ q = q \circ p = 0$. Conversely, if h is an involution, we put $p = \frac{1}{2}(1 + h)$ and $q = \frac{1}{2}(1 - h)$. Then clearly, $p - q = h$. It is easy to see that this decomposition is unique:

$$h = p - q, \quad p + q = 1 \Rightarrow p = \tfrac{1}{2}(1 + h)$$

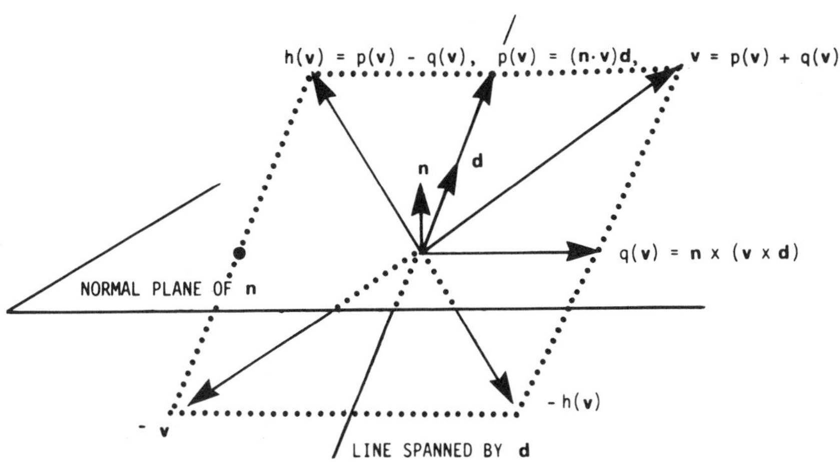

A reflection in the line spanned by the unit vector **u** is a special case of involutions in a line: here $\mathbf{n}=\mathbf{d}=\mathbf{u}$ and $A=a$, $B=b$, $C=c$ are just the direction cosines of the line. For this reason, the involution in a line along a plane is sometimes called a *skew reflection* in the line (or, a *skew rotation* around the line through 180°). Notice that trace $h=2(Aa+Bb+Cc)-3=2-3=-1$. If h is the involution in the line with direction vector **d** along the plane with normal vector **n**, then $-h$ is the involution in the plane with normal vector **n** along the line with direction vector **d**, and vice versa, and $\text{trace}(-h)=1$. We shall see in Chap. 10 that these two types of involutions, together with the identity operator 1 and the reflection in the origin -1, exhaust the set of involutions.

Examples

1. Write the matrix of the involution in the plane $x+2y+3z=0$ along the line $x/2=y=-z$.

Answer: $Aa+Bb+Cc=2+2-3=1$,

$$h=\begin{pmatrix}-3 & -8 & -12\\ -2 & -3 & -6\\ 2 & 4 & 7\end{pmatrix}$$

2. Write the matrix of the involution in the NE & SW bisector axis, along the normal plane of the UNE & LSW equiangular line.

Answer: NE & SW bisector axis: $x/1=y/1=z/0$, normal plane of the UNE & LSW equiangular line: $x+y+z=0$. Therefore, $Aa+Bb+Cc=2$;

$$p=\begin{pmatrix}\tfrac12 & \tfrac12 & \tfrac12\\ \tfrac12 & \tfrac12 & \tfrac12\\ 0 & 0 & 0\end{pmatrix}\qquad h=2p-1=\begin{pmatrix}0 & 1 & 1\\ 1 & 0 & 1\\ 0 & 0 & -1\end{pmatrix}$$

3. Dilatations

Let p be a principal projection, i.e., $p(\mathbf{v})=(\mathbf{n}\cdot\mathbf{v})\mathbf{d}$ $(\mathbf{n}\cdot\mathbf{d}=1)$. Then the linear operator

$$f=1+(\lambda-1)p \qquad \lambda\in\mathbb{R},\ \lambda\neq 0$$

is called a *dilatation* in ratio λ along the line with direction vector **d**, relative to the plane with normal vector **n**. f leaves unchanged every vector in the plane with normal vector **n** and stretches every vector parallel to the line of **d** in ratio λ (provided that $\lambda>1$; contracts in ratio λ if $0<\lambda<1$, and a reflection in the origin is also involved if $\lambda < 0$). In the special case of $\lambda = 1$, the dilatation becomes the identity operator; if $\lambda = 0$, f is the projection onto the normal plane of **n** along the line spanned by **d**; and if $\lambda = -1$, then f is the involution relative to the normal plane of **n** along the line spanned by **d**.

$$f = \begin{pmatrix} 1+(\lambda-1)Aa & (\lambda-1)Ba & (\lambda-1)Ca \\ (\lambda-1)Ab & 1+(\lambda-1)Bb & (\lambda-1)Cb \\ (\lambda-1)Ac & (\lambda-1)Bc & 1+(\lambda-1)Cc \end{pmatrix} \qquad Aa+Bb+Cc=1$$

is the matrix of the dilatation f along the line $x/a = y/b = z/c$ relative to the plane $Ax + By + Cz = 0$.

THE MATRIX PRODUCT RULE†

Let us now turn to the general case where f, g are two arbitrary linear operators and find the matrix of their product:

$$g \circ f = \begin{pmatrix} b_{11} & b_{12} & b_{12} \\ b_{21} & b_{22} & b_{23} \\ b_{31} & b_{32} & b_{33} \end{pmatrix} \circ \begin{pmatrix} a_{11} & a_{12} & a_{13} \\ a_{21} & a_{22} & a_{23} \\ a_{31} & a_{32} & a_{33} \end{pmatrix} = \begin{pmatrix} c_{11} & c_{12} & c_{13} \\ c_{21} & c_{22} & c_{23} \\ c_{31} & c_{32} & c_{33} \end{pmatrix}$$

Recall that the column vectors of the matrix of $g \circ f$ are just the vectors $g(f(\mathbf{i}))$, $g(f(\mathbf{j}))$, $g(f(\mathbf{k}))$, e.g.,

$$\begin{aligned} c_{11}\mathbf{i} + c_{21}\mathbf{j} + c_{31}\mathbf{k} &= g(f(\mathbf{i})) = g(a_{11}\mathbf{i} + a_{21}\mathbf{j} + a_{31}\mathbf{k}) \\ &= a_{11}g(\mathbf{i}) + a_{21}g(\mathbf{j}) + a_{31}g(\mathbf{k}) \\ &= a_{11}(b_{11}\mathbf{i} + b_{21}\mathbf{j} + b_{31}\mathbf{k}) \\ &\quad + a_{21}(b_{12}\mathbf{i} + b_{22}\mathbf{j} + b_{32}\mathbf{k}) \\ &\quad + a_{31}(b_{13}\mathbf{i} + b_{23}\mathbf{j} + b_{33}\mathbf{k}) = (b_{11}a_{11} + b_{12}a_{21} + b_{13}a_{31})\mathbf{i} \\ &\qquad + (b_{21}a_{11} + b_{22}a_{21} + b_{23}a_{31})\mathbf{j} \\ &\qquad + (b_{31}a_{11} + b_{32}a_{21} + b_{33}a_{31})\mathbf{k} \end{aligned}$$

so that we have $c_{11} = b_{11}a_{11} + b_{12}a_{21} + b_{13}a_{31}$, etc. In general,

$$c_{mn} = b_{m1}a_{1n} + b_{m2}a_{2n} + b_{m3}a_{3n} \qquad (m,\ n = 1,\ 2,\ 3)$$

Thus the entry c_{mn} in the mth row and the nth column of the product matrix is calculated as the dot product of the mth row of g with the nth column of f.

†*Note*: The matrix product rule is valid in terms of skew affine coordinates also. Accordingly, the basis in this section may be arbitrary.

$$\begin{pmatrix} \cdot & \cdot & \cdot \\ b_{m1} & b_{m2} & b_{m3} \\ \cdot & \cdot & \cdot \end{pmatrix} \circ \begin{pmatrix} \cdot & a_{1n} & \cdot \\ \cdot & a_{2n} & \cdot \\ \cdot & a_{3n} & \cdot \end{pmatrix} = \begin{pmatrix} \cdot & \cdot & \cdot \\ \cdot & c_{mn} & \cdot \\ \cdot & \cdot & \cdot \end{pmatrix}$$

Examples

1. $$\begin{pmatrix} -26 & -18 & -27 \\ 21 & 15 & 21 \\ 12 & 8 & 13 \end{pmatrix} \circ \begin{pmatrix} -26 & -18 & -27 \\ 21 & 15 & 21 \\ 12 & 8 & 13 \end{pmatrix} = \begin{pmatrix} -26 & -18 & -27 \\ 21 & 15 & 21 \\ 12 & 8 & 13 \end{pmatrix}$$

This operator is a projection.

2. $$\begin{pmatrix} 3 & 2 & -2 \\ 8 & 3 & -4 \\ 12 & 6 & -7 \end{pmatrix} \circ \begin{pmatrix} 3 & 2 & -2 \\ 8 & 3 & -4 \\ 12 & 6 & -7 \end{pmatrix} = \begin{pmatrix} 1 & 0 & 0 \\ 0 & 1 & 0 \\ 0 & 0 & 1 \end{pmatrix}$$

This is an involution.

3. $$\begin{pmatrix} 4 & -8 & 12 \\ -1 & 2 & -3 \\ -2 & 4 & -6 \end{pmatrix}^2 = \begin{pmatrix} 0 & 0 & 0 \\ 0 & 0 & 0 \\ 0 & 0 & 0 \end{pmatrix}$$

This operator is a lift.

4. $$\begin{pmatrix} -26 & -18 & -27 \\ 21 & 15 & 21 \\ 12 & 8 & 13 \end{pmatrix} \circ \begin{pmatrix} 27 & 18 & 27 \\ -21 & -14 & -21 \\ -12 & -8 & -12 \end{pmatrix} = \begin{pmatrix} 0 & 0 & 0 \\ 0 & 0 & 0 \\ 0 & 0 & 0 \end{pmatrix}$$

This is the product of a pair of complementary projections.

5. $$\begin{pmatrix} 0 & 1 & 0 \\ 0 & 0 & 0 \\ 0 & 0 & 0 \end{pmatrix} \circ \begin{pmatrix} 0 & 0 & 0 \\ 0 & 0 & 1 \\ 0 & 0 & 0 \end{pmatrix} = \begin{pmatrix} 0 & 0 & 1 \\ 0 & 0 & 0 \\ 0 & 0 & 0 \end{pmatrix}$$

6. $$\begin{pmatrix} 0 & 0 & 0 \\ 0 & 0 & 1 \\ 0 & 0 & 0 \end{pmatrix} \circ \begin{pmatrix} 0 & 1 & 0 \\ 0 & 0 & 0 \\ 0 & 0 & 0 \end{pmatrix} = \begin{pmatrix} 0 & 0 & 0 \\ 0 & 0 & 0 \\ 0 & 0 & 0 \end{pmatrix}$$

A left zero divisor of a linear operator is not necessarily a right zero divisor.

For linear operators of the plane, the product rule reads

$$\begin{pmatrix} b_{11} & b_{12} \\ b_{21} & b_{22} \end{pmatrix} \circ \begin{pmatrix} a_{11} & a_{12} \\ a_{21} & a_{22} \end{pmatrix} = \begin{pmatrix} b_{11}a_{11} + b_{12}a_{21} & b_{11}a_{12} + b_{12}a_{22} \\ b_{21}a_{11} + b_{22}a_{21} & b_{21}a_{12} + b_{22}a_{22} \end{pmatrix}$$

APPLICATIONS

4. Rotations of the Plane

Let r denote the rotation of the plane around the origin through ϕ. Then

$$r(\mathbf{i}) = \cos\phi\,\mathbf{i} + \sin\phi\,\mathbf{j} \qquad r(\mathbf{j}) = -\sin\phi\,\mathbf{i} + \cos\phi\,\mathbf{j}$$

and, by the column formulas,

$$r = \begin{pmatrix} \cos\phi & -\sin\phi \\ \sin\phi & \cos\phi \end{pmatrix}$$

Using the matrix product rule for 2×2 matrices, we can now prove the

Addition Formulas

$$\cos(\phi + \phi') = \cos\phi\cos\phi' - \sin\phi\sin\phi'$$
$$\sin(\phi + \phi') = \sin\phi\cos\phi' + \cos\phi\sin\phi'$$

Proof:

$$r' \circ r = \begin{pmatrix} \cos\phi' & -\sin\phi' \\ \sin\phi' & \cos\phi' \end{pmatrix} \circ \begin{pmatrix} \cos\phi & -\sin\phi \\ \sin\phi & \cos\phi \end{pmatrix}$$
$$= \begin{pmatrix} \cos\phi\cos\phi' - \sin\phi\sin\phi' & -\sin\phi\cos\phi' - \cos\phi\sin\phi' \\ \sin\phi\cos\phi' + \cos\phi\sin\phi' & \cos\phi\cos\phi' - \sin\phi\sin\phi' \end{pmatrix}$$
$$= \begin{pmatrix} \cos(\phi + \phi') & -\sin(\phi + \phi') \\ \sin(\phi + \phi') & \cos(\phi + \phi') \end{pmatrix}$$

because the product of the rotations through ϕ and ϕ' is the rotation through $\phi + \phi'$.

5. Decomposition of a Rotation into the Product of Two Reflections

Let h denote the reflection of the plane in the line passing through the origin and making an angle $\frac{1}{2}\phi$ with the x axis. Then

$$h(\mathbf{i}) = \cos\phi\,\mathbf{i} + \sin\phi\,\mathbf{j} \qquad h(\mathbf{j}) = \sin\phi\,\mathbf{i} - \cos\phi\,\mathbf{j}$$

and, by the column formulas,

$$h = \begin{pmatrix} \cos\phi & \sin\phi \\ \sin\phi & -\cos\phi \end{pmatrix}$$

Note that h is a symmetric matrix. Moreover, h is an involution: $h^2 = 1$. (By contrast, a rotation r is not a symmetric involution in general, as we have seen in Application 4 above. Exception: $\phi = 0°$ or $180°$.)

Using the rational parametric form of the circle (see Chap. 2, Application 12), the reflection h can also be written in the form

$$h = \begin{pmatrix} \dfrac{1-\omega^2}{1+\omega^2} & \dfrac{2\omega}{1+\omega^2} \\ \dfrac{2\omega}{1+\omega^2} & -\dfrac{1-\omega^2}{1+\omega^2} \end{pmatrix}$$

where $\omega = \tan \tfrac{1}{2}\phi$ is the slope of the axis of reflection.

Let us now consider the matrix product

$$\begin{pmatrix} \cos \tfrac{1}{2}\phi & \sin \tfrac{1}{2}\phi \\ \sin \tfrac{1}{2}\phi & -\cos \tfrac{1}{2}\phi \end{pmatrix} \circ \begin{pmatrix} \cos \tfrac{1}{2}\phi & -\sin \tfrac{1}{2}\phi \\ -\sin \tfrac{1}{2}\phi & -\cos \tfrac{1}{2}\phi \end{pmatrix} = \begin{pmatrix} \cos\phi & -\sin\phi \\ \sin\phi & \cos\phi \end{pmatrix}$$

The geometric interpretation of this product is as follows. The rotation through ϕ can be decomposed into the product of two reflections in lines making an angle $\tfrac{1}{2}\phi$ with each other. (This decomposition is certainly not unique: any pair of lines making an angle $\tfrac{1}{2}\phi$ will do.)

Complete the table:

f	Geometric description of f	f	Geometric description of f
$\begin{pmatrix} \cos\phi & -\sin\phi \\ \sin\phi & \cos\phi \end{pmatrix}$	Rotation through ϕ	$\begin{pmatrix} \cos\phi & \sin\phi \\ \sin\phi & -\cos\phi \end{pmatrix}$	Reflection in the line making an angle $\tfrac{1}{2}\phi$ with the x axis
$\begin{pmatrix} \sin\phi & -\cos\phi \\ \cos\phi & \sin\phi \end{pmatrix}$		$\begin{pmatrix} \sin\phi & \cos\phi \\ \cos\phi & -\sin\phi \end{pmatrix}$	
$\begin{pmatrix} -\sin\phi & -\cos\phi \\ \cos\phi & -\sin\phi \end{pmatrix}$		$\begin{pmatrix} -\sin\phi & \cos\phi \\ \cos\phi & \sin\phi \end{pmatrix}$	
$\begin{pmatrix} -\cos\phi & -\sin\phi \\ \sin\phi & -\cos\phi \end{pmatrix}$		$\begin{pmatrix} -\cos\phi & \sin\phi \\ \sin\phi & \cos\phi \end{pmatrix}$	
	Rotation through $180° + \phi$		Reflection in the line making an angle $90° + \tfrac{1}{2}\phi$ with the x axis

(continued)

	Rotation through $270° - \phi$		Reflection in the line making an angle $135° - ½\phi$ with the x axis
	Rotation through $270° + \phi$		Reflection in the line making an angle $135° + ½\phi$ with the x axis
	Rotation through $-\phi$		Reflection in the ine making an angle $-½\phi$ with the x axis

6. Invections

An *invection* j is defined as a linear operator satisfying $j^4 = 1$. Involutions are special invections, but there exist noninvolutory invections as well, the most important of which are

$$i = \begin{pmatrix} 0 & -1 \\ 1 & 0 \end{pmatrix} \qquad -i = \begin{pmatrix} 0 & 1 \\ -1 & 0 \end{pmatrix}$$

the rotations through 90° and –90°, which are *antisymmetric*.

There are infinitely many other noninvolutory invections which fail to be antisymmetric, e.g.,

$$j = \begin{pmatrix} -1 & 2 \\ -1 & 1 \end{pmatrix}$$

also satisfying $j^4 = 1$. In Chap. 10 we shall see a method that can be used to find all the invections.

The matrix products

$$\begin{pmatrix} -1 & 0 \\ 0 & 1 \end{pmatrix} \circ \begin{pmatrix} 0 & 1 \\ 1 & 0 \end{pmatrix} = \begin{pmatrix} 0 & -1 \\ 1 & 0 \end{pmatrix}$$

and

$$\begin{pmatrix} 0 & 1 \\ 1 & 0 \end{pmatrix} \circ \begin{pmatrix} -1 & 0 \\ 0 & 1 \end{pmatrix} = \begin{pmatrix} 0 & 1 \\ -1 & 0 \end{pmatrix}$$

can be interpreted geometrically by saying that the products of the reflections in the NW & SE bisector axis and in the y axis yield the rotations through $\pm 90°$ (depending on the order in which the product is taken). Later we shall see that every invection can be decomposed into the product of two involutions.

We have seen (Application 2) that

$$h \text{ involution} \Rightarrow p = \tfrac{1}{2}(1+h) \text{ projection}$$

Let us now prove that

$$j \text{ invection} \Rightarrow p = \tfrac{1}{4}(1+j+j^2+j^3) \text{ projection}$$

Indeed,

$$\begin{aligned} p &= \frac{1}{16}(1+j^2+j^4+j^6+2j+2j^2+4j^3+2j^4+2j^5) \\ &= \frac{1}{16}(4+4j+4j^2+4j^3) = p \end{aligned}$$

because $j^5 = j$, on the strength of $j^4 = 1$. These are special cases of the ergodic formula (see Application 10 below).

7. Lifts

A *lift* ℓ is a linear operator satisfying $\ell^2 = 0$, $\ell \neq 0$. Earlier (Chap. 6, Application 7) a lift was defined in terms of the dot product; later (Chap. 10, Application 1) we shall see that the two definitions are, in fact, equivalent.

Consider the matrix products

$$\begin{pmatrix} 0 & 1 \\ 0 & 0 \end{pmatrix} \circ \begin{pmatrix} 0 & 0 \\ 1 & 0 \end{pmatrix} = \begin{pmatrix} 1 & 0 \\ 0 & 0 \end{pmatrix} \qquad \begin{pmatrix} 0 & 0 \\ 1 & 0 \end{pmatrix} \circ \begin{pmatrix} 0 & 1 \\ 0 & 0 \end{pmatrix} = \begin{pmatrix} 0 & 0 \\ 0 & 1 \end{pmatrix}$$

They can be interpreted geometrically by saying that the principal projections

$$p = \begin{pmatrix} 1 & 0 \\ 0 & 0 \end{pmatrix} \qquad q = \begin{pmatrix} 0 & 0 \\ 0 & 1 \end{pmatrix}$$

which are mutual zero divisors: $p \circ q = 0 = q \circ p$, can be decomposed into the product of a pair of lifts

$$\ell = \begin{pmatrix} 0 & 1 \\ 0 & 0 \end{pmatrix} \qquad \ell' = \begin{pmatrix} 0 & 0 \\ 1 & 0 \end{pmatrix}$$

i.e., $\ell^2 = 0 = \ell'^2$, $(\ell \circ \ell')^2 = \ell \circ \ell'$, $(\ell' \circ \ell)^2 = \ell' \circ \ell$. Later (Chap. 9, Application 3) we shall see that principal projections that are mutual zero divisors can always be decomposed into the product of lifts.

8. Twists

A linear operator t satisfying $t^3 = -t$ is called a *twist*. In Chap. 4 we have seen that, for every unit vector **u**, the formula $t(\mathbf{v}) = \mathbf{u} \times \mathbf{v}$ defines a twist and obviously $-t$ is also a twist; in fact, they are both *antisymmetric* twists. Thus there are infinitely many antisymmetric twists. There is an even greater abundance of twists that fail to be antisymmetric. As we shall see (Chap. 9, Application 4) for every pair of lists ℓ, ℓ' such that $\ell' \circ \ell$, $\ell \circ \ell'$ are projections, $t = \ell' - \ell$ is a twist. In fact, every twist can be decomposed in this way (Chap. 10, Application 3).

Example

$$\ell = \begin{pmatrix} -1 & 2 & -1 \\ -1 & 2 & -1 \\ -1 & 2 & -1 \end{pmatrix} \qquad \ell' = \begin{pmatrix} 0 & -2 & 4 \\ 0 & -2 & 4 \\ 0 & -1 & 2 \end{pmatrix}$$

satisfy $\ell^2 = 0 = \ell'^2$;

$$t = \ell' - \ell = \begin{pmatrix} 1 & -4 & 5 \\ 1 & -4 & 5 \\ 1 & -3 & 3 \end{pmatrix}$$

satisfies $t^3 = -t$; and

$$p = \ell \circ \ell' = \begin{pmatrix} 0 & -1 & 2 \\ 0 & -1 & 2 \\ 0 & -1 & 2 \end{pmatrix}$$

satisfies $p^2 = p$.

We leave it to the reader to establish the following:

$$t \text{ twist} \Rightarrow j = 1 + t + t^2 \text{ invection and } j^{-1} = 1 - t + t^2$$

$$j \text{ invection} \Rightarrow t = \tfrac{1}{2}(j - j^3) = \tfrac{1}{2}(j - j^{-1}) \text{ twist}$$

9. Product of Projections

The product of projections is not necessarily a projection. A sufficient condition is that they commute:

$$p^2 = p,\ q^2 = q,\ p \circ q = q \circ p \Rightarrow (p \circ q)^2 = p \circ q$$

If this is the case and p, q are projections onto planes, then $p \circ q$ is a projection onto the line of intersection of those planes.

The sum of two projections is not necessarily a projection. A necessary and sufficient condition is that they be mutual zero divisors:

$$p^2 = p, \quad q^2 = q, \quad p \circ q = 0 = q \circ p \Leftrightarrow (p + q)^2 = p + q$$

The condition is clearly sufficient. To show that it is also necessary we write

$$(p+q)^2 = p+q \Rightarrow p\circ q + q\circ p = 0$$
$$\Rightarrow q\circ p\circ q + q^2\circ p = 0 = p\circ q^2 + q\circ p\circ q$$
$$\Rightarrow q^2\circ p = p\circ q^2 \Rightarrow q\circ p = p\circ q$$

But then $p\circ q + q\circ p = 0 \Rightarrow 2p\circ q = 0 \Rightarrow p\circ q = 0$.

In what follows p and q will be principal projections along the same plane. We want to show that then $p\circ q = p$ and $q\circ p = q$. Indeed, let $p(\mathbf{v}) = (\mathbf{n}\cdot\mathbf{v})\mathbf{d}$, $q(\mathbf{v}) = (\mathbf{n}\cdot\mathbf{v})\mathbf{e}$ with $\mathbf{n}\cdot\mathbf{d} = 1 = \mathbf{n}\cdot\mathbf{e}$. Then

$$(p\circ q)(\mathbf{v}) = p(q(\mathbf{v})) = (\mathbf{n}\cdot q(\mathbf{v}))\mathbf{d} = (\mathbf{n}\cdot(\mathbf{n}\cdot\mathbf{v})\mathbf{e})\mathbf{d} = (\mathbf{n}\cdot\mathbf{v})\mathbf{d} = p(\mathbf{v})$$

for all $\mathbf{v}$, proving that $p\circ q = p$. The proof of $q\circ p = q$ is similar.

With p, q as above, we now wish to calculate $(p-q)^2$.

$$(p-q)^2 = p^2 + q^2 - p\circ q - q\circ p = p + q - p - q = 0$$

i.e., $\ell = p - q$ is a lift: $\ell^2 = 0$. The converse is also true. Let $\ell(\mathbf{v}) = (\mathbf{n}\cdot\mathbf{v})\mathbf{e}'$ with $\mathbf{n}\cdot\mathbf{e}' = 0$. We may assume that $\ell \neq 0$, then for some $\mathbf{e}''$, $\mathbf{n}\cdot\mathbf{e}'' \neq 0$. Then set $\mathbf{d} = \mathbf{e}''/\mathbf{n}\cdot\mathbf{e}''$ and $\mathbf{e} = \mathbf{e}' + \mathbf{d}$. It follows that $\mathbf{n}\cdot\mathbf{d} = 1 = \mathbf{n}\cdot\mathbf{e}$ and therefore $p(\mathbf{v}) = (\mathbf{n}\cdot\mathbf{v})\mathbf{d}$, $q(\mathbf{v}) = (\mathbf{n}\cdot\mathbf{v})\mathbf{e}$ are principal projections along the same plane; moreover, $\ell = p - q$ because

$$p(\mathbf{v}) - q(\mathbf{v}) = (\mathbf{n}\cdot\mathbf{v})\mathbf{d} - (\mathbf{n}\cdot\mathbf{v})\mathbf{e} = (\mathbf{n}\cdot\mathbf{v})(\mathbf{d}-\mathbf{e}) = (\mathbf{n}\cdot\mathbf{v})\mathbf{e}' = \ell(\mathbf{v}) \qquad \text{for all } \mathbf{v}$$

To summarize: *ℓ is a lift if, and only if, ℓ is the difference of two principal projections along the same plane,*

$$\ell^2 = 0 \Leftrightarrow \ell = p - q, \quad p^2 = p = p\circ q, \quad q^2 = q = q\circ p$$

This representation of ℓ is, of course, not unique.

10. The Ergodic Formula

We have seen that

$$h^2 = 1 \Rightarrow p = \tfrac{1}{2}(1+h)$$

and

$$j^4 = 1 \Rightarrow p = \tfrac{1}{4}(1 + j + j^2 + j^3)$$

where p is a projection. These are special cases of a more general result, the ergodic formula, which we shall state for rotations: *the arithmetic mean of the first n powers of a rotation through the angle 360°/n is equal to the perpendicular projection onto the axis of the rotation*:

$$p = \frac{1 + r + r^2 + \cdots + r^{n-1}}{n} \qquad r^n = 1, \quad p^2 = p$$

The proof of this fascinating formula is left as an exercise. The hint that $p(\mathbf{v})$ is the centroid of the vertices $\mathbf{v}, r(\mathbf{v}), r^2(\mathbf{v}), \ldots, r^{n-1}(\mathbf{v})$ of a regular polygon of n vertices should make this exercise a particularly easy one. In Chap. 12 we shall prove a generalization of this result, called the ergodic theorem, from which the restriction $r^n = 1$ has been removed.

Example Let r denote the rotation through 120° around the UNE & LSW equiangular line. Then $r^3 = 1$, and

$$p = \frac{1 + r + r^2}{3} = \frac{1}{3}\left\{\begin{pmatrix}1 & 0 & 0\\ 0 & 1 & 0\\ 0 & 0 & 1\end{pmatrix} + \begin{pmatrix}0 & 0 & 1\\ 1 & 0 & 0\\ 0 & 1 & 0\end{pmatrix} + \begin{pmatrix}0 & 1 & 0\\ 0 & 0 & 1\\ 1 & 0 & 0\end{pmatrix}\right\}$$

$$= \begin{pmatrix}1/3 & 1/3 & 1/3\\ 1/3 & 1/3 & 1/3\\ 1/3 & 1/3 & 1/3\end{pmatrix}$$

is the perpendicular projection onto the UNE & LSW equiangular line.

EXERCISES

7.1 Show by matrix multiplication that each of the following is a projection.

(i) $\begin{pmatrix}4 & 8 & -12\\ -3 & -6 & 9\\ -1 & -2 & 3\end{pmatrix}$ (ii) $\begin{pmatrix}-1 & -1 & -1\\ -1 & -1 & -1\\ 3 & 3 & 3\end{pmatrix}$

(iii) $\begin{pmatrix}-3 & -8 & 12\\ 3 & 7 & -9\\ 1 & 2 & -2\end{pmatrix}$ (iv) $\begin{pmatrix}14 & 2 & -8\\ 21 & 3 & -12\\ 28 & 4 & -16\end{pmatrix}$

(v) $\begin{pmatrix}-13 & -2 & 8\\ -21 & -2 & 12\\ -28 & -4 & 17\end{pmatrix}$ (vi) $\begin{pmatrix}-8 & -12 & -21\\ -15 & -19 & -35\\ 12 & 16 & 29\end{pmatrix}$

(vii) $\begin{pmatrix}-26 & -18 & -27\\ 21 & 15 & 21\\ 12 & 8 & 13\end{pmatrix}$ (viii) $\begin{pmatrix}-24 & 48 & -56\\ 9 & -18 & 21\\ 18 & -36 & 42\end{pmatrix}$

(ix) $\begin{pmatrix}32 & -16 & 20\\ -8 & 4 & -5\\ -56 & 28 & -35\end{pmatrix}$

7.2 Show by matrix multiplication that each of the following is an involution.

(i) $\begin{pmatrix} -1 & -2 & -2 \\ -2 & -1 & -2 \\ 2 & 2 & 3 \end{pmatrix}$ (ii) $\begin{pmatrix} 3 & 2 & 2 \\ 2 & 3 & 1 \\ -6 & -6 & -4 \end{pmatrix}$

(iii) $\begin{pmatrix} -27 & -4 & 16 \\ -42 & -5 & 24 \\ -56 & -8 & 33 \end{pmatrix}$ (iv) $\begin{pmatrix} 5 & -6 & 0 \\ 4 & -5 & 0 \\ 2 & -2 & -1 \end{pmatrix}$

(v) $\begin{pmatrix} -7 & -2 & -24 \\ 6 & 13 & -18 \\ 2 & 4 & -5 \end{pmatrix}$ (vi) $\begin{pmatrix} -53 & -36 & -54 \\ 42 & 29 & 42 \\ 24 & 16 & 25 \end{pmatrix}$

(vii) $\begin{pmatrix} 2/3 & 1/3 & 2/3 \\ 1/3 & 2/3 & -2/3 \\ 2/3 & -2/3 & -1/3 \end{pmatrix}$

7.3 Show by matrix multiplication that each of the following is a lift.

(i) $\begin{pmatrix} 1 & 1 & 1 \\ -1 & -1 & -1 \\ 0 & 0 & 0 \end{pmatrix}$ (ii) $\begin{pmatrix} -4 & 8 & -12 \\ 1 & -2 & 3 \\ 2 & -4 & 6 \end{pmatrix}$

(iii) $\begin{pmatrix} 1 & -2 & 1 \\ 1 & -2 & 1 \\ 1 & -2 & 1 \end{pmatrix}$ (iv) $\begin{pmatrix} 0 & 2 & -4 \\ 0 & 2 & -4 \\ 0 & 1 & -2 \end{pmatrix}$

7.4 Show by matrix multiplication that each of the following is a twist.

(i) $\begin{pmatrix} 1 & -2 & 1 \\ 1 & -1 & 0 \\ 0 & 0 & 0 \end{pmatrix}$ (ii) $\begin{pmatrix} -1 & 4 & -5 \\ -1 & 4 & -5 \\ -1 & 3 & -3 \end{pmatrix}$

(iii) $\begin{pmatrix} -15 & 12 & 21 \\ -8 & 6 & 10 \\ -6 & 5 & 9 \end{pmatrix}$

7.5 Show by matrix multiplication that each of the following is an invection.

(i) $\begin{pmatrix} -4 & 7 & -5 \\ -3 & 6 & -5 \\ -2 & 4 & -3 \end{pmatrix}$ (ii) $\begin{pmatrix} -1 & 2 & -2 \\ -1 & 1 & -1 \\ 0 & 0 & -1 \end{pmatrix}$

(iii) $\begin{pmatrix} 11 & -9 & -15 \\ -4 & 3 & 8 \\ 10 & -8 & -15 \end{pmatrix}$

7.6 Prove that

$$t \text{ twist} \Rightarrow 1+t+t^2;\quad 1-t+t^2 \text{ invections}$$

in fact, if $j=1+t+t^2$, then $j^{-1}=1-t+t^2=j^3$.

7.7 Prove that

$$t \text{ twist} \Rightarrow 1+t^2 \text{ projection}$$

7.8 Prove that

$$j \text{ invection} \Rightarrow \tfrac{1}{2}(j-j^{-1}) \text{ twist}$$

7.9 Show that

$$p=\begin{pmatrix} Aa & Ba \\ Ab & Bb \end{pmatrix}$$

is a projection of $\mathbb{R}^2$, provided that $Aa+Bb=1$. Describe p geometrically.

7.10 Describe

$$q=\begin{pmatrix} 1-Aa & -Ba \\ -Ab & 1-Bb \end{pmatrix}$$

geometrically, if $Aa+Bb=1$.

7.11 Show that a lift of $\mathbb{R}^2$ can be written as a difference of projections along the same line.

7.12 Show that

$$j \text{ invection} \Rightarrow p=\tfrac{1}{4}(1+j+j^2+j^3) \text{ projection}$$

7.13 In each of the following, calculate the Lie product

$$[f,g]=f\circ g-g\circ f$$

(i) $f=\begin{pmatrix} 1 & 2 & 1 \\ 2 & 1 & 2 \\ 1 & 2 & 3 \end{pmatrix}$ $\quad g=\begin{pmatrix} 4 & 1 & 1 \\ -4 & 2 & 0 \\ 1 & 2 & 1 \end{pmatrix}$

(ii) $f=\begin{pmatrix} 2 & 1 & 0 \\ 1 & 1 & 2 \\ -1 & 2 & 1 \end{pmatrix}$ $\quad g=\begin{pmatrix} 3 & 1 & -2 \\ 3 & -2 & 4 \\ -3 & 5 & -1 \end{pmatrix}$

7.14 Calculate the nth power of each of the following operators.

(i) $\begin{pmatrix} \cos\phi & -\sin\phi \\ \sin\phi & \cos\phi \end{pmatrix}$ $\quad$ (ii) $\begin{pmatrix} 1 & 1 \\ 0 & 1 \end{pmatrix}$

(iii) $\begin{pmatrix} 0 & 1 & 0 \\ 0 & 0 & 1 \\ 0 & 0 & 0 \end{pmatrix}$ $\quad$ (iv) $\begin{pmatrix} 1 & 0 & 1 \\ 0 & 1 & 0 \\ 0 & 0 & 1 \end{pmatrix}$

(v) $\begin{pmatrix} -1 & -2 & -2 \\ -2 & -1 & -2 \\ 2 & 2 & 3 \end{pmatrix}$ (vi) $\begin{pmatrix} 1 & -2 & 1 \\ 1 & -1 & 0 \\ 0 & 0 & 0 \end{pmatrix}$

(vii) $\begin{pmatrix} -1 & 2 & -2 \\ -1 & 1 & -1 \\ 0 & 0 & -1 \end{pmatrix}$ (viii) $\begin{pmatrix} 1/2 & 1/2 & 1/2 \\ 1/2 & 1/2 & 1/2 \\ 0 & 0 & 0 \end{pmatrix}$

(ix) $\begin{pmatrix} -1 & 2 & -1 \\ -1 & 2 & -1 \\ -1 & 2 & -1 \end{pmatrix}$ (x) $\begin{pmatrix} \lambda & 0 & 0 \\ 0 & 1 & 0 \\ 0 & 0 & 1 \end{pmatrix}$

(xi) $\begin{pmatrix} \lambda & 0 & 0 \\ 0 & 0 & 0 \\ 0 & 0 & 0 \end{pmatrix}$ (xii) $\begin{pmatrix} \kappa & 0 & 0 \\ 0 & \lambda & 0 \\ 0 & 0 & \mu \end{pmatrix}$

7.15 Write the lifts

$$\begin{pmatrix} 0 & 0 & 1 \\ 0 & 0 & 0 \\ 0 & 0 & 0 \end{pmatrix}, \quad \begin{pmatrix} 4 & -8 & 12 \\ -1 & 2 & -3 \\ -2 & 4 & -6 \end{pmatrix}$$

as a difference of two projections.

8

Invertible Linear Operators†

NECESSARY AND SUFFICIENT CONDITION FOR INVERTIBILITY

Given a linear operator f, we shall say that it is *invertible* if there exists another linear operator f^{-1}, called the *inverse* of f, such that

$$f^{-1} \circ f = 1 \qquad \text{and} \qquad f \circ f^{-1} = 1$$

If this is the case, the matrix of f is also called invertible, and the matrix of f^{-1}, in terms of the same coordinate system, is called the *inverse matrix*.

Not every linear operator is invertible: projections other than the identity, lifts, and twists of $\mathbb{R}^3$ are examples of noninvertible linear operators. Obviously, a necessary condition for a linear operator f to be invertible is that it be *bijective*. It is true, although not so obvious, that this condition is also sufficient. This is to say that if a linear operator f is bijective, then its set-theoretic inverse f^{-1} is also linear; i.e., it preserves the sum and scalar multiple of vectors. Indeed,

$$\begin{aligned} f^{-1}(\mathbf{v} + \mathbf{v}') &= f^{-1}(1(\mathbf{v}) + 1(\mathbf{v}')) = f^{-1}((f \circ f^{-1})(\mathbf{v}) + (f \circ f^{-1})(\mathbf{v}')) \\ &= f^{-1}(f(f^{-1}(\mathbf{v})) + f(f^{-1}(\mathbf{v}'))) \\ &= f^{-1}(f(f^{-1}(\mathbf{v}) + f^{-1}(\mathbf{v}'))) = (f^{-1} \circ f)(f^{-1}(\mathbf{v}) + f^{-1}(\mathbf{v}')) \\ &= 1(f^{-1}(\mathbf{v}) + f^{-1}(\mathbf{v}')) = f^{-1}(\mathbf{v}) + f^{-1}(\mathbf{v}') \\ f^{-1}(\lambda \mathbf{v}) &= f^{-1}(\lambda 1(\mathbf{v})) = f^{-1}(\lambda (f \circ f^{-1})(\mathbf{v})) = f^{-1}(\lambda f(f^{-1}(\mathbf{v})) \end{aligned}$$

†*Note*: In this chapter all bases are orthonormal and all coordinate systems are rectangular metric, except where otherwise stipulated.

$$= f^{-1}(f(\lambda f^{-1}(\mathbf{v}))) = (f^{-1} \circ f)(\lambda f^{-1}(\mathbf{v})) = 1(\lambda f^{-1}(\mathbf{v}))$$
$$= \lambda f^{-1}(\mathbf{v}) \qquad \lambda \in \mathbb{R}$$

Further necessary and sufficient conditions for invertibility are given in Chaps. 9 and 10.

Examples

1. A strain in ratio $\lambda \neq 0$ is invertible; namely, its inverse is a strain in ratio $1/\lambda$:

$$\begin{pmatrix} \lambda & 0 & 0 \\ 0 & \lambda & 0 \\ 0 & 0 & \lambda \end{pmatrix}^{-1} = \begin{pmatrix} 1/\lambda & 0 & 0 \\ 0 & 1/\lambda & 0 \\ 0 & 0 & 1/\lambda \end{pmatrix}$$

2. A reflection (in a plane, line, or in the origin) is invertible; namely, it is its own inverse. More generally, any involution is invertible and is its own inverse:

$$h^2 = 1 \Leftrightarrow h^{-1} = h$$

For example, the reflection h in the line spanned by the unit vector

$$\mathbf{u} = \frac{1}{\sqrt{14}}\mathbf{i} + \frac{2}{\sqrt{14}}\mathbf{j} + \frac{3}{\sqrt{14}}\mathbf{k}$$

satisfies

$$h^{-1} = \begin{pmatrix} -6/7 & 2/7 & 3/7 \\ 2/7 & -3/7 & 6/7 \\ 3/7 & 6/7 & 2/7 \end{pmatrix}^{-1} = \begin{pmatrix} -6/7 & 2/7 & 3/7 \\ 2/7 & -3/7 & 6/7 \\ 3/7 & 6/7 & 2/7 \end{pmatrix} = h$$

3. If we observe that

$$\begin{pmatrix} 1 & 2 & 2 \\ 0 & -1 & 0 \\ 0 & 0 & -1 \end{pmatrix}^2 = \begin{pmatrix} 1 & 0 & 0 \\ 0 & 1 & 0 \\ 0 & 0 & 1 \end{pmatrix}$$

then we may conclude that the matrix represents an involution, and therefore

$$\begin{pmatrix} 1 & 2 & 2 \\ 0 & -1 & 0 \\ 0 & 0 & -1 \end{pmatrix}^{-1} = \begin{pmatrix} 1 & 2 & 2 \\ 0 & -1 & 0 \\ 0 & 0 & -1 \end{pmatrix}$$

4. If we notice that

$$\begin{pmatrix} 1 & -2 & -2 \\ 1 & -1 & -1 \\ 0 & 0 & 1 \end{pmatrix}^4 = \begin{pmatrix} 1 & 0 & 0 \\ 0 & 1 & 0 \\ 0 & 0 & 1 \end{pmatrix}$$

then we may conclude that the matrix represents an invection, and therefore

$$\begin{pmatrix} 1 & -2 & -2 \\ 1 & -1 & -1 \\ 0 & 0 & 1 \end{pmatrix}^{-1} = \begin{pmatrix} 1 & -2 & -2 \\ 1 & -1 & -1 \\ 0 & 0 & 1 \end{pmatrix}^{3} = \begin{pmatrix} -1 & 2 & 0 \\ -1 & 1 & -1 \\ 0 & 0 & 1 \end{pmatrix}$$

FORMAL RULES OF THE INVERSE

$$f \text{ invertible} \Leftrightarrow \det f \neq 0$$

The reason for this is that the system of linear equations obtained as the scalar version of $f(\mathbf{v}) = \mathbf{w}$ has a unique solution for $\mathbf{v}$; namely, $\mathbf{v} = f^{-1}(\mathbf{w})$ if, and only if, $\det f \neq 0$. If the linear operators f and g are both invertible, then so is their product $g \circ f$; namely:

Product Rule

$$(g \circ f)^{-1} = f^{-1} \circ g^{-1}$$

Notice that the inverse of the product is the product of the inverses taken in the *opposite order*. The inverse f^{-1} of an invertible operator f is also invertible; namely,

$$(f^{-1})^{-1} = f$$

The identity operator 1 is invertible; namely,

$$1^{-1} = 1$$

Finally, for the determinant of the inverse operator we have (see Chap. 13)

$$\det f^{-1} = (\det f)^{-1}$$

The rules governing the inverse of the product and the inverse follow from the set-theoretical properties of the inverse of a bijective function.

THE GENERAL LINEAR GROUP

Let us denote the set of invertible linear operators by the symbol $GL(\mathbb{R}^3)$. It clearly satisfies the following properties:

(0°) Closure

$$f, g \in GL(\mathbb{R}^3) \Rightarrow g \circ f \in GL(\mathbb{R}^3)$$

(2°) Associative Law

$$(h \circ g) \circ f = h \circ (g \circ f)$$

(3°) Neutrality of the Identity

$$1 \in GL(\mathbb{R}^3) \quad \text{and} \quad 1 \circ f = f = f \circ 1$$

(4°) Inverse Operator

$$f \in GL(\mathbb{R}^3) \Rightarrow f^{-1} \in GL(\mathbb{R}^3) \qquad \text{and} \qquad f^{-1} \circ f = 1 = f \circ f^{-1}$$

A set in which an operation $\circ$ is defined subject to these four rules is called a *group*. Note the absence of the commutative law, which, however, may be present in some special cases; then we talk about a *commutative group*. $GL(\mathbb{R}^3)$ is called the *general linear group* of the vector space $\mathbb{R}^3$.

It is convenient to collect the foregoing rules, called the *group axioms*, in the following table, for ready reference:

0°	CLOSURE
[redacted]	
2°	ASSOCIATIVITY
3°	NEUTRAL
4°	INVERSE

APPLICATIONS

1. Shears

A shear $f = 1 + \ell$ where ℓ is a lift (i.e., $\ell^2 = 0$) is invertible, namely, $f^{-1} = 1 - \ell$. Indeed, $(1 + \ell) \circ (1 - \ell) = 1 - \ell^2 = 1$. The positive and negative powers of the shear f can be calculated by the formula

$$(1 + \ell)^n = 1 + n\ell$$

Suppose now that f and g are shears parallel to the same plane and along the same line, i.e., $f = 1 + \alpha\ell$, $g = 1 + \beta\ell$, $\alpha, \beta \in \mathbb{R}$. Then the product $g \circ f$ is also a shear parallel to the same plane and along the same line; namely:

$$(1 + \alpha\ell) \circ (1 + \beta\ell) = 1 + (\alpha + \beta)\ell$$

We see that *the shears parallel to the same plane along the same line form a commutative group* (which is the same as the additive group of $\mathbb{R}$).

2. Dilatations

A dilatation $f = 1 + (\lambda - 1)p$, where $\lambda \neq 0$ and p is a principal projection, is invertible; namely, $f^{-1} = 1 + (\lambda^{-1} - 1)p$. If g is another dilatation relative to the same plane and along the same line, i.e., $g = 1 + (\mu - 1)p$, then the product $g \circ f$ is also a dilatation of the same type:

$$(1 + (\lambda - 1)p) \circ (1 + (\mu - 1)p) = 1 + (\lambda\mu - 1)p$$

We see that *the dilatations relative to the same plane along the same line form a commutative group* (which is the same as the multiplicative group of $\mathbb{R}$).

3. Rotations

Every rotation of the plane can be written as

$$r = \begin{pmatrix} \cos\phi & -\sin\phi \\ \sin\phi & \cos\phi \end{pmatrix}$$

where ϕ is the angle of rotation. Matrix multiplication yields

$$\begin{pmatrix} \cos\phi & -\sin\phi \\ \sin\phi & \cos\phi \end{pmatrix} \circ \begin{pmatrix} \cos\phi & \sin\phi \\ -\sin\phi & \cos\phi \end{pmatrix} = \begin{pmatrix} 1 & 0 \\ 0 & 1 \end{pmatrix}$$

We conclude that

$$r^{-1} = \begin{pmatrix} \cos\phi & \sin\phi \\ -\sin\phi & \cos\phi \end{pmatrix}$$

If we also take into account that the inverse of the rotation through ϕ is the rotation through $-\phi$, i.e.,

$$r^{-1} = \begin{pmatrix} \cos(-\phi) & -\sin(-\phi) \\ \sin(-\phi) & \cos(-\phi) \end{pmatrix}$$

then we also get, incidentally, the trigonometric formulas

$$\cos(-\phi) = \cos\phi$$
$$\sin(-\phi) = -\sin\phi$$

Note that the matrix of r^{-1} can be obtained by "transposition," i.e., by reflecting the entries of the matrix of r in the main diagonal. In Chap. 11 we shall see the reason for this rule, which is also applicable to rotations of the space.

Our results, together with those of Chap. 7, Application 4, imply that *the rotations of the plane form a commutative group*. This group is the same as the additive group $\mathbb{R}$ modulo 360° (because a rotation through 360° is the same as the rotation through 0°, i.e., the identity operator 1). More on rotations can be found in Chap. 12.

THE INVERSE MATRIX FORMULA[†]

Recall the important result:

$$f^{-1} \text{ exists} \Leftrightarrow \det f \neq 0$$

We shall write $D = \det f$ for the determinant of the linear operator

$$f = \begin{pmatrix} a_{11} & a_{12} & a_{13} \\ a_{21} & a_{22} & a_{23} \\ a_{31} & a_{32} & a_{33} \end{pmatrix}$$

Let us further write $A_{11}, A_{12}, \ldots, A_{33}$ for the cofactors of the entries $a_{11}, a_{12}, \ldots, a_{33}$; i.e.,

$$A_{11} = \begin{vmatrix} a_{22} & a_{23} \\ a_{32} & a_{33} \end{vmatrix}, A_{12} = -\begin{vmatrix} a_{21} & a_{23} \\ a_{31} & a_{33} \end{vmatrix}, \ldots, A_{33} = \begin{vmatrix} a_{11} & a_{12} \\ a_{21} & a_{22} \end{vmatrix}$$

where the signatures are assigned according to the Chessboard Rule: $\operatorname{sgn} A_{mn} = (-1)^{m+n}$.

The Inverse Matrix Formula

$$\begin{pmatrix} a_{11} & a_{12} & a_{13} \\ a_{21} & a_{22} & a_{23} \\ a_{31} & a_{32} & a_{33} \end{pmatrix}^{-1} = \begin{pmatrix} \frac{A_{11}}{D} & \frac{A_{21}}{D} & \frac{A_{31}}{D} \\ \frac{A_{12}}{D} & \frac{A_{22}}{D} & \frac{A_{32}}{D} \\ \frac{A_{13}}{D} & \frac{A_{23}}{D} & \frac{A_{33}}{D} \end{pmatrix}$$

Accordingly, the calculation of the inverse matrix f^{-1} of the matrix f involves 4 steps:

1. Calculate the determinant $D = \det f$. If $D = 0$, stop and conclude that there is no inverse f^{-1}.
2. Replace each entry of the matrix f by its cofactor. Do not forget to attach the appropriate signature, in accordance with the chessboard rule!
3. Transpose the matrix (i.e., interchange its rows and columns).
4. Divide each entry of the matrix by D.

[†] *Note*: The inverse matrix formula is valid in terms of skew affine coordinates as well. Accordingly, the basis in this section may be arbitrary.

In order to emphasize the importance of the inverse matrix formula, we shall give two different proofs.

First Proof: This one is based on Cramer's rule. We wish to solve the equation $f(\mathbf{v}) = \mathbf{v}'$, where $\mathbf{v}$ is the unknown and $\mathbf{v}'$ is given. This means solving the regular system of linear equations

$$\begin{aligned} a_{11}x + a_{12}y + a_{13}z &= x' \\ a_{21}x + a_{22}y + a_{23}z &= y' \qquad D \neq 0 \\ a_{31}x + a_{32}y + a_{33}z &= z' \end{aligned}$$

The unique solution is

$$x = \frac{\begin{vmatrix} x' & a_{12} & a_{13} \\ y' & a_{22} & a_{23} \\ z' & a_{32} & a_{32} \end{vmatrix}}{\mathrm{D}} = \frac{A_{11}}{D}x' + \frac{A_{21}}{D}y' + \frac{A_{31}}{D}z'$$

$$y = \frac{\begin{vmatrix} a_{11} & x' & a_{13} \\ a_{21} & y' & a_{23} \\ a_{31} & z' & a_{33} \end{vmatrix}}{\mathrm{D}} = \frac{A_{12}}{D}x' + \frac{A_{22}}{D}y' + \frac{A_{32}}{D}z'$$

$$z = \frac{\begin{vmatrix} a_{11} & a_{12} & x' \\ a_{21} & a_{22} & y' \\ a_{31} & a_{32} & z' \end{vmatrix}}{\mathrm{D}} = \frac{A_{13}}{D}x' + \frac{A_{23}}{D}y' + \frac{A_{33}}{D}z'$$

Since the solution to the vector equation $f(\mathbf{v}) = \mathbf{v}'$ is the vector $\mathbf{v} = f^{-1}(\mathbf{v}')$, the system above is the scalar version of the latter, yielding the desired matrix f^{-1} as given by the inverse matrix formula.

Second Proof: This one is based on the expansion and the vanishing properties of determinants. Given the matrix

$$f = \begin{pmatrix} a_{11} & a_{12} & a_{13} \\ a_{21} & a_{22} & a_{23} \\ a_{31} & a_{32} & a_{33} \end{pmatrix}$$

we are looking for another

$$g = \begin{pmatrix} b_{11} & b_{12} & b_{13} \\ b_{21} & b_{22} & b_{23} \\ b_{31} & b_{32} & b_{33} \end{pmatrix}$$

such that

$$\begin{pmatrix} b_{11} & b_{12} & b_{13} \\ b_{21} & b_{22} & b_{23} \\ b_{31} & b_{32} & b_{33} \end{pmatrix} \circ \begin{pmatrix} a_{11} & a_{12} & a_{13} \\ a_{21} & a_{22} & a_{23} \\ a_{31} & a_{32} & a_{33} \end{pmatrix} = \begin{pmatrix} D & 0 & 0 \\ 0 & D & 0 \\ 0 & 0 & D \end{pmatrix}$$

i.e.,

$$b_{11}a_{11} + b_{12}a_{21} + b_{13}a_{31} = D$$
$$b_{11}a_{12} + b_{12}a_{22} + b_{13}a_{32} = 0$$
$$\dots\dots\dots\dots\dots\dots\dots\dots\dots\dots$$
$$b_{31}a_{13} + b_{32}a_{23} + b_{33}a_{33} = D$$

But by the expansion property and by the vanishing property of determinants, we know that the cofactors of the determinant D satisfy this condition, i.e., $b_{11} = A_{11}$, $b_{12} = A_{21}$, ..., $b_{33} = A_{33}$. It follows that

$$\begin{pmatrix} A_{11}/D & A_{21}/D & A_{31}/D \\ A_{12}/D & A_{22}/D & A_{32}/D \\ A_{13}/D & A_{23}/D & A_{33}/D \end{pmatrix} \circ \begin{pmatrix} a_{11} & a_{12} & a_{13} \\ a_{21} & a_{22} & a_{32} \\ a_{31} & a_{32} & a_{33} \end{pmatrix} = \begin{pmatrix} 1 & 0 & 0 \\ 0 & 1 & 0 \\ 0 & 0 & 1 \end{pmatrix}$$

Example Find the inverse, if any, of

$$f = \begin{pmatrix} 1 & 0 & 1 \\ 2 & 1 & 0 \\ 3 & 1 & 4 \end{pmatrix}$$

Answer:

$$D = \begin{vmatrix} 1 & 0 & 1 \\ 2 & 1 & 0 \\ 3 & 1 & 4 \end{vmatrix} = 3 \neq 0$$

Hence f^{-1} exists, namely:

$$f^{-1} = \frac{1}{3}\begin{pmatrix} 4 & 1 & -1 \\ -8 & 1 & 2 \\ -1 & -1 & 1 \end{pmatrix} = \begin{pmatrix} 4/3 & 1/3 & -1/3 \\ -8/3 & 1/3 & 2/3 \\ -1/3 & -1/3 & 1/3 \end{pmatrix}$$

For the invertible linear operators of the plane, the inverse matrix formula reads

$$\begin{pmatrix} a_{11} & a_{12} \\ a_{21} & a_{22} \end{pmatrix}^{-1} = \begin{pmatrix} a_{22}/D & -a_{12}/D \\ -a_{21}/D & a_{11}/D \end{pmatrix}$$

where

$$D = \begin{vmatrix} a_{11} & a_{12} \\ a_{21} & a_{22} \end{vmatrix} = a_{11}a_{22} - a_{12}a_{21} \neq 0$$

SECOND METHOD FOR CALCULATING THE INVERSE MATRIX

The problem of finding the inverse matrix can be described—and solved—as a problem of finding the solutions of nonhomogeneous systems of linear equations. Let

$$\begin{pmatrix} a_{11} & a_{12} & a_{13} \\ a_{21} & a_{22} & a_{23} \\ a_{31} & a_{32} & a_{33} \end{pmatrix}^{-1} = \begin{pmatrix} b_{11} & b_{12} & b_{13} \\ b_{21} & b_{22} & b_{23} \\ b_{31} & b_{32} & b_{33} \end{pmatrix}$$

This means, according to the product rule of matrices, that

$$\begin{aligned} a_{11}b_{11} + a_{12}b_{21} + a_{13}b_{31} &= 1 \\ a_{21}b_{11} + a_{22}b_{21} + a_{23}b_{31} &= 0 \\ a_{31}b_{11} + a_{32}b_{21} + a_{33}b_{31} &= 0 \end{aligned}$$

$$\begin{aligned} a_{11}b_{12} + a_{12}b_{22} + a_{13}b_{32} &= 0 \\ a_{21}b_{12} + a_{22}b_{22} + a_{23}b_{32} &= 1 \\ a_{31}b_{12} + a_{32}b_{22} + a_{33}b_{32} &= 0 \end{aligned}$$

$$\begin{aligned} a_{11}b_{13} + a_{12}b_{23} + a_{13}b_{33} &= 0 \\ a_{21}b_{13} + a_{22}b_{23} + a_{23}b_{33} &= 0 \\ a_{31}b_{13} + a_{32}b_{23} + a_{33}b_{33} &= 1 \end{aligned}$$

where the coefficients $a_{11}, \ldots, a_{33}$ are known, and $b_{11}, \ldots, b_{33}$ are unknown. Thus we have to solve three nonhomogeneous systems of linear equations. The work on these three systems can be done simultaneously, and instead of the three systems of equations we shall be working on one single matrix, as follows.

We wish to bring the triply augmented matrix

$$\left(\begin{array}{ccc|ccc} a_{11} & a_{12} & a_{13} & 1 & 0 & 0 \\ a_{21} & a_{22} & a_{23} & 0 & 1 & 0 \\ a_{31} & a_{32} & a_{33} & 0 & 0 & 1 \end{array}\right)$$

to the form of

$$\left(\begin{array}{ccc|ccc} 1 & 0 & 0 & b_{11} & b_{12} & b_{13} \\ 0 & 1 & 0 & b_{21} & b_{22} & b_{23} \\ 0 & 0 & 1 & b_{31} & b_{32} & b_{33} \end{array}\right)$$

by operating on its rows, i.e., by interchanging rows, by multiplying rows by a nonzero scalar, and by adding (subtracting) multiples of a row to (from) another row. If we succeed in doing this, then the b matrix on the right side is the inverse of the a matrix. If, however, we come across an all-zero row on the left side, the work cannot be completed, and we conclude that no inverse matrix exists. The following examples should clarify the process, which is essentially a gaussian elimination.

Examples

1. $$\begin{pmatrix} 0 & 0 & 1 \\ 1 & 0 & 0 \\ 0 & 1 & 0 \end{pmatrix}^{-1} = \begin{pmatrix} 0 & 1 & 0 \\ 0 & 0 & 1 \\ 1 & 0 & 0 \end{pmatrix}$$

   because

$$\left(\begin{array}{ccc|ccc} 0 & 0 & 1 & 1 & 0 & 0 \\ 1 & 0 & 0 & 0 & 1 & 0 \\ 0 & 1 & 0 & 0 & 0 & 1 \end{array}\right) \Rightarrow \left(\begin{array}{ccc|ccc} 1 & 0 & 0 & 0 & 1 & 0 \\ 0 & 1 & 0 & 0 & 0 & 1 \\ 0 & 0 & 1 & 1 & 0 & 0 \end{array}\right)$$

2. $$\begin{pmatrix} 2 & 0 & 0 \\ 0 & 3 & 0 \\ 0 & 0 & 4 \end{pmatrix}^{-1} = \begin{pmatrix} 1/2 & 0 & 0 \\ 0 & 1/3 & 0 \\ 0 & 0 & 1/4 \end{pmatrix}$$

   because

$$\left(\begin{array}{ccc|ccc} 2 & 0 & 0 & 1 & 0 & 0 \\ 0 & 3 & 0 & 0 & 1 & 0 \\ 0 & 0 & 4 & 0 & 0 & 1 \end{array}\right) \Rightarrow \left(\begin{array}{ccc|ccc} 1 & 0 & 0 & 1/2 & 0 & 0 \\ 0 & 1 & 0 & 0 & 1/3 & 0 \\ 0 & 0 & 1 & 0 & 0 & 1/4 \end{array}\right)$$

3. $$\begin{pmatrix} 1 & 1 & 1 \\ 1 & 1 & 1 \\ 1 & 1 & 1 \end{pmatrix}^{-1}$$

   does not exist because

$$\left(\begin{array}{ccc|ccc} 1 & 1 & 1 & 1 & 0 & 0 \\ 1 & 1 & 1 & 0 & 1 & 0 \\ 1 & 1 & 1 & 0 & 0 & 1 \end{array}\right) \Rightarrow \left(\begin{array}{ccc|ccc} 1 & 1 & 1 & 1 & 0 & 0 \\ 0 & 0 & 0 & -1 & 1 & 0 \\ 0 & 0 & 0 & -1 & 0 & 1 \end{array}\right)$$

and the process cannot be completed because of the presence of an all-zero row on the left side.

4. $$\begin{pmatrix} 3 & 0 & 1 \\ 2 & 2 & 3 \\ 1 & -1 & -1 \end{pmatrix}^{-1} = \begin{pmatrix} -1 & 1 & 2 \\ -5 & 4 & 7 \\ 4 & -3 & -6 \end{pmatrix}$$

because

$$\left(\begin{array}{ccc|ccc} 3 & 0 & 1 & 1 & 0 & 0 \\ 2 & 2 & 3 & 0 & 1 & 0 \\ 1 & -1 & -1 & 0 & 0 & 1 \end{array}\right) \Rightarrow \left(\begin{array}{ccc|ccc} 1 & -1 & -1 & 0 & 0 & 1 \\ 2 & 2 & 3 & 0 & 1 & 0 \\ 3 & 0 & 1 & 1 & 0 & 0 \end{array}\right)$$

$$\Rightarrow \left(\begin{array}{ccc|ccc} 1 & -1 & -1 & 0 & 0 & 1 \\ 0 & 4 & 5 & 0 & 1 & -2 \\ 0 & 3 & 4 & 1 & 0 & -3 \end{array}\right) \Rightarrow \left(\begin{array}{ccc|ccc} 1 & -1 & -1 & 0 & 0 & 1 \\ 0 & 1 & 1 & -1 & 1 & 1 \\ 0 & 0 & 1 & 4 & -3 & -6 \end{array}\right)$$

$$\Rightarrow \left(\begin{array}{ccc|ccc} 1 & -1 & -1 & 0 & 0 & 1 \\ 0 & 1 & 1 & -1 & 1 & 1 \\ 0 & 0 & 1 & 4 & -3 & -6 \end{array}\right) \Rightarrow \left(\begin{array}{ccc|ccc} 1 & 0 & 0 & -1 & 1 & 2 \\ 0 & 1 & 0 & -5 & 4 & 7 \\ 0 & 0 & 1 & 4 & -3 & -6 \end{array}\right)$$

5. $$\begin{pmatrix} 1 & 0 & 1 \\ 2 & 1 & 0 \\ 3 & 1 & 4 \end{pmatrix}^{-1} = \begin{pmatrix} 4/3 & 1/3 & -1/3 \\ -8/3 & 1/3 & 2/3 \\ -1/3 & -1/3 & 1/3 \end{pmatrix}$$

because

$$\left(\begin{array}{ccc|ccc} 1 & 0 & 1 & 1 & 0 & 0 \\ 2 & 1 & 0 & 0 & 1 & 0 \\ 3 & 1 & 4 & 0 & 0 & 1 \end{array}\right) \Rightarrow \left(\begin{array}{ccc|ccc} 1 & 0 & 1 & 1 & 0 & 0 \\ 0 & 1 & -2 & -2 & 1 & 0 \\ 0 & 1 & 1 & -3 & 0 & 1 \end{array}\right) \Rightarrow \left(\begin{array}{ccc|ccc} 1 & 0 & 1 & 1 & 0 & 0 \\ 0 & 1 & -2 & -2 & 1 & 0 \\ 0 & 0 & 3 & -1 & -1 & 1 \end{array}\right)$$

$$\Rightarrow \left(\begin{array}{ccc|ccc} 1 & 0 & 0 & 4/3 & 1/3 & -1/3 \\ 0 & 1 & 0 & -8/3 & 1/3 & 2/3 \\ 0 & 0 & 1 & -1/3 & -1/3 & 1/3 \end{array}\right) \Rightarrow \left(\begin{array}{ccc|ccc} 1 & 0 & 1 & 1 & 0 & 0 \\ 0 & 1 & 0 & -8/3 & 1/3 & 2/3 \\ 0 & 0 & 1 & -1/3 & -1/3 & 1/3 \end{array}\right)$$

APPLICATIONS

4. Conjugate of an Operator

Let g be an invertible operator. $g \circ f \circ g^{-1}$ is called the *conjugate* of the operator f by g. The conjugate of a projection is a projection:

$$p^2 = p \Rightarrow (g \circ p \circ g^{-1})^2 = (g \circ p \circ g^{-1}) \circ (g \circ p \circ g^{-1}) = g \circ p^2 \circ g^{-1} = g \circ p \circ g^{-1}$$

In the same way one can see that the conjugate of an involution (lift, shear, dilatation, twist, invection) is an involution (lift, shear, dilatation, twist, invection, resp.) In Chap. 11 (Application 7) we shall see in more detail that if p is the projection onto the line ℓ along the plane π, then its conjugate $g \circ p \circ g^{-1}$ is the projection onto the line $g(\ell)$ along the plane $g(\pi)$. Similar statements can be made about the conjugate of an involution, lift, etc.

The algebraic properties of conjugation are treated more fully in Chap. 14.

5. Decomposition of a Shear into the Product of Two Involutions

Let p, q be a pair of principal projections along the same plane (onto different lines). Then

$$h = 1 - 2p \qquad k = 1 - 2q$$

are involutions in the same plane (along different lines), by Chap. 7, Application 2. Then

$$\begin{aligned} h \circ k &= (1-2p) \circ (1-2q) \\ &= 1 - 2p - 2q + 4p \circ q \\ &= 1 - 2p - 2q + 4p \\ &= 1 + 2(p-q) = 1 + 2\ell \end{aligned}$$

because $p \circ q = p$ by Chap. 7, Application 9. Moreover, $\ell = p - q$ is a lift, and hence $h \circ k$ is a shear.

The converse is also true: Every shear can be written in the form $f = 1 + 2\ell$, where ℓ is a lift. Therefore, $\ell = p - q$ for some principal projections p, q along the same plane. Then $h = 1 - 2p$, $k = 1 - 2q$ are involutions and the calculation above shows that $f = h \circ k$. We have proved that *f is a shear if, and only if, f is the product of two involutions in the same plane*. This decomposition is not necessarily unique.

EXERCISES

8.1 Calculate the inverse of the given matrix in one way, then check the result in another way, for each of the following matrices.

(i) $\begin{pmatrix} 1 & 0 & 0 \\ 0 & 0 & -1 \\ 0 & 1 & 0 \end{pmatrix}$ (ii) $\begin{pmatrix} 1 & 0 & n \\ 0 & 1 & 0 \\ 0 & 0 & 1 \end{pmatrix}$ (iii) $\begin{pmatrix} \kappa & 0 & 0 \\ 0 & \lambda & 0 \\ 0 & 0 & \mu \end{pmatrix}$

(iv) $\begin{pmatrix} 1 & 2 & 2 \\ 0 & -1 & 0 \\ 0 & 0 & -1 \end{pmatrix}$ (v) $\begin{pmatrix} 2 & 0 & 1 \\ 0 & -7 & 3 \\ 1 & -6 & 3 \end{pmatrix}$ (vi) $\begin{pmatrix} 1 & -1 & 1 \\ 2 & 1 & 1 \\ 1 & -6 & 3 \end{pmatrix}$

(vii) $\begin{pmatrix} 3 & 0 & 1 \\ 2 & 2 & 3 \\ 1 & -1 & -1 \end{pmatrix}$ (viii) $\begin{pmatrix} -3 & -6 & 7 \\ 3 & 5 & -6 \\ 7 & 12 & -14 \end{pmatrix}$ (ix) $\begin{pmatrix} 1 & 4 & 3 \\ 2 & 5 & 4 \\ 1 & -3 & -2 \end{pmatrix}$

(x) $\begin{pmatrix} 1 & 2 & 3 \\ 0 & 4 & 5 \\ 0 & 0 & 6 \end{pmatrix}$ (xi) $\begin{pmatrix} 1 & 1 & 1 \\ 0 & 1 & 1 \\ 1 & 0 & 2 \end{pmatrix}$ (xii) $\begin{pmatrix} -2 & 5 & 1 \\ 3 & -1 & 4 \\ 2 & 0 & 7 \end{pmatrix}$

8.2 Show that the rotations of $\mathbb{R}^3$ with the same axis form a group.

8.3 Define what is meant by a dilatation in $\mathbb{R}^2$. Show that the dilatations of $\mathbb{R}^2$ along the same line form a group.

8.4 Show that the shears of $\mathbb{R}^2$ along a fixed line, which are parallel to another given line, form a group.

8.5 Show that if every element of a group is an involution, then the group is commutative.

8.6 Show that the invertible triangular matrices

$$\begin{pmatrix} \kappa & \alpha & \beta \\ 0 & \lambda & \gamma \\ 0 & 0 & \mu \end{pmatrix}$$

form a group.

8.7 Show that $(1-f)^{-1} = 1 + f + f^2$, $(1+f)^{-1} = 1 - f + f^2$ for the linear operator

$$f = \begin{pmatrix} 0 & 1 & 1 \\ 0 & 0 & 1 \\ 0 & 0 & 0 \end{pmatrix}$$

8.8 Calculate f^{-1} for

$$f = \begin{pmatrix} \cos\phi & -\sin\phi \\ \sin\phi & \cos\phi \end{pmatrix}$$

8.9 Calculate f^{-1} for each of the following.

(i) $\begin{pmatrix} 1 & 2 \\ 2 & 5 \end{pmatrix}$ (ii) $\begin{pmatrix} 1 & 2 & -3 \\ 0 & 1 & 2 \\ 0 & 0 & 1 \end{pmatrix}$ (iii) $\begin{pmatrix} 1 & 3 & -5 & 7 \\ 0 & 1 & 2 & -3 \\ 0 & 0 & 1 & 2 \\ 0 & 0 & 0 & 1 \end{pmatrix}$

(iv) $\begin{pmatrix} 2 & 2 & 3 \\ 1 & -1 & 0 \\ -1 & 2 & 1 \end{pmatrix}$ (v) $\begin{pmatrix} 1 & 1 & 1 & 1 \\ 1 & 1 & -1 & -1 \\ 1 & -1 & 1 & -1 \\ 1 & -1 & -1 & 1 \end{pmatrix}$ (vi) $\begin{pmatrix} 2 & 1 & 0 & 0 \\ 3 & 2 & 0 & 0 \\ 1 & 1 & 3 & 4 \\ 2 & -1 & 2 & 3 \end{pmatrix}$

(vii) $\begin{pmatrix} 0 & 1 & 1 & 1 \\ 1 & 0 & 1 & 1 \\ 1 & 1 & 0 & 1 \\ 1 & 1 & 1 & 0 \end{pmatrix}$

8.10 Find f in each of the following:

(i) $\begin{pmatrix} 2 & 5 \\ 1 & 3 \end{pmatrix} \circ f = \begin{pmatrix} 4 & -6 \\ 2 & 1 \end{pmatrix}$ (ii) $f \circ \begin{pmatrix} 1 & 1 & -1 \\ 2 & 1 & 0 \\ 1 & -1 & 1 \end{pmatrix} = \begin{pmatrix} 1 & -1 & 3 \\ 4 & 3 & 2 \\ 1 & -2 & 5 \end{pmatrix}$

(iii) $\begin{pmatrix} 2 & 1 \\ 2 & 1 \end{pmatrix} \circ f = \begin{pmatrix} 2 & 1 \\ 2 & 1 \end{pmatrix}$ (iv) $f \circ \begin{pmatrix} 2 & 1 \\ 2 & 1 \end{pmatrix} = \begin{pmatrix} 1 & 0 \\ 0 & 1 \end{pmatrix}$

8.11 Decompose the shears

$$\begin{pmatrix} 1 & 0 & 2 \\ 0 & 1 & 0 \\ 0 & 0 & 1 \end{pmatrix}, \quad \begin{pmatrix} 9 & -16 & 24 \\ -2 & 5 & -6 \\ -4 & 8 & -11 \end{pmatrix}$$

into the product of two involutions.

8.12 Show that the conjugate of an involution (lift, shear, twist, invection) is an involution (lift, shear, twist, invection, resp.)

8.13 Show that the conjugate of a dilatation is a dilatation with the same ratio.

8.14 The linear operator $f \circ g \circ f^{-1} \circ g^{-1}$ is called the *commutator* of f,g. Show that every shear is the commutator of a certain pair of operators.

8.15 Show that the commutator of g,f is the inverse of the commutator of f,g.

8.16 Show that the commutator of f,g^{-1} is the same as the conjugate of the commutator of g,f by g^{-1}.

9
Kernel and Image: The Classification of Linear Maps†

KERNEL AND IMAGE

We must distinguish between linear maps and linear operators. A linear map is more general in that its domain and codomain are not necessarily the same (as are those of a linear operator) but may be two distinct vector spaces. The symbol $\mathbb{R}^n$ will denote the n-dimensional vector space with scalars in $\mathbb{R}$, the field of real numbers. Then a *linear map f* is a function $f: \mathbb{R}^n \to \mathbb{R}^m$ that preserves the sum and the scalar multiple of vectors:

$$f(\mathbf{v} + \mathbf{v}') = f(\mathbf{v}) + f(\mathbf{v}')$$

$$f(\lambda \mathbf{v}) = \lambda f(\mathbf{v}) \qquad (\lambda \in \mathbb{R})$$

In this chapter we tackle the important problem of classifying linear maps (the problem of classifying linear operators is an entirely different, and paradoxically more difficult and complex problem, which is treated in Chap. 10). The classification of linear maps turns on the concept of rank f, a nonnegative integer having the property that two linear maps with the same rank are geometrically indistinguishable.

The discussion depends on the construction of the *kernel* and the *image* of a linear map, denoted by the symbols $\ker f$ and $\operatorname{im} f$, respectively.

Definition

$$\ker f = \{\mathbf{v} \in \mathbb{R}^n \mid f(\mathbf{v}) = \mathbf{0}\}$$

i.e., the kernel of f is the set of those vectors of the domain $\mathbb{R}^n$ which "get killed" by f.

† *Note*: The bases in this chapter are not necessarily orthonormal, and hence the coordinates are skew affine.

Examples

1. The kernel of a projection (other than the identity) is nontrivial. Let p be the perpendicular projection onto the horizontal plane and let q be the perpendicular projection onto the horizontal axis. Then $\ker p$ is the vertical axis and $\ker q$ is the horizontal coordinate plane.
2. The kernel of rotations, reflections, invections, involutions, strains, shears is trivial, i.e., it consists of the null vector and nothing else.

   More generally,

$$\ker f = \mathbf{0} \Leftrightarrow f \text{ injective}$$

Since the sum and scalar multiple of vectors in $\ker f$ is again in $\ker f$, we see that $\ker f$ is a subspace of the domain $\mathbb{R}^n$ of f, so its dimension n' is subject to $0 \leq n' \leq n$. However, not every n' satisfying this condition may actually occur as the dimension of $\ker f$ for some linear map f: if $m < n$, then dim $\ker f$ is at least $n - m$ or, to put the matter a little differently,

$$0 \leq n - \dim \ker f \leq m \tag{1}$$

The reason for this is that if the codomain $\mathbb{R}^m$ is "smaller" than the domain $\mathbb{R}^n$, then some vectors, other than the null vector, must get killed by every linear map from $\mathbb{R}^n$ to $\mathbb{R}^m$.

Definition

$$\operatorname{im} f = \{\mathbf{w} \in \mathbb{R}^m | \mathbf{w} = f(\mathbf{v}) \text{ for some } \mathbf{v} \in \operatorname{dom} f\}$$

i.e., the image of f is the set of those vectors of the codomain $\mathbb{R}^m$ which "get hit" by f.

Examples

1. Let p be the perpendicular projection onto the horizontal plane, and let q be the perpendicular projection onto the vertical axis. Then, obviously, $\operatorname{im} p$ is the horizontal plane, and $\operatorname{im} q$ is the vertical axis.
2. The image of rotations, reflections, involutions, invections, strains, and shears is the whole of the codomain $\mathbb{R}^3$. More generally,

$$\operatorname{im} f = \operatorname{cod} f \Leftrightarrow f \text{ surjective}$$

APPLICATIONS

1. Exact Composition

We have the following necessary and sufficient condition:

$$g \circ f = 0 \Leftrightarrow \operatorname{im} f \subseteq \ker g$$

Proof:

$$\begin{aligned} g \circ f = 0 &\Leftrightarrow g(f(\mathbf{v})) = \mathbf{0} \text{ for all } \mathbf{v} \\ &\Leftrightarrow f(\mathbf{v}) \in \ker g \text{ for all } \mathbf{v} \\ &\Leftrightarrow (\mathbf{w} \in \operatorname{im} f \Rightarrow \mathbf{w} \in \ker g) \Leftrightarrow \operatorname{im} f \subseteq \ker g \end{aligned}$$

Definition We say that $g \circ f$ is an *exact composition* if $\operatorname{im} f = \ker g$ (i.e., g kills exactly those vectors which are hit by f). In particular, an exact composition is trivial: $g \circ f = 0$.

Example Let p, q be a pair of complementary projections (i.e., $q = 1 - p$). Then both compositions $q \circ p$ and $p \circ q$ are exact: $\ker q = \operatorname{im} p$, $\operatorname{im} q = \ker p$. In other words, when switching from a projection to its complementary projection, we must switch around the kernel and image.

Proof: $\ker(1-p) \supseteq \operatorname{im} p$ follows from $(1-p) \circ p = 0$. On the other hand,

$$\begin{aligned} \mathbf{v} \in \ker(1-p) &\Rightarrow (1-p)(\mathbf{v}) = \mathbf{0} \\ &\Rightarrow \mathbf{v} = p(\mathbf{v}) \Rightarrow \mathbf{v} \in \operatorname{im} p \end{aligned}$$

Therefore $\ker(1-p) \subseteq \operatorname{im} p$. Combining the two, $\ker q = \ker(1-p) = \operatorname{im} p$ follows. $\operatorname{im} q = \ker p$ is immediate from this, on the strength of $1-(1-p) = p$.

2. Lifts

Let $\mathbf{n} \perp \mathbf{e}$, $\mathbf{n} \neq \mathbf{0}$, $\mathbf{e} \neq \mathbf{0}$ and put $\ell(\mathbf{v}) = (\mathbf{n} \cdot \mathbf{v})\mathbf{e}$ for all $\mathbf{v} \in \mathbb{R}^3$. Then $\ell^2 = 0$, $\ell \neq 0$: ℓ is a lift. Clearly, $\operatorname{im} \ell$ is the line spanned by $\mathbf{e}$; we want to find $\ker \ell$.

$$\mathbf{v} \in \ker \ell \Leftrightarrow \mathbf{n} \cdot \mathbf{v} = 0 \Leftrightarrow \mathbf{n} \perp \mathbf{v}$$

Case of $\mathbb{R}^2$ $\ker \ell$ is the normal line of $\mathbf{n}$. $\ker \ell \perp \mathbf{n} \Rightarrow \ker \ell \parallel \operatorname{im} \ell \Rightarrow \operatorname{im} \ell = \ker \ell$. Hence $\ell^2 = 0$ is an exact composition. Note that $\dim \operatorname{im} \ell + \dim \ker \ell = 2 = \dim \mathbb{R}^2$.

Case of $\mathbb{R}^3$ $\ker \ell$ is the normal plane of $\mathbf{n}$. Note that $\dim \operatorname{im} \ell + \dim \ker \ell = 3 = \dim \mathbb{R}^3$.

3. Projections

Let $\ell'(\mathbf{v}) = (\mathbf{n}' \cdot \mathbf{v})\mathbf{e}'$ be another lift such that $\mathbf{n}' \cdot \mathbf{e}' = 0$,

$\mathbf{n} \cdot \mathbf{e}' = 1 = \mathbf{n}' \cdot \mathbf{e}$. It can be seen that $\ell \circ \ell'$, $\ell' \circ \ell$ are a pair of principal projections which are mutual zero divisors. In fact,

$$\operatorname{im} \ell \circ \ell' = \operatorname{im} \ell \qquad \ker \ell \circ \ell' = \ker \ell'$$
$$\operatorname{im} \ell' \circ \ell = \operatorname{im} \ell' \qquad \ker \ell' \circ \ell = \ker \ell$$

We shall show below (Application 5) that the converse is also true: if p, p' are two principal projections which are mutual zero divisors, then they can be decomposed into the product of a pair of lifts.

We put $p(\mathbf{v}) = (\mathbf{n} \cdot \mathbf{v})\mathbf{e}'$ for all $\mathbf{v}$ $(\mathbf{n} \cdot \mathbf{e}' = 1)$. Then $p^2 = p$: p is a principal projection. Clearly, $\operatorname{im} p$ is the line spanned by $\mathbf{e}'$, and $\ker p$ is the normal plane of $\mathbf{n}$. Again,

$$\dim \operatorname{im} p + \dim \ker p = 3 = \dim \mathbb{R}^3$$

Let $p'(\mathbf{v}) = (\mathbf{n}' \cdot \mathbf{v})\mathbf{e}$ be another principal projection $(\mathbf{n}' \cdot \mathbf{e} = 1 = \mathbf{n} \cdot \mathbf{e}'$, $\mathbf{n} \cdot \mathbf{e} = 0 = \mathbf{n}' \cdot \mathbf{e}')$. Then $p \circ p' = 0 = p' \circ p$; i.e., p and p' are mutual zero divisors. Put

$$q = p + p'$$

It is easy to see that q is also a projection: $q^2 = q$; moreover, $\operatorname{im} q$ is the plane spanned by the lines $\operatorname{im} p$ and $\operatorname{im} p'$, and

$$\ker q = \ker p \cap \ker p'$$

the intersection of the planes $\ker p$ and $\ker p'$. Note that, once more,

$$\dim \operatorname{im} q + \dim \ker q = 3 = \dim \mathbb{R}^3$$

These examples motivate the law of nullity, which we shall now discuss.

THE LAW OF NULLITY. RANK

Since the sum and scalar multiple of vectors in $\operatorname{im} f$ is again in $\operatorname{im} f$, we see that $\operatorname{im} f$ is a subspace of the codomain $\mathbb{R}^m$ of f, so its dimension m' is subject to $0 \leq m' \leq m$. However, not every m' satisfying this condition may actually occur as the dimension of $\operatorname{im} f$ for some linear map f: if $n < m$, then $\dim \operatorname{im} f$ is at most n:

$$0 \leq \dim \operatorname{im} f \leq n \tag{2}$$

The reason for this is that if the codomain $\mathbb{R}^m$ is "larger" than the domain $\mathbb{R}^n$, then no linear map can hit every vector; a linear map can, at best, hit a target as large as $\mathbb{R}^n$, its domain.

Furthermore, the dimensions of $\ker f$ and $\operatorname{im} f$ are not independent of one another, in fact, either one determines the other, as the next formula shows:

$$\dim \operatorname{im} f = n - \dim \ker f \tag{3}$$

This formula is a nontrivial theorem, called the law of nullity. Intuitively, the more vectors f kills, the fewer it can hit, and vice versa.

Proof for f: $\mathbb{R}^3 \to \mathbb{R}^3$:

Case 1: $\dim \ker f = 3 \Rightarrow f(\mathbf{v}) = \mathbf{0}$ for all $\mathbf{v} \Rightarrow \operatorname{im} f = \mathbf{0}$

$$\Rightarrow \dim \operatorname{im} f = 0$$

Case 2: $\dim \ker f = 2$, $\ker f$ is spanned by $\mathbf{e}_1$, $\mathbf{e}_2$. Complete this to a basis for $\mathbb{R}^3$, $\{\mathbf{e}_1, \mathbf{e}_2, \mathbf{e}_3\}$. Then

$$\mathbf{w} \in \operatorname{im} f \Rightarrow \mathbf{w} = f(\mathbf{v}) = f(\alpha_1\mathbf{e}_1 + \alpha_2\mathbf{e}_2 + \alpha_3\mathbf{e}_3) = \alpha_3 f(\mathbf{e}_3)$$

because $f(\mathbf{e}_1) = \mathbf{0} = f(\mathbf{e}_2)$. Therefore, $\operatorname{im} f$ is a line spanned by $f(\mathbf{e}_3) \neq \mathbf{0}$, $\dim \operatorname{im} f = 1$.

Case 3: $\dim \ker f = 1$, $\ker f$ is spanned by $\mathbf{e}_1$. Complete this to a basis for $\mathbb{R}^3$, $\{\mathbf{e}_1, \mathbf{e}_2, \mathbf{e}_3\}$. Then

$$\mathbf{w} \in \operatorname{im} f \Rightarrow \mathbf{w} = f(\mathbf{v}) = f(\alpha_1\mathbf{e}_1 + \alpha_2\mathbf{e}_2 + \alpha_3\mathbf{e}_3) = \alpha_2 f(\mathbf{e}_2) + \alpha_3 f(\mathbf{e}_3)$$

because $f(\mathbf{e}_1) = \mathbf{0}$. We want to show that $\dim \operatorname{im} f = 2$ by showing that $\{f(\mathbf{e}_2), f(\mathbf{e}_3)\}$ is linearly independent. Indeed,

$$\begin{aligned}\beta_2 f(\mathbf{e}_2) + \beta_3 f(\mathbf{e}_3) = \mathbf{0} &\Rightarrow f(\beta_2\mathbf{e}_2 + \beta_3\mathbf{e}_3) = \mathbf{0} \Rightarrow \beta_2\mathbf{e}_2 + \beta_3\mathbf{e}_3 \in \ker f\\ &\Rightarrow \beta_2 = \beta_3 = 0\end{aligned}$$

because $\ker f$ is spanned by $\mathbf{e}_1$ and $\{\mathbf{e}_1, \mathbf{e}_2, \mathbf{e}_3\}$ is linearly independent.

Case 4: $\dim \ker f = 0$. Let $\{\mathbf{e}_1, \mathbf{e}_2, \mathbf{e}_3\}$ be a basis for $\mathbb{R}^3$; we want to show that $\{f(\mathbf{e}_1), f(\mathbf{e}_2), f(\mathbf{e}_3)\}$ is linearly independent. Indeed,

$$\begin{aligned}\beta_1 f(\mathbf{e}_1) + \beta_2 f(\mathbf{e}_2) + \beta_3 f(\mathbf{e}_3) = \mathbf{0} &\Rightarrow f(\beta_1\mathbf{e}_1 + \beta_2\mathbf{e}_2 + \beta_3\mathbf{e}_3) = \mathbf{0}\\ &\Rightarrow \beta_1\mathbf{e}_1 + \beta_2\mathbf{e}_2 + \beta_3\mathbf{e}_3 \in \ker f\\ &\Rightarrow \beta_1\mathbf{e}_1 + \beta_2\mathbf{e}_2 + \beta_3\mathbf{e}_3 = \mathbf{0} \qquad (\text{because } \ker f = \mathbf{0})\\ &\Rightarrow \beta_1 = \beta_2 = \beta_3 = 0\end{aligned}$$

It follows that $\dim \operatorname{im} f = 3$.

In each case the equation $\dim \ker f + \dim \operatorname{im} f = 3$ is satisfied. The proof for f: $\mathbb{R}^3 \to \mathbb{R}^3$ is complete. The proof for the general case f: $\mathbb{R}^n \to \mathbb{R}^m$ can be given along the same lines.

The appearance of the nonnegative integer (3) in (1) and (2) points to its importance, calling for a definition. We define the *rank* of a linear map f: $\mathbb{R}^n \to \mathbb{R}^m$ as $\operatorname{rank} f = \dim \operatorname{im} f$. We then have

The Law of Nullity

$$\text{rank } f = \dim \text{im} f = n - \dim \ker f$$

APPLICATIONS

4. Twists

It follows from the Law of Nullity that a lift $\ell \neq 0$ of $\mathbb{R}^3$ has rank $\ell = 1$. Indeed,

$$\ell^2 = 0 \Rightarrow \text{im}\,\ell \subseteq \ker \ell \Rightarrow \dim \text{im}\,\ell \leqq \dim \ker \ell$$

Hence rank $\ell \geqq 2$ would contradict the law of nullity, limiting the sum of dim im ℓ and dim ker ℓ to 3. We conclude that all lifts are of the type $\ell(\mathbf{v}) = \lambda(\mathbf{n} \cdot \mathbf{v})\mathbf{e}$, where $\mathbf{n} \cdot \mathbf{e} = 0$ and $\mathbf{n}$ is the normal vector of ker ℓ, $\mathbf{e}$ is the direction vector of im ℓ, and $\lambda \in \mathbb{R}$.

Let ℓ, ℓ' be lifts: $\ell^2 = 0 = \ell'^2$, $\ell \neq 0$, $\ell' \neq 0$ such that $\ell \circ \ell'$, $\ell' \circ \ell$ are projections. Then it follows from the foregoing, and from Application 2 above, that im $\ell \circ \ell'$ = im ℓ and ker $\ell \circ \ell'$ = ker ℓ', so that $\ell \circ \ell'$ is a principal projection onto the line im ℓ along the plane ker ℓ'. $1 - \ell \circ \ell'$ is the complementary projection onto ker $\ell' = \text{im}(1 - \ell \circ \ell')$ along im ℓ, so we have an exact composition $\ell' \circ (1 - \ell \circ \ell') = 0 \Rightarrow \ell' \circ \ell \circ \ell' = \ell'$. Similarly, we may conclude that $\ell \circ \ell' \circ \ell = \ell$. Let us calculate the powers of $t = \ell - \ell'$:

$$t^2 = -\ell \circ \ell' - \ell' \circ \ell$$

$$t^3 = \ell' \circ \ell \circ \ell' - \ell \circ \ell' \circ \ell = \ell' - \ell = -t$$

Therefore, t is a twist. The converse is also true: every twist t can be decomposed into the difference of lifts whose products are projections which are mutual zero divisors, but we shall not be able to prove this before Chap. 10. Granting this result for the moment, we shall show—with the aid of the law of nullity—that im t is the plane spanned by im ℓ and im ℓ', and

$$\ker t = \ker \ell \cap \ker \ell'$$

the intersection of the planes ker ℓ and ker ℓ'.

$$\mathbf{v} \in \ker t \Leftrightarrow t(\mathbf{v}) = \mathbf{0} \Leftrightarrow \ell(\mathbf{v}) = \ell'(\mathbf{v})$$

$$\Leftrightarrow \ell(\mathbf{v}) = \mathbf{0} = \ell'(\mathbf{v}) \qquad \text{(because im } \ell \text{ and im } \ell' \text{ are } \textit{distinct} \text{ lines)}$$

$$\Leftrightarrow \mathbf{v} \in \ker \ell \text{ and } \mathbf{v} \in \ker \ell' \Leftrightarrow \mathbf{v} \in \ker \ell \cap \ker \ell'$$

Case of $\mathbb{R}^2$ ker $t = \mathbf{0}$ for any twist $t \neq 0$, because ker ℓ, ker ℓ' are distinct lines intersecting at the origin. Then by the law of nullity, rank $t = 2$,

$\operatorname{im} t = \mathbb{R}^2$ so that t is bijective and hence t^{-1} exists. Calculating t^{-1} yields

$$t^3 = -t \Rightarrow t^2 = -1 \Rightarrow t^4 = 1 \Rightarrow t^{-1} = t^3 = -t$$

every nontrivial twist of $\mathbb{R}^2$ is a noninvolutory invection. As we shall see in Chap. 10, the converse is also true: in $\mathbb{R}^2$ every noninvolutory invection is a twist.

Case of $\mathbb{R}^3$ The twists of $\mathbb{R}^3$ are very different in that they fail to be bijective and hence have no inverse. In fact, for $t \neq 0$, $\ker t$ is the line of intersection of the planes $\ker \ell$, $\ker \ell'$. By the law of nullity, $\operatorname{rank} t = 3 - 1 = 2$, hence $\operatorname{im} t$ is a plane. Obviously, it is the plane spanned by the lines $\operatorname{im} \ell$, $\operatorname{im} \ell'$.

5. Principal Projections That Are Mutual Zero Divisors

Let p, p' be principal projections which are mutual zero divisors: $p \circ p' = 0 = p' \circ p$. Then $\operatorname{im} p$, $\operatorname{im} p'$ are lines and, by virtue of the law of nullity, $\ker p$, $\ker p'$ are planes. Let $\mathbf{e}$, $\mathbf{e}'$ denote the vectors spanning the lines $\operatorname{im} p$, $\operatorname{im} p'$, respectively. Furthermore, let $\mathbf{n}$ be the normal vector of the plane $\ker p'$, and let $\mathbf{n}'$ be the normal vector of the plane $\ker p$. Then

$$\begin{aligned} p' \circ p = 0 = p \circ p' &\Rightarrow \operatorname{im} p \subseteq \ker p', \quad \operatorname{im} p' \subseteq \ker p \\ &\Rightarrow \mathbf{e} \perp \mathbf{n},\ \mathbf{e}' \perp \mathbf{n}' \Rightarrow \mathbf{n} \cdot \mathbf{e}' = 0 = \mathbf{n}' \cdot \mathbf{e} \end{aligned}$$

Moreover, $\mathbf{n} \pm \mathbf{e}'$, $\mathbf{n}' \pm \mathbf{e}$ (otherwise, p or p' would be a lift) and hence $\mathbf{n} \cdot \mathbf{e}' \neq 0$, $\mathbf{n}' \cdot \mathbf{e} \neq 0$. We shall assume that the lengths of these vectors have been fixed such that $\mathbf{n} \cdot \mathbf{e}' = 1 = \mathbf{n}' \cdot \mathbf{e}$. We now put $\ell(\mathbf{v}) = (\mathbf{n} \cdot \mathbf{v})\mathbf{e}$, $\ell'(\mathbf{v}) = (\mathbf{n}' \cdot \mathbf{v})\mathbf{e}'$ for all $\mathbf{v} \in \mathbb{R}^3$. Then clearly $\ell^2 = 0 = \ell'^2$ and $\ell \circ \ell' = p$, $\ell' \circ \ell = p'$ (see Application 3 above), and we have $\ker p = \ker \ell'$, $\operatorname{im} p = \operatorname{im} \ell$, $\ker p' = \ker \ell$, $\operatorname{im} p' = \operatorname{im} \ell'$.

Recall that $p^2 = p \Rightarrow \operatorname{rank} p = \operatorname{trace} p$ (Chap. 7, Application 1). This rule makes the calculation of rank, *for projections only*, a very simple matter. The calculation of the rank in general is a more arduous process.

Example

$$p = \begin{pmatrix} 32 & -16 & 20 \\ -8 & 4 & -5 \\ -56 & 28 & -35 \end{pmatrix}$$

is a principal projection. Indeed, $p^2 = p$ by matrix multiplication; $\operatorname{rank} p = \dim \operatorname{im} p = \operatorname{trace} p = 32 + 4 - 35 = 1$, hence p is a projection onto a line, i.e., a principal projection.

CLASSIFICATION OF LINEAR MAPS

The nullity of f is just dim ker f, and it measures the size of the set of vectors that get killed by f. The rank of f is just dim im f, and it measures the size of the set of vectors that get hit by f. The law of nullity states that the sum of the rank and the nullity is always the same: it is the dimension of the domain.

We are now ready to proceed with the task of classifying the linear maps from $\mathbb{R}^n$ to $\mathbb{R}^m$. Let us rewrite the inequalities (1) and (2) in terms of the rank: $0 \leq \text{rank} f \leq m$, $0 \leq \text{rank} f \leq n$. But since these inequalities must be satisfied simultaneously, we have that rank f cannot be greater than the *smaller* of the dimensions of the domain and the codomain of f:

$$0 \leq \text{rank} f \leq \min(n,m)$$

On the other hand, given any integer r between 0 and $\min(n,m)$, these numbers included, we can construct a linar map f: $\mathbb{R}^n \to \mathbb{R}^m$ such that $\text{rank} f = r$. Two linear maps from $\mathbb{R}^n$ to $\mathbb{R}^m$ having the same rank exhibit identical geometric properties. Therefore, we have a complete classification of linear maps from $\mathbb{R}^n$ to $\mathbb{R}^m$, effected by the rank function. *There are altogether*

$$\min(n,m) + 1$$

different types of linear maps from $\mathbb{R}^n$ *to* $\mathbb{R}^m$ according as $\text{rank} f = 0, 1, 2, \ldots, \min(n,m)$.

Important Special Cases of f: $\mathbb{R}^n \to \mathbb{R}^m$

1. $\text{rank} f = 0 \Leftrightarrow f = 0$. There is one, and only one, linear map of zero rank, namely, the trivial map.
2. Let $n > m$. Then f surjective $\Leftrightarrow \text{rank} f = m$ (maximal). Conversely, if there exists a surjective linear map, then $n \geq m$. There are, of course, infinitely many linear maps with maximal rank (which is equal to the dimension of the codomain), but they are all surjective, hitting every vector in cod f.
3. Let $m < n$. Then f injective $\Leftrightarrow \text{rank} f = n$ (maximal). Conversely, if there is an injective linear map, then $m \leq n$. There are, of course, infinitely many linear maps with maximal rank (which is equal to the dimension of the domain) but they are all injective, killing no vector in dom f (other than the null vector).
4. Let $n = m$. Then f bijective (isomorphism) $\Leftrightarrow \text{rank} f = n = m$(maximal). Conversely, if there is a bijective linear map, then $n = m$. There are, of course, infinitely many linear maps with maximal rank, and they are all both injective and surjective, hence bijective. A bijective linear map is

called an *isomorphism*, and its domain and codomain are said to be *isomorphic* vector spaces. It follows from the law of nullity that two vector spaces are isomorphic if, and only if, they have the same dimension.

We shall summarize these results in the form of a

Theorem Let $L(\mathbb{R}^n, \mathbb{R}^m)$ denote the set of all linear maps f: $\mathbb{R}^n \to \mathbb{R}^m$.

$$\text{Injective} f \in L(\mathbb{R}^n, \mathbb{R}^m) \text{ exists} \Leftrightarrow n \leq m$$
$$\text{Surjective} f \in L(\mathbb{R}^n, \mathbb{R}^m) \text{ exists} \Leftrightarrow n \geq m$$
$$\text{Bijective} f \in L(\mathbb{R}^n, \mathbb{R}^m) \text{ exists} \Leftrightarrow n = m$$

and, in each case, rank f will be maximal. Conversely, if rank f is maximal, then f is injective or surjective, according as $n \leq m$ or $n \geq m$.

If $n = m$, then

$$f \text{ injective} \Leftrightarrow f \text{ surjective} \Leftrightarrow f \text{ bijective}$$

Remark It will help to retain the contents of this theorem if we recall the following elementary fact about a function f: $X \to Y$ between *finite* sets, X having n elements, and Y having m elements:

$$\text{Injective} f \text{ exists} \Leftrightarrow n \leq m$$
$$\text{Surjective} f \text{ exists} \Leftrightarrow n \geq m$$
$$\text{Bijective} f \text{ exists} \Leftrightarrow n = m$$

If $n = m$, then f injective $\Leftrightarrow f$ surjective $\Leftrightarrow f$ bijective.

NECESSARY AND SUFFICIENT CONDITIONS FOR THE INVERTIBILITY OF LINEAR MAPS AND OPERATORS

Any two of the following conditions on the linear map f: $\mathbb{R}^n \to \mathbb{R}^n$ are equivalent:

1. f is invertible.
2. f is bijective.
3. f is injective.
4. f is surjective.
5. f preserves linear independence (see Application 7 below).
6. $\ker f = \mathbf{0}$.
7. $\text{rank} f = n$.

Another necessary and sufficient condition for the invertibility of a linear operator will be added in Chap. 10.

STANDARD FORM OF A MATRIX. CALCULATION OF THE RANK

We shall now turn to the matrix representation of a linear map f: $\mathbb{R}^n \to \mathbb{R}^m$. Elements of $\mathbb{R}^n$ will be represented as column vectors:

$$\begin{pmatrix} x_1 \\ x_2 \\ \vdots \\ x_n \end{pmatrix}$$

$x_i \in \mathbb{R}$ being the coordinates with respect to the standard basis:

$$\begin{pmatrix} 1 \\ 0 \\ 0 \\ \cdot \\ \cdot \\ \cdot \\ 0 \end{pmatrix}, \begin{pmatrix} 0 \\ 1 \\ 0 \\ \cdot \\ \cdot \\ \cdot \\ 0 \end{pmatrix}, \begin{pmatrix} 0 \\ 0 \\ 1 \\ \cdot \\ \cdot \\ \cdot \\ 0 \end{pmatrix}, \ldots, \begin{pmatrix} 0 \\ 0 \\ 0 \\ \cdot \\ \cdot \\ \cdot \\ 1 \end{pmatrix}$$

Then the linear map f appears as an $n \times m$ matrix:

$$f = \begin{pmatrix} a_{11} & a_{12} & a_{13} & \cdots & a_{1n} \\ a_{21} & a_{22} & a_{23} & \cdots & a_{2n} \\ \cdots & \cdots & \cdots & \cdots & \cdots \\ a_{m1} & a_{m2} & a_{m3} & \cdots & a_{mn} \end{pmatrix}$$

in which the column vectors are just the images of the basis vectors. Note that the number of columns is the same as the dimension of the domain, and the number of rows, the dimension of the codomain of f.

Much useful information can be obtained from the matrix of the linear map f, some of it directly, and some more, after further calculations. Thus rank$f$ can be calculated, and from it, the injectivity or surjectivity of f established. Moreover, ker$f$ and im$f$ can be calculated from the matrix of f. In the rest of this chapter we shall survey these techniques.

We say that the matrix of $f \in L(\mathbb{R}^n, \mathbb{R}^m)$ is in *standard form* if

$$f = \underbrace{\left.\begin{pmatrix} \overbrace{\begin{matrix}1 & 0 & 0 & 0\end{matrix}}^{r} & \cdots & 0 \\ 0 \quad 1 \quad 0 \quad 0 & \cdots & 0 \\ \cdots \ddots & & \\ 0 \quad \cdots \quad\quad 1 & \cdots & 0 \\ \hline 0 \quad \cdots \quad\quad 0 & \cdots & 0 \\ & \ddots & \\ 0 \quad \cdots \quad\quad 0 & \cdots & 0 \end{pmatrix}\right\}}_{n} m$$

is a matrix having n columns with $a_{11} = a_{22} = \cdots = a_{rr} = 1$, and all other entries are 0, where $r = \operatorname{rank} f$ is the number of 1's in the matrix. With an appropriate choice for bases in dom f and cod f, the matrix of every linear map can be brought to standard form.

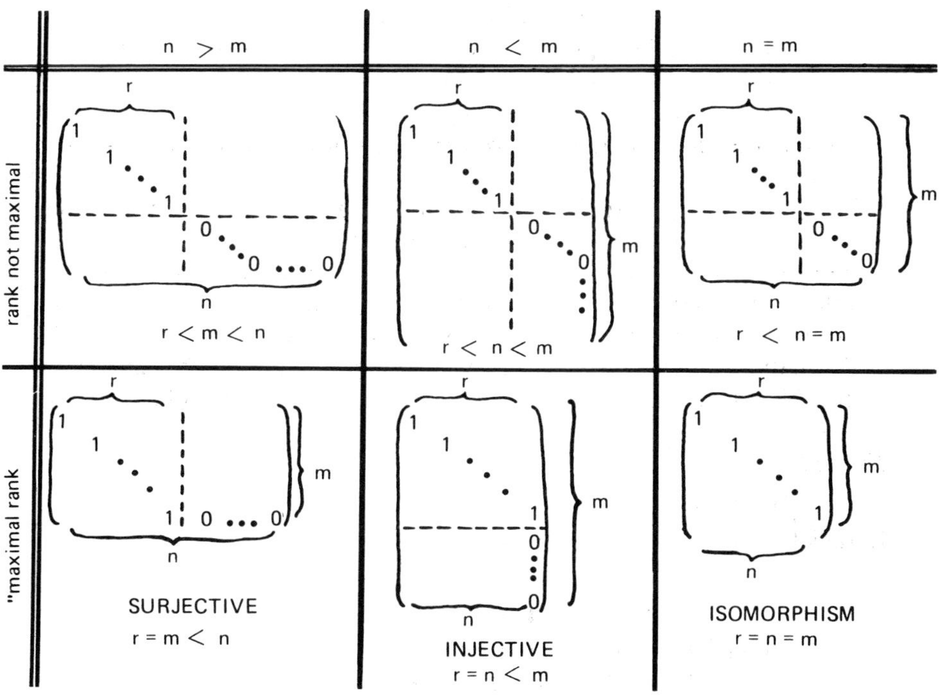

We must guard against the confusion which may arise when looking at the standard form of an $n \times n$ matrix of maximal rank:

$$f = \begin{pmatrix} 1 & 0 & 0 & \cdots & 0 \\ 0 & 1 & 0 & \cdots & 0 \\ 0 & 0 & 1 & \cdots & 0 \\ \vdots & & & \ddots & \\ 0 & 0 & 0 & \cdots & 1 \end{pmatrix} \qquad \operatorname{rank} f = n$$

This should not be confused with the identity operator, which also has maximal rank: rank $1 = n$. But whereas the identity operator is unique, there are infinitely many isomorphisms all having the same standard form as f above. The possibility of this and similar confusions warns us that the matrix language, unsupported by the language of linear maps and linar operators, is a very deficient and inadequate language in which to describe linear algebra.

We calculate rank f by bringing the matrix to standard form. The rank of the matrix will remain unchanged if we perform any of the following operations on the matrix:

1. Interchange two rows of the matrix.
2. Multiply a row of the matrix by a nonzero scalar.
3. Add (subtract) a multiple of a row to (from) another row of the matrix.

1′. Interchange two columns of the matrix.
2′. Multiply a column of the matrix by a nonzero scalar.
3′. Add (subtract) a multiple of a column to (from) another column of the matrix.

The purpose of performing these operations is to "kill off" (=to reduce to zero) as many elements of the matrix as possible. It will not be possible to kill off all the elements of the matrix in this way: a residue will always "survive" and the number of the "survivors" is independent of the manner of killing; in fact, this number is the same as the rank of the matrix. We have reached the position in which no further killing is possible when the survivors belong to different rows and columns.

Examples

1. $\operatorname{rank} \begin{pmatrix} 1 & 1 & 1 & 1 & 1 & 1 & 1 & 1 & 1 \\ 1 & 1 & 1 & 1 & 1 & 1 & 1 & 1 & 1 \\ 1 & 1 & 1 & 1 & 1 & 1 & 1 & 1 & 1 \end{pmatrix}$

$$= \operatorname{rank} \begin{pmatrix} 1 & 0 & 0 & 0 & 0 & 0 & 0 & 0 & 0 \\ 1 & 0 & 0 & 0 & 0 & 0 & 0 & 0 & 0 \\ 1 & 0 & 0 & 0 & 0 & 0 & 0 & 0 & 0 \end{pmatrix}$$

$$= \operatorname{rank}\begin{pmatrix} 1 & 0 & 0 & 0 & 0 & 0 & 0 & 0 & 0 \\ 0 & 0 & 0 & 0 & 0 & 0 & 0 & 0 & 0 \\ 0 & 0 & 0 & 0 & 0 & 0 & 0 & 0 & 0 \end{pmatrix} = 1$$

2. $\operatorname{rank}\begin{pmatrix} 1 & 2 & 3 \\ 4 & 5 & 6 \end{pmatrix} = \operatorname{rank}\begin{pmatrix} 1 & 0 & 0 \\ 4 & -3 & -6 \end{pmatrix} = \operatorname{rank}\begin{pmatrix} 1 & 0 & 0 \\ 0 & -3 & -6 \end{pmatrix}$

$$= \operatorname{rank}\begin{pmatrix} 1 & 0 & 0 \\ 0 & 1 & 0 \end{pmatrix} = 2$$

Since the rank is maximal, the matrix represents a surjective linear map.

3. $\operatorname{rank}\begin{pmatrix} 1 & 0 & 1 \\ 2 & 1 & -1 \\ 1 & 3 & 2 \\ -2 & 4 & 5 \end{pmatrix} = \operatorname{rank}\begin{pmatrix} 1 & 0 & 1 \\ 0 & 1 & -3 \\ 0 & 3 & 1 \\ 0 & 4 & 7 \end{pmatrix}$

$$= \operatorname{rank}\begin{pmatrix} 1 & 0 & 1 \\ 0 & 1 & -3 \\ 0 & 0 & 10 \\ 0 & 0 & 19 \end{pmatrix} = \operatorname{rank}\begin{pmatrix} 1 & 0 & 0 \\ 0 & 1 & -3 \\ 0 & 0 & 1 \\ 0 & 0 & 19 \end{pmatrix}$$

$$= \operatorname{rank}\begin{pmatrix} 1 & 0 & 0 \\ 0 & 1 & 0 \\ 0 & 0 & 1 \\ 0 & 0 & 0 \end{pmatrix} = 3$$

The rank is maximal; hence the matrix represents an injective linear map from $\mathbb{R}^3$ to $\mathbb{R}^4$.

4. $\operatorname{rank}\begin{pmatrix} 1 & -2 & -3 \\ 0 & 1 & -4 \\ 2 & -2 & -2 \\ 3 & -4 & 1 \end{pmatrix} = \operatorname{rank}\begin{pmatrix} 1 & -2 & 3 \\ 0 & 1 & -4 \\ 0 & 2 & -8 \\ 0 & 2 & -8 \end{pmatrix}$

$$= \operatorname{rank}\begin{pmatrix} 1 & 0 & 0 \\ 0 & 1 & -4 \\ 0 & 2 & -8 \\ 0 & 2 & -8 \end{pmatrix} = \operatorname{rank}\begin{pmatrix} 1 & 0 & 0 \\ 0 & 1 & -4 \\ 0 & 0 & 0 \\ 0 & 0 & 0 \end{pmatrix}$$

$$= \operatorname{rank}\begin{pmatrix} 1 & 0 & 0 \\ 0 & 1 & 0 \\ 0 & 0 & 0 \\ 0 & 0 & 0 \end{pmatrix} = 2$$

5. If the column vectors (or the row vectors) of a matrix are pairwise

parallel, then the rank of the matrix is equal to 1; e.g.,

$$\operatorname{rank}\begin{pmatrix} -14 & 7 & -35 & 21 \\ 2 & -1 & 5 & -3 \\ -10 & 5 & -25 & 15 \end{pmatrix} = 1$$

HOMOGENEOUS SYSTEMS OF LINEAR EQUATIONS

A system of linear equations

$$\begin{aligned} a_{11}x_1 + a_{12}x_2 + \cdots + a_{1n}x_n &= b_1 \\ a_{21}x_1 + a_{22}x_2 + \cdots + a_{2n}x_n &= b_2 \\ &\ \ \vdots \\ a_{m1}x_1 + a_{m2}x_2 + \cdots + a_{mn}x_n &= b_m \end{aligned}$$

is called *homogeneous* or *nonhomogeneous* according as all b_k's are 0 or not. The *matrix of the system* is the matrix

$$\begin{pmatrix} a_{11} & a_{12} & \cdots & a_{1n} \\ a_{21} & a_{22} & \cdots & a_{2n} \\ \cdots & \cdots & \cdots & \cdots \\ a_{m1} & a_{m2} & \cdots & a_{mn} \end{pmatrix}$$

consisting of the coefficients on the left-hand side of the equations, and we shall also refer to the *augmented matrix*

$$\left(\begin{array}{cccc|c} a_{11} & a_{12} & \cdots & a_{1n} & b_1 \\ a_{21} & a_{22} & \cdots & a_{2n} & b_2 \\ \cdots & \cdots & \cdots & \cdots & \cdots \\ a_{m1} & a_{m2} & \cdots & a_{mn} & b_m \end{array}\right)$$

where the coefficients on the right-hand side are also included as the "augmentation." However, for a homogeneous system this distinction will not be made and we shall write

$$\begin{pmatrix} a_{11} & a_{12} & \cdots & a_{1n} \\ a_{21} & a_{22} & \cdots & a_{2n} \\ \cdots & \cdots & \cdots & \cdots \\ a_{m1} & a_{m2} & \cdots & a_{mn} \end{pmatrix} \quad \text{instead of} \quad \left(\begin{array}{cccc|c} a_{11} & a_{12} & \cdots & a_{1n} & 0 \\ a_{21} & a_{22} & \cdots & a_{2n} & 0 \\ \cdots & \cdots & \cdots & \cdots & \cdots \\ a_{m1} & a_{m2} & \cdots & a_{mn} & 0 \end{array}\right)$$

Solving a homogeneous system of linear equations is equivalent to the problem of finding ker f.

A homogeneous system of linear equations can always be solved: $x_1 = x_2 = \cdots = x_n = 0$, called the *trivial solution*, always satisfies it. This corresponds to the fact that ker f is never empty: the null vector, at least, belongs to it.

The interesting question, however, is whether a nontrivial solution exists. This is the same question whether ker f is at least one dimensional.

Theorem The homogeneous system of m linear equations with n unknowns has a nontrivial solution if, and only if,

$$r < n$$

i.e., the rank of the matrix is less than the number of unknowns.

This is just the scalar version of the evident statement that ker f is at least one dimensional if, and only if, rank $f < n$. Two important special cases must be stressed.

A homogeneous system of linear equations consisting of fewer equations than unknowns always has a nontrivial solution. This is just the scalar version of the evident statement that f: $\mathbb{R}^n \to \mathbb{R}^m$ with $n > m$ must kill some "live" vectors.

A homogeneous system of n *linear equations with the same number of unknowns has a nontrivial solution if, and only if, the determinant of its matrix vanishes.* Otherwise, the linear map is bijective, having a trivial kernel.

In solving a homogenous system of linear equations, we use the method of gaussian† elimination. The idea is to eliminate the unknown x_1 from all but the first, the unknown x_2 from all but the second, ..., the unknown x_r from all but the rth equation. In terms of the matrix, this means row reduction, i.e., bringing the matrix to *row-echelon form*:

$$\begin{pmatrix} * & * & * & \cdots & * \\ * & * & * & \cdots & * \\ * & * & * & \cdots & * \\ \cdots & \cdots & \cdots & \cdots & \cdots \end{pmatrix} \Rightarrow \begin{pmatrix} 1 & 0 & 0 & \cdots & * & \cdots & * \\ 0 & 1 & 0 & \cdots & * & \cdots & * \\ 0 & 0 & 1 & \cdots & * & \cdots & * \\ \cdots & \cdots & \cdots & \cdots & \cdots & \cdots & \cdots \end{pmatrix}$$

where all the entries are 0 except $a_{11} = a_{22} = \cdots = a_{rr} = 1$. This may be achieved by successive subtractions of rows, *in preference to division of a row by a scalar. Such a division may introduce fractions making the calculation*

†Carl Friedrich Gauss (1777-1855), professor at Göttingen, the greatest mathematician of all times.

more cumbersome. In a homogeneous system, fractions in the solution can be avoided altogether (provided that all the coefficients are rational numbers) through a judicious choice of the free parameters.

In the process of row reduction we may lose a row (i.e., a row may become all zero). This means that the corresponding equation of the system, being redundant, disappears. Such an occurrence will increase the degrees of freedom, i.e., the number of free parameters, by one. An all-zero row may be deleted from the matrix: neither the rank nor the solution is being changed thereby.

By contrast, we cannot lose a column of the matrix by row reduction. The worst that can happen is that "we lose the 1 in a column." In trying to eliminate x_k from all but the kth equation, we may have inadvertently eliminated x_{k+1} from the $(k+1)$st and all subsequent equations. This means that x_{k+1} can be given any arbitrary value; i.e., x_{k+1} is one of the free parameters. The way this shows up in the matrix is $a_{k+1,k+1} = a_{k+1,k+2} = \cdots = 0$ and we cannot have $a_{k+1,k+1} = 1$. When we lose the 1 from a column, this column may be moved to the last place. When doing this we may be well advised to label this column by the corresponding unknown x_{k+1} as a reminder that the order of the unknowns has been changed.

Thus the process of gaussian elimination, when carried out in terms of the matrix, will yield the row-echelon form

$$\begin{pmatrix} 1 & 0 & 0 & \cdots & 0 & * & \cdots & * \\ 0 & 1 & 0 & \cdots & 0 & * & \cdots & * \\ 0 & 0 & 1 & \cdots & 0 & * & \cdots & * \\ \cdots & \cdots & \cdots & \cdots & \cdots & \cdots & \cdots & \cdots \\ 0 & 0 & 0 & \cdots & 1 & * & \cdots & * \\ 0 & 0 & 0 & \cdots & 0 & 0 & \cdots & 0 \\ \cdots & \cdots & \cdots & \cdots & \cdots & \cdots & \cdots & \cdots \end{pmatrix}$$
$$\underbrace{\hphantom{1\;0\;0\;\cdots\;0}}_{r}\qquad\underbrace{\hphantom{*\;\cdots\;*}}_{n-r}$$

From the row-echelon form, the solution of the homogeneous system can be read off directly. The degree of freedom, i.e., the number of free parameters, is the same as the number of $*$'s in the last nonvanishing row, which is the same as the difference between the number of columns and the number of surviving rows: $n-r$. Since the number of surviving rows is rank f, the number of free parameters is the same as the nullity of f. In fact, the solution of the homogeneous system of linear equations can be considered as the parametric form of the kernel of its matrix. Hence the number of free parameters is just dim ker$f$.

Examples

1. Solve

$$\begin{aligned} 2x_1 + 3x_2 - x_3 + 5x_4 &= 0 \\ 3x_1 - x_2 + 2x_3 - 7x_4 &= 0 \\ x_1 - 2x_2 + 4x_3 - 7x_4 &= 0 \end{aligned}$$

Answer:

$$\begin{pmatrix} 2 & 3 & -1 & 5 \\ 3 & -1 & 2 & -7 \\ 1 & -2 & 4 & -7 \end{pmatrix} \Rightarrow \begin{pmatrix} 1 & -2 & 4 & -7 \\ 0 & 7 & -9 & 19 \\ 0 & 5 & -10 & 14 \end{pmatrix}$$

$$\Rightarrow \begin{pmatrix} 1 & -2 & 4 & -7 \\ 0 & 2 & 1 & 5 \\ 0 & 1 & -12 & 4 \end{pmatrix} \Rightarrow \begin{pmatrix} 1 & -2 & 4 & -7 \\ 0 & 1 & -12 & 4 \\ 0 & 0 & 25 & -3 \end{pmatrix}$$

$$\Rightarrow \begin{pmatrix} 1 & 0 & 0 & -35/25 \\ 0 & 1 & 0 & 64/25 \\ 0 & 0 & 1 & -3/25 \end{pmatrix} \Rightarrow \begin{cases} x_1 = 35t \\ x_2 = -64t \\ x_3 = 3t \\ x_4 = 25t \end{cases}$$

2. Solve

$$\begin{aligned} 2x_1 + x_2 + x_3 - 2x_4 &= 0 \\ x_1 + 3x_2 + x_3 - 5x_4 &= 0 \\ x_1 + x_2 + 5x_3 + 7x_4 &= 0 \\ 2x_1 + 3x_2 - 3x_3 - 14x_4 &= 0 \end{aligned}$$

Answer:

$$\begin{pmatrix} 2 & 1 & 1 & -2 \\ 1 & 3 & 1 & -5 \\ 1 & 1 & 5 & 7 \\ 2 & 3 & -3 & -14 \end{pmatrix} \Rightarrow \begin{pmatrix} 1 & 3 & 1 & -5 \\ 1 & 1 & 5 & 7 \\ 2 & 3 & -3 & -14 \\ 2 & 1 & 1 & -2 \end{pmatrix}$$

$$\Rightarrow \begin{pmatrix} 1 & 3 & 1 & -5 \\ 0 & -2 & 4 & 12 \\ 0 & -3 & -5 & -4 \\ 0 & -5 & -1 & 8 \end{pmatrix} \Rightarrow \begin{pmatrix} 1 & 3 & 1 & -5 \\ 0 & 1 & -2 & -6 \\ 0 & 0 & -11 & -22 \\ 0 & 0 & -11 & -22 \end{pmatrix}$$

$$\Rightarrow\begin{pmatrix}1&3&1&-5\\0&1&-2&-6\\0&0&1&2\\0&0&0&0\end{pmatrix}\Rightarrow\begin{pmatrix}1&0&0&-1\\0&1&0&-2\\0&0&1&2\\0&0&0&0\end{pmatrix}\Rightarrow\begin{cases}x_1=t\\x_2=2t\\x_3=-2t\\x_4=t\end{cases}$$

3. Solve

$$\begin{aligned}3x_1+\ 4x_2-\ 5x_3+\ 7x_4&=0\\2x_1-\ 3x_2+\ 3x_3-\ 2x_4&=0\\4x_1+11x_2-13x_3+16x_4&=0\\7x_1-\ 2x_2+\ x_3+\ 3x_4&=0\\x_1+24x_2-27x_3+29x_4&=0\end{aligned}$$

Answer:

$$\begin{pmatrix}3&4&-5&7\\2&-3&3&-2\\4&11&-13&16\\7&-2&1&3\\1&24&-27&29\end{pmatrix}\Rightarrow\begin{pmatrix}1&7&-8&9\\0&-17&19&-20\\0&-17&19&-20\\0&-5&57&-60\\0&17&-19&20\end{pmatrix}$$

$$\Rightarrow\begin{pmatrix}1&7&-8&9\\0&17&-19&20\\0&0&0&0\\0&0&0&0\\0&0&0&0\end{pmatrix}\Rightarrow\begin{pmatrix}1&0&-3/17&13/17\\0&1&-19/17&20/17\\0&0&0&0\\0&0&0&0\\0&0&0&0\end{pmatrix}\Rightarrow\begin{cases}x_1=3s-13t\\x_2=19s-20t\\x_3=17s\\x_4=17t\end{cases}$$

APPLICATIONS

6. Finding the Kernel

Examples

1. Find ker f if

$$f=\begin{pmatrix}1&5&11\\2&3&8\\-1&2&3\end{pmatrix}$$

Answer:

$$f(\mathbf{v})=\mathbf{0}\Rightarrow\begin{pmatrix}1&5&11\\0&-7&-14\\0&7&14\end{pmatrix}\Rightarrow\begin{pmatrix}1&5&11\\0&1&2\end{pmatrix}\Rightarrow\begin{pmatrix}1&0&1\\0&1&2\end{pmatrix}\Rightarrow\begin{cases}x=\ -t\\y=-2t\\z=\ \ t\end{cases}$$

ker f is the line $-x=y/-2=z$.

2. Find ker f if

$$f=\begin{pmatrix} 6 & -1 & 4 \\ -1 & 3 & -1 \\ 2 & 7 & 0 \end{pmatrix}$$

Answer:

$$f(\mathbf{v})=\mathbf{0}\Rightarrow\begin{pmatrix} 1 & -3 & 1 \\ 0 & 17 & -2 \\ 0 & 13 & -2 \end{pmatrix}\Rightarrow\begin{pmatrix} 1 & -3 & 1 \\ 0 & 4 & 0 \\ 0 & 13 & -2 \end{pmatrix}\Rightarrow\begin{pmatrix} 1 & -3 & 1 \\ 0 & 1 & 0 \\ 0 & 0 & -2 \end{pmatrix}$$

$$\Rightarrow \det f\neq 0\Rightarrow \ker f=\mathbf{0}$$

f is bijective.

3. Find ker f if

$$f=\begin{pmatrix} 49 & 42 & 63 \\ 56 & 48 & 72 \\ 63 & 54 & 81 \end{pmatrix}$$

Answer:

$$f(\mathbf{v})=\mathbf{0}\Rightarrow\begin{pmatrix} 49 & 42 & 63 \\ 7 & 6 & 9 \\ 14 & 12 & 18 \end{pmatrix}\Rightarrow(7 \quad 6 \quad 9)\Rightarrow 7x+6y+9z=0$$

is the equation of ker f, which is a plane.

7. Linear Independence and Dependence

We say that the vectors $\mathbf{v}_1, \mathbf{v}_2, \ldots, \mathbf{v}_k$ are *linearly independent* if

$$\lambda_1\mathbf{v}_1+\lambda_2\mathbf{v}_2+\cdots+\lambda_k\mathbf{v}_k=\mathbf{0}\Rightarrow\lambda_1=\lambda_2=\cdots=\lambda_k=0$$

i.e., the only way the null vector can be written as a linear combination is the trivial way. If the vectors $\mathbf{v}_1, \mathbf{v}_2, \ldots, \mathbf{v}_k$ are *not* linearly independent, we call them *linearly dependent*. This means that there is a nontrivial linear combination $\lambda_1\mathbf{v}_1+\lambda_2\mathbf{v}_2+\cdots+\lambda_k\mathbf{v}_k=\mathbf{0}$, i.e., at least for one coefficient $\lambda_\nu\neq 0$. In determining whether a set of vectors is linearly independent or not, we ask the question whether the homogeneous system of linear equations formed of the coordinates of the vectors has not, or has, a nontrivial solution.

We shall now prove that

$$f \text{ preserves linear independence} \Leftrightarrow \ker f=\mathbf{0}$$

If $\ker f = \mathbf{0}$, then for every linearly independent set $\{\mathbf{v}_1, \ldots, \mathbf{v}_k\}$ we have

$$\begin{aligned} \lambda_1 f(\mathbf{v}_1) + \cdots + \lambda_k f(\mathbf{v}_k) = \mathbf{0} &\Rightarrow f(\lambda_1 \mathbf{v}_1 + \cdots + \lambda_k \mathbf{v}_k) = \mathbf{0} \\ &\Rightarrow \lambda_1 \mathbf{v}_1 + \cdots + \lambda_k \mathbf{v}_k = \mathbf{0} \Rightarrow \lambda_1 = \cdots = \lambda_k = 0 \end{aligned}$$

We conclude that f preserves linear independence. Conversely, assume that f preserves linear independence, and let $\{\mathbf{e}_1, \ldots, \mathbf{e}_n\}$ be a basis. Then

$$\begin{aligned} f(\mathbf{v}) = \mathbf{0} &\Rightarrow f(\lambda_1 \mathbf{e}_1 + \cdots + \lambda_n \mathbf{e}_n) = \mathbf{0} \\ &\Rightarrow \lambda_1 f(\mathbf{e}_1) + \cdots + \lambda_n f(\mathbf{e}_n) = \mathbf{0} \\ &\Rightarrow \lambda_1 = \cdots = \lambda_n = 0 \Rightarrow \mathbf{v} = \lambda_1 \mathbf{e}_1 + \cdots + \lambda_n \mathbf{e}_n = \mathbf{0} \end{aligned}$$

Since $\mathbf{v}$ is arbitrary, we conclude that $\ker f = 0$.

Examples

1. Are the vectors

$$\begin{pmatrix} 1 \\ 1 \\ 1 \\ 1 \end{pmatrix}, \begin{pmatrix} 1 \\ 2 \\ 3 \\ 4 \end{pmatrix}, \begin{pmatrix} 1 \\ 3 \\ 6 \\ 10 \end{pmatrix}, \begin{pmatrix} 1 \\ 4 \\ 10 \\ 20 \end{pmatrix}$$

linearly independent or dependent?

Answer:

$$\begin{pmatrix} 1 & 1 & 1 & 1 \\ 1 & 2 & 3 & 4 \\ 1 & 3 & 6 & 10 \\ 1 & 4 & 10 & 20 \end{pmatrix} \Rightarrow \begin{pmatrix} 1 & 1 & 1 & 1 \\ 0 & 1 & 2 & 3 \\ 0 & 2 & 5 & 9 \\ 0 & 3 & 9 & 19 \end{pmatrix}$$

$$\Rightarrow \begin{pmatrix} 1 & 1 & 1 & 1 \\ 0 & 1 & 2 & 3 \\ 0 & 0 & 1 & 3 \\ 0 & 0 & 3 & 10 \end{pmatrix} \Rightarrow \begin{pmatrix} 1 & 1 & 1 & 1 \\ 0 & 1 & 2 & 3 \\ 0 & 0 & 1 & 3 \\ 0 & 0 & 0 & 1 \end{pmatrix} \Rightarrow \text{no nontrivial solution}$$

The vectors are linearly independent.

2. Are the vectors

$$\begin{pmatrix} 1 \\ 3 \\ 1 \\ -2 \end{pmatrix}, \begin{pmatrix} 3 \\ 2 \\ 4 \\ 2 \end{pmatrix}, \begin{pmatrix} 2 \\ -1 \\ 3 \\ 4 \end{pmatrix} \text{ linearly independent or dependent?}$$

Answer:

$$\begin{pmatrix} 1 & 3 & 2 \\ 3 & 2 & -1 \\ 1 & 4 & 3 \\ -2 & 2 & 4 \end{pmatrix} \Rightarrow \begin{pmatrix} 1 & 3 & 2 \\ 0 & -7 & -7 \\ 0 & 1 & 1 \\ 0 & 8 & 8 \end{pmatrix}$$

$$\Rightarrow \begin{pmatrix} 1 & 3 & 2 \\ 0 & 1 & 1 \\ 0 & 0 & 0 \\ 0 & 0 & 0 \end{pmatrix} \Rightarrow \begin{cases} x = t \\ y = -t \\ z = t \end{cases}$$

Nontrivial solutions exist, e.g.,

$$\begin{pmatrix} 1 \\ 3 \\ 1 \\ -2 \end{pmatrix} - \begin{pmatrix} 3 \\ 2 \\ 4 \\ 2 \end{pmatrix} + \begin{pmatrix} 2 \\ -1 \\ 3 \\ 4 \end{pmatrix} = \begin{pmatrix} 0 \\ 0 \\ 0 \\ 0 \end{pmatrix}$$

The vectors are linearly dependent.

NONHOMOGENEOUS SYSTEMS OF LINEAR EQUATIONS

The problem of deciding whether a vector does or does not belong to the image of a linear map is the same as that of solving a nonhomogeneous system of linear equations.

Let $f: \mathbb{R}^n \to \mathbb{R}^m$ be a linear map; we wish to find all the solutions in the unknown $\mathbf{v} \in \mathbb{R}^n$ of the vector equation $f(\mathbf{v}) = \mathbf{b}$, where $\mathbf{b} \in \mathbb{R}^m$ is a fixed vector. If we translate this vector equation into scalar equations for the coordinates, we see that this is the same problem as that of solving the nonhomogeneous system of linear equations:

$$
\begin{aligned}
a_{11}x_1 + a_{12}x_2 + \cdots + a_{1n}x_n &= b_1 \\
a_{21}x_1 + a_{22}x_2 + \cdots + a_{2n}x_n &= b_2 \\
&\ \vdots \\
a_{m1}x_1 + a_{m2}x_2 + \cdots + a_{mn}x_n &= b_m
\end{aligned}
$$

or, what is the same, finding solutions of the matrix equation

$$
\begin{pmatrix} a_{11} & a_{12} & \cdots & a_{1n} \\ a_{21} & a_{22} & \cdots & a_{2n} \\ \cdots & \cdots & \cdots & \cdots \\ a_{m1} & a_{m2} & \cdots & a_{mn} \end{pmatrix} \circ \begin{pmatrix} x_1 \\ x_2 \\ \cdot \\ \cdot \\ \cdot \\ x_n \end{pmatrix} = \begin{pmatrix} b_1 \\ b_2 \\ \cdot \\ \cdot \\ \cdot \\ b_m \end{pmatrix}
$$

for the unknown

$$
\begin{pmatrix} x_1 \\ x_2 \\ \cdot \\ \cdot \\ \cdot \\ x_n \end{pmatrix} = \mathbf{v} \in \mathbb{R}^n
$$

Clearly, solutions may or may not exist. We now give a necessary and sufficient condition for the existence of a solution to the equation.

Theorem The vector equation $f(\mathbf{v}) = \mathbf{b}$, where $f\colon \mathbb{R}^n \to \mathbb{R}^m$ is a linear map and $\mathbf{b} \in \mathbb{R}^m$ is a given fixed vector, can be solved for $\mathbf{v} \in \mathbb{R}^n$ if, and only if, $\mathbf{b} \in \operatorname{im} f$.

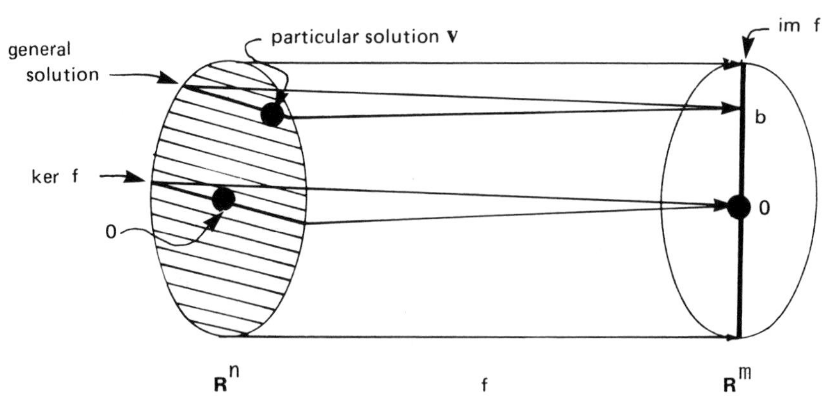

In practice, we decide solvability by checking the ranks of certain matrices. Since the column vectors of the matrix are $f(\mathbf{e}_1), f(\mathbf{e}_2), \ldots, f(\mathbf{e}_n) \in \mathbb{R}^m$, where $\mathbf{e}_1, \mathbf{e}_2, \ldots, \mathbf{e}_n \in \mathbb{R}^n$ are the basis vectors for $\mathbb{R}^n$, and since those column vectors span $\operatorname{im} f$, the condition $\mathbf{b} \in \operatorname{im} f$ can be replaced by another condition, namely: the span of the set of vectors $\{f(\mathbf{e}_1), f(\mathbf{e}_2), \ldots, f(\mathbf{e}_n), \mathbf{b}\}$ is no larger than the span of the set $\{f(\mathbf{e}_1), f(\mathbf{e}_2), \ldots, f(\mathbf{e}_n)\}$, in fact, both spans are the same, i.e., $\operatorname{im} f$.

This is the same as to say that

$$\mathbf{b} \in \operatorname{im} f \Leftrightarrow \operatorname{rank} \begin{pmatrix} a_{11} & a_{12} & \cdots & a_{1n} \\ a_{21} & a_{22} & \cdots & a_{2n} \\ \cdots & \cdots & \cdots & \cdots \\ a_{m1} & a_{m2} & \cdots & a_{mn} \end{pmatrix}$$

$$= \operatorname{rank} \left(\begin{array}{cccc|c} a_{11} & a_{12} & \cdots & a_{1n} & b_1 \\ a_{21} & a_{22} & \cdots & a_{2n} & b_2 \\ \cdots & \cdots & \cdots & \cdots & \cdots \\ a_{m1} & a_{m2} & \cdots & a_{mn} & b_m \end{array}\right)$$

i.e., *a solution for the nonhomogeneous system of linear equations*

$$\begin{aligned} a_{11}x_1 + a_{12}x_2 + \cdots + a_{1n}x_n &= b_1 \\ a_{21}x_1 + a_{22}x_2 + \cdots + a_{2n}x_n &= b_2 \\ &\vdots \\ a_{m1}x_1 + a_{m2}x_2 + \cdots + a_{mn}x_n &= b_m \end{aligned}$$

exists if, and only if, the rank of the augmented matrix:

$$\left(\begin{array}{cccc|c} a_{11} & a_{12} & \cdots & a_{1n} & b_1 \\ a_{21} & a_{22} & \cdots & a_{2n} & b_2 \\ \cdots & \cdots & \cdots & \cdots & \cdots \\ a_{m1} & a_{m2} & \cdots & a_{mn} & b_m \end{array}\right)$$

is no greater than the rank of the matrix of the system:

$$\begin{pmatrix} a_{11} & a_{12} & \cdots & a_{1n} \\ a_{21} & a_{22} & \cdots & a_{2n} \\ \cdots & \cdots & \cdots & \cdots \\ a_{m1} & a_{m2} & \cdots & a_{mn} \end{pmatrix}$$

Suppose now that this condition is satisfied; we want to find all the solutions of the equation $f(\mathbf{v}) = \mathbf{b}$. It is clear that the *general solution* of the nonhomogeneous system will have the form

$$\mathbf{v} = \tilde{\mathbf{v}} + \ker f$$

where

$$\tilde{\mathbf{v}} = \begin{pmatrix} \tilde{x}_1 \\ \tilde{x}_2 \\ \cdot \\ \cdot \\ \cdot \\ \tilde{x}_n \end{pmatrix}$$

is a *particular solution* of the nonhomogeneous system and $\ker f$ stands for the general solution of the corresponding homogeneous system.

In practice, the calculation of the ranks of the two matrices, the calculation of the general solution of the underlying homogeneous system, and the finding of a particular solution of the nonhomogeneous system is compressed into the same procedure of bringing the augmented matrix to row-echelon form via gaussian elimination, as in the homogeneous case. The important difference is that, unlike in the homogeneous case, *existence of solution is not guaranteed in the nonhomogeneous case*. If a solution does not exist, it will show up in the row-echelon form of the augmented matrix, with the augmenting column having one extra survivor:

$$\left(\begin{array}{cccccccc|c} 1 & 0 & 0 & \cdots & 0 & * & * & * & * \\ 0 & 1 & 0 & \cdots & 0 & * & * & * & * \\ 0 & 0 & 1 & \cdots & 0 & * & * & * & * \\ \cdots & \cdots & \cdots & \cdots & \cdots & \cdots & \cdots & \cdots & \cdots \\ 0 & 0 & 0 & \cdots & 1 & * & * & * & * \\ 0 & 0 & 0 & \cdots & 0 & 0 & 0 & 0 & 1 \\ \cdots & \cdots & \cdots & \cdots & \cdots & \cdots & \cdots & \cdots & \cdots \\ 0 & 0 & 0 & \cdots & 0 & 0 & 0 & 0 & 0 \end{array}\right) \Rightarrow \text{no solution}$$

because the rank of the augmented matrix is greater than the rank of the matrix of the system.

If, on the other hand, the row-echelon form has the shape

$$\left(\begin{array}{cccccccc|c} 1 & 0 & 0 & \cdots & 0 & * & * & * & * \\ 0 & 1 & 0 & \cdots & 0 & * & * & * & * \\ 0 & 0 & 1 & \cdots & 0 & * & * & * & * \\ \cdots & \cdots & \cdots & \cdots & \cdots & \cdots & \cdots & \cdots & \cdots \\ 0 & 0 & 0 & \cdots & 1 & * & * & * & * \\ 0 & 0 & 0 & \cdots & 0 & 0 & 0 & 0 & 0 \\ \cdots & \cdots & \cdots & \cdots & \cdots & \cdots & \cdots & \cdots & \cdots \\ 0 & 0 & 0 & \cdots & 0 & 0 & 0 & 0 & 0 \end{array}\right)$$

then a solution does exist, because the rank of the augmented matrix agrees with the rank of the matrix of the system. If this rank is r, and if the number of unknowns is n, then the nullity of the matrix of the system is $n-r$, the same as the number of the *'s in the last surviving row, apart from the augmenting column. This number is the degree of freedom, i.e., the number of free parameters in the general solution; it is just the nullity of the matrix of the system, $\dim \ker f$.

Sometimes it will be found advantageous (and, sometimes, necessary) to interchange columns of the matrix (although never with the augmenting column!). In such a case we may be well advised to label each column by the corresponding unknown, as a reminder that the order of unknowns has been changed. Fractions in the particular solution may be unavoidable, but we can always get away with integral coefficients for the free parameters through their judicious choice, provided that the original system has rational coefficients.

Examples

1. Find all the solutions, if any, of the system

$$\begin{aligned} x_1+3x_2+x_3-x_4-3x_5 &= 3 \\ 4x_1-\ x_2+x_3-x_4-\ x_5 &= 4 \\ -7x_1+5x_2-x_3+x_4-\ x_5 &= -5 \\ x_1-5x_2-\ x_3-x_4+\ x_5 &= -2 \end{aligned}$$

Answer:

$$\begin{array}{c} \begin{array}{ccccc} x_4 & x_2 & x_3 & x_4 & x_5 \end{array} \\ \left(\begin{array}{rrrrr|r} 1 & 3 & 1 & -1 & -3 & 3 \\ 4 & -1 & 1 & -1 & -1 & 4 \\ -7 & 5 & -1 & 1 & -1 & -5 \\ 1 & -5 & -1 & -1 & 1 & -2 \end{array}\right) \end{array} \Rightarrow \begin{array}{c} \begin{array}{ccccc} x_3 & x_4 & x_5 & x_1 & x_2 \end{array} \\ \left(\begin{array}{rrrrr|r} 1 & -1 & -3 & 1 & 1 & 3 \\ 0 & 0 & 2 & 3 & -4 & 1 \\ 0 & 0 & -4 & -6 & 8 & -2 \\ 0 & -2 & -2 & 2 & -2 & 1 \end{array}\right) \end{array}$$

$$\Rightarrow \begin{array}{c} \begin{array}{ccccc} x_3 & x_4 & x_5 & x_1 & x_2 \end{array} \\ \left(\begin{array}{rrrrr|r} 1 & -1 & -3 & 1 & 3 & 3 \\ 0 & -2 & -2 & 2 & -2 & 1 \\ 0 & 0 & 2 & 3 & -4 & 1 \\ 0 & 0 & 0 & 0 & 0 & 0 \end{array}\right) \end{array} \Rightarrow \begin{array}{c} \begin{array}{ccccc} x_3 & x_4 & x_5 & x_4 & x_2 \end{array} \\ \left(\begin{array}{rrrrr|r} 1 & 0 & 0 & 3 & 0 & 7/2 \\ 0 & 1 & 0 & -5/2 & 3 & -1 \\ 0 & 0 & 1 & 3/2 & -2 & 1/2 \\ 0 & 0 & 0 & 0 & 0 & 0 \end{array}\right) \end{array}$$

$$\Rightarrow \begin{cases} x_1 = 2s \\ x_2 = 2t \\ x_3 = 7/2 - 6s \\ x_4 = -\ 1 + 5s - 6t \\ x_5 = 1/2 - 3s + 4t \end{cases}$$

We see from this that the rank of both the matrix of the system as well as the augmented matrix is 3, and the nullity is $5-3=2$. A particular solution is

$$\begin{pmatrix} 0 \\ 0 \\ 7/2 \\ -1 \\ 1/2 \end{pmatrix},$$

and $\ker f$ is spanned by the vectors

$$\begin{pmatrix} 2 \\ 0 \\ -6 \\ 5 \\ -3 \end{pmatrix}, \begin{pmatrix} 0 \\ 2 \\ 0 \\ -6 \\ 4 \end{pmatrix}$$

2. Find all the solutions, if any, of the system

$$\begin{aligned} x_1 + x_2 - 3x_3 &= -1 \\ 2x_1 + x_2 - 2x_3 &= 1 \\ x_1 + x_2 + x_3 &= 3 \\ x_1 + 2x_2 - 3x_3 &= 1 \end{aligned}$$

Answer:

$$\left(\begin{array}{ccc|c} 1 & 1 & -3 & -1 \\ 2 & 1 & -2 & 1 \\ 1 & 1 & 1 & 3 \\ 1 & 2 & -3 & 1 \end{array}\right) \Rightarrow \left(\begin{array}{ccc|c} 1 & 1 & -3 & -1 \\ 0 & -1 & 4 & 3 \\ 0 & 0 & 4 & 4 \\ 0 & 1 & 0 & 2 \end{array}\right)$$

$$\Rightarrow \left(\begin{array}{ccc|c} 1 & 1 & -3 & -1 \\ 0 & 1 & 0 & 2 \\ 0 & 0 & 4 & 5 \\ 0 & 0 & 4 & 4 \end{array}\right) \Rightarrow \left(\begin{array}{ccc|c} 1 & 1 & -3 & -3 \\ 0 & 1 & 0 & 2 \\ 0 & 0 & 1 & 5/4 \\ 0 & 0 & 0 & 1 \end{array}\right)$$

implies no solution exists, since the rank of the augmented matrix exceeds that of the matrix of the system.

3. Find all the solutions, if any, of the system

$$\begin{aligned} 2x_1 + x_2 + x_3 &= 2 \\ x_1 + 3x_2 + x_3 &= 5 \\ x_1 + x_2 + 5x_3 &= -7 \\ 2x_1 + 3x_2 - 3x_3 &= 14 \end{aligned}$$

$$\left(\begin{array}{ccc|c} 2 & 1 & 1 & 2 \\ 1 & 3 & 1 & 5 \\ 1 & 1 & 5 & -7 \\ 2 & 3 & -3 & 14 \end{array}\right) \Rightarrow \left(\begin{array}{ccc|c} 1 & 3 & 1 & 5 \\ 0 & -2 & 4 & -12 \\ 0 & -3 & -5 & 4 \\ 0 & -5 & -1 & -8 \end{array}\right)$$

$$\Rightarrow \left(\begin{array}{ccc|c} 1 & 3 & 1 & 5 \\ 0 & 1 & 9 & -16 \\ 0 & -3 & -5 & 4 \\ 0 & -5 & -1 & -8 \end{array}\right) \Rightarrow \left(\begin{array}{ccc|c} 1 & 0 & 0 & 1 \\ 0 & 1 & 0 & 2 \\ 0 & 0 & 1 & -2 \\ 0 & 0 & 0 & 0 \end{array}\right)$$

$$\Rightarrow \begin{cases} x_1 = 1 \\ x_2 = 2 \\ x_3 = -2 \end{cases}$$

Notice that rank f is *maximal*; hence f is injective and the solution is unique.

4. Solve

$$\begin{aligned} 4x_1 + 2x_2 - x_3 + x_4 - x_5 &= 2 \\ x_1 - 3x_2 + 2x_3 + x_4 + x_5 &= 1 \\ 3x_1 - x_2 + 3x_3 - x_4 - x_5 &= 0 \end{aligned}$$

Answer:

$$\left(\begin{array}{ccccc|c} 4 & 2 & -1 & 1 & -1 & 2 \\ 1 & -3 & 2 & 1 & 1 & 1 \\ 3 & -1 & 3 & -1 & -1 & 0 \end{array}\right) \Rightarrow \begin{array}{c} \begin{array}{ccccc} x_5 & x_2 & x_3 & x_4 & x_1 \end{array} \\ \left(\begin{array}{ccccc|c} 1 & -2 & 1 & -1 & -4 & -2 \\ 1 & -3 & 2 & 1 & 1 & 1 \\ -1 & -1 & 3 & -1 & 3 & 0 \end{array}\right) \end{array}$$

$$\Rightarrow \begin{array}{c} \begin{array}{ccccc} x_5 & x_2 & x_3 & x_4 & x_1 \end{array} \\ \left(\begin{array}{ccccc|c} 1 & -2 & 1 & -1 & -4 & -2 \\ 0 & -1 & 1 & 2 & 5 & 3 \\ 0 & -3 & 4 & -2 & -1 & -2 \end{array}\right) \end{array} \Rightarrow \begin{array}{c} \begin{array}{ccccc} x_5 & x_2 & x_3 & x_4 & x_1 \end{array} \\ \left(\begin{array}{ccccc|c} 1 & -2 & 1 & -1 & -4 & -2 \\ 0 & 1 & -1 & -2 & -5 & -3 \\ 0 & 0 & 1 & -8 & -16 & -11 \end{array}\right) \end{array}$$

$$\Rightarrow \begin{array}{c} \begin{array}{ccccc} x_5 & x_2 & x_3 & x_4 & x_1 \end{array} \\ \left(\begin{array}{ccccc|c} 1 & -2 & 0 & 7 & 12 & 9 \\ 0 & 1 & 0 & -10 & -21 & -14 \\ 0 & 0 & 1 & -8 & -16 & -11 \end{array}\right) \end{array} \Rightarrow \begin{array}{c} \begin{array}{ccccc} x_5 & x_2 & x_3 & x_4 & x_1 \end{array} \\ \left(\begin{array}{ccccc|c} 1 & 0 & 0 & -13 & -30 & -19 \\ 0 & 1 & 0 & -10 & -21 & -14 \\ 0 & 0 & 1 & -8 & -16 & -11 \end{array}\right) \end{array}$$

$$\Rightarrow \begin{cases} x_1 = t \\ x_2 = -14 + 10s + 21t \\ x_3 = -11 + 8s + 16t \\ x_4 = s \\ x_5 = -19 + 13s + 30t \end{cases}$$

Notice that the rank is *maximal*; hence f is *surjective*.

APPLICATIONS

8. Finding the Image

Finding $\ker f$ is the same as solving a homogeneous system of linear equations, as we have seen. On the face of it, the problem of finding $\operatorname{im} f$ presents its own solution, since the column vectors of the matrix of f span $\operatorname{im} f$. However, they may not do this most economically, and the problem of finding $\operatorname{im} f$ actually consists in picking a maximal subset of linearly independent column vectors from the matrix of f. This can be done by reducing the matrix to *column-echelon form*:

$$\begin{pmatrix} 1 & 0 & 0 & 0 & 0 & 0 & \cdots & 0 \\ * & 1 & 0 & 0 & 0 & 0 & \cdots & 0 \\ * & * & 1 & 0 & 0 & 0 & \cdots & 0 \\ * & * & * & 1 & 0 & 0 & \cdots & 0 \\ * & * & * & * & 1 & 0 & \cdots & 0 \\ * & * & * & * & * & 0 & \cdots & 0 \\ \cdots & \cdots & \cdots & \cdots & \cdots & \cdots & \cdots & \cdots \\ * & * & * & * & * & 0 & \cdots & 0 \end{pmatrix}$$

The procedure is, once more, gaussian elimination, but this time we keep adding and subtracting columns, rather than rows. The surviving columns will mark off those column vectors of the matrix which form a maximal subset of linearly independent vectors.

Examples

1. Find $\operatorname{im} f$ if

$$f = \begin{pmatrix} 1 & 2 & 7 \\ -2 & 1 & -4 \\ 1 & -1 & 1 \end{pmatrix}$$

Answer:

$$\begin{pmatrix} 1 & 2 & 7 \\ -2 & 1 & -4 \\ 1 & -1 & 1 \end{pmatrix} \Rightarrow \begin{pmatrix} 1 & 0 & 0 \\ -2 & 5 & 10 \\ 1 & -3 & -6 \end{pmatrix} \Rightarrow \begin{pmatrix} 1 & 0 & 0 \\ -2 & 5 & 0 \\ 1 & -3 & 0 \end{pmatrix}$$

$\operatorname{im} f$ is the plane spanned by the first two column vectors:

$$\operatorname{im} f = s\begin{pmatrix}1\\-2\\1\end{pmatrix} + t\begin{pmatrix}2\\1\\-1\end{pmatrix}$$

2. Find im f if

$$f = \begin{pmatrix}1 & 1 & 2\\2 & -3 & -1\\-3 & 2 & 5\end{pmatrix}$$

Answer:

$$\begin{pmatrix}1 & 1 & 2\\2 & -3 & -1\\-3 & 2 & 5\end{pmatrix} \Rightarrow \begin{pmatrix}1 & 0 & 0\\2 & -5 & -5\\-3 & 5 & -8\end{pmatrix} \Rightarrow \begin{pmatrix}1 & 0 & 0\\2 & 1 & 0\\-3 & -1 & 1\end{pmatrix}$$

im $f = \mathbb{R}^3$, f is surjective and hence bijective; it is an isomorphism.

3. Find im f if

$$f = \begin{pmatrix}42 & -54 & 36\\-49 & 63 & -42\\56 & -72 & 48\end{pmatrix}$$

Answer:

$$\begin{pmatrix}42 & -54 & 36\\-49 & 63 & -42\\56 & -72 & 48\end{pmatrix} \Rightarrow \begin{pmatrix}6 & 0 & 0\\-7 & 0 & 0\\8 & 0 & 0\end{pmatrix}$$

im f is the line

$$\begin{pmatrix}6\\-7\\8\end{pmatrix}t$$

4. Show that $\mathbf{b} \in \operatorname{im} f$, where

$$f = \begin{pmatrix}3 & 11\\1 & 7\end{pmatrix} \quad \text{and} \quad \mathbf{b} = \begin{pmatrix}2\\1\end{pmatrix}$$

Answer:

$$\left(\begin{array}{cc|c}3 & 11 & 2\\1 & 7 & 1\end{array}\right) \Rightarrow \left(\begin{array}{cc|c}1 & 7 & 1\\0 & -10 & -1\end{array}\right) \Rightarrow \left(\begin{array}{cc|c}1 & 0 & 0.3\\0 & 1 & 0.1\end{array}\right)$$

i.e., $\mathbf{b} = f(\mathbf{v})$ for

$$\mathbf{v} = \begin{pmatrix}0.3\\0.1\end{pmatrix}$$

Hence $\mathbf{b} \in \operatorname{im} f$.

5. Does $\mathbf{b}$ belong to $\operatorname{im} f$ if

$$f = \begin{pmatrix} 2 & 3 & 4 \\ 1 & -2 & 3 \end{pmatrix} \qquad \text{and} \qquad \mathbf{b} = \begin{pmatrix} 5 \\ 4 \end{pmatrix}?$$

Answer:

$$\left(\begin{array}{ccc|c} 2 & 3 & 4 & 5 \\ 1 & -2 & 3 & 4 \end{array}\right) \Rightarrow \left(\begin{array}{ccc|c} 1 & 0 & 1 & 2 \\ 0 & 1 & 2 & 3 \end{array}\right)$$

i.e., $\mathbf{b} = f(\mathbf{v})$ for

$$\mathbf{v} = \begin{pmatrix} 2-t \\ 3-2t \\ t \end{pmatrix} \qquad t \in \mathbb{R}$$

Therefore, $\mathbf{b} \in \operatorname{im} f$.

6. Does $\mathbf{b}$ belong to $\operatorname{im} f$ if

$$f = \begin{pmatrix} 2 & 5 \\ -3 & 1 \\ 4 & 2 \end{pmatrix} \qquad \text{and} \qquad \mathbf{b} = \begin{pmatrix} 0 \\ 1 \\ 0 \end{pmatrix}?$$

Answer:

$$\left(\begin{array}{cc|c} 2 & 5 & 0 \\ -3 & 1 & 1 \\ 4 & 2 & 0 \end{array}\right) \Rightarrow \left(\begin{array}{cc|c} 1 & 0 & -5 \\ 0 & 1 & 2 \\ 0 & 0 & 1 \end{array}\right) \Rightarrow \mathbf{b} \notin \operatorname{im} f$$

7. Can the vector

$$\begin{pmatrix} 4 \\ 11 \\ 11 \end{pmatrix}$$

be obtained as a linear combination of the vectors

$$\begin{pmatrix} -1 \\ 4 \\ -2 \end{pmatrix}, \quad \begin{pmatrix} -1 \\ -2 \\ 4 \end{pmatrix}, \quad \begin{pmatrix} 2 \\ 3 \\ 3 \end{pmatrix}$$

and if so, how?

Answer:

$$\left(\begin{array}{ccc|c} -1 & -1 & 2 & 4 \\ 4 & -2 & 3 & 11 \\ -2 & 4 & 3 & 11 \end{array}\right) \Rightarrow \left(\begin{array}{ccc|c} 1 & 1 & -2 & -4 \\ 0 & -6 & 11 & 27 \\ 0 & 6 & 1 & 3 \end{array}\right) \Rightarrow \left(\begin{array}{ccc|c} 1 & 1 & 0 & 2 \\ 0 & -6 & 0 & -6 \\ 0 & 0 & 1 & 3 \end{array}\right) \Rightarrow \left(\begin{array}{ccc|c} 1 & 0 & 0 & 1 \\ 0 & 1 & 0 & 1 \\ 0 & 0 & 1 & 3 \end{array}\right)$$

$$\Rightarrow \begin{cases} x = 1 \\ y = 1 \\ z = 3 \end{cases}$$

The unique solution is

$$\begin{pmatrix}-1\\4\\-2\end{pmatrix}+\begin{pmatrix}-1\\-2\\4\end{pmatrix}+3\begin{pmatrix}2\\3\\3\end{pmatrix}=\begin{pmatrix}4\\11\\11\end{pmatrix}$$

8. Pick a maximal subset of linearly independent vectors from the set

$$\begin{pmatrix}1\\1\\1\\2\end{pmatrix},\ \begin{pmatrix}5\\3\\4\\6\end{pmatrix},\ \begin{pmatrix}4\\5\\3\\13\end{pmatrix},\ \begin{pmatrix}-2\\1\\2\\-1\end{pmatrix},\ \begin{pmatrix}3\\4\\4\\7\end{pmatrix}$$

Answer:

$$\begin{pmatrix}1&5&4&-2&3\\1&3&5&1&4\\1&4&3&2&4\\2&6&13&-1&7\end{pmatrix}\Rightarrow\begin{pmatrix}1&0&0&0&0\\1&-2&1&3&1\\1&-1&-1&4&1\\2&-4&5&3&1\end{pmatrix}$$

$$\Rightarrow\begin{matrix}1^{st}&5^{th}&2^{nd}&3^{rd}&4^{th}\\ \end{matrix}\begin{pmatrix}1&0&0&0&0\\1&1&0&0&0\\1&1&1&-2&1\\2&1&-2&4&0\end{pmatrix}\Rightarrow\begin{matrix}1^{st}&5^{th}&2^{nd}&3^{rd}&4^{th}\\ \end{matrix}\begin{pmatrix}1&0&0&0&0\\1&1&0&0&0\\1&1&1&0&0\\2&1&-2&0&1\end{pmatrix}$$

and we may conclude that

$$\overset{1^{st}}{\begin{pmatrix}1\\1\\1\\2\end{pmatrix}}\ \overset{5^{th}}{\begin{pmatrix}3\\4\\4\\7\end{pmatrix}},\ \overset{2^{nd}}{\begin{pmatrix}5\\3\\4\\6\end{pmatrix}},\ \overset{4^{th}}{\begin{pmatrix}-2\\1\\2\\-2\end{pmatrix}}$$

is a maximal subset of linearly independent vectors in the set of five vectors.

9. Spanning Sets

Examples

1. Do the vectors

$$\begin{pmatrix}1\\2\end{pmatrix}, \quad \begin{pmatrix}2\\3\end{pmatrix}, \quad \begin{pmatrix}3\\4\end{pmatrix}$$

form a spanning set for $\mathbb{R}^2$?

Answer: Apply column reduction to find a subset of linearly independent vectors in $\mathbb{R}^2$:

$$\begin{pmatrix}1 & 2 & 3\\2 & 3 & 4\end{pmatrix} \Rightarrow \begin{pmatrix}1 & 0 & 0\\2 & 1 & 0\end{pmatrix}$$

We conclude that the first two vectors are linearly independent in $\mathbb{R}^2$; therefore, the given set spans $\mathbb{R}^2$.

2. Show that the vectors

$$\begin{pmatrix}1\\-2\\3\end{pmatrix}, \quad \begin{pmatrix}2\\3\\4\end{pmatrix}, \quad \begin{pmatrix}1\\5\\1\end{pmatrix}, \quad \begin{pmatrix}3\\1\\7\end{pmatrix}$$

do not span $\mathbb{R}^3$. Also, show that they span a plane of $\mathbb{R}^3$ and find the equation of that plane.

Answer:

$$\left(\begin{array}{cccc|c}1 & 2 & 1 & 3 & x\\-2 & 3 & 5 & 1 & y\\3 & 4 & 1 & 7 & z\end{array}\right) \Rightarrow \left(\begin{array}{cccc|c}1 & 2 & 1 & 3 & x\\0 & 7 & 7 & 7 & 2x+y\\0 & 0 & 0 & 0 & 17x-2y-7z\end{array}\right)$$

i.e., the given vectors span the plane $17x-2y-7z=0$.

3. Given the vectors

$$\begin{pmatrix}1\\1\\1\\2\end{pmatrix}, \quad \begin{pmatrix}5\\3\\4\\6\end{pmatrix}, \quad \begin{pmatrix}4\\5\\3\\13\end{pmatrix}, \quad \begin{pmatrix}-2\\1\\2\\-1\end{pmatrix}, \quad \begin{pmatrix}3\\4\\4\\7\end{pmatrix}$$

pick a minimal spanning set for $\mathbb{R}^4$.

Answer: A minimal spanning set of $\mathbb{R}^4$ is also a maximal set of linearly independent vectors in $\mathbb{R}^4$, and we may use the method of Application 7 above.

EXERCISES

9.1 Find $\ker f$ if

$$f = \begin{pmatrix} 1 & 2 & -3 \\ 2 & -1 & 4 \\ 4 & 3 & -2 \end{pmatrix}$$

9.2 Find

$$\operatorname{rank} \begin{pmatrix} 1 & 2 & 3 \\ 4 & 5 & 6 \\ 7 & 8 & 9 \end{pmatrix}$$

9.3 Find $\ker f$ and $\operatorname{im} f$ for

$$f = \begin{pmatrix} 4 & 8 & -12 \\ -3 & -6 & 9 \\ -1 & -2 & 3 \end{pmatrix}$$

and check the law of nullity in this example.

9.4 Find all the vectors $\mathbf{v}$ (if any) satisfying $f(\mathbf{v}) = \mathbf{b}$ if f, $\mathbf{b}$ are given as follows:

(i) $f = \begin{pmatrix} 2 & 8 & 6 \\ 4 & 2 & -2 \\ 3 & -1 & 1 \end{pmatrix}$ and $\mathbf{b} = \begin{pmatrix} 20 \\ -2 \\ 11 \end{pmatrix}$

(ii) $f = \begin{pmatrix} 1 & 1 & 2 \\ 2 & 4 & -3 \\ 3 & 6 & -5 \end{pmatrix}$ and $\mathbf{b} = \begin{pmatrix} 9 \\ 1 \\ 0 \end{pmatrix}$

(iii) $f = \begin{pmatrix} 2 & 1 & 1 \\ 3 & -2 & -1 \\ 4 & -7 & 3 \end{pmatrix}$ and $\mathbf{b} = \begin{pmatrix} 8 \\ 1 \\ 10 \end{pmatrix}$

(iv) $f = \begin{pmatrix} 2 & 1 & 4 \\ 3 & 2 & 1 \\ 1 & 3 & 3 \end{pmatrix}$ and $\mathbf{b} = \begin{pmatrix} 16 \\ 10 \\ 16 \end{pmatrix}$

(v) $f = \begin{pmatrix} 2 & 1 & -3 \\ 3 & -2 & 2 \\ 5 & -3 & -1 \end{pmatrix}$ and $\mathbf{b} = \begin{pmatrix} 5 \\ 5 \\ 16 \end{pmatrix}$

(vi) $f = \begin{pmatrix} 3 & 6 & -6 \\ 2 & -5 & 4 \\ -1 & 16 & -14 \end{pmatrix}$ and $\mathbf{b} = \begin{pmatrix} 9 \\ 6 \\ 3 \end{pmatrix}$

(vii) $f = \begin{pmatrix} 1 & 5 & 11 \\ 2 & 3 & 8 \\ -1 & 2 & 3 \end{pmatrix}$ and $\mathbf{b} = \begin{pmatrix} -5 \\ 4 \\ -9 \end{pmatrix}$

(viii) $f = \begin{pmatrix} 1 & 2 & -3 \\ 2 & -1 & 4 \\ 4 & 3 & -2 \end{pmatrix}$ and $\mathbf{b} = \begin{pmatrix} 6 \\ 2 \\ 14 \end{pmatrix}$

(ix) $f = \begin{pmatrix} 1 & -2 \\ 2 & -4 \end{pmatrix}$ and $\mathbf{b} = \begin{pmatrix} 3 \\ 5 \end{pmatrix}$

(x) $f = \begin{pmatrix} 2 & -3 \\ 3 & -4 \end{pmatrix}$ and $\mathbf{b} = \begin{pmatrix} 4 \\ -1 \end{pmatrix}$

9.5 In each case state whether the given set of vectors is linearly independent or dependent in the vector space indicated, and give reasons.

(i) $\begin{pmatrix} -1 \\ 2 \\ 1 \end{pmatrix}, \begin{pmatrix} 2 \\ 4 \\ -3 \end{pmatrix}, \begin{pmatrix} 4 \\ 0 \\ -5 \end{pmatrix}$ in $\mathbb{R}^4$

(ii) $\begin{pmatrix} 3 \\ 4 \\ 1 \\ 2 \end{pmatrix}, \begin{pmatrix} 5 \\ 7 \\ 1 \\ 9 \end{pmatrix}, \begin{pmatrix} 2 \\ 5 \\ -4 \\ 6 \end{pmatrix}$ in $\mathbb{R}^4$

(iii) $\begin{pmatrix} 1 \\ 0 \\ -1 \\ 0 \end{pmatrix}, \begin{pmatrix} 0 \\ 1 \\ 0 \\ -1 \end{pmatrix}, \begin{pmatrix} 0 \\ 1 \\ -1 \\ 0 \end{pmatrix}, \begin{pmatrix} 1 \\ 0 \\ 0 \\ -1 \end{pmatrix}$ in $\mathbb{R}^4$

(iv) $\begin{pmatrix} 1 \\ 2 \\ 3 \end{pmatrix}, \begin{pmatrix} 2 \\ 3 \\ 4 \end{pmatrix}, \begin{pmatrix} 3 \\ 4 \\ 5 \end{pmatrix}$ in $\mathbb{R}^3$

(v) $\begin{pmatrix} 3 \\ 1 \\ 2 \end{pmatrix}, \begin{pmatrix} 4 \\ 2 \\ 2 \end{pmatrix}, \begin{pmatrix} -2 \\ -1 \\ -1 \end{pmatrix}$ in $\mathbb{R}^3$

(vi) $\begin{pmatrix} 6 \\ -1 \\ 2 \end{pmatrix}, \begin{pmatrix} -1 \\ 3 \\ 7 \end{pmatrix}, \begin{pmatrix} 4 \\ -1 \\ 0 \end{pmatrix}$ in $\mathbb{R}^3$

9.6 In each of the following, find out whether the first vector can be obtained as a linear combination of the rest. If so, write such a linear combination and state whether it is unique.

(i) $\begin{pmatrix}3\\0\\0\\-6\end{pmatrix}$; $\begin{pmatrix}-1\\1\\-1\\3\end{pmatrix}, \begin{pmatrix}3\\-5\\1\\-13\end{pmatrix}, \begin{pmatrix}2\\3\\4\\1\end{pmatrix}$

(ii) $\begin{pmatrix}-5\\-4\\12\\5\end{pmatrix}$; $\begin{pmatrix}0\\1\\3\\4\end{pmatrix}, \begin{pmatrix}1\\0\\2\\3\end{pmatrix}, \begin{pmatrix}-3\\-2\\0\\-5\end{pmatrix}, \begin{pmatrix}4\\3\\-5\\0\end{pmatrix}$

(iii) $\begin{pmatrix}31\\29\\10\end{pmatrix}$; $\begin{pmatrix}1\\5\\3\end{pmatrix}, \begin{pmatrix}2\\1\\-1\end{pmatrix}, \begin{pmatrix}4\\2\\1\end{pmatrix}$

(iv) $\begin{pmatrix}1\\2\\-1\\4\end{pmatrix}$; $\begin{pmatrix}1\\-2\\1\\-1\end{pmatrix}, \begin{pmatrix}-1\\3\\-1\\1\end{pmatrix}, \begin{pmatrix}1\\-3\\2\\-3\end{pmatrix}, \begin{pmatrix}2\\3\\5\\2\end{pmatrix}$

(v) $\begin{pmatrix}1\\2\\-3\\-6\end{pmatrix}$; $\begin{pmatrix}-1\\-1\\0\\2\end{pmatrix}, \begin{pmatrix}1\\0\\-1\\-2\end{pmatrix}, \begin{pmatrix}-1\\-3\\1\\5\end{pmatrix}, \begin{pmatrix}2\\2\\3\\2\end{pmatrix}$

(vi) $\begin{pmatrix}1\\4\\0\end{pmatrix}$; $\begin{pmatrix}1\\3\\1\end{pmatrix}, \begin{pmatrix}1\\-1\\5\end{pmatrix}, \begin{pmatrix}-2\\1\\-9\end{pmatrix}, \begin{pmatrix}-1\\4\\-8\end{pmatrix}, \begin{pmatrix}1\\3\\2\end{pmatrix}$

9.7 Find all the solutions, if any, of the nonhomogeneous system

$$\begin{aligned} x + y - z &= a \\ x - y + z &= b \\ -x + y + z &= c \end{aligned}$$

9.8 What conditions must be put on a, b, c so that the system

$$\begin{aligned} x + 2y - 3z &= a \\ 2x + 6y - 11z &= b \\ x - 2y + 7z &= c \end{aligned}$$

have a solution?

9.9 What conditions must be put on a, b, c so that the system

$$\begin{aligned} x + 2y - 3z &= a \\ 3x - y + 2z &= b \\ -x + 5y - 8z &= c \end{aligned}$$

have a solution?

9.10 For what values of k does the system

$$\begin{aligned} kx + y + z &= 1 \\ x + ky + z &= 1 \\ x + y + kz &= 1 \end{aligned}$$

have (i) no solution, (ii) a unique solution, and (iii) infinitely many solutions?

9.11 A man has a pocketful of pennies, nickels, and dimes. If he has 14 coins in the value of 89 cents, how many coins of each type has he got?

9.12 Show, by matrix multiplication, that each of the following is a projection. Write the equation of the kernel and the equation of the image of the projection. Calculate the rank and the trace of each.

(i) $\begin{pmatrix} 1 & 0 & 0 \\ 2 & 3 & 2 \\ -3 & -3 & -2 \end{pmatrix}$ (ii) $\begin{pmatrix} 4 & 8 & -12 \\ -3 & -6 & 9 \\ -1 & -2 & 3 \end{pmatrix}$ (iii) $\begin{pmatrix} 7 & -9 & 15 \\ -2 & 4 & -5 \\ -4 & 6 & -9 \end{pmatrix}$

(iv) $\begin{pmatrix} 4 & 1 & 2 \\ 6 & 3 & 4 \\ -9 & -3 & -5 \end{pmatrix}$ (v) $\begin{pmatrix} -26 & -18 & -27 \\ 21 & 15 & 21 \\ 12 & 8 & 13 \end{pmatrix}$ (vi) $\begin{pmatrix} -8 & -12 & -21 \\ -15 & -19 & -35 \\ 12 & 16 & 29 \end{pmatrix}$

9.13 Show that $\ker f$ is a subspace of $\operatorname{dom} f$ (i.e., show that $\ker f$ is closed under the sum and scalar multiple of vectors).

9.14 Show that $\operatorname{im} f$ is a subspace of $\operatorname{cod} f$.

9.15 Show that every principal projection can be decomposed into the product of two lifts. Is the decomposition unique?

9.16 Prove that f is injective $\Leftrightarrow \ker f = 0$.

9.17 Let p, q be principal projections along the same plane. Show that

$$\frac{\alpha p + \beta q}{\alpha + \beta} \qquad (\alpha + \beta \neq 0)$$

is also a principal projection along the same plane, and

$$\operatorname{im} \frac{\alpha p + \beta q}{\alpha + \beta} \quad \text{is spanned by} \quad \frac{\alpha \mathbf{a} + \beta \mathbf{b}}{\alpha + \beta}$$

where $\mathbf{a}$, $\mathbf{b}$ are the vectors spanning $\operatorname{im} p$ and $\operatorname{im} q$, respectively.

9.18 Show, in two different ways, that $\operatorname{im} f \subseteq \ker f$ if

$$f = \begin{pmatrix} 4 & -8 & 12 \\ -1 & 2 & -3 \\ -2 & 4 & -6 \end{pmatrix}$$

9.19 Show that

$$\begin{pmatrix} 1 & 4 & 6 \\ 1 & 1 & 3 \\ -1 & -2 & -4 \end{pmatrix} \circ \begin{pmatrix} 2 & 4 & 6 \\ 1 & 2 & 3 \\ -1 & -2 & -3 \end{pmatrix}$$

is an exact composition.

9.20 In each of the following show that the given set of vectors spans a plane in $\mathbb{R}^3$. Write the equation of that plane.

(i) $$\begin{pmatrix} 2 \\ -3 \\ 5 \end{pmatrix}, \begin{pmatrix} 7 \\ 1 \\ 0 \end{pmatrix}, \begin{pmatrix} 5 \\ 4 \\ -5 \end{pmatrix}, \begin{pmatrix} 3 \\ 7 \\ -10 \end{pmatrix}$$

(ii) $$\begin{pmatrix} 1 \\ -1 \\ 0 \end{pmatrix}, \begin{pmatrix} 0 \\ 1 \\ -1 \end{pmatrix}, \begin{pmatrix} 1 \\ 0 \\ -1 \end{pmatrix}, \begin{pmatrix} 2 \\ 1 \\ -3 \end{pmatrix}, \begin{pmatrix} 3 \\ 2 \\ -5 \end{pmatrix}$$

9.21 Show that the set of vectors

$$\begin{pmatrix} 24 \\ -42 \\ 18 \end{pmatrix}, \begin{pmatrix} -31 \\ 56 \\ -24 \end{pmatrix}, \begin{pmatrix} 28 \\ -49 \\ 21 \end{pmatrix}$$

spans a line in $\mathbb{R}^3$. Write the equation of that line in symmetric form.

9.22 Let p, q be projections. Show that

(i) $q \circ p = p \Leftrightarrow \operatorname{im} p \subseteqq \operatorname{im} q$,
(ii) $q \circ p = q \Leftrightarrow \ker p \subseteqq \ker q$.

9.23 Let p, q be projections. Show that

$$p \circ q = q \circ p \Leftrightarrow q(\operatorname{im} p) \subseteqq \operatorname{im} p \qquad q(\ker p) \subseteqq \ker p$$

9.24 Let V, W be subspaces of the vector space $\mathbb{R}^n$. Define $V + W$ such that

$$V + W = \{\mathbf{v} + \mathbf{w} \mid \mathbf{v} \in V,\ \mathbf{w} \in W\}$$

Show that $V + W$ is a subspace of $\mathbb{R}^n$ and

$$\dim(V + W) = \dim V + \dim W - \dim(V \cap W)$$

9.25 Let p, q be projections such that $p \circ q = q \circ p$. Show that $p' = p \circ q$, $q' = p + q - p \circ q$ are also projections. Moreover,

$$\operatorname{im} p' = \operatorname{im} p \cap \operatorname{im} q \qquad \operatorname{im} q' = \operatorname{im} p + \operatorname{im} q$$

10

Eigenvalues and Eigenvectors: The Classification of Linear Operators†

EXISTENCE AND UNIQUENESS OF EIGENVECTORS

One of the most important problems in linear algebra is that of identifying linear operators geometrically, if their matrix is given. This problem is usually solved by finding the eigenvectors (also called characteristic vectors) of the linear operator.

Let f be a linear operator and suppose that for some vector $\mathbf{e} \neq 0$, and for some scalar λ,

$$f(\mathbf{e}) = \lambda \mathbf{e}$$

holds. We shall say that $\mathbf{e}$ is an *eigenvector* of f belonging to the *eigenvalue* λ. Note that the null vector shall not be considered as an eigenvector, although the scalar zero may occur as an eigenvalue.

Examples (*Note*: In these examples orthonormal bases are used.)

1. $p(\mathbf{v}) = (\mathbf{k} \cdot \mathbf{v})\mathbf{k}$ is the perpendicular projection onto the vertical axis,

$$p = \begin{pmatrix} 0 & 0 & 0 \\ 0 & 0 & 0 \\ 0 & 0 & 1 \end{pmatrix}$$

Then $p(\mathbf{k}) = \mathbf{k} = 1\mathbf{k}$, $p(\mathbf{i}) = \mathbf{0} = 0\mathbf{i}$, $p(\mathbf{j}) = \mathbf{0} = 0\mathbf{j}$; hence $\mathbf{k}$ is an eigenvector of p belonging to $\lambda = 1$, whereas $\mathbf{i}$ and $\mathbf{j}$ are eigenvectors of p belonging to $\lambda = 0$. Notice that, in fact, every horizontal vector is an eigenvector of p belonging to $\lambda = 0$.

†*Note*: The bases in this chapter are not necessarily orthonormal, and hence the coordinates are skew affine, except where stipulated otherwise.

2. $h(\mathbf{v}) = -(\mathbf{i}\cdot\mathbf{v})\mathbf{i} - (\mathbf{j}\cdot\mathbf{v})\mathbf{j} + (\mathbf{k}\cdot\mathbf{v})\mathbf{k}$ is the reflection in the vertical axis, or what is the same to say, rotation through 180° around the vertical axis:

$$h = \begin{pmatrix} -1 & 0 & 0 \\ 0 & -1 & 0 \\ 0 & 0 & 1 \end{pmatrix}$$

Then $h(\mathbf{i}) = -\mathbf{i} = (-1)\mathbf{i}$, $h(\mathbf{j}) = -\mathbf{j} = (-1)\mathbf{j}$, $h(\mathbf{k}) = \mathbf{k} = 1\mathbf{k}$. Therefore, **i** and **j** are both eigenvectors belonging to $\lambda = -1$ and **k** is an eigenvector of h belonging to $\lambda = 1$. Notice that, in fact, every horizontal vector is an eigenvector of h belonging to $\lambda = -1$.

3. Let $t(\mathbf{v}) = (\mathbf{i}\cdot\mathbf{v})\mathbf{j} - (\mathbf{j}\cdot\mathbf{v})\mathbf{i}$ define a linear operator of $\mathbb{R}^2$ where **i**, **j** are perpendicular unit vectors. Then t is the rotation of the plane through 90°:

$$t = \begin{pmatrix} 0 & -1 \\ 1 & 0 \end{pmatrix}$$

It is clear from the geometric meaning of t that $t(\mathbf{e}) = \lambda\mathbf{e} \Rightarrow \mathbf{e} = \mathbf{0}$ so that t has no eigenvectors.

4. $r(\mathbf{v}) = (\mathbf{k}\cdot\mathbf{v})\mathbf{k} + \mathbf{k} \times \mathbf{v}$ is the rotation around the vertical axis through 90°:

$$r = \begin{pmatrix} 0 & -1 & 0 \\ 1 & 0 & 0 \\ 0 & 0 & 1 \end{pmatrix}$$

It is clear from the geometric meaning of r that all the eigenvectors of r are vertical and they belong to $\lambda = 1$. Thus the eigenvectors of a rotation populate the axis of the rotation; in fact, they are the fixed vectors of the rotation satisfying $r(\mathbf{e}) = \mathbf{e}$.

The examples above suggest that eigenvectors may or may not exist, but if a linear operator f has an eigenvector, then it has infinitely many (namely, every vector parallel to it, i.e., every scalar multiple of it). There is no sense in which the length and the orientation of an eigenvector could be considered as unique. However, under certain circumstances the alignment of an eigenvector can be unique.

5. $f(\mathbf{v}) = \tfrac{1}{2}(\mathbf{i}\cdot\mathbf{v})\mathbf{i} + 2(\mathbf{j}\cdot\mathbf{v})\mathbf{j} + 3(\mathbf{k}\cdot\mathbf{v})\mathbf{k}$ has the matrix form

$$f = \begin{pmatrix} \tfrac{1}{2} & 0 & 0 \\ 0 & 2 & 0 \\ 0 & 0 & 3 \end{pmatrix}$$

f transforms the unit cube spanned by **i**, **j**, **k** into a standing matchbox of dimensions $\tfrac{1}{2} \times 2 \times 3$. We have $f(\mathbf{i}) = \tfrac{1}{2}\mathbf{i}$, $f(\mathbf{j}) = 2\mathbf{j}$, $f(\mathbf{k}) = 3\mathbf{k}$. It is clear from the geometric meaning of f that it has three uniquely determined eigenvectors, namely **i**, **j**, **k** belonging to the respective eigenvalues ½, 2, 3. Eigenvectors are considered to within a scalar multiple; i.e., two eigenvectors having the same alignment but possibly different length and orientation are considered to be the same. So **i** and 99**i** are both eigenvectors, but we identify them as eigenvectors, although, of course, they are different *qua* vectors.

NECESSARY AND SUFFICIENT CONDITIONS FOR THE INVERTIBILITY OF A LINEAR OPERATOR

The first example above suggests that *if a linear operator f has a nontrivial kernel*: $\ker f \neq 0$, *then the eigenvectors of f belonging to* $\lambda = 0$ *are just the nontrivial elements of* $\ker f$. The converse is also true, so that we have

$$\lambda = 0 \text{ is an eigenvalue of } f \Leftrightarrow \ker f \neq 0$$

Any two of the following conditions on the linear operator f of $\mathbb{R}^3$ *are equivalent*:

1. f is invertible.
2. f is bijective.
3. f is injective.
4. f is surjective.
5. f preserves linear independence.
6. $\{\mathbf{a},\mathbf{b},\mathbf{c}\}$ linearly independent $\Rightarrow [f(\mathbf{a}),f(\mathbf{b}),f(\mathbf{c})] \neq 0$.
7. $\det f \neq 0$.
8. $\ker f = 0$.
9. $\operatorname{rank} f = 3$.
10. $\lambda = 0$ is *not* an eigenvalue of f.

APPLICATIONS

1. Lifts

A lift ℓ was defined as a linear operator such that, for all **v**,

$$\ell(\mathbf{v}) = (\mathbf{n} \cdot \mathbf{v})\mathbf{e}$$

where $\mathbf{n} \perp \mathbf{e}$ are some fixed vectors. We have also seen that every lift satisfies $\ell^2 = 0$. Now we shall prove that the converse is also true, i.e.,

$$\ell^2 = 0 \Leftrightarrow \ell(\mathbf{v}) = (\mathbf{n} \cdot \mathbf{v})\mathbf{e} \quad (\mathbf{n} \cdot \mathbf{e} = 0)$$

Thus lifts can be described geometrically as rank 1 operators with image contained in the kernel: $\operatorname{im} \ell \subseteq \ker \ell$.

Let **e** be an eigenvector of ℓ, $\ell^2 = 0$. Then

$$\ell(\mathbf{e}) = \lambda\mathbf{e} \Rightarrow \ell^2(\mathbf{e}) = \ell(\ell(\mathbf{e})) = \ell(\lambda\mathbf{e}) = \lambda\ell(\mathbf{e}) = \lambda^2\mathbf{e}$$
$$\Rightarrow \lambda^2\mathbf{e} = \mathbf{0} \Rightarrow \lambda^2 = 0 \qquad (\text{because } \mathbf{e} \neq \mathbf{0})$$
$$\Rightarrow \lambda = 0$$

Therefore, ℓ can only have eigenvalue 0.

If $\ell \neq 0$, then for some $\mathbf{v} \neq \mathbf{0}$, $\ell(\mathbf{v}) = \mathbf{e} \neq \mathbf{0}$. Therefore, $\ell(\mathbf{e}) = \ell^2(\mathbf{v}) = \mathbf{0} = 0\mathbf{e}$, so that any vector $\mathbf{e} \neq \mathbf{0}$ in $\operatorname{im} \ell$ is an eigenvector of ℓ belonging to $\lambda = 0$. It follows that rank ℓ is at least 1; we want to show that it is exactly 1. Suppose that rank $\ell = 2$ or more. Then we would have vectors $\mathbf{e} = \ell(\mathbf{v})$, $\mathbf{e}' = \ell(\mathbf{v}')$, linearly independent, but $\ell(x\mathbf{e} + x'\mathbf{e}') = x\ell(\mathbf{e}) + x'\ell(\mathbf{e}') = x\ell^2(\mathbf{v}) + x'\ell^2(\mathbf{v}') = \mathbf{0}$. This means that dim ker ℓ is at least 2. This is impossible, however, as it would contradict the law of nullity. We conclude that rank $\ell = 1$ and dim ker $\ell = 2$. Let $\mathbf{n}$ be a normal vector of the plane $\ker \ell$. Since $\mathbf{e} \in \ker \ell$, $\mathbf{e} \perp \mathbf{n}$. We claim that $\ell(\mathbf{v}) = (\mathbf{n} \cdot \mathbf{v})\mathbf{e}$ for all $\mathbf{v}$, where $\mathbf{e}$ is a direction vector of the line $\operatorname{im} \ell$. Indeed, $\ell(\mathbf{n}) \neq \mathbf{0}$ and we may assume that $\ell(\mathbf{n}) = \mathbf{e}$. Then we have two linear operators assigning the same value to each of the basis vectors $\mathbf{e}$, $\mathbf{n}$, $\mathbf{e} \times \mathbf{n}$. But then these linear operators must coincide, i.e., $\ell(\mathbf{v}) = (\mathbf{n} \cdot \mathbf{v})\mathbf{e}$; ℓ is a lift. Observe that $\mathbf{e}$ is not the only eigenvector belonging to 0 (any vector perpendicular to $\mathbf{n}$ is also one), but $\mathbf{e}$ can be characterized as the only eigenvector of ℓ that belongs to both $\operatorname{im} \ell$ and $\ker \ell$.

Example Find the geometric meaning of the matrix

$$\ell = \begin{pmatrix} 4 & -8 & 12 \\ -1 & 2 & -3 \\ -2 & 4 & -6 \end{pmatrix}$$

Answer: It can be checked by calculation that $\ell^2 = 0$ and hence ℓ is a lift. It follows that $\operatorname{im} \ell$ is a line spanned by an eigenvector $\mathbf{e}$ belonging to $\lambda = 0$; for example, take the first column of the matrix: $\mathbf{e} = \ell(\mathbf{i}) = 4\mathbf{i} - \mathbf{j} - 2\mathbf{k}$. Calculating $\ker \ell$, we find the plane $x - 2y + 3z = 0$ with normal vector $\mathbf{n} = \mathbf{i} - 2\mathbf{j} + 3\mathbf{k}$. Thus ℓ is a $\pm\mathbf{n}$ lift to $\mathbf{e}$.

2. Shears

A shear was defined as a linear operator $f = 1 + \ell$, where $\ell^2 = 0$ (ℓ is a lift). Recall that ℓ can have eigenvalue 0 only. Suppose that $\mathbf{e}$ is an eigenvector of f; then

$$f(\mathbf{e}) = \lambda\mathbf{e} \Rightarrow (f-1)(\mathbf{e}) = (\lambda - 1)\mathbf{e}$$
$$\Rightarrow \ell(\mathbf{e}) = (\lambda - 1)\mathbf{e} \Rightarrow \lambda - 1 = 0 \Rightarrow \lambda = 1$$

Thus a shear can have eigenvalue $\lambda = 1$ only.

In order to identify the shear f geometrically, it will suffice to identify the lift $\ell = f - 1$.

Example Show that

$$f = \begin{pmatrix} 5 & -8 & 12 \\ -1 & 3 & -3 \\ -2 & 4 & -5 \end{pmatrix}$$

is a shear and describe it geometrically.

Answer: f is a shear because $\ell = f - 1$ is a lift; in fact, it is the lift described in the preceding example. Therefore, f is the shear in the direction of the vector $\mathbf{e} = 4\mathbf{i} - \mathbf{j} - 2\mathbf{k}$ parallel to the plane $x - 2y + 3z = 0$.

3. Twists

A twist t was defined as a linear operator satisfying $t^3 = -t$. We shall exclude the trivial operator $t = 0$ from the discussion.

Every twist can be represented as a difference of two lifts whose product is a projection (this representation is not unique), i.e.,

$$t^3 = -t \Leftrightarrow t = \ell - \ell', \quad \ell^2 = 0 = \ell'^2, \quad (\ell \circ \ell')^2 = \ell \circ \ell', \quad (\ell' \circ \ell)^2 = \ell' \circ \ell$$

The necessity of this condition was proved in Chap. 9 (Application 4). We shall now prove sufficiency. We consider the cases of $\mathbb{R}^2$ and $\mathbb{R}^3$ separately.

Case of $\mathbb{R}^2$ We shall see that a twist t in $\mathbb{R}^2$ has no (real) eigenvalues and eigenvectors. First let us show that $\operatorname{rank} t = 2$ so that t is bijective and hence $t^{-1} = t^3 = -t$ (in fact, t is an invection: $t^4 = 1$ with $t^2 = -1$). Indeed,

$$\begin{aligned} \operatorname{rank} t = 1 &\Rightarrow t(\mathbf{e}) \parallel \mathbf{e} \\ &\Rightarrow t(\mathbf{e}) = \lambda \mathbf{e} \qquad \text{for all } \mathbf{e} \in \operatorname{im} t, \mathbf{e} \neq \mathbf{0} \end{aligned}$$

i.e., $\mathbf{e}$ is an eigenvector of t. However,

$$\begin{aligned} t^3 = -t &\Rightarrow \lambda^3 = -\lambda \Rightarrow \lambda(\lambda^2 + 1) = 0 \Rightarrow \lambda = 0 \\ &\Rightarrow t(\mathbf{e}) = \mathbf{0} \Rightarrow \operatorname{im} t \subseteq \ker t \Rightarrow t^2 = 0 \\ &\Rightarrow t^3 = 0 = -t \Rightarrow t = 0 \Rightarrow \operatorname{rank} t = 0 \end{aligned}$$

which is impossible. We conclude that $\operatorname{rank} t = 2$. By the law of nullity, $\ker t = 0$, so there are no eigenvectors belonging to $\lambda = 0$. But $\lambda = 0$ is the only possible eigenvalue, so a twist $t \neq 0$ in $\mathbb{R}^2$ has no (real) eigenvectors.

In order to find a decomposition $t = \ell - \ell'$, let $\mathbf{d}' = a'\mathbf{i} + b'\mathbf{j}$ be the direction vector and let $\mathbf{n}' = A'\mathbf{i} + B'\mathbf{j}$ be the normal vector of a line in $\mathbb{R}^2$.

Further, let $\mathbf{d} = a\mathbf{i} + b\mathbf{j}$ be the direction vector, and let $\mathbf{n} = A\mathbf{i} + B\mathbf{j}$ be the normal vector of the line spanned by $t(\mathbf{d}') = \mathbf{d}$. Then $\mathbf{n} \cdot \mathbf{d} = Aa + Bb = 0$, $\mathbf{n}' \cdot \mathbf{d}' = A'a' + B'b' = 0$. Moreover,

$$\begin{aligned} \mathbf{d}' \nparallel t(\mathbf{d}') &\Rightarrow \mathbf{n} \not\perp \mathbf{d}', \quad \mathbf{n}' \not\perp \mathbf{d} \\ &\Rightarrow \mathbf{n} \cdot \mathbf{d}' \neq 0, \quad \mathbf{n}' \cdot \mathbf{d} \neq 0 \end{aligned}$$

In fact, we shall assume that the lengths of these vectors have been so fixed that $\mathbf{n} \cdot \mathbf{d}' = 1 = \mathbf{n}' \cdot \mathbf{d}$. Then clearly

$$t(\mathbf{v}) = (\mathbf{n} \cdot \mathbf{v})\mathbf{d} - (\mathbf{n}' \cdot \mathbf{v})\mathbf{d}' \qquad \text{for all } \mathbf{v} \in \mathbb{R}^2$$

i.e., $t = \ell - \ell'$, where $\ell(\mathbf{v}) = (\mathbf{n} \cdot \mathbf{v})\mathbf{d}$, $\ell'(\mathbf{v}) = (\mathbf{n}' \cdot \mathbf{v})\mathbf{d}'$ are lifts and $\ell \circ \ell'(\mathbf{v}) = (\mathbf{n}' \cdot \mathbf{v})\mathbf{d}$, $\ell' \circ \ell(\mathbf{v}) = (\mathbf{n} \cdot \mathbf{v})\mathbf{d}'$ are principal projections. In matrix form,

$$t = \begin{pmatrix} Aa - A'a' & Ba - B'a' \\ Ab - A'b' & Bb - B'b' \end{pmatrix}$$

where $Aa + Bb = 0 = A'a' + B'b'$, $Aa' + Bb' = 1 = A'a + B'b$.

We have thus found all the twists of $\mathbb{R}^2$. Let us observe that trace $t = 0$ for every twist in $\mathbb{R}^2$.

Case of $\mathbb{R}^3$ Let $\mathbf{e} \in \mathbb{R}^3$ be an eigenvector of the twist $t \neq 0$. Then

$$\begin{aligned} t^3(\mathbf{e}) = -t(\mathbf{e}) &\Rightarrow \lambda^3\mathbf{e} = -\lambda\mathbf{e} \\ &\Rightarrow \lambda(\lambda^2 + 1)\mathbf{e} = \mathbf{0} \Rightarrow \lambda = 0 \qquad (\text{since } \mathbf{e} \neq \mathbf{0}) \end{aligned}$$

Thus the eigenvector $\mathbf{e}$ belongs to $\lambda = 0$ and t has no other eigenvalue. In other words, the eigenvectors of t are precisely the vectors $\mathbf{e} \in \ker t$, $\mathbf{e} \neq \mathbf{0}$. Clearly, dim ker t is at least 1 but it cannot be 3. We shall now see that it is not 2 either, so it must be 1. Indeed,

$$\begin{aligned} \dim \ker t = 2 &\Rightarrow \dim \operatorname{im} t = 1 \qquad (\text{law of nullity}) \\ &\Rightarrow \mathbf{e} \in \operatorname{im} t \quad \text{with } t(\mathbf{e}) \neq \mathbf{0} \text{ exists} \end{aligned}$$

(otherwise, $t^2 = 0 \Rightarrow t = -t^3 = 0$, a contradiction). Hence

$$t(\mathbf{e}) \parallel \mathbf{e} \Rightarrow t(\mathbf{e}) = \lambda\mathbf{e} \text{ with } \lambda \neq 0$$

which is impossible, because t has no eigenvalue other than 0.

We have proved that ker t is a line and im t is a plane not containing the line. We can determine both ker t and im t from the matrix of t using the methods of Chapter 9. However, this may not determine t uniquely.

Here we follow another route. Let $\mathbf{m}' = t(\mathbf{m}_0) \neq \mathbf{0}$ and $\mathbf{m} = t(\mathbf{m}')$. Then

$$t(\mathbf{m}) = -t^3(\mathbf{m}) = -t^4(\mathbf{m}') = -t^5(\mathbf{m}_0) = t^3(\mathbf{m}_0) = -t(\mathbf{m}_0) = -\mathbf{m}' \neq \mathbf{0}$$

Thus $\mathbf{m}$, $\mathbf{m}'$ span the plane $\operatorname{im} t$. Let $\mathbf{n}$ be the normal vector of the plane spanned by $\mathbf{m}$ and $\ker t$, and let $\mathbf{n}'$ be the plane spanned by $\mathbf{m}'$ and $\ker t$. Clearly, $\mathbf{n} \perp \mathbf{m}$, $\mathbf{n}' \perp \mathbf{m}'$; but $\mathbf{n} \nparallel \mathbf{n}'$, so that $\mathbf{n} \not\perp \mathbf{m}'$ and $\mathbf{n}' \not\perp \mathbf{m}$. Therefore, $\mathbf{n} \cdot \mathbf{m} = 0 = \mathbf{n}' \cdot \mathbf{m}'$ but $\mathbf{n} \cdot \mathbf{m}' \neq 0$, $\mathbf{n}' \cdot \mathbf{m} \neq 0$. In fact, we shall assume that the lengths of these vectors have been so fixed that $\mathbf{n} \cdot \mathbf{m}' = 1 = \mathbf{n}' \cdot \mathbf{m}$. We put

$$\ell(\mathbf{v}) = (\mathbf{n} \cdot \mathbf{v})\mathbf{m}, \quad \ell'(\mathbf{v}) = (\mathbf{n}' \cdot \mathbf{v})\mathbf{m}' \qquad \text{for all } \mathbf{v} \in \mathbb{R}^3$$

It is left as an exercise to show that $t = \ell - \ell'$ is the desired decomposition. To summarize, every twist $t \neq 0$ of $\mathbb{R}^3$ is the difference of two lifts ℓ, ℓ' such that the lines $\operatorname{im} \ell$, $\operatorname{im} \ell'$ span the plane $\operatorname{im} t$ and the intersection of the planes $\ker \ell$, $\ker \ell'$ is the line $\ker t$. We have thereby found all the twists of $\mathbb{R}^3$ and conclude that trace $t = 0$.

Example Show that

$$t = \begin{pmatrix} 1 & -4 & 5 \\ 1 & -4 & 5 \\ 1 & -3 & 3 \end{pmatrix}$$

is a twist and decompose it into a difference of lifts.

Answer: t satisfies $t^3 = -t$; hence it is a twist. We want to find $\ker t$:

$$\begin{pmatrix} 1 & -4 & 5 \\ 1 & -3 & 3 \end{pmatrix} \Rightarrow \begin{pmatrix} 1 & -4 & 5 \\ 0 & 1 & -2 \end{pmatrix} \Rightarrow \begin{pmatrix} 1 & 0 & -3 \\ 0 & 1 & -2 \end{pmatrix} \Rightarrow \begin{cases} x = 3s \\ y = 2s \\ z = \ s \end{cases}$$

so that $\ker t$ is the line $x/3 = y/2 = z/1$ with direction vector $\mathbf{e} = 3\mathbf{i} + 2\mathbf{j} + \mathbf{k}$. $\operatorname{im} t$ is spanned, say, by $\mathbf{m}' = t(\mathbf{i}) = \mathbf{i} + \mathbf{j} + \mathbf{k}$ and $\mathbf{m} = t(\mathbf{m}') = 2\mathbf{i} + 2\mathbf{j} + \mathbf{k}$. Then $\mathbf{m} \times \mathbf{e} = \mathbf{j} - 2\mathbf{k}$, $\mathbf{m}' \times \mathbf{e} = -\mathbf{i} + 2\mathbf{j} - \mathbf{k}$. Put $\mathbf{n} = -\mathbf{j} + 2\mathbf{k}$, $\mathbf{n}' = -\mathbf{i} + 2\mathbf{j} - \mathbf{k}$; then $\mathbf{m} \cdot \mathbf{n} = 0 = \mathbf{m}' \cdot \mathbf{n}'$, $\mathbf{m} \cdot \mathbf{n}' = 1 = \mathbf{m}' \cdot \mathbf{n}$. We have lifts $\ell(\mathbf{v}) = (\mathbf{n} \cdot \mathbf{v})\mathbf{m}$, $\ell'(\mathbf{v}) = (\mathbf{n}' \cdot \mathbf{v})\mathbf{m}'$; in matrix form,

$$\ell = \begin{pmatrix} 0 & -2 & 4 \\ 0 & -2 & 4 \\ 0 & -1 & 2 \end{pmatrix} \qquad \ell' = \begin{pmatrix} -1 & 2 & -1 \\ -1 & 2 & -1 \\ -1 & 2 & -1 \end{pmatrix}$$

The required decomposition is

$$t = \ell - \ell'$$

4. Decomposition of an Invection into the Product of Two Involutions

An invection j was defined as a linear operator such that $j^4 = 1$. If j also satisfies $j^2 = 1$, then j is an involution; otherwise, we shall speak of a noninvolutory invection. Every noninvolutory invection j is the product of two involutions:

$$j^4 = 1 \Rightarrow j = k \circ h, \quad h^2 = 1 = k^2$$

Indeed, j invection $\Rightarrow t = \frac{1}{2}(j - j^3)$ twist, satisfying

$$1 + t + t^2 = j \qquad 1 - t + t^2 = j^3 = j^{-1}$$

Let $t = \ell - \ell'$ be the decomposition of the twist t as a difference of lifts: $\ell^2 = 0 = \ell'^2$, $(\ell \circ \ell')^2 = \ell \circ \ell'$ (see Application 3 above). Put

$$h = 1 - 2\ell \circ \ell' \qquad k = 1 - \ell \circ \ell' - \ell' \circ \ell - \ell - \ell'$$

Then (as a simple calculation, not reproduced here, shows) $h^2 = 1 = k^2$ and $j = k \circ h$, $h \circ k = j^{-1} = j^3$.

In $\mathbb{R}^2$, $\ell \circ \ell'$ and $\ell' \circ \ell$ are complementary projections: $\ell \circ \ell' + \ell' \circ \ell = 1$, so that $h = \ell' \circ \ell - \ell \circ \ell'$ and $k = -(\ell + \ell')$. It follows immediately that $j = k \circ h = \ell - \ell' = t$, $j^{-1} = j^3 = t^3 = -t = -j$, because $\ell \circ \ell' \circ \ell = \ell$ (why?). We conclude that in $\mathbb{R}^2$ (but not in $\mathbb{R}^3$!) every noninvolutory invection is a nontrivial twist, and vice versa. Thus a noninvolutory invection of $\mathbb{R}^2$ has no (real) eigenvalues and eigenvectors. The most important example is the rotation of the plane through 90° and –90°,

$$\begin{pmatrix} 0 & -1 \\ 1 & 0 \end{pmatrix} \text{ and } \begin{pmatrix} 0 & 1 \\ -1 & 0 \end{pmatrix}$$

However, in $\mathbb{R}^3$ a noninvolutory invection j has a unique eigenvalue $\lambda = 1$ or $\lambda = -1$ (but not both), since

$$j^4 = 1 \Rightarrow \lambda^4 = 1 \Rightarrow \lambda^4 - 1 = 0 \Rightarrow (\lambda - 1)(\lambda + 1)(\lambda^2 + 1) = 0$$
$$\Rightarrow \lambda = 1 \text{ or } \lambda = -1$$

5. The Cayley† Transform of a Twist

If the invection j has eigenvalue $\lambda = 1$, it is called *orientation preserving*; in this case j satisfies $(j - 1) \circ (j^2 + 1) = 0$ or

$$j^3 - j^2 + j - 1 = 0$$

If j has eigenvalue $\lambda = -1$, it is called *orientation reversing*; in this case j satisfies $(j + 1) \circ (j^2 + 1) = 0$ or

$$j^3 + j^2 + j + 1 = 0$$

We conclude that in $\mathbb{R}^3$,

j orientation-preserving invection $\Leftrightarrow -j$ orientation-reversing invection

†Arthur Cayley (1821 – 1895), professor of mathematics at Cambridge.

Examples

1. Let $\mathbf{u}$ be a unit vector in $\mathbb{R}^3$; then j defined by the formula

$$j(\mathbf{v}) = (\mathbf{u} \cdot \mathbf{v})\mathbf{u} + \mathbf{u} \times \mathbf{v}$$

   is a rotation through 90° around the line spanned by $\mathbf{u}$ (see Chap. 6, Application 6). In fact, j is an orientation-preserving invection; the eigenvector $\mathbf{u}$ belongs to $\lambda = 1$.
2. The formula $j(\mathbf{v}) = -(\mathbf{u} \cdot \mathbf{v})\mathbf{u} - \mathbf{u} \times \mathbf{v}$, where $\mathbf{u}$ is a unit vector, defines an orientation-reversing invection, $\mathbf{u}$ being the eigenvector belonging to $\lambda = -1$. j is called a *rotary reflection* through 90°; it could be described as the product of a rotation through 90° around the line spanned by $\mathbf{u}$ and a reflection in the normal plane of $\mathbf{u}$.

Recall that t twist $\Rightarrow j = 1 + t + t^2$ invection; j is called the *Cayley transform* of the twist t. It is left as an exercise to verify that $j = 1 + t + t^2$ satisfies $j^3 - j^2 + j - 1 = 0$, so that the Cayley transform of a twist is an orientation-preserving invection.

The Cayley transform establishes a bijection between the set of twists and the set of orientation-preserving invections. Indeed, given an orientation-preserving invection j, its Cayley transform is the twist

$$t = \tfrac{1}{2}(j - j^3) = \tfrac{1}{2}(j - j^{-1})$$

satisfying $1 + t + t^2 = j$. Thus the two Cayley transforms are inverses of one another. We may observe also that the Cayley transform of 0 is 1, and the Cayley transform of $-t$ is j^{-1}, and vice versa.

The unique eigenvector of a twist t and that of its Cayley transform j coincide (why?). This brings out the geometric meaning of an orientation-preserving invection; it is a "skew rotation" through 90° around the line $\ker t$. In Application 9 below, we shall look at "skew rotations" through ϕ. In Chap. 11, Application 6, we shall consider other aspects of the Cayley transform. In Chap. 13 we shall discuss its relation to the exponential functor.

FINDING THE EIGENVECTORS IF THE EIGENVALUES ARE KNOWN

If we know the eigenvalues of f, then the corresponding eigenvectors can be found as the kernel of a certain linear operator:

$$f(\mathbf{e}) = \lambda\mathbf{e} \Leftrightarrow \mathbf{e} \in \ker(f - \lambda)$$

provided that the kernel is nontrivial. In this way the problem of finding the eigenvectors of a linear operator (with known eigenvalues) has been reduced to the problem of solving a homogeneous system of linear equations.

If dim $\ker(f-\lambda)=1$, then the eigenvector $\mathbf{e}$ of f belonging to λ is uniquely determined (up to a scalar multiple). If dim $\ker(f-\lambda)=m>1$, then there are m linearly independent eigenvectors $\mathbf{e}_1, \mathbf{e}_2, \ldots, \mathbf{e}_m$ of f belonging to λ. At the same time every linear combination $\mathbf{e}=x_1\mathbf{e}_1+x_2\mathbf{e}_2+\cdots+x_m\mathbf{e}_m$ is also an eigenvector of f belonging to λ, so we may speak of an m-dimensional *eigenspace* of f belonging to λ.

APPLICATIONS

6. Projections

We have seen (in Chap. 7, Application 1, and in Chap. 9, Application 5) that, in $\mathbb{R}^3$, there exist projections p with trace $p=\operatorname{rank} p=0, 1, 2, 3$. In fact, $p=0$ is a projection of rank 0; $p=1$ is a projection of rank 3; and $p(\mathbf{v})=(\mathbf{n}\cdot\mathbf{v})\mathbf{e}$, where $\mathbf{n}\cdot\mathbf{e}=1$ is a projection of rank 1 (i.e., a principal projection) onto the line $\operatorname{im} p$ (spanned by $\mathbf{e}$) along the plane $\ker p$ (the normal plane of $\mathbf{n}$); and $q(\mathbf{v})=(\mathbf{d}\times\mathbf{v})\times\mathbf{m}$, where $\mathbf{m}\cdot\mathbf{d}=1$ is a projection of rank 2 onto the plane $\operatorname{im} q$ (the normal plane of $\mathbf{m}$) along the line $\ker q$ (the line spanned by $\mathbf{d}$). We shall now see that these examples actually exhaust the set of projections in $\mathbb{R}^3$.

Let p be a linear operator satisfying $p^2=p$, and let $\mathbf{e}$ be an eigenvector of p. Then

$$\begin{aligned} p^2(\mathbf{e})=p(\mathbf{e}) &\Rightarrow \lambda^2\mathbf{e}=\lambda\mathbf{e} \\ &\Rightarrow (\lambda^2-\lambda)\mathbf{e}=\mathbf{0} \\ &\Rightarrow \lambda^2-\lambda=0 \qquad (\text{because } \mathbf{e}\neq\mathbf{0}) \\ &\Rightarrow \lambda(\lambda-1)=0 \Rightarrow \lambda=0 \text{ or } \lambda=1 \end{aligned}$$

i.e., an eigenvalue of p must be equal to 0 or 1. Since there are at most three linearly independent eigenvectors of p, we have four possibilities:

1. $\operatorname{rank} p=0 \Leftrightarrow \lambda_1=\lambda_2=\lambda_3=0 \Leftrightarrow p=0$.
2. $\operatorname{rank} p=1 \Leftrightarrow \lambda_1=1,\ \lambda_2=\lambda_3=0$; put $\mathbf{e}_1=\mathbf{e},\ \mathbf{e}_2\times\mathbf{e}_3=\mathbf{n}$ and fix the lengths of $\mathbf{e}$, $\mathbf{n}$ such that $\mathbf{e}\cdot\mathbf{n}=1$ (this is possible because $\mathbf{e}\not\perp\mathbf{n}\Rightarrow\mathbf{e}\cdot\mathbf{n}\neq 0$) and find that $p(\mathbf{v})=(\mathbf{n}\cdot\mathbf{v})\mathbf{e}$ for all $\mathbf{v}\in\mathbb{R}^3$; i.e., p is a projection onto the line spanned by $\mathbf{e}$ along the normal plane of $\mathbf{n}$.
3. $\operatorname{rank} p=2 \Leftrightarrow \lambda_1=\lambda_2=1,\ \lambda_3=0$; put $\mathbf{e}_1\times\mathbf{e}_2=\mathbf{m},\ \mathbf{e}_3=\mathbf{d}$ and fix the lengths of $\mathbf{d}$, $\mathbf{m}$ such that $\mathbf{d}\cdot\mathbf{m}=1$ (this is possible because $\mathbf{d}\not\perp\mathbf{m}\Rightarrow\mathbf{d}\cdot\mathbf{m}\neq 0$) and find that $p(\mathbf{v})=(\mathbf{d}\times\mathbf{v})\times\mathbf{m}$ for all $\mathbf{v}\in\mathbb{R}^3$; i.e., p is a projection onto the normal plane of $\mathbf{m}$ along the line spanned by $\mathbf{d}$.
4. $\operatorname{rank} p=3 \Leftrightarrow \lambda_1=\lambda_2=\lambda_3=1 \Leftrightarrow p=1$.

Example Find the eigenvalues and eigenvectors of the linear operator

$$f=\begin{pmatrix}-3 & -2 & -10\\ 16 & 9 & 40\\ -2 & -1 & -4\end{pmatrix}$$

and check the results by substituting into $f(\mathbf{e})=\lambda\mathbf{e}$. (*Hint*: Calculate f^2 first.)

Answer: By matrix multiplication, $f^2=f$, hence f is a projection. $\operatorname{rank} f=\operatorname{trace} f=2$, hence f is a projection onto a plane having eigenvalues $\lambda_1=\lambda_2=1$, $\lambda_3=0$ and eigenvectors $\mathbf{e}_1$, $\mathbf{e}_2$, $\mathbf{e}_3$. $\mathbf{e}_3$ is the vector spanning $\ker f$:

$$\begin{pmatrix}-3 & -2 & -10\\ 16 & 9 & 40\\ -2 & -1 & -4\end{pmatrix}\Rightarrow\begin{pmatrix}1 & 0 & -2\\ 0 & 1 & 8\end{pmatrix}\Rightarrow\begin{cases}x=2t\\ y=-8t\\ z=t\end{cases}\qquad \text{i.e., } \mathbf{e}_3=\begin{pmatrix}2\\ -8\\ 1\end{pmatrix}$$

To find $\mathbf{e}_1$ and $\mathbf{e}_2$, we may pick any two linearly independent vectors in $\operatorname{im} f$; our choice is the first two column vectors of f, i.e.,

$$\mathbf{e}_1=\begin{pmatrix}-3\\ 16\\ -2\end{pmatrix}\qquad \mathbf{e}_2=\begin{pmatrix}-2\\ 9\\ -1\end{pmatrix}$$

Check:

$$f(\mathbf{e}_3)=\begin{pmatrix}-3 & -2 & -10\\ 16 & 9 & 40\\ -2 & -1 & -4\end{pmatrix}\begin{pmatrix}2\\ -8\\ 1\end{pmatrix}=\begin{pmatrix}0\\ 0\\ 0\end{pmatrix}=0\mathbf{e}_3$$

Similarly, $f(\mathbf{e}_1)=\mathbf{e}_1$, $f(\mathbf{e}_2)=\mathbf{e}_2$. Note that $\mathbf{e}_1$ and $\mathbf{e}_2$ span a two-dimensional eigenspace of f.

7. Involutions

We have seen (in Chap. 7, Application 2) that in $\mathbb{R}^3$ there exist involutions h with trace $h=3, 1, -1, -3$. In fact, $h=1$ is an involution of trace 3; $h=-1$ is an involution of trace -3; $h(\mathbf{v})=\mathbf{v}-2(\mathbf{n}\cdot\mathbf{v})\mathbf{e}=\mathbf{n}\times(\mathbf{v}\times\mathbf{e})-(\mathbf{n}\cdot\mathbf{v})\mathbf{e}$, where $\mathbf{n}\cdot\mathbf{e}=1$ is an involution of trace 1; and $h(\mathbf{v})=2(\mathbf{n}\cdot\mathbf{v})\mathbf{e}-\mathbf{v}=(\mathbf{n}\cdot\mathbf{v})\mathbf{e}-\mathbf{n}\times(\mathbf{v}\times\mathbf{e})$ $(\mathbf{n}\cdot\mathbf{e}=1)$ is an involution of trace -1. We shall now see that these examples actually exhaust the set of involutions in $\mathbb{R}^3$.

Let h be a linear operator satisfying $h^2=1$ and let $\mathbf{e}$ be an eigenvector of h. Then $h^2(\mathbf{e})=\mathbf{e}\Rightarrow\lambda^2-1=0\Rightarrow\lambda=1$ or $\lambda=-1$; i.e., an eigenvalue of h must be equal to 1 or -1. We can see, by using three linearly independent eigenvectors as a basis for $\mathbb{R}^3$, that there are four possibilities (since in this case trace $h=\lambda_1+\lambda_2+\lambda_3$):

1. $\operatorname{trace} h = 3 \Leftrightarrow \lambda_1 = \lambda_2 = \lambda_3 = 1 \Leftrightarrow h = 1$.
2. $\operatorname{trace} h = -3 \Leftrightarrow \lambda_1 = \lambda_2 = \lambda_3 = -1 \Leftrightarrow h = -1$.
3. $\operatorname{trace} h = 1 \Leftrightarrow \lambda_1 = \lambda_2 = 1,\ \lambda_3 = -1$; put $\mathbf{e}_1 \times \mathbf{e}_2 = \mathbf{n}$, $\mathbf{e}_3 = \mathbf{e}$ and fix the lengths of $\mathbf{n}$, $\mathbf{e}$ such that $\mathbf{n} \cdot \mathbf{e} = 1$, and find that $h(\mathbf{v}) = \mathbf{v} - 2(\mathbf{n} \cdot \mathbf{v})\mathbf{e}$ is the involution in the plane $\ker \tfrac{1}{2}(1-h)$ along the line $\operatorname{im} \tfrac{1}{2}(1-h)$.
4. $\operatorname{trace} h = -1 \Leftrightarrow \lambda_1 = 1,\ \lambda_2 = \lambda_3 = -1$; this is the involution in the line $\ker \tfrac{1}{2}(1-h)$ along the plane $\operatorname{im} \tfrac{1}{2}(1-h)$.

Example Find the eigenvalues and eigenvectors of the linear operator

$$f = \begin{pmatrix} 3 & 8 & 12 \\ 2 & 3 & 6 \\ -2 & -4 & -7 \end{pmatrix}$$

and check the results by substituting into $f(\mathbf{e}) = \lambda \mathbf{e}$. (*Hint*: Calculate f^2 first.)

Answer: By matrix multiplication, $f^2 = 1 \Rightarrow f$ involution. Since $\operatorname{trace} f = -1$, f is the involution in the line $\ker \tfrac{1}{2}(1-f)$ along the plane $\operatorname{im} \tfrac{1}{2}(1-f)$ with eigenvalues $\lambda_1 = 1$, $\lambda_2 = \lambda_3 = -1$ and eigenvectors $\mathbf{e}_1$, spanning the line $\ker \tfrac{1}{2}(1-f)$; $\mathbf{e}_2$, $\mathbf{e}_3$, spanning the plane $\operatorname{im} \tfrac{1}{2}(1-f)$.

Calculating $\mathbf{e}_1$:

$$\tfrac{1}{2}(1-f) = \begin{pmatrix} -1 & -4 & -6 \\ -1 & -1 & 3 \\ 1 & 2 & 4 \end{pmatrix}$$

$\ker \tfrac{1}{2}(1-f)$:

$$\begin{pmatrix} -1 & -4 & -6 \\ -1 & -1 & 3 \\ 1 & 2 & 4 \end{pmatrix} \Rightarrow \begin{pmatrix} 1 & 0 & 2 \\ 0 & 1 & 1 \end{pmatrix} \Rightarrow \begin{cases} x = -2t \\ y = -t \\ z = t \end{cases} \qquad \text{i.e., } \mathbf{e}_1 = \begin{pmatrix} -2 \\ -1 \\ 1 \end{pmatrix}$$

Calculating $\mathbf{e}_2$, $\mathbf{e}_3$: Any two linearly independent vectors in $\operatorname{im} \tfrac{1}{2}(1-f)$ can be taken as $\mathbf{e}_2$, $\mathbf{e}_3$; our choice is the first two column vectors of the matrix, i.e.,

$$\mathbf{e}_2 = \begin{pmatrix} -1 \\ -1 \\ 1 \end{pmatrix} \qquad \mathbf{e}_3 = \begin{pmatrix} -4 \\ -1 \\ 2 \end{pmatrix}$$

Check: $f(\mathbf{e}_1) = \mathbf{e}_1$, etc. Note that $\mathbf{e}_2$ and $\mathbf{e}_3$ span a two-dimensional eigenspace of f.

FINDING THE EIGENVALUES OF A LINEAR OPERATOR

We have seen that the eigenvectors of a linear operator f can be determined by solving a homogeneous system of linear equations, provided that the eigenvalues of f are known. Therefore, our problem reduces to finding the eigenvalues of a linear operator. The key to this is the following important result:

$$f(\mathbf{e}) = \lambda\mathbf{e}, \quad \mathbf{e} \neq \mathbf{0} \Leftrightarrow \det(f-\lambda) = 0$$

$\det(f-\lambda) = 0$ is called the *characteristic equation* of the linear operator f. Thus our result states that the eigenvalues of f can be obtained as the roots of the characteristic equation of f. Indeed,

$$\begin{aligned} f(\mathbf{e}) = \lambda\mathbf{e}, \ \mathbf{e} \neq \mathbf{0} &\Leftrightarrow f(\mathbf{e}) - \lambda\mathbf{e} = \mathbf{0} \\ &\Leftrightarrow (f-\lambda)(\mathbf{e}) = \mathbf{0} \\ &\Leftrightarrow \mathbf{e} \in \ker(f-\lambda), \ \mathbf{e} \neq \mathbf{0} \\ &\Leftrightarrow \ker(f-\lambda) \neq 0 \\ &\Leftrightarrow f-\lambda \text{ is } \textit{not} \text{ injective} \\ &\Leftrightarrow \det(f-\lambda) = 0 \end{aligned}$$

Examples

1. The characteristic equation of the projection

$$p = \begin{pmatrix} -3 & -1 & -2 \\ -6 & -2 & -4 \\ 9 & 3 & 6 \end{pmatrix}$$

is

$$\begin{vmatrix} -3-\lambda & -1 & -2 \\ -6 & -2-\lambda & -4 \\ 9 & 3 & 6-\lambda \end{vmatrix} = 0$$

or $\lambda^3 - \lambda^2 = 0$. Every projection of $\mathbb{R}^3$ onto a line has the same characteristic equation. (Why?)

2. The characteristic equation of the projection

$$1 - p = \begin{pmatrix} 4 & 1 & 2 \\ 6 & 3 & 4 \\ -9 & -3 & -5 \end{pmatrix}$$

is

$$\begin{vmatrix} 4-\lambda & 1 & 2 \\ 6 & 3-\lambda & 4 \\ 9 & 3 & -5-\lambda \end{vmatrix} = 0$$

or $\lambda^3 - 2\lambda^2 + \lambda = 0$. Every projection of $\mathbb{R}^3$ onto a plane has the same characteristic equation. (Why?)

3. The characteristic equation of the involution

$$h = \begin{pmatrix} 3 & 2 & -2 \\ 8 & 3 & -4 \\ 12 & 6 & -7 \end{pmatrix}$$

is $\lambda^3 - \lambda^2 - \lambda + 1 = 0$. h is an involution in a line; every other involution in a line has the same characteristic equation. (Why?)
4. The characteristic equation of the involution

$$-h = \begin{pmatrix} -3 & -2 & 2 \\ -8 & -3 & 4 \\ -12 & -6 & 7 \end{pmatrix}$$

$\lambda^3 + \lambda^2 - \lambda - 1 = 0$. h is an involution in a plane; every other involution in a plane has the same characteristic equation. (Why?)
5. The characteristic equation of a twist of $\mathbb{R}^3$ is $\lambda^3 - \lambda = 0$.
6. The characteristic equation of an invection of $\mathbb{R}^3$ is $\lambda^3 - \lambda^2 + \lambda - 1 = 0$ or $\lambda^3 + \lambda^2 + \lambda + 1 = 0$, according as the invection is orientation preserving or orientation reversing.

The characteristic equation of a linear operator f of $\mathbb{R}^3$ is a *cubic* equation (i.e., an equation of degree 3)

$$\lambda^3 + A\lambda^2 + B\lambda + C = 0$$

and (as a simple calculation, not reproduced here, shows)

$$A = -\operatorname{trace} f$$
$$C = -\det f$$

In the case there are three real eigenvalues, the characteristic equation factorizes:

$$(\lambda - \lambda_1)(\lambda - \lambda_2)(\lambda - \lambda_3) = 0$$

If the three eigenvalues are distinct, f is called a *hyperbolic* operator; we shall see below that the three eigenvectors of a hyperbolic operator are linearly independent.

It may happen that f has one (real) eigenvalue only, and the characteristic equation of f has an irreducible quadratic factor. In this case f is called an *elliptic* operator (see Application 9 below). In addition, there are several other types of linear operators in $\mathbb{R}^3$ (those having double/triple eigenvalues or double/triple eigenvectors). The story in $\mathbb{R}^2$ is quite different, as we shall now see.

The characteristic equation of the linear operator

$$f = \begin{pmatrix} a & b \\ c & d \end{pmatrix}$$

is the quadratic equation

$$\det(f - \lambda) = \lambda^2 + A\lambda + B = 0$$

where

$$A = -(a + d) = -\operatorname{trace} f$$
$$B = ad - bc = \det f$$

f is *elliptic* (no eigenvalues and eigenvectors) or *hyperbolic* (two distinct eigenvalues and two linearly independent eigenvectors) according as the discriminant of its characteristic equation

$$\operatorname{trace}^2 f - 4 \det f$$

is negative or positive. In addition, there are two other types of linear operators in $\mathbb{R}^2$ (having double eigenvectors or double eigenvectors), as we shall see later in this chapter.

Examples

1. $f = \begin{pmatrix} 0 & -1 \\ 1 & 0 \end{pmatrix}$ has characteristic equation $\lambda^2 + 1 = 0$. f has no eigenvalues or eigenvectors. f is elliptic.
2. $f = \begin{pmatrix} 0 & 1 \\ 1 & 0 \end{pmatrix}$ has characteristic equation $\lambda^2 - 1 = 0$. Its eigenvalues are $\lambda_1 = 1$, $\lambda_2 = -1$ corresponding to the eigenvectors $\mathbf{e}_1 = \mathbf{i} + \mathbf{j}$, $\mathbf{e}_2 = \mathbf{i} - \mathbf{j}$. f is hyperbolic.

The eigenvalues of a linear operator of $\mathbb{R}^2$ can thus be found by using the quadratic formula. In the case of $\mathbb{R}^3$, however, the direct method of finding the roots of a cubic equation would entail an inordinate amount of calculations, and one tries to find the eigenvalues by trial and error. For example, the *rational* roots of an algebraic equation with rational coefficients can be determined by surveying the factors of the constant term and of the leading coefficient. In particular, if the entries of the matrix are all integers, then the leading coefficient is 1 and all rational roots of the characteristic equation must in fact be integers, and they will come from among the factors of the constant term. (It goes without saying that the characteristic equation need not have rational roots, and testing the factors of the constant term does not decide conclusively whether real eigenvalues exist or not.)

In this connection the *tabular method* of finding the substitution value of polynomials at specific places can make the calculation very simple. It is based on the formula

$$P(\lambda) = A\lambda^3 + B\lambda^2 + C\lambda + D = \lambda[\lambda(\lambda A + B) + C] + D$$

We shall, accordingly, evaluate $P(\lambda)$ through the following tabular arrangement:

	A	B	C	D
λ	A	$\lambda A + B$	$\lambda(\lambda A + B) + C$	$\lambda[\lambda(\lambda A + B) + C] + D = P(\lambda)$

That is, we prepare a table with columns headed by the successive coefficients of the cubic polynomial. In the rows we enter λ_n, i.e., the rational numbers that are our candidates (the factors of the constant term of the polynomial, if the leading coefficient is 1 or –1) in order to check whether they qualify as a root. In the first column, headed by A, we enter A once more, and in the second and every subsequent column of the same row we enter the number we get by multiplying the number in the previous column by λ_n and add the number heading our column. When we reach the last column, headed by the constant term D, the calculation will yield $P(\lambda_n)$. If this number is 0, we conclude that λ_n is a root of the polynomial; otherwise, λ_n is not a root.

Examples

1. Find all the rational roots, if any, of the cubic polynomial $-\lambda^3 + 2\lambda^2 + 5\lambda - 6$ and put the polynomial in factorized form.

   Answer: The candidates are the divisors of the constant term, i.e., 1, 2, 3, 6, –1, –2, –3, –6. The tabular method yields the following table:

	–1	2	5	–6
1	–1	1	6	0
2	–1	0	5	4
3	–1	–1	2	0
6	–1	–4	–19	–120
–1	–1	3	2	–8
–1	–1	4	–3	0

This means that the roots of the polynomial $-\lambda^3 + 2\lambda^2 + 5\lambda - 6 = 0$ are $\lambda_1 = 1$, $\lambda_2 = 3$, $\lambda_3 = -2$. Thus the factorized form of the cubic polynomial is

$$-\lambda^3 + 2\lambda^2 + 5\lambda - 6 = -(\lambda - 1)(\lambda - 3)(\lambda + 2)$$

The tabular method of finding the substitution value of a polynomial works for polynomials of any degree.

2. Find the values $P(2)$ and $P(3)$ for the polynomial $P(x) = x^4 + 2x^2 - 9x - 6$.

Answer: The tabular method yields

	1	0	2	–9	–6
2	1	2	6	3	0
3	1	3	11	24	66

Therefore, $P(2) = 0$ and $x = 2$ is a root, and $P(3) = 66$, $x = 3$ is not a root of the polynomial.

The tabular method has one further advantage. Suppose we find that the rational number x_0 is a root of the polynomial $P(x)$; i.e., $P(x_0) = 0$. Then, as we know, the linear polynomial $x - x_0$ is a factor of $P(x)$; that is, $P(x)$ is divisible by $x - x_0$ and the quotient $P(x)/(x - x_0)$ is a polynomial of degree one less, which may be obtained by long division. However, if we found x_0 via the tabular method, there is no need to perform the long division: the coefficients of the quotient are precisely the numbers entering the row of x_0 in the table.

3. Factor out $x - 2$ from the polynomial $x^4 + 2x^2 - 9x - 6$.

Answer: From the table above, we get the coefficients of the quotient that yields the factorization

$$x^4 + 2x^2 - 9x - 6 = (x - 2)(x^3 + 2x^2 + 6x + 3)$$

This remark is important because it will also help cut down on the amount of computation involved in the tabular method once a rational root of the original polynomial is found. Since the quotient is likely to have fewer and smaller coefficients, yet the same roots, we are better off if we search for the rational roots of the quotient.

In case the polynomial is cubic and we have found one rational root via the tabular method, the missing roots can be found quickly even if they are not rational, as the quotient is a quadratic polynomial that can be solved by the quadratic formula.

The following will illustrate how to find the eigenvectors of linear operators by finding the eigenvalues first.

Examples

1. Find the eigenvalues and eigenvectors of the linear operator

$$f = \begin{pmatrix} 1 & -1 & 4 \\ 3 & 2 & -1 \\ 2 & 1 & -1 \end{pmatrix}$$

and check by substituting into $f(\mathbf{e}) = \lambda\mathbf{e}$.

Answer: The characteristic equation is

$$\det(f-\lambda) = \begin{vmatrix} 1-\lambda & -1 & 4 \\ 3 & 2-\lambda & -1 \\ 2 & 1 & -1-\lambda \end{vmatrix} = -(\lambda^3 - 2\lambda^2 - 5\lambda + 6) = 0$$

This is the same equation whose roots we determined above by trial and error using the tabular method: $\lambda_1 = 1$, $\lambda_2 = -2$, $\lambda_3 = 3$.

To find $\mathbf{e}_1$, we determine $\ker(f-1)$:

$$\begin{pmatrix} 0 & -1 & 4 \\ 3 & 1 & -1 \\ 2 & 1 & -2 \end{pmatrix} \Rightarrow \begin{pmatrix} 1 & 0 & 1 \\ 0 & 1 & -4 \end{pmatrix} \Rightarrow \begin{cases} x = -t \\ y = 4t \\ z = t \end{cases} \qquad \text{i.e., } \mathbf{e}_1 = \begin{pmatrix} -1 \\ 4 \\ 1 \end{pmatrix}$$

Similarly, $\ker(f+2)$ yields

$$\mathbf{e}_2 = \begin{pmatrix} -1 \\ 1 \\ 1 \end{pmatrix}$$

and $\ker(f-3)$,

$$\mathbf{e}_3 = \begin{pmatrix} 1 \\ 2 \\ 1 \end{pmatrix}$$

Check: $f(\mathbf{e}_1) = 1\mathbf{e}_1$, $f(\mathbf{e}_2) = -2\mathbf{e}_2$, $f(\mathbf{e}_3) = 3\mathbf{e}_3$.

2. Find the eigenvalues and eigenvectors of the linear operator

$$f = \begin{pmatrix} 3 & 2 & 4 \\ 2 & 0 & 2 \\ 4 & 2 & 3 \end{pmatrix}$$

Answer: The characteristic equation of f is $-\lambda^3 + 6\lambda^2 + 15\lambda + 8 = 0$. The rational roots, if any, must come from among the factors of the constant term, i.e., from among 1, 2, 4, 8, –1, –2, –4, –8. Using the tabular method, we have

	–1	6	15	8
1	–1	5	20	28
–1	–1	7	8	0

This tells us that $\lambda_1 = -1$ is a root and if we factor out $\lambda + 1$ from the cubic polynomial, we get the factorization

$$-\lambda^3 + 6\lambda^2 + 15\lambda + 8 = (\lambda + 1)(-\lambda^2 + 7\lambda + 8)$$

At this stage we may use the quadratic formula to find the other two eigenvalues, or we may simply continue with the tabular method, this time applying to the quadratic:

	–1	7	8
8	–1	–1	0

The other two roots are $\lambda_2 = 8$ and $\lambda_3 = -1$.

$\ker(f + 1)$:

$$\begin{pmatrix} 4 & 2 & 4 \\ 2 & 1 & 2 \\ 4 & 2 & 4 \end{pmatrix} \Rightarrow (2 \quad 1 \quad 2) \Rightarrow \begin{cases} x = s \\ y = -2s - 2t \\ z = t \end{cases}$$

the eigenvectors belonging to the eigenvalue –1 form the population of a plane; we may pick a pair of linearly independent vectors in that plane, e.g.,

$$\mathbf{e}_1 = \begin{pmatrix} 1 \\ -2 \\ 0 \end{pmatrix} \qquad \mathbf{e}_2 = \begin{pmatrix} 0 \\ -2 \\ 1 \end{pmatrix}$$

spanning the two-dimensional eigenspace.

$\ker(f - 8)$:

$$\begin{pmatrix} -5 & 2 & 4 \\ 2 & -8 & 2 \\ 4 & 2 & -5 \end{pmatrix} \Rightarrow \begin{pmatrix} 1 & 0 & -1 \\ 0 & 2 & -1 \end{pmatrix} \Rightarrow \begin{cases} x = 2t \\ y = t \\ z = 2t \end{cases} \qquad \mathbf{e}_3 = \begin{pmatrix} 2 \\ 1 \\ 2 \end{pmatrix}$$

We shall classify linear operators below, and we shall call the type of this operator *shear type*.

Two linear operators may have the same characteristic equation and yet have quite different character, as the following example will show.

Example Find the eigenvalues and the eigenvectors of the linear operator

$$f = \begin{pmatrix} 8 & -8 & -1 \\ 0 & 0 & -1 \\ 0 & 1 & -2 \end{pmatrix}$$

Answer: The characteristic equation in factorized form is

$$(\lambda - 8)(\lambda + 1)(\lambda + 1) = 0$$

i.e., the same as in the preceding example, where the linear operator had three linearly independent eigenvectors, two of which spanned a two-dimensional eigenspace belonging to the double eigenvalue −1. The present linear operator, however, has only two linearly independent eigenvectors, one belonging to $\lambda = -1$ and the other to $\lambda = 8$.

$\ker(f + 1)$:

$$\begin{pmatrix} 9 & -8 & -1 \\ 0 & 1 & -1 \end{pmatrix} \Rightarrow \begin{pmatrix} 9 & 0 & -9 \\ 0 & 1 & -1 \end{pmatrix} \Rightarrow \begin{pmatrix} 1 & 0 & -1 \\ 0 & 1 & -1 \end{pmatrix} \Rightarrow \begin{cases} x = t \\ y = t \\ z = t \end{cases} \qquad \mathbf{e}_1 = \begin{pmatrix} 1 \\ 1 \\ 1 \end{pmatrix}$$

$\ker(f - 8)$:

$$\begin{pmatrix} 0 & -8 & -1 \\ 0 & 1 & -10 \end{pmatrix} \Rightarrow \begin{cases} x = t \\ y = 0 \\ z = 0 \end{cases} \qquad \mathbf{e}_2 = \begin{pmatrix} 1 \\ 0 \\ 0 \end{pmatrix}$$

Later in this chapter we shall see that the eigenvector $\mathbf{e}_1$ is a "double eigenvector" of f. Linear operators of this type are called *hyperbolic-parabolic*.

These examples demonstrate that the name "characteristic equation" is a misnomer: it is far from characterizing the linear operator in general. However, in the special case when the linear operator f has three distinct eigenvalues, the characteristic equation actually characterizes f. It is only in the case of double or triple eigenvalues that the characteristic equation fails to characterize f and the phenomenon of a double of triple eigenvector, and that of two- and three-dimensional eigenspaces (or both) occur.

APPLICATION

8. Changing the Basis (Changing the Coordinate System)

Here "basis" will mean three linearly independent vectors in $\mathbb{R}^3$ which may or may not form an orthonormal frame. Suppose that we want to change the "old" basis $\{\mathbf{i},\mathbf{j},\mathbf{k}\}$ to a "new" one $\{\mathbf{i}',\mathbf{j}',\mathbf{k}'\}$. Two questions arise:

1. How are the old coordinates of a vector $\mathbf{v}$ being changed into the new coordinates (of the same vector $\mathbf{v}$)?
2. How is the old matrix of a linear operator f being changed into the new matrix (of the same linear operator f)?

There is a linear operator g, called the *transition operator*, such that

$$\mathbf{i}' = g(\mathbf{i}) \qquad \mathbf{j}' = g(\mathbf{j}) \qquad \mathbf{k}' = g(\mathbf{k})$$

(In other words, the column vectors of the transition matrix are just the old coordinates of the new basis vectors.) Moreover, since $\{\mathbf{i}',\mathbf{j}',\mathbf{k}'\}$ is a basis, g must be invertible (bijective) so that

$$\mathbf{i} = g^{-1}(\mathbf{i}') \qquad \mathbf{j} = g^{-1}(\mathbf{j}') \qquad \mathbf{k} = g^{-1}(\mathbf{k}')$$

Let x', y', z' denote the new coordinates of the vector $\mathbf{v}$; then $\mathbf{v} = x'\mathbf{i}' + y'\mathbf{j}' + z'\mathbf{k}'$ and hence

$$x'\mathbf{i} + y'\mathbf{j} + z'\mathbf{k} = g^{-1}(\mathbf{v})$$

This answers the first question: *In order to find the new coordinates of the vector* $\mathbf{v}$ *one calculates the coordinates of the vector* $g^{-1}(\mathbf{v})$ *in the old basis*, where g is the transition operator.

Let

$$f = \begin{pmatrix} a'_{11} & a'_{12} & a'_{13} \\ a'_{21} & a'_{22} & a'_{23} \\ a'_{31} & a'_{32} & a'_{33} \end{pmatrix}$$

be the new matrix of the linear operator f; then

$$f(\mathbf{i}') = a'_{11}\mathbf{i}' + a'_{21}\mathbf{j}' + a'_{31}\mathbf{k}' \qquad \text{and} \qquad g^{-1} \circ f(\mathbf{i}') = a'_{12}\mathbf{i} + a'_{21}\mathbf{j} + a'_{31}\mathbf{j}$$

Since $\mathbf{i}' = g(\mathbf{i})$, we have that

$$a'_{11}\mathbf{i} + a'_{21}\mathbf{j} + a'_{31}\mathbf{k} = (g^{-1} \circ f \circ g)(\mathbf{i})$$

and similar formulas apply for $\mathbf{j}$ and $\mathbf{k}$. This answers the second question: *In order to find the new matrix of the linear operator* f, *one calculates the matrix of the linear operator* $g^{-1} \circ f \circ g$ *in the old basis*, where g is the transition operator.

Examples

1. Given that

$$f = \begin{pmatrix} 3 & 2 & -2 \\ 8 & 3 & -4 \\ 12 & 6 & -7 \end{pmatrix}$$

is an involution in a line (i.e., $f^2 = 1$, $\operatorname{trace} f = -1$), find a new basis in which the matrix of f becomes

$$\begin{pmatrix} 1 & 0 & 0 \\ 0 & -1 & 0 \\ 0 & 0 & -1 \end{pmatrix}$$

Write the transition equation.

Answer: The elements of the new basis will be the eigenvectors of f. Calculating $\ker(f-1)$, we find $\mathbf{i}' = \mathbf{i} + 2\mathbf{j} + 3\mathbf{k}$; and from $\ker(f+1)$, $\mathbf{j}' = -\mathbf{i} + 2\mathbf{j}$, $\mathbf{k}' = \mathbf{i} + 2\mathbf{k}$. The column vectors of the transition matrix are the coordinates of the new basis vectors:

$$g = \begin{pmatrix} 1 & -1 & 1 \\ 2 & 2 & 0 \\ 3 & 0 & 2 \end{pmatrix}$$

We shall also need the inverse:

$$g^{-1} = \begin{pmatrix} 2 & 1 & -1 \\ -2 & -\frac{1}{2} & 1 \\ -3 & -\frac{3}{2} & 2 \end{pmatrix}$$

The transition equation is

$$\begin{pmatrix} 2 & 1 & -1 \\ -2 & -\frac{1}{2} & 1 \\ -3 & -\frac{3}{2} & 2 \end{pmatrix} \circ \begin{pmatrix} 3 & 2 & -2 \\ 8 & 3 & -4 \\ 12 & 6 & -7 \end{pmatrix} \circ \begin{pmatrix} 1 & -1 & 1 \\ 2 & 2 & 0 \\ 3 & 0 & 2 \end{pmatrix} = \begin{pmatrix} 1 & 0 & 0 \\ 0 & -1 & 0 \\ 0 & 0 & -1 \end{pmatrix}$$

2. The linear operator

$$p = \begin{pmatrix} -3 & -8 & 12 \\ 3 & 7 & -9 \\ 1 & 2 & -2 \end{pmatrix}$$

satisfies $p^2 = p$ and $\operatorname{rank} p = \operatorname{trace} p = 2$, so that p is projection onto a plane. Find a new basis in which the matrix of p will assume the form

$$\begin{pmatrix} 1 & 0 & 0 \\ 0 & 1 & 0 \\ 0 & 0 & 0 \end{pmatrix}$$

Write the transition equation.

Answer: The new basis vectors will be $\mathbf{i}'$, $\mathbf{j}'$ spanning $\operatorname{im} p$ and $\mathbf{k}'$ spanning $\ker p$. We shall use the first two column vectors of p as $\mathbf{i}' = -3\mathbf{i} + 3\mathbf{j} + \mathbf{k}$, $\mathbf{j}' = -8\mathbf{i} + 7\mathbf{j} + 2\mathbf{k}$. Calculating $\ker p$ yields $\mathbf{k}' = -4\mathbf{i} + 3\mathbf{j} + \mathbf{k}$. Hence the transition operator is

$$g = \begin{pmatrix} -3 & -8 & -4 \\ 3 & 7 & 3 \\ 1 & 2 & 1 \end{pmatrix} \quad \text{and} \quad g^{-1} = \begin{pmatrix} 1 & 0 & 4 \\ 0 & 1 & -3 \\ -1 & -2 & 3 \end{pmatrix}$$

The transition equation is

$$\begin{pmatrix} 1 & 0 & 4 \\ 0 & 1 & -3 \\ -1 & -2 & 3 \end{pmatrix} \circ \begin{pmatrix} -3 & -8 & 12 \\ 3 & 7 & -9 \\ 1 & 2 & -2 \end{pmatrix} \circ \begin{pmatrix} -3 & -8 & -4 \\ 3 & 7 & 3 \\ 1 & 2 & 1 \end{pmatrix} = \begin{pmatrix} 1 & 0 & 0 \\ 0 & 1 & 0 \\ 0 & 0 & 0 \end{pmatrix}$$

3. The linear operator

$$t = \begin{pmatrix} 1 & -2 \\ 1 & -1 \end{pmatrix}$$

is a twist. Change the basis such that its new matrix will have the form

$$\begin{pmatrix} 0 & -1 \\ 1 & 0 \end{pmatrix}$$

Write the transition equation.

Answer: The new basis vectors are, say, $\mathbf{i}' = \mathbf{i}$, $\mathbf{j}' = t(\mathbf{i}) = \mathbf{i} + \mathbf{j}$. The transition matrix is

$$g = \begin{pmatrix} 1 & 1 \\ 0 & 1 \end{pmatrix} \qquad g^{-1} = \begin{pmatrix} 1 & -1 \\ 0 & 1 \end{pmatrix}$$

The transition equation is

$$\begin{pmatrix} 1 & -1 \\ 0 & 1 \end{pmatrix} \circ \begin{pmatrix} 1 & -2 \\ 1 & -1 \end{pmatrix} \circ \begin{pmatrix} 1 & 1 \\ 0 & 1 \end{pmatrix} = \begin{pmatrix} 0 & -1 \\ 1 & 0 \end{pmatrix}$$

4. Find a new basis in which

$$f = \begin{pmatrix} 1 & -1 & 4 \\ 3 & 2 & -1 \\ 2 & 1 & -1 \end{pmatrix}$$

takes the form of a diagonal matrix. Write the transition equation.

Answer: The eigenvalues and eigenvectors of f have been determined earlier in this chapter as $\lambda_1 = 1$, $\lambda_2 = -2$, $\lambda_3 = 3$; $\mathbf{e}_1 = -\mathbf{i} + 4\mathbf{j} + \mathbf{k}$, $\mathbf{e}_2 = -\mathbf{i} + \mathbf{j} + \mathbf{k}$, $\mathbf{e}_3 = \mathbf{i} + 2\mathbf{j} + \mathbf{k}$. We conclude that in terms of the new basis consisting of these eigenvectors the matrix of f will take the form of a diagonal matrix with the eigenvalues as the diagonal entries. The column vectors of the transition matrix are the coordinates of the eigenvectors in the old basis, and the transition equation is

$$\begin{pmatrix} -1/6 & 1/3 & -1/2 \\ -1/3 & -1/3 & 1 \\ 1/2 & 0 & 1/2 \end{pmatrix} \circ \begin{pmatrix} 1 & -1 & 4 \\ 3 & 2 & -1 \\ 2 & 1 & -1 \end{pmatrix} \circ \begin{pmatrix} -1 & -1 & 1 \\ 4 & 1 & 2 \\ 1 & 1 & 1 \end{pmatrix} = \begin{pmatrix} 1 & 0 & 0 \\ 0 & -2 & 0 \\ 0 & 0 & 3 \end{pmatrix}$$

Remark A new basis in which f takes the form of a diagonal matrix exists if, and only if, f has three linearly independent eigenvectors. For example, the matrix of a twist cannot be "diagonalized." We shall also see other types of linear operators whose matrix cannot be diagonalized. However, even for these operators a new basis can be found in terms of which f is "standardized." This is the problem of classifying linear operators. In older books on linear algebra the classification problem of linear operators is ignored and, instead, a pseudoproblem of "matrix diagonalization" is treated at length—a bad distortion of the correct perspective of the underlying ideas.

CLASSIFICATION OF LINEAR OPERATORS

Our classification of linear maps (from one vector space to *another*) depended on one single integer rank f and, for that reason, it was a simple affair. The classification of linear operators (from a vector space to the *same* vector space), on the other hand, is more complicated because it depends on *two* integers: the number of eigenvalues and the number of eigenvectors of f. A further complicating factor is that the reason for having fewer than the maximum number of eigenvalues and eigenvectors is twofold: on the one hand, certain eigenvalues and eigenvectors may be "missing" (due to the characteristic equation having an irreducible quadratic factor); on the other hand, some of the eigenvectors and some of the eigenvalues may in fact coincide, and the eigenvalues may coincide independently of the eigenvectors.

We shall simplify language by saying that f has one, two, or three *distinct* eigenvectors if the maximum number of its linearly independent eigenvectors is one, two, or three, respectively. Furthermore, we shall say that the eigenvectors of a linear operator f of $\mathbb{R}^2$ coincide (or that the multiplicity of its eigenvector is 2, or that it has a double eigenvector) if f has an eigenvalue with multiplicity 2 but it has no eigenspace of dimension 2. In $\mathbb{R}^3$ there are linear operators having an eigenvector with multiplicity 2 or 3. The advantage of this convention is obvious: Now each linear operator has the same number of eigenvectors as eigenvalues, if we count the eigenvalues and the *distinct* eigenvectors with their respective multiplicity.

The difference between an eigenspace of dimension 2 and an eigenvector of multiplicity 2 is this: The former is a subspace of dimension 2, every element of which is an eigenvector belonging to the same eigenvalue. The latter is an isolated eigenvector that lives in an invariant subspace of dimension 2 *containing no other eigenvector*. If there is an eigenvector of multiplicity 3, there is no other (distinct) eigenvector, but there is an *invariant plane* containing the triple eigenvector. An invariant plane is one that is mapped into itself as a whole, although not pointwise, by the linear operator.

The classification of linear operators is based on the following

Theorem Let $\mathbf{e}_1$, $\mathbf{e}_2$ be eigenvectors of f belonging to λ_1, λ_2. Then

$$\lambda_1 \neq \lambda_2 \Rightarrow \mathbf{e}_1 \nparallel \mathbf{e}_2$$

(The condition is sufficient but not necessary.)

Indeed, $\mathbf{e}_1 \parallel \mathbf{e}_2 \Rightarrow \mathbf{e}_2 = x\mathbf{e}_1$ $(x \neq 0)$, so that

$$x\lambda_2\mathbf{e}_1 = \lambda_2(x\mathbf{e}_1) = \lambda_2\mathbf{e}_2 = f(\mathbf{e}_2) = f(x\mathbf{e}_1) = xf(\mathbf{e}_1) = x\lambda_1\mathbf{e}_1$$

and

$$\begin{aligned} x\lambda_2\mathbf{e}_1 = x\lambda_1\mathbf{e}_1 &\Rightarrow x(\lambda_2 - \lambda_1)\mathbf{e}_1 = \mathbf{0} \\ &\Rightarrow \lambda_2 - \lambda_1 = 0 \qquad \text{(because } x\mathbf{e}_1 \neq \mathbf{0}\text{)} \\ &\Rightarrow \lambda_2 = \lambda_1, \quad \text{a contradiction} \end{aligned}$$

We conclude that $\mathbf{e}_1 \nparallel \mathbf{e}_2$.

Corollary Let $\mathbf{e}_1$, $\mathbf{e}_2$, $\mathbf{e}_3$ be eigenvectors of f belonging to λ_1, λ_2, λ_3. Then

$$\lambda_1, \lambda_2, \lambda_3 \text{ distinct} \Rightarrow \mathbf{e}_1, \mathbf{e}_2, \mathbf{e}_3 \text{ linearly independent}$$

Proof of the Corollary: $\mathbf{e}_1$, $\mathbf{e}_2$, $\mathbf{e}_3$ linearly dependent $\Rightarrow \mathbf{e}_3 = x_1\mathbf{e}_1 + x_2\mathbf{e}_2$ and we may assume that $x_1 \neq 0$, $x_2 \neq 0$ and $\mathbf{e}_1 \nparallel \mathbf{e}_2$ (otherwise, there would be equal eigenvalues among λ_1, λ_2, λ_3 by the theorem). Then

$$\begin{aligned} \lambda_3x_1\mathbf{e}_1 + \lambda_3x_2\mathbf{e}_2 = \lambda_3(x_1\mathbf{e}_1 + x_2\mathbf{e}_2) &= \lambda_3\mathbf{e}_3 = f(\mathbf{e}_3) \\ &= f(x_1\mathbf{e}_1 + x_2\mathbf{e}_2) = x_1f(\mathbf{e}_1) + x_2f(\mathbf{e}_2) = \lambda_1x_1\mathbf{e}_1 + \lambda_2x_2\mathbf{e}_2 \\ &\Rightarrow (\lambda_3 - \lambda_1)x_1\mathbf{e}_1 + (\lambda_3 - \lambda_2)x_2\mathbf{e}_2 = 0 \\ &\Rightarrow \lambda_3 - \lambda_1 = 0 \text{ and } \lambda_3 - \lambda_2 = 0 \qquad \text{(because } \mathbf{e}_1 \nparallel \mathbf{e}_2\text{)} \\ &\Rightarrow \lambda_1 = \lambda_2 = \lambda_3 \end{aligned}$$

which is impossible.

We conclude that:

1. Eigenvectors belonging to distinct eigenvalues are always distinct.
2. Distinct eigenvectors may belong to eigenvalues that are not distinct; if this is the case, they span an eigenspace.
3. The multiplicity of an eigenvalue could be greater than the dimension of the corresponding eigenspace. If this is the case, there is an eigenvector in the eigenspace with multiplicity at least 2. In more detail, if the multiplicity of λ is 2 and there is no eigenspace of dimension 2, there is a double eigenvector belonging to λ. If the multiplicity of λ is 3 and there is no eigenspace of dimension 2, there is a triple eigenvector. If the multiplicity of λ is 3 and there is an eigenspace of dimension 2 but not 3, there is a double eigenvector in the eigenspace belonging to λ. If we add up the multiplicities of the eigenvectors belonging to λ, we get the multiplicity of the eigenvalue λ.

Case of $\mathbb{R}^2$ The number of solutions of the characteristic equation is 2 or 0. In the first case, the eigenvalues may be distinct or they may coincide. If they are distinct, then (by the theorem) the eigenvectors will also be distinct and we have a *hyperbolic* operator, as indicated by

$$\begin{bmatrix} \lambda_1, & \lambda_2 \\ \mathbf{e}_1, & \mathbf{e}_2 \end{bmatrix}$$

If the two eigenvalues coincide, there are two possibilities, according as the corresponding eigenvectors coincide or not. If f has two distinct eigenvectors, then they span an eigenspace of dimension 2, i.e., f is a *strain*, as indicated by

$$\begin{bmatrix} \lambda\ , & \lambda \\ \mathbf{e}_1, & \mathbf{e}_2 \end{bmatrix}$$

If the eigenvalues as well as the eigenvectors coincide (the case of a *double* eigenvector), then f is a *parabolic* operator (see Application 10 below), as indicated by

$$\begin{bmatrix} \lambda, & \lambda \\ \mathbf{e}, & \mathbf{e} \end{bmatrix}$$

Finally, if the characteristic equation has no (real) roots, then f has neither eigenvalues nor eigenvectors, and is called *elliptic*, as indicated by

$$\begin{bmatrix} -, & - \\ -, & - \end{bmatrix}$$

That all four cases actually do occur will be seen in the exercises.

Classification of Linear Operators of $\mathbb{R}^2$

Number of distinct eigenvectors	No eigenvalues	Two eigenvalues		
		Distinct eigenvalues	Eigenvalues coincide	
			No eigenspace	No double eigenvector
0	Elliptic	–	–	–
1	–	–	Parabolic	–
2	–	Hyperbolic	–	Strain

Case of $\mathbb{R}^3$ The characteristic equation can have either 3 (not necessarily distinct) roots, or it can have one root (the other two roots being complex conjugate). In the first case, the possibilities are:

(1) There are three distinct eigenvalues.
(2) Exactly two of the three eigenvalues coincide.
(3) All three eigenvalues coincide.

(a) There are three distinct eigenvectors.
(b) Exactly two of the three eigenvectors coincide, leaving two distinct eigenvectors, one of them a double eigenvector.
(c) All three eigenvectors coincide, leaving one triple eigenvector.

The first three possibilities, as well as the last three, are mutually exclusive. If all the remaining combinations of these possibilities could occur independently, then we would have $3 \times 3 = 9$ types. However, our theorem rules out (1)(b), (1)(c), and (2)(c). This leaves $6 + 1 = 7$ types as follows:

(1)(a) $\begin{bmatrix} \lambda_1, & \lambda_2, & \lambda_3 \\ \mathbf{e}_1, & \mathbf{e}_2, & \mathbf{e}_3 \end{bmatrix}$ called *hyperbolic type* (three distinct eigenvalues and three distinct eigenvectors)

(2)(a) $\begin{bmatrix} \lambda_1, & \lambda_2, & \lambda_2 \\ \mathbf{e}_1, & \mathbf{e}_2, & \mathbf{e}_3 \end{bmatrix}$ called *dilatation type* (two eigenvalues, one with multiplicity 2 to which an eigenspace of dimension 2 belongs)

(2)(b) $\begin{bmatrix} \lambda_1, & \lambda_2, & \lambda_2 \\ \mathbf{e}_1, & \mathbf{e}_2, & \mathbf{e}_2 \end{bmatrix}$ called *hyperbolic-parabolic type* (two eigenvalues, one with multiplicity 2, to which an eigenvector with multiplicity 2 belongs)

(3)(a) $\begin{bmatrix} \lambda, & \lambda, & \lambda \\ \mathbf{e}_1, & \mathbf{e}_2, & \mathbf{e}_3 \end{bmatrix}$ is a *strain* (one eigenvalue with multiplicity 3, to which an eigenspace of dimension 3 belongs)

(3)(b) $\begin{bmatrix} \lambda, & \lambda, & \lambda \\ \mathbf{e}_1, & \mathbf{e}_2, & \mathbf{e}_2 \end{bmatrix}$ called *shear type* (one eigenvalue with multiplicity 3, two eigenvectors, one with multiplicity 2 and an eigenspace of dimension 2)

(3)(c) $\begin{bmatrix} \lambda, & \lambda, & \lambda \\ \mathbf{e}, & \mathbf{e}, & \mathbf{e} \end{bmatrix}$ called *parabolic type* (one eigenvalue with multiplicity 3 and one eigenvector with multiplicity 3)

Finally, if the characteristic equation has only one solution, we have

$\begin{bmatrix} \lambda, — , — \\ \mathbf{e}, — , — \end{bmatrix}$ called *elliptic* type (one eigenvalue and one eigenvector, both with multiplicity 1, two eigenvalues and two eigenvectors are "missing")

That all seven cases actually do occur will be seen in the examples.

Examples

1. For each matrix, write the characteristic equation; find all eigenvalues and eigenvectors with their multiplicity. Find eigenspaces. State the type of operator. ($\lambda_\nu = \lambda_\mu \Leftrightarrow \nu = \mu$.)

(a) $\begin{pmatrix} \lambda_1 & 0 & 0 \\ 0 & \lambda_2 & 0 \\ 0 & 0 & \lambda_3 \end{pmatrix}$ char. equation: $(\lambda-\lambda_1)(\lambda-\lambda_2)(\lambda-\lambda_3)=0$

λ_1, λ_2, λ_3 eigenvalues; **i**, **j**, **k** corresponding eigenvectors. Type: hyperbolic.

(b) $\begin{pmatrix} \lambda_1 & 0 & 0 \\ 0 & \lambda_1 & 0 \\ 0 & 0 & \lambda_2 \end{pmatrix}$ char. equation: $(\lambda-\lambda_1)^2(\lambda-\lambda_2)=0$

λ_1, λ_2 eigenvalues, λ_1 double eigenvalue.

i, **j** eigenvectors belonging to λ_1 spanning an eigenspace of dimension 2; **k** is an eigenvector belonging to λ_2. Type: dilatation.

(c) $\begin{pmatrix} \lambda_1 & 0 & 0 \\ 0 & \lambda_1 & 0 \\ 0 & 0 & \lambda_1 \end{pmatrix}$ char. equation: $(\lambda-\lambda_1)^3=0$

λ_1 triple eigenvalue; **i**, **j**, **k** eigenvectors belonging to λ_1, spanning a three-dimensional eigenspace. Type: strain.

Classification of Linear Operators of $\mathbb{R}^3$

Number of distinct eigenvectors	One eigenvalue	Three eigenvalues					
		Distinct eigenvalues	Not all eigenvalues are distinct				
			No eigenspace		No multiple eigenvector		Eigenspace & multiple eigenvector
			Triple eigenvector	Double eigenvector	Eigenspace of dimension 3	Eigenspace of dimension 2	
1	Elliptic	–	Parabolic	–	–	–	–
2	–	–	–	Hyperbolic-parabolic	–	–	Shear type
3	–	Hyperbolic	–	–	Strain	Dilatation type	–

(d) $\begin{pmatrix} \lambda_1 & 1 & 0 \\ 0 & \lambda_1 & 0 \\ 0 & 0 & \lambda_1 \end{pmatrix}$ char. equation: $(\lambda-\lambda_1)^3=0$

λ_1 triple eigenvalue; **i**, **k** eigenvectors belonging to λ_1, spanning an eigenspace of dimension 2; **i** is a double eigenvector. Type: shear.

(e) $\begin{pmatrix} \lambda_1 & 1 & 0 \\ 0 & \lambda_1 & 1 \\ 0 & 0 & \lambda_1 \end{pmatrix}$ char. equation: $(\lambda-\lambda_1)^3=0$

λ_1 triple eigenvalue; **i** triple eigenvector belonging to λ_1. Type: parabolic.

(f) $\begin{pmatrix} \lambda_1 & 1 & 0 \\ 0 & \lambda_1 & 0 \\ 0 & 0 & \lambda_2 \end{pmatrix}$ char. equation: $(\lambda-\lambda_1)^2(\lambda-\lambda_2)=0$

λ_1, λ_2 eigenvalues, λ_1 double eigenvalue; **i**, **k** eigenvectors, **i** is a double eigenvector. Type: hyperbolic-parabolic.

(g) $\begin{pmatrix} \lambda_1 & 0 & 0 \\ 0 & \rho\cos\phi & -\rho\sin\phi \\ 0 & \rho\sin\phi & \rho\cos\phi \end{pmatrix}$ $(\rho\neq 0;\ \phi\neq 0,\ 180°)$

Characteristic equation: $(\lambda-\lambda_1)(\lambda^2-2\rho\cos\phi\lambda+\rho^2)=0$.
λ_1 single eigenvalue, **i** single eigenvector. Type: elliptic.

2. Find the type of the linear operator

$$f=\begin{pmatrix} 1 & 3 & 2 \\ -5 & 17 & 10 \\ 7 & -21 & -12 \end{pmatrix}$$

Answer: $\det(f-\lambda)=0 \Rightarrow (\lambda-2)^3=0 \Rightarrow \lambda=2$ is the only eigenvalue with multiplicity 3.

$\ker(f-2)$:

$$\begin{pmatrix} -1 & 3 & 2 \\ -5 & 15 & 10 \\ 7 & -21 & -14 \end{pmatrix} \Rightarrow (1 \quad -3 \quad -2) \Rightarrow \begin{cases} x=3s+2t \\ y=s \\ z=t \end{cases}$$

is an eigenspace of dimension 2,

$$\mathbf{e}_1=\begin{pmatrix} 3 \\ 1 \\ 0 \end{pmatrix} \qquad \mathbf{e}_2=\begin{pmatrix} 2 \\ 0 \\ 1 \end{pmatrix}$$

Since f has a triple eigenvalue and an eigenspace of dimension 2 but not 3, f is shear type.

Let us find the double eigenvector of f. A shear-type operator can be written as $f = \lambda(1 + \ell)$, where $\ell^2 = 0$. In our case

$$f = 2 + 2\ell \Rightarrow \ell = \tfrac{1}{2}f - 1 = \begin{pmatrix} -1/2 & 3/2 & 1 \\ -5/2 & 15/2 & 5 \\ 7/2 & -21/2 & -7 \end{pmatrix}$$

is a lift and im ℓ is spanned by

$$\mathbf{e} = \begin{pmatrix} 1 \\ 5 \\ -7 \end{pmatrix}$$

the double eigenvector of f.

3. Find the type of the linear operator

$$f = \begin{pmatrix} 2 & -1 & 6 \\ 0 & 2 & 1 \\ 0 & 0 & 2 \end{pmatrix}$$

Answer: $\det(f - \lambda) = 0 \Rightarrow (\lambda - 2)^3 = 0 \Rightarrow \lambda = 2$ is an eigenvalue with multiplicity 3.

$\ker(f - 2)$:

$$\begin{pmatrix} 0 & -1 & 6 \\ 0 & 0 & 1 \\ 0 & 0 & 0 \end{pmatrix} \Rightarrow \begin{cases} x = t \\ y = 0 \\ z = 0 \end{cases}$$

$\mathbf{e} = \mathbf{i}$ is a triple eigenvector; f is parabolic.

4. Find the type of

$$f = \begin{pmatrix} 2 & 0 & 0 \\ 0 & 2 & 0 \\ 0 & 0 & 2 \end{pmatrix}$$

Answer: $\det(f - \lambda) = 0 \Rightarrow (\lambda - 2)^3 = 0 \Rightarrow \lambda = 2$ is an eigenvalue with multiplicity 3; clearly, **i**, **j**, **k** are eigenvectors spanning an eigenspace of dimension 3. f is a strain.

APPLICATIONS

9. Elliptic Operators

Case of $\mathbb{R}^2$ We want to find all the elliptic operators of the plane, i.e., all those operators that have no (real) eigenvalues and eigenvectors. Let f be such an operator and let $\mathbf{v}, \mathbf{w} \neq \mathbf{0}$ satisfy $\mathbf{w} = f(\mathbf{v})$. Since $\mathbf{v}$ is no eigenvector of f, $\mathbf{w} \nparallel \mathbf{v}$ and they span $\mathbb{R}^2$. Then $f(\mathbf{w})$ can be written as a linear combination of $\mathbf{v}$, $\mathbf{w}$ which we shall find it expedient to put in the form

$$f(\mathbf{v}) = \mathbf{w}$$
$$f(\mathbf{w}) = \alpha\mathbf{v} + 2\beta\mathbf{w}$$

or, what is the same in terms of $g = f - \beta$,

$$g(\mathbf{v}) = -\beta\mathbf{v} + \mathbf{w}$$
$$g(\mathbf{w}) = \alpha\mathbf{v} + \beta\mathbf{w}$$

This implies that

$$g^2(\mathbf{v}) = (\alpha + \beta^2)\mathbf{v}, \quad g^2(\mathbf{w}) = (\alpha + \beta^2)\mathbf{w}$$

i.e., $g^2 = \alpha + \beta^2$ is a strain. Moreover, $\alpha + \beta^2 \in \mathbb{R}$ cannot be positive; otherwise, we would have a scalar γ and a vector $\mathbf{e}$ such that $\alpha + \beta^2 = \gamma^2$ and

$$\begin{aligned}\mathbf{e} = (\gamma - \beta)\mathbf{v} + \mathbf{w} \Rightarrow g(\mathbf{e}) &= (\gamma - \beta)g(\mathbf{v}) + g(\mathbf{w})\\ &= (\gamma - \beta)(-\beta\mathbf{v} + \mathbf{w}) + \alpha\mathbf{v} + \beta\mathbf{w}\\ &= (\gamma^2 - \beta\gamma + \alpha)\mathbf{v} + \gamma\mathbf{w} = (\gamma^2 + \beta\gamma)\mathbf{v} + \gamma\mathbf{w}\\ &= \gamma[(\gamma - \beta)\mathbf{v} + \mathbf{w}] = \gamma\mathbf{e}\\ &\Rightarrow (f - \beta)(\mathbf{e}) = \gamma\mathbf{e} \Rightarrow f(\mathbf{e}) = (\beta + \gamma)\mathbf{e}\end{aligned}$$

and $\mathbf{e}$ would be an eigenvector of f, which is impossible.

Therefore, $\alpha + \beta^2 = -\delta^2$, $\delta \in \mathbb{R}$, and we may define a linear operator t by setting $g = \delta t$. Then

$$g^2(\mathbf{v}) = (\alpha + \beta^2)\mathbf{v}, \; g^2(\mathbf{w}) = (\alpha + \beta^2)\mathbf{w}$$

i.e., t is a twist and $f = \beta + g = \beta + \delta t$. Set $\rho^2 = \beta^2 + \delta^2$, $\rho > 0$; then $(\beta/\rho)^2 + (\delta/\rho)^2 = 1$ and there is an angle $\phi \neq 180°$, $0 < \phi < 360°$, such that $\cos\phi = \beta/\rho$, $\sin\phi = \delta/\rho$, and the linear operator f can be written in the

Standard Form of Elliptic Operators in $\mathbb{R}^2$

$$f = \rho(\cos\phi + t\sin\phi) \qquad (t^3 = -t)$$

Conversely, a linear operator of this form is elliptic, provided that $\phi \neq 0°$, $180°$ because the discriminant of its characteristic equation

$$\lambda^2 - 2\rho\cos\phi\lambda + \rho^2 = 0$$

is negative: $-4\rho^2\sin^2\phi < 0$. Thus we have found all the elliptic operators f of the plane. Note that f determines the twist t up to signature, and the scalars $\rho > 0$, $0 < \phi < 360°$ ($\phi \neq 180°$) uniquely.

In the special case of $\rho = 1$, f is a "skew rotation" through ϕ. If, in addition,

$$t = i = \begin{pmatrix} 0 & -1 \\ 1 & 0 \end{pmatrix}$$

is the rotation of the plane through 90°, then f is the rotation of the plane through ϕ.

Case of $\mathbb{R}^3$ A linear operator f of $\mathbb{R}^3$ is called elliptic if its characteristic equation has exactly one real root λ (i.e., it has a pair of complex conjugate roots). Let the eigenvector corresponding to λ be $\mathbf{e}$; then

$$\dim\ker(f-\lambda) = 1 \qquad \text{and} \qquad \dim\operatorname{im}(f-\lambda) = 2$$

by the law of nullity. We now show that $\operatorname{im}(f-\lambda)$ is an invariant plane for f:

$$\begin{aligned} \mathbf{v} \in \operatorname{im}(f-\lambda) \Rightarrow \mathbf{v} &= f(\mathbf{w}) - \lambda\mathbf{w} \qquad \text{for some } \mathbf{w} \in \mathbb{R}^3 \\ \Rightarrow f(\mathbf{v}) &= f(f(\mathbf{w}) - \lambda\mathbf{w}) \\ &= f^2(\mathbf{w}) - \lambda f(\mathbf{w}) \\ &= (f-\lambda)(f(\mathbf{w})) \Rightarrow f(\mathbf{v}) \in \operatorname{im}(f-\lambda) \end{aligned}$$

This means that f is acting on the plane $\operatorname{im}(f-\lambda)$ and its action is elliptic; that is, f has no eigenvectors in that plane. Moreover, in the notation of above, there are scalars β, δ and linearly independent vectors $\mathbf{m}$, $\mathbf{m}' \in \operatorname{im}(f-\lambda)$ such that

$$\begin{aligned} f(\mathbf{m}) &= \beta\mathbf{m} + \delta\mathbf{m}' \\ f(\mathbf{m}') &= -\delta\mathbf{m} + \beta\mathbf{m}' \end{aligned}$$

Let $\mathbf{n}$ be the normal vector of the plane spanned by $\mathbf{m}$, $\mathbf{e}$ and $\mathbf{n}'$ the normal vector of the plane spanned by $\mathbf{m}'$, $\mathbf{e}$ satisfying $\mathbf{n}\cdot\mathbf{m} = 0 = \mathbf{n}'\cdot\mathbf{m}'$, $\mathbf{n}\cdot\mathbf{m}' = 1 = \mathbf{n}'\cdot\mathbf{m}$. Then a twist t is defined by the formula $t(\mathbf{v}) = (\mathbf{n}\cdot\mathbf{v})\mathbf{m} - (\mathbf{n}'\cdot\mathbf{v})\mathbf{m}'$ and the linear operator f can be written in the

Standard Form of Elliptic Operators in $\mathbb{R}^3$

$$f = \lambda(1 + t^2) + \rho(t^4\cos\phi + t\sin\phi) \qquad (t^3 = -t;\ \phi \neq 0°, 180°)$$

where $\cos\phi = \beta/\rho$, $\sin\phi = \delta/\rho$, $\rho^2 = \beta^2 + \delta^2 \neq 0$. The verification of this is an exercise. Conversely, a linear operator of this form is elliptic, as can be seen from the characteristic equation

$$(\lambda - \lambda_0)(\lambda^2 - 2\rho\cos\phi\,\lambda + \rho^2) = 0$$

where the quadratic factor is irreducible. Thus every elliptic operator f determines a twist t (up to signature) and unique scalars $\rho > 0$; $0 < \phi < 360°$, $\phi \neq 180°$; $\lambda = \lambda_0$.

If $\rho = \lambda = 1$, then f is a skew rotation through ϕ around the axis $\ker t$. If, in addition, t is an antisymmetric twist (see Chap. 11), f is a rotation through ϕ around $\ker t$.

10. Parabolic Operators

We treat the case of $\mathbb{R}^2$ only. Our main result is that

$$f \text{ parabolic} \Leftrightarrow f = \lambda + \ell \qquad (\ell^2 = 0,\ \lambda \in \mathbb{R})$$

or, in matrix form,

$$f \text{ parabolic} \Leftrightarrow f = \begin{pmatrix} Aa + \lambda & Ba \\ Ab & Bb + \lambda \end{pmatrix} \qquad (Aa + Bb = 0)$$

The condition is necessary, because

$$\det f = (Aa + \lambda)(Bb + \lambda) - AaBb = \lambda^2$$
$$\operatorname{trace} f = Aa + Bb + 2\lambda = 2\lambda$$

implies that the characteristic equation of f is $(\lambda - \tfrac{1}{2}\operatorname{trace} f)^2 = 0$, hence f has a double eigenvalue $\lambda = \tfrac{1}{2}\operatorname{trace} f$. Moreover,

$$f(\mathbf{e}) = \lambda\mathbf{e} \Leftrightarrow \mathbf{e} \in \ker(f - \lambda) = \ker \ell$$
$$\Rightarrow \mathbf{e} = a\mathbf{i} + b\mathbf{j}$$

so f has a double eigenvector, namely $\mathbf{e}$.

Conversely,

$$f \text{ parabolic} \Rightarrow 4 \det f = \operatorname{trace}^2 f$$

and hence the characteristic equation of f is

$$\lambda^2 - 2(\tfrac{1}{2}\operatorname{trace} f)\lambda + (\tfrac{1}{2}\operatorname{trace} f)^2 = (\lambda - \tfrac{1}{2}\operatorname{trace} f)^2 = 0$$

i.e., the double eigenvalue is $\lambda = \tfrac{1}{2}\operatorname{trace} f$. If

$$f = \begin{pmatrix} \alpha & \beta \\ \gamma & \delta \end{pmatrix}$$

then $\operatorname{trace} f = \alpha + \delta$ and

$$\ell = f - \lambda = \begin{pmatrix} \tfrac{1}{2}(\alpha - \delta) & \beta \\ \gamma & \tfrac{1}{2}(\delta - \alpha) \end{pmatrix} \neq 0$$

(otherwise, $f = \lambda$ strain, not parabolic). $\ell^2 = 0$ can be checked by matrix multiplication. We conclude that $f = \lambda + \ell$, where ℓ is a lift. A shear is a special parabolic operator with eigenvalue $\lambda = 1$.

TYPE OF LINEAR OPERATORS IN $\mathbb{R}^n$

It is not the task of this book to discuss the classification of linear operators of $\mathbb{R}^n$ for $n \geqq 4$. It may, however, be helpful if we review the ideas on which the classification rests and call attention to the new types that arise. An eigenvector $\mathbf{e}$ of f with multiplicity k ($2 \leqq k \leqq n$) is an eigenvector that lives in a "nested set" of invariant subspaces: $\mathbf{e} \in E_1 \subset E_2 \subset \cdots \subset E_k$ of dimension $1, 2, \ldots, k$, respectively, and there is no eigenvector of f in E_k, other than those of E_1. If $n \geqq 4$, then f may have several eigenvectors with multiplicity $\geqq 2$ which may belong to the same or to different eigenvalues. In $\mathbb{R}^3$ there is not enough room for this to happen. If we add the multiplicities of the eigenvectors that belong to λ, we get the multiplicity of the eigenvalue λ.

If f is "missing" some eigenvectors (due to the characteristic equation having some irreducible quadratic factors), it will be missing an *even* number of them, i.e., pairs of eigenvectors. Such missing pairs of eigenvectors will nevertheless determine invariant planes of f; in each of those invariant planes f will act by way of a rank 2 operator

$$\rho(t^4 \cos \phi + t \sin \phi)$$

($\rho > 0$; $\phi \neq 0°$, $180°$; $t^3 = -t$, rank $t = 2$; such an operator t is called a *principal* twist.) Therefore, we may regard t as a substitute for the pair of missing eigenvectors corresponding to the pair of complex conjugate eigenvalues

$$\rho(\cos \phi \pm i \sin \phi) \qquad (i = \sqrt{-1})$$

which we may regard as a pair of missing roots of an irreducible quadratic factor of the characteristic equation of f.

With this new convention the count of eigenvalues and eigenvectors is complete: Every linear operator f of $\mathbb{R}^n$ has exactly n eigenvalues and n eigenvectors, each counted with multiplicity, including pairs of complex conjugate eigenvalues, and pairs of complex conjugate eigenvectors or principal twists belonging to them. The pairs of complex conjugate eigenvalues may also have multiplicity $\geqq 2$. In this case there is either an eigenspace of even dimension, every plane of which is invariant under f, or there is a principal twist with multiplicity $\geqq 2$, whose image is contained in an invariant subspace containing the image of no other principal twist of f. To the same pair of complex conjugate eigenvalues with multiplicity, an eigenspace and a principal twist with multiplicity may also belong. These possibilities will give rise to new types of linear operators which are unknown in $\mathbb{R}^3$.

Two linear operators of $\mathbb{R}^n$ are said to have the same *type* if there are bijective correspondences between their respective sets of eigenvalues, as well as between their respective sets of eigenvectors, which preserve the multiplicities of eigenvalues and eigenvectors, preserve complex conjugate pairs, and respect the relationship that makes eigenvectors belong to specific eigenvalues.

If we denote the number of types of linear operators in $\mathbb{R}^n$ by the symbol $\tau(n)$, then these values, up to $n = 11$, are given by the following table:

n	1	2	3	4	5	6	7	8	9	10	11
$\tau(n)$	1	4	7	20	36	87	162	355	666	1367	2557

EXERCISES

10.1 Find, without solving the characteristic equation, all the eigenvalues and eigenvectors of f in each of the following cases. (*Hint*: Calculate f^2 first.)

(i) $f = \begin{pmatrix} -1 & 0 & 0 \\ 6 & 3 & -2 \\ -6 & -4 & -3 \end{pmatrix}$ (ii) $f = \begin{pmatrix} 1 & 0 & 0 \\ 2 & 3 & 2 \\ -3 & -3 & -2 \end{pmatrix}$

(iii) $f = \begin{pmatrix} 7 & -9 & 15 \\ -2 & 4 & -5 \\ -4 & 6 & -9 \end{pmatrix}$ (iv) $f = \begin{pmatrix} -23 & -40 & -56 \\ -12 & -19 & -28 \\ 18 & 30 & 42 \end{pmatrix}$

(v) $f = \begin{pmatrix} -26 & -18 & -22 \\ 21 & 15 & 21 \\ 12 & 8 & 13 \end{pmatrix}$ (vi) $f = \begin{pmatrix} 7 & 16 & -24 \\ -6 & -13 & 18 \\ -2 & -4 & -5 \end{pmatrix}$

(vii) $f = \begin{pmatrix} -14 & 20 & -18 \\ 21 & -30 & 27 \\ 35 & -50 & 45 \end{pmatrix}$ (viii) $f = \begin{pmatrix} 1 & -4 & 5 \\ 1 & -4 & 5 \\ 1 & -3 & 3 \end{pmatrix}$

(ix) $f = \begin{pmatrix} 1 & 0 & 0 \\ 4 & 5 & 4 \\ -6 & -6 & -5 \end{pmatrix}$ (x) $f = \begin{pmatrix} -8 & -12 & -21 \\ -15 & -19 & -35 \\ 12 & 16 & 29 \end{pmatrix}$

(xi) $f=\begin{pmatrix} 4 & -8 & 12 \\ -1 & 2 & -3 \\ -2 & 4 & -6 \end{pmatrix}$ (xii) $f=\begin{pmatrix} 3 & 8 & 12 \\ 2 & 3 & 6 \\ -2 & -4 & -7 \end{pmatrix}$

(xiii) $f=\begin{pmatrix} 5 & 12 \\ -2 & -5 \end{pmatrix}$ (xiv) $f=\begin{pmatrix} -1 & 2 \\ -1 & 1 \end{pmatrix}$

(xv) $f=\begin{pmatrix} 25 & 52 & -78 \\ -6 & -13 & 18 \\ 4 & 8 & -13 \end{pmatrix}$ (xvi) $f=\begin{pmatrix} -9 & -41 \\ 2 & 9 \end{pmatrix}$

10.2 Show that every projection $p \neq 0, 1$ of $\mathbb{R}^3$ can be written in one of the forms $p(\mathbf{v}) = (\mathbf{n} \cdot \mathbf{v})\mathbf{e}$ or $p(\mathbf{v}) = (\mathbf{e} \times \mathbf{v}) \times \mathbf{n}$, where $\mathbf{e}$, $\mathbf{n}$ are fixed vectors subject to $\mathbf{e} \cdot \mathbf{n} = 1$. Study the special case $\mathbf{e} = \mathbf{n} = \mathbf{u}$. (See Chap. 4, Exercise 13, and Chap. 11, Exercises 26 and 27.)

10.3 Given that

$$f=\begin{pmatrix} -15 & 12 & 21 \\ -8 & 6 & 10 \\ -6 & 5 & 9 \end{pmatrix}$$

is a twist, find its unique eigenvector (without using the characteristic equation).

10.4 Given that

$$f=\begin{pmatrix} 4 & -7 & 5 \\ 3 & -6 & 5 \\ 2 & -4 & 3 \end{pmatrix}$$

is an invection, find its unique eigenvector (without using the characteristic equation). Is f orientation preserving or reversing?

10.5 Show that if $t^3 = -t$ and $j = 1 + t + t^2$, then $j^3 - j^2 + j - 1 = 0$, $t = \tfrac{1}{2}(j - j^3)$. Conversely, if $j^3 - j^2 + j - 1 = 0$ and $t = \tfrac{1}{2}(j - j^3)$, then $t^3 = -t$ and $j = 1 + t + t^2$. (In other words, the Cayley transform between the set of twists and the set of orientation-preserving invections is bijective.)

10.6 Show that the unique eigenvector of a twist $t \neq 0$ and that of its Cayley transform coincide.

10.7 Show that $j(\mathbf{v}) = (\mathbf{u} \cdot \mathbf{v})\mathbf{u} + \mathbf{u} \times \mathbf{v}$ $(|\mathbf{u}| = 1)$ is an orientation-preserving invection. Find the Cayley transform t of j, and calculate $1 + t + t^2$.

10.8 Let $t = \ell - \ell'$ be a twist (where ℓ, ℓ' are lifts and $\ell \circ \ell'$, $\ell' \circ \ell$ are principal projections). Find the Cayley transform of t in terms of ℓ, ℓ'.

10.9 Show that every twist t of $\mathbb{R}^3$ can be written in the form

$$t(\mathbf{v}) = (\mathbf{m}' \times \mathbf{v}) \times \mathbf{n}' - (\mathbf{m} \times \mathbf{v}) \times \mathbf{n}$$

where $\mathbf{m} \cdot \mathbf{n} = 0 = \mathbf{m}' \cdot \mathbf{n}'$, $\mathbf{m} \cdot \mathbf{n}' = 1 = \mathbf{m}' \cdot \mathbf{n}$. Study the special case $\mathbf{m} = \mathbf{n}'$,

$\mathbf{m}' = \mathbf{n}$ (the case of an antisymmetric twist). Calculate the Cayley transform of t. (See Chap. 4, Exercise 11, and Chap. 11, Exercise 25.)

10.10 Show that if $t \neq 0$ is a twist of $\mathbb{R}^3$, then $p = 1 + t^2$, $q = t^4$ are projections, in more detail; p is the principal projection onto the line $\ker t$ along the plane $\operatorname{im} t$; and q is the projection onto the plane $\operatorname{im} t$ along the line $\ker t$.

10.11 Decompose the twist

$$t = \begin{pmatrix} 5 & 13 \\ -2 & -5 \end{pmatrix}$$

into the difference of lifts.

10.12 In each of the following, decompose the invection into the product of involutions.

(i) $j = \begin{pmatrix} 4 & -7 & 5 \\ 3 & -6 & 5 \\ 2 & -4 & 3 \end{pmatrix}$ (ii) $j = \begin{pmatrix} 5 & 13 \\ -2 & -5 \end{pmatrix}$

10.13 Calculate the Cayley transform of the invection

$$j = \begin{pmatrix} 4 & -7 & 5 \\ 3 & -6 & 5 \\ 2 & -4 & 3 \end{pmatrix}$$

and show that it is a twist.

10.14 Calculate the Cayley transform of the twist

$$t = \begin{pmatrix} 1 & -4 & 5 \\ 1 & -4 & 5 \\ 1 & -3 & 3 \end{pmatrix}$$

and show that it is an orientation-preserving invection.

10.15 In each of the following, find the characteristic equation of the linear operator f, and hence find all the eigenvalues and eigenvectors. Check by substituting into the equation $f(\mathbf{e}) = \lambda \mathbf{e}$. State the type of f.

(i) $f = \begin{pmatrix} 1 & 1 & 1 \\ 1 & 1 & -1 \\ 1 & -1 & -1 \end{pmatrix}$ (ii) $f = \begin{pmatrix} 3 & 4 & -2 \\ 3 & 2 & 0 \\ 0 & 4 & 1 \end{pmatrix}$

(iii) $f = \begin{pmatrix} -7 & -7 & 5 \\ -8 & -8 & -5 \\ 0 & -5 & 0 \end{pmatrix}$ (iv) $f = \begin{pmatrix} 3 & 2 & 4 \\ 2 & 0 & 2 \\ 4 & 2 & 3 \end{pmatrix}$

(v) $f = \begin{pmatrix} 100 & 99 & 99 \\ 99 & 100 & 99 \\ -99 & -99 & -98 \end{pmatrix}$ (vi) $f = \begin{pmatrix} 8 & -8 & -1 \\ 0 & 0 & -1 \\ 0 & 1 & -2 \end{pmatrix}$

(vii) $f=\begin{pmatrix}-2 & 1 & 0\\ -2 & 1 & -1\\ -1 & 1 & -2\end{pmatrix}$ (viii) $f=\begin{pmatrix}-1 & -1 & 2\\ -2 & -1 & 3\\ -1 & -1 & 2\end{pmatrix}$

(ix) $f=\begin{pmatrix}-1 & 4 & -5\\ -1 & 4 & -5\\ -1 & 3 & -3\end{pmatrix}$ (x) $f=\begin{pmatrix}-1 & 18 & 7\\ -1 & -13 & -4\\ 1 & 25 & 8\end{pmatrix}$

(xi) $f=\begin{pmatrix}3 & 1 & 4\\ 0 & -2 & 7\\ 0 & 0 & 5\end{pmatrix}$ (xii) $f=\begin{pmatrix}1 & 1 & 1\\ 1 & 1 & 1\\ 1 & 1 & 1\end{pmatrix}$

10.16 In each of the following, write the characteristic equation, and find all the eigenvalues and eigenvectors with their multiplicity. Find eigenspaces, if any. State the type of operator. ($\lambda_\nu=\lambda_\mu \Leftrightarrow \nu=\mu$.)

(i) $\begin{pmatrix}\lambda_1 & 0\\ 0 & \lambda_2\end{pmatrix}$ (ii) $\begin{pmatrix}\lambda_1 & 1\\ 0 & \lambda_1\end{pmatrix}$ (iii) $\begin{pmatrix}\lambda_1 & 0\\ 0 & \lambda_1\end{pmatrix}$

(iv) $\begin{pmatrix}\rho\cos\phi & -\rho\sin\phi\\ \rho\sin\phi & \rho\cos\phi\end{pmatrix}$ $(\rho>0,\ \phi\neq 0°,\ 180°)$

10.17 What is the type of each of the following linear operators? (i) projection; (ii) involution; (iii) invection; (iv) twist; (v) lift; (vi) shear; (vii) rotation; (viii) reflection; (ix) dilatation; (x) strain.

10.18 In each of the following, find eigenvalues and eigenvectors, and state the type of f.

(i) $f=\begin{pmatrix}2 & 1\\ 1 & 2\end{pmatrix}$ (ii) $f=\begin{pmatrix}3 & 4\\ 5 & 2\end{pmatrix}$

(iii) $f=\begin{pmatrix}5 & 6 & -3\\ -1 & 0 & 1\\ 1 & 2 & -1\end{pmatrix}$ (iv) $f=\begin{pmatrix}2 & -1 & 2\\ 5 & -3 & 3\\ -1 & 0 & -2\end{pmatrix}$

10.19 Given that

$$f=\begin{pmatrix}1 & 0 & 0\\ 4 & 5 & 4\\ -6 & -6 & -5\end{pmatrix}$$

is an involution in a plane (i.e., $f^2=1$, trace $f=1$), find a new basis in which the matrix of f becomes

$$\begin{pmatrix}1 & 0 & 0\\ 0 & 1 & 0\\ 0 & 0 & -1\end{pmatrix}$$

Write the transition equation.

10.20 Given that

$$f=\begin{pmatrix} 4 & 8 & -12 \\ -3 & -6 & 9 \\ -1 & -2 & 3 \end{pmatrix}$$

is a projection onto a line, find a new basis in which the matrix of f becomes

$$\begin{pmatrix} 1 & 0 & 0 \\ 0 & 0 & 0 \\ 0 & 0 & 0 \end{pmatrix}$$

Write the transition equation.

10.21 In each of the following, find a new basis in terms of which the matrix of f is a diagonal matrix. Write the transition equation.

(i) $f=\begin{pmatrix} 5 & -6 & -6 \\ -1 & 4 & 2 \\ 3 & -6 & -4 \end{pmatrix}$ (ii) $f=\begin{pmatrix} 4 & -2 & 1 \\ -2 & 1 & 2 \\ 1 & 2 & 4 \end{pmatrix}$

(iii) $f=\begin{pmatrix} 3 & 6 & 6 \\ 0 & 2 & 0 \\ -3 & -12 & -6 \end{pmatrix}$ (iv) $f=\begin{pmatrix} 5 & -12 \\ 4 & 9 \end{pmatrix}$

(v) $f=\begin{pmatrix} -1 & 4 & -2 \\ -3 & 4 & 0 \\ -3 & 1 & 3 \end{pmatrix}$ (vi) $f=\begin{pmatrix} 1 & 1 & -1 \\ 0 & 2 & 1 \\ 0 & 0 & -2 \end{pmatrix}$

10.22 Show that

$$f=\begin{pmatrix} 4 & -8 & 12 \\ -1 & 2 & -3 \\ -2 & 4 & -6 \end{pmatrix}$$

is a lift and find a new basis in which the matrix of f becomes

$$\begin{pmatrix} 0 & 0 & 1 \\ 0 & 0 & 0 \\ 0 & 0 & 0 \end{pmatrix}$$

Write the transition equation.

10.23 Given that

$$f=\begin{pmatrix} 1 & -4 & 5 \\ 1 & -4 & 5 \\ 1 & -3 & 3 \end{pmatrix}$$

is a twist, find a new basis in which the matrix of f takes the form

$$\begin{pmatrix} 0 & 0 & 0 \\ 0 & 0 & -1 \\ 0 & 1 & 0 \end{pmatrix}$$

Write the transition equation.

10.24 Show that if $\mathbf{e}$ is an eigenvector of f belonging to λ, then $\mathbf{e}$ is also an eigenvector of f^2. What is the corresponding eigenvalue?

10.25 Let f be a bijective linear operator. Show that if $\mathbf{e}$ is an eigenvector of f belonging to λ, then $\mathbf{e}$ is also an eigenvector of f^{-1}. What is the corresponding eigenvalue?

10.26 Prove that a linear operator f is bijective if, and only if, it has no zero eigenvalue.

10.27 Write the characteristic equation of

$$f = \begin{pmatrix} \lambda_1 & \alpha & \beta \\ 0 & \lambda_2 & \gamma \\ 0 & 0 & \lambda_3 \end{pmatrix}$$

10.28 Show that $g \circ f$ and $f \circ g$ have the same characteristic equation.

10.29 Show that a volume-preserving elliptic operator can be decomposed into the product of two involutions.

10.30 Write the matrix version of the decomposition in Exercise 10.29, for $\mathbb{R}^2$.

10.31 Show that

$$j \text{ orientation-preserving invection} \Rightarrow \operatorname{trace} j = 1$$
$$j \text{ orientation-reversing invection} \Rightarrow \operatorname{trace} j = -1$$

10.32 True or false?
"Given a plane π and a line ℓ not in π, both passing through the origin, there are exactly two twists $\pm t$ such that $\pi = \operatorname{im} t$ and $\ell = \ker t$".

10.33 Show that

$$\left\{ \begin{pmatrix} 1 \\ 2 \\ 1 \end{pmatrix}, \begin{pmatrix} 2 \\ 3 \\ 3 \end{pmatrix}, \begin{pmatrix} 3 \\ 7 \\ 1 \end{pmatrix} \right\} \quad \text{and} \quad \left\{ \begin{pmatrix} 3 \\ 1 \\ 4 \end{pmatrix}, \begin{pmatrix} 5 \\ 2 \\ 3 \end{pmatrix}, \begin{pmatrix} 1 \\ 1 \\ -6 \end{pmatrix} \right\}$$

are both bases for $\mathbb{R}^3$. Find the transition operator from the first to the second basis.

10.34 Let g be invertible. Show that if $\mathbf{e}$ is the eigenvector of the operator f belonging to eigenvalue λ, then $g(\mathbf{e})$ is the eigenvector of the conjugate operator $g \circ f \circ g^{-1}$ belonging to eigenvalue λ.

10.35 Show that the operator f has the same type as its conjugate $g \circ f \circ g^{-1}$, for any invertible g.

10.36 Let f be a linear operator of $\mathbb{R}^3$ with rank $f = 1$. Show that either f is a lift, or $f = \lambda p$, $\lambda \in \mathbb{R}$, where p is a principal projection.

10.37 Show that any two (nontrivial) lifts are conjugate.

10.38 Show that any two (nonidentity) shears are conjugate.

10.39 Show that any two (nontrivial) twists are conjugate in $\mathbb{R}^3$.

10.40 Show that any two (nonidentity) invections are conjugate in $\mathbb{R}^3$.

10.41 Show that two dilatations are conjugate if, and only if, they have the same ratio.

10.42 Show that every linear operator of $\mathbb{R}^2$ can be decomposed into the product of not more than two shears and one dilatation (in any order).

10.43 Show that every linear operator of $\mathbb{R}^3$ can be decomposed into the product of not more than three shears and one dilatation (in any order).

11

Symmetric, Antisymmetric, and Orthogonal Operators[†]

THE STAR-DOT FORMULA

The *transpose* of a matrix is obtained by interchanging its rows and columns:

$$\begin{pmatrix} a_{11} & a_{12} & a_{13} \\ \cdots & \cdots & \cdots \\ \cdots & \cdots & \cdots \end{pmatrix}^* = \begin{pmatrix} a_{11} & \vdots & \vdots \\ a_{12} & \vdots & \vdots \\ a_{13} & \vdots & \vdots \end{pmatrix}$$

The operator f^* whose matrix is the transpose of the matrix of f (in terms of the same rectangular metric coordinate system) is also called the transpose of f, and the operation * is called *transposition*.

Examples

1. Let ℓ be a lift (i.e., $\ell^2 = 0$). Then

$$\ell = \begin{pmatrix} Aa & Ba & Ca \\ Ab & Bb & Cb \\ Ac & Bc & Cc \end{pmatrix}$$

where $Aa + Bb + Cc = 0$. Hence

$$\ell^* = \begin{pmatrix} Aa & Ab & Ac \\ Ba & Bb & Bc \\ Ca & Cb & Cc \end{pmatrix}$$

and we conclude that the transpose of a lift is a lift.

[†] *Note*: In this chapter all bases are assumed to be orthonormal and all coordinate systems are assumed to be rectangular metric.

2. Let p be a principal projection (i.e., $p^2 = p$ and rank $p = 1$). Then

$$p = \begin{pmatrix} Aa & Ba & Ca \\ Ab & Bb & Cb \\ Ac & Bc & Cc \end{pmatrix}$$

where $Aa + Bb + Cc = 1$. Again, it follows that the transpose of a principal projection is a principal projection.

We shall now see that the relationship between f and f^* remains undisturbed under a change of the coordinate system (to another rectangular metric coordinate system). This is not immediately obvious, but will follow from the important

Star-Dot Formula

$$f(\mathbf{v}) \cdot \mathbf{w} = \mathbf{v} \cdot f^*(\mathbf{w})$$

which holds for any pair of vectors $\mathbf{v}$, $\mathbf{w}$.

Proof: $f(\mathbf{i}) \cdot \mathbf{v} = \mathbf{i} \cdot f^*(\mathbf{v})$ because both sides are equal to the same expression $a_{11}x + a_{21}y + a_{31}z$. Similarly, $f(\mathbf{j}) \cdot \mathbf{v} = \mathbf{j} \cdot f^*(\mathbf{v})$ and $f(\mathbf{k}) \cdot \mathbf{v} = \mathbf{k} \cdot f^*(\mathbf{v})$. Therefore,

$$\begin{aligned} f(\mathbf{v}) \cdot \mathbf{w} &= f(x\mathbf{i} + y\mathbf{j} + z\mathbf{k}) \cdot \mathbf{w} = (xf(\mathbf{i}) + yf(\mathbf{j}) + zf(\mathbf{k})) \cdot \mathbf{w} \\ &= x(f(\mathbf{i}) \cdot \mathbf{w}) + y(f(\mathbf{j}) \cdot \mathbf{w}) + z(f(\mathbf{k}) \cdot \mathbf{w}) \\ &= x(\mathbf{i} \cdot f^*(\mathbf{w})) + y(\mathbf{j} \cdot f^*(\mathbf{w})) + z(\mathbf{k} \cdot f^*(\mathbf{w})) \\ &= (x\mathbf{i} + y\mathbf{j} + z\mathbf{k}) \cdot f^*(\mathbf{w}) = \mathbf{v} \cdot f^*(\mathbf{w}) \end{aligned}$$

The star-dot formula can be used as an equivalent definition of the transpose f^* of a linear operator f. This definition does not have any reference to matrices or coordinates. It follows that transposition is independent of coordinates.

APPLICATIONS

1. Invariant Planes of a Linear Operator

We say that a plane is invariant under the linear operator f if every vector $\mathbf{v}$ parallel to the plane is mapped into a vector $f(\mathbf{v})$ also parallel to the plane. Let $\mathbf{n} \neq \mathbf{0}$ be a normal vector of the plane; then

$$\mathbf{n} \cdot \mathbf{v} = 0 \Rightarrow \mathbf{n} \cdot f(\mathbf{v}) = 0 \Rightarrow f^*(\mathbf{n}) \cdot \mathbf{v} = 0$$

by the star-dot formula and we may conclude that $f^*(\mathbf{n}) \parallel \mathbf{n}$, i.e., $f^*(\mathbf{n}) = \lambda\mathbf{n}$. The converse is also true, so that we have the result: *the eigenvectors of the transpose f^* are precisely the normal vectors of the invariant planes of f.*

Example Given the parabolic operator

$$f = \begin{pmatrix} -2 & 1 & 0 \\ -2 & 1 & -1 \\ -1 & 1 & -2 \end{pmatrix}$$

find its unique invariant plane.

Answer:

$$f^* = \begin{pmatrix} -2 & -2 & -1 \\ 1 & 1 & 1 \\ 0 & -1 & -2 \end{pmatrix}$$

and

$$\det(f^* - \lambda) = 0 \Rightarrow (\lambda + 1)^3 = 0 \Rightarrow \lambda = -1$$

is the only eigenvalue, and it has multiplicity 3.

$\ker(f^* + 1)$:

$$\begin{pmatrix} -1 & -2 & -1 \\ 1 & 2 & 1 \\ 0 & -1 & -2 \end{pmatrix} \Rightarrow \begin{pmatrix} 1 & 2 & 1 \\ 0 & 1 & 1 \end{pmatrix} \Rightarrow \begin{pmatrix} 1 & 0 & -1 \\ 0 & 1 & 1 \end{pmatrix} \Rightarrow \begin{cases} x = t \\ y = -t \\ z = t \end{cases}$$

$$\Rightarrow \mathbf{n} = \mathbf{i} - \mathbf{j} + \mathbf{k}$$

So $x - y + z = 0$ is the equation of the unique invariant plane of the parabolic operator f. Note that the unique eigenvector $\mathbf{e} = \mathbf{i} + \mathbf{j}$ of f, which is a triple eigenvector, belongs to the invariant plane.

FORMAL PROPERTIES OF TRANSPOSITION

Transposition is bilinear, i.e.,

$$(\lambda f)^* = \lambda f^* \qquad (\lambda \in \mathbb{R})$$
$$(f + g)^* = f^* + g^*$$

This is obvious if we think of the rules how to add and transpose matrices: it makes no difference whether we add them first and transpose afterward, or vice versa. The same is true of the order of performing the multiplication of the matrix by a scalar and transposing. A special case of this is the transpose of the negative:

$$(-f)^* = -f^*$$

The transpose of the identity operator is itself:

$$1^* = 1$$

The transpose of the trivial operator is itself:

$$0^* = 0$$

The important rule that repeating transposition twice yields the same linear operator:

$$(f^*)^* = f$$

is also self-evident if we think of the transposition of a matrix.

The transpose of the product is the same as the product of the transposes, taken in the *opposite order*:

Product Rule

$$(g \circ f)^* = f^* \circ g^*$$

This is a consequence of the star-dot formula:

$$\begin{aligned} \mathbf{v} \cdot (g \circ f)^* (\mathbf{w}) &= (g \circ f)(\mathbf{v}) \cdot \mathbf{w} = g(f(\mathbf{v})) \cdot \mathbf{w} = f(\mathbf{v}) \cdot g^* (\mathbf{w}) \\ &= \mathbf{v} \cdot f^* (g^* (\mathbf{w})) = \mathbf{v} \cdot (f^* \circ g^*)(\mathbf{w}) \qquad \text{for all } \mathbf{v}, \mathbf{w} \end{aligned}$$

The transpose of the inverse of an invertible operator is just the inverse of the transpose, which is also invertible:

$$(f^{-1})^* = (f^*)^{-1}$$

as can be seen if we apply the previous formula to the product $f \circ f^{-1} = 1$.

Example Show that the transpose of an involution is an involution.

Answer: $h^2 = 1 \Rightarrow (h^2)^* = 1^* \Rightarrow (h^*)^2 = 1$.

The concept of transposition is introduced in order to distinguish three most important classes of linear operators, the symmetric operators ($f^* = f$), the antisymmetric operators ($f^* = -f$), and the orthogonal operators ($f^* = f^{-1}$).

SYMMETRIC OPERATORS. THE PRINCIPAL AXIS THEOREM

A linear operator s is called *symmetric* if $s^* = s$. This notion is independent of the choice of coordinates, although we determine whether s is symmetric by inspecting its matrix. Indeed, the term "symmetric" was chosen in this context because the matrix of s satisfying $s^* = s$ is symmetric with respect to its main diagonal, i.e., its rows and columns may be interchanged without altering the matrix:

$$a_{nm} = a_{mn}$$

We also have, from the star-dot formula, that

$$s \text{ symmetric} \Leftrightarrow s(\mathbf{v}) \cdot \mathbf{w} = \mathbf{v} \cdot s(\mathbf{w})$$

for all **v**, **w**.

Examples

1. Let p be a perpendicular projection onto a line. Show that p is symmetric.

   We have $p(\mathbf{v}) = (\mathbf{u} \cdot \mathbf{v})\mathbf{u}$ for all **v**, where $|\mathbf{u}| = 1$. Then

   $$p(\mathbf{v}) \cdot \mathbf{w} = (\mathbf{u} \cdot \mathbf{v})(\mathbf{u} \cdot \mathbf{w}) = \mathbf{v} \cdot p(\mathbf{w}) \qquad \text{for all } \mathbf{v}, \mathbf{w}$$

   By the star-dot formula, we may conclude that $p^* = p$.
   The matrix of perpendicular projections was given in Chap. 6, Applications 1 and 2.
2. Let q be a perpendicular projection onto a plane. Show that q is symmetric.

   $q = 1 - p$, where p is the perpendicular projection onto the normal line of $\operatorname{im} q$. Then $q^* = (1-p)^* = 1^* - p^* = 1 - p = q$.

   It is left as an exercise to show that if h is a reflection (in a plane, line, or in the origin), then h is symmetric. The converse of these statements is also true:

   $$p^2 = p = p^* \Rightarrow p \text{ perpendicular projection}$$

   $$h^{-1} = h = h^* \Rightarrow h \text{ reflection}$$

However, the proof of these results requires a deeper theorem, the principal axis theorem, which we shall now discuss.

The principal axis theorem provides a geometric characterization of symmetric operators, asserting that precisely those operators are symmetric which have three *mutually perpendicular* eigenvectors. In particular, the type of a symmetric operator is either hyperbolic, or dilatational, or strain (depending on the multiplicity of the eigenvalues). It can have an eigenspace of dimension greater than 1, but it cannot have an eigenvector with multiplicity greater than 1, nor can it be elliptic with missing eigenvectors. The three mutually perpendicular eigenvectors $\mathbf{e}_1$, $\mathbf{e}_2$, $\mathbf{e}_3$ determine the three principal axes of the symmetric operator. The corresponding eigenvalues λ_1, λ_2, λ_3 indicate the ratio in which the principal axis is being stretched ($\lambda > 1$) contracted ($0 < \lambda < 1$), or if a reflection in the normal plane is also involved ($\lambda < 0$). Thus, if s is symmetric and none of its eigenvalues is 0, s deforms the unit cube determined by its eigenvectors into a *rectangular* parallelepiped (matchbox), or what is the same to say, s deforms a sphere centered at the origin into an ellipsoid whose axes of symmetry are the principal axes of s.

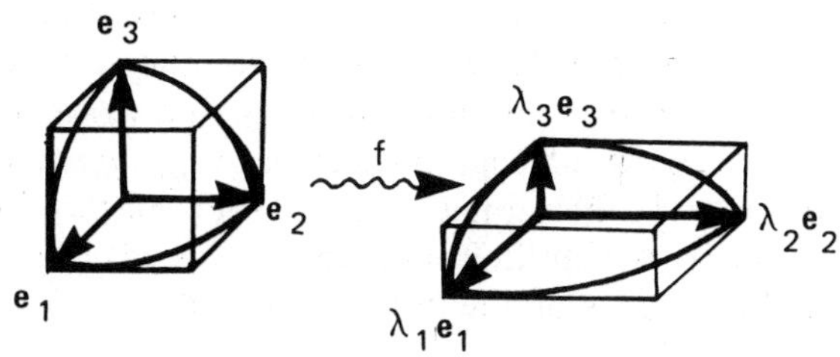

The Principal Axis Theorem

$$s^* = s \Leftrightarrow s(\mathbf{v}) = \lambda_1(\mathbf{e}_1 \cdot \mathbf{v})\mathbf{e}_1 + \lambda_2(\mathbf{e}_2 \cdot \mathbf{v})\mathbf{e}_2 + \lambda_3(\mathbf{e}_3 \cdot \mathbf{v})\mathbf{e}_3$$

for every $\mathbf{v}$ *where* $\{\mathbf{e}_1,\mathbf{e}_2,\mathbf{e}_3\}$ *is some orthonormal basis.*

It is clear that $s(\mathbf{e}_1) = \lambda_1\mathbf{e}_1$, $s(\mathbf{e}_2) = \lambda_2\mathbf{e}_2$, $s(\mathbf{e}_3) = \lambda_3\mathbf{e}_3$, so that $\mathbf{e}_1$, $\mathbf{e}_2$, $\mathbf{e}_3$ are eigenvectors of s belonging to the eigenvalues λ_1, λ_2, λ_3, respectively. In particular, $s^* = s \Rightarrow \operatorname{im} s \perp \ker s$.

If the principal axes of the symmetric operator s are chosen as coordinate axes, then the matrix of s becomes particularly simple:

$$s = \begin{pmatrix} \lambda_1 & 0 & 0 \\ 0 & \lambda_2 & 0 \\ 0 & 0 & \lambda_3 \end{pmatrix}$$

This may be called the "diagonalized form" of s. In older books on linear algebra the principal axis theorem is quoted as saying that "a matrix is orthogonally diagonizable if, and only if, it is symmetric." However, this formulation badly distorts the correct perspective of the underlying ideas, which can be stated without any reference to coordinates or to changes in coordinates.

The principal axis theorem does not assert that the principal axes of a symmetric operator s are uniquely determined. They may not be, as in the case of the strain in ratio λ:

$$s = \begin{pmatrix} \lambda & 0 & 0 \\ 0 & \lambda & 0 \\ 0 & 0 & \lambda \end{pmatrix}$$

In this case, every orthonormal basis will furnish a set of pairwise perpendicular eigenvectors. However, it will follow from Proposition 1 below that, in the case when the eigenvalues λ_1, λ_2, λ_3 are all distinct, the principal axes are uniquely determined (apart from orientation).

The formula given by the principal axis theorem can also be written as $s = \lambda_1 p_1 + \lambda_2 p_2 + \lambda_3 p_3$, where p_1, p_2, p_3 are the perpendicular projections onto the pairwise perpendicular principal axes. This immediately shows that the condition $s^* = s$ for the existence of the (pairwise perpendicular) principal axes is *necessary*, because if s can be written as a linear combination of three perpendicular projections p_1, p_2, p_3 which are known to be symmetric, then by the formal properties of transposition, s will also be symmetric. The proof that the condition $s^* = s$ is also *sufficient* is more complicated and it depends on the following results:

Proposition 1 Let s be a symmetric operator: $s^* = s$. Suppose that $\mathbf{e}_1$ and $\mathbf{e}_2$ are two eigenvectors of s belonging to the eigenvalues $\lambda_1 \neq \lambda_2$. Then the two eigenvectors $\mathbf{e}_1$ and $\mathbf{e}_2$ are perpendicular: $\mathbf{e}_1 \perp \mathbf{e}_2$.

Proof: By assumption, $s(\mathbf{e}_1) = \lambda_1 \mathbf{e}_1$, $s(\mathbf{e}_2) = \lambda_2 \mathbf{e}_2$. As the eigenvalues are distinct, we may assume that $\lambda_2 \neq 0$. Then

$$\mathbf{e}_1 \cdot \mathbf{e}_2 = \frac{1}{\lambda_2}\mathbf{e}_1 \cdot (\lambda_2 \mathbf{e}_2) = \frac{1}{\lambda_2}\mathbf{e}_1 \cdot s(\mathbf{e}_2) = \frac{1}{\lambda_2}s^*(\mathbf{e}_1) \cdot \mathbf{e}_2 \qquad \text{(star-dot formula)}$$

$$= \frac{1}{\lambda_2}s(\mathbf{e}_1) \cdot \mathbf{e}_2 = \frac{\lambda_1}{\lambda_2}\mathbf{e}_1 \cdot \mathbf{e}_2 \Rightarrow \left(1 - \frac{\lambda_1}{\lambda_2}\right)\mathbf{e}_1 \cdot \mathbf{e}_2 = 0$$

$$\Rightarrow \mathbf{e}_1 \cdot \mathbf{e}_2 = 0 \Rightarrow \mathbf{e}_1 \perp \mathbf{e}_2$$

This completes the proof of Proposition 1.

Proposition 2 Let s be a symmetric operator: $s^* = s$, and let $\mathbf{e}$ be an eigenvector of s. Then

$$\mathbf{v} \perp \mathbf{e} \Rightarrow s(\mathbf{v}) \perp \mathbf{e}$$

Since any vector perpendicular to $\mathbf{e}$ is taken into a vector perpendicular to $\mathbf{e}$ by the symmetric operator s, we see that the normal plane of the eigenvector $\mathbf{e}$ is left unchanged (although not necessarily pointwise). This means that the normal plane of an eigenvector $\mathbf{e}$ of the symmetric operator s is an invariant plane of s.

Proof: We have that $s^* = s$, $s(\mathbf{e}) = \lambda \mathbf{e}$, $\mathbf{e} \neq \mathbf{0}$. Then

$$\begin{aligned}\mathbf{v} \perp \mathbf{e} &\Rightarrow \mathbf{v} \cdot \mathbf{e} = 0 \Rightarrow \mathbf{v} \cdot \lambda \mathbf{e} = 0 \\ &\Rightarrow \mathbf{v} \cdot s(\mathbf{e}) = 0 \\ &\Rightarrow s^*(\mathbf{v}) \cdot \mathbf{e} = 0 \qquad \text{(by the star-dot formula)} \\ &\Rightarrow s(\mathbf{v}) \cdot \mathbf{e} = 0 \Rightarrow s(\mathbf{v}) \perp \mathbf{e}\end{aligned}$$

This completes the proof of Proposition 2.

Now we are ready to prove sufficiency of the condition $s^* = s$ in the principal axis theorem.

Case of $\mathbb{R}^2$

$$s^* = s \Rightarrow s(\mathbf{v}) = \lambda_1(\mathbf{e}_1 \cdot \mathbf{v})\mathbf{e}_1 + \lambda_2(\mathbf{e}_2 \cdot \mathbf{v})\mathbf{e}_2$$

where $\{\mathbf{e}_1, \mathbf{e}_2\}$ is an orthonormal basis.

We want to exhibit the existence of an eigenvector $\mathbf{e}_1$ of s. Let $\mathbf{w}$ be an arbitrary vector; if $s(\mathbf{w}) = \lambda\mathbf{w}$, then $\mathbf{w} = \mathbf{e}_1$ is an eigenvector. Otherwise, $s(\mathbf{w}) \neq \lambda\mathbf{w}$; i.e., $s(\mathbf{w}) \nparallel \mathbf{w}$ and $\{\mathbf{w}, s(\mathbf{w})\}$ span $\mathbb{R}^2$. It follows that $s^2(\mathbf{w})$ can be written as a linear combination:

$$s^2(\mathbf{w}) = \alpha\mathbf{w} + \beta s(\mathbf{w})$$

It also follows that $s^2(s(\mathbf{w})) = \alpha s(\mathbf{w}) + \beta s(s(\mathbf{w}))$, i.e., $s^2 - \beta s - \alpha$ kills both vectors $\mathbf{w}$, $s(\mathbf{w})$. This can be only if $s^2 - \beta s - \alpha = 0$. By completing the square, we get

$$s^2 - 2\left(\frac{\beta}{2}\right)s + \left(\frac{\beta}{2}\right)^2 = \frac{\beta^2}{4} + \alpha$$

or, what is the same,

$$(s - \tfrac{1}{2}\beta)^2 = \gamma \qquad \text{with } \gamma = \frac{\beta^2}{4} + \alpha$$

The real number γ cannot be negative because it is the square of the vector $(s - \frac{1}{2}\beta)(\mathbf{u})$, where $\mathbf{u}$ is a unit vector:

$$\begin{aligned} |(s - \tfrac{1}{2}\beta)(\mathbf{u})|^2 &= (s - \tfrac{1}{2}\beta)\mathbf{u} \cdot (s - \tfrac{1}{2}\beta)\mathbf{u} \\ &= \mathbf{u} \cdot (s - \tfrac{1}{2}\beta)^* \circ (s - \tfrac{1}{2}\beta)\mathbf{u} \qquad \text{(star-dot formula)} \\ &= \mathbf{u} \cdot (s - \tfrac{1}{2}\beta) \circ (s - \tfrac{1}{2}\beta)\mathbf{u} \\ &= \mathbf{u} \cdot \gamma\mathbf{u} = \gamma(\mathbf{u} \cdot \mathbf{u}) = \gamma|\mathbf{u}|^2 = \gamma \end{aligned}$$

Therefore, we may write $\gamma = \delta^2$ for some $\delta \in \mathbb{R}$. Then

$$(s - \tfrac{1}{2}\beta)^2 - \delta^2 = 0 \Rightarrow (s - \tfrac{1}{2}\beta + \delta) \circ (s - \tfrac{1}{2}\beta - \delta) = 0$$

This, however, implies that either $s - \frac{1}{2}\beta + \delta$ or $s - \frac{1}{2}\beta - \delta$ must have a nontrivial kernel (otherwise, they would both be bijective, so their product could not be 0). We shall assume that $\ker(s - \frac{1}{2}\beta + \delta) \neq 0$. Then, for some $\mathbf{e}_1 \neq 0$,

$$\mathbf{e}_1 \in \ker(s - \tfrac{1}{2}\beta + \delta)$$

so that $\mathbf{e}_1$ is an eigenvector of s belonging to $\lambda_1 = \frac{1}{2}\beta - \delta$. We may, of course, assume that $\mathbf{e}_1$ is a unit vector. Let $\mathbf{e}_2$ be another unit vector in $\mathbb{R}^2$

which is perpendicular to $\mathbf{e}_1$. Then, by Proposition 2, $s(\mathbf{e}_2)$ is also perpendicular to $\mathbf{e}_1$; i.e., $s(\mathbf{e}_2)$ is parallel to $\mathbf{e}_2$ or, what is the same, $s(\mathbf{e}_2) = \lambda_2\mathbf{e}_2$ for some $\lambda_2 \in \mathbb{R}$. This means that $\mathbf{e}_2$ is another eigenvector of s and we have, for all $\mathbf{v}$,

$$s(\mathbf{v}) = \lambda_1(\mathbf{e}_1 \cdot \mathbf{v})\mathbf{e}_1 + \lambda_2(\mathbf{e}_2 \cdot \mathbf{v})\mathbf{e}_2$$

because the two sides of the equation describe linear operators which assign the same values $\lambda_1\mathbf{e}_1$ and $\lambda_2\mathbf{e}_2$ to the basis vectors $\mathbf{e}_1$ and $\mathbf{e}_2$, respectively. But then the two operators must coincide.

Case of $\mathbb{R}^3$ Now let s be a symmetric operator of $\mathbb{R}^3$. Then the characteristic equation of s, $\det(s-\lambda)=0$, is a cubic equation in λ with real coefficients and therefore it must have at least one real root λ_1, which is an eigenvalue of s, yielding an eigenvector $\mathbf{e}_1 \in \ker(s-\lambda_1)$. We consider the normal plane of $\mathbf{e}_1$ which we identify with $\mathbb{R}^2$. By Proposition 2, s takes every vector of the normal plane of $\mathbf{e}_1$ into a vector of the same normal plane, so that s can also be considered as a linear operator of the normal plane $\mathbb{R}^2$. Moreover, by the star-dot formula one can see that s as a linear operator of $\mathbb{R}^2$ is also symmetric. It follows that there are two perpendicular unit vectors $\mathbf{e}_2$ and $\mathbf{e}_3$ in the normal plane of $\mathbf{e}_1$ such that $s(\mathbf{e}_2) = \lambda_2\mathbf{e}_2$ and $s(\mathbf{e}_3) = \lambda_3\mathbf{e}_3$. Then

$$s(\mathbf{v}) = \lambda_1(\mathbf{e}_1 \cdot \mathbf{v})\mathbf{e}_1 + \lambda_2(\mathbf{e}_2 \cdot \mathbf{v})\mathbf{e}_2 + \lambda_3(\mathbf{e}_3 \cdot \mathbf{v})\mathbf{e}_3$$

for all $\mathbf{v}$, and the proof of the principal axis theorem is now complete.

APPLICATIONS

2. Perpendicular Projections

$$p \text{ perpendicular projection} \Leftrightarrow p^* = p = p^2$$

Indeed, $p^* = p = p^2 \Leftrightarrow p(\mathbf{v}) = \lambda_1(\mathbf{e}_1 \cdot \mathbf{v})\mathbf{e}_1 + \lambda_2(\mathbf{e}_2 \cdot \mathbf{v})\mathbf{e}_2 + \lambda_3(\mathbf{e}_3 \cdot \mathbf{v})\mathbf{e}_3$ where $\{\mathbf{e}_1,\mathbf{e}_2,\mathbf{e}_3\}$ is an orthonormal basis (because $p^* = p$), and λ_1, λ_2, λ_3 can only be 0 or 1 (because $p^2 = p$).

There are four cases, according to the rank of p:

(1) $\operatorname{rank} p = 0 \Leftrightarrow \lambda_1 = \lambda_2 = \lambda_3 = 0 \Leftrightarrow p = 0$ (projection into the origin).

(2) $\operatorname{rank} p = 1 \Leftrightarrow \lambda_1 = \lambda_2 = 0,\ \lambda_3 = 1 \Leftrightarrow p(\mathbf{v}) = (\mathbf{e}_3 \cdot \mathbf{v})\mathbf{e}_3$ (perpendicular projection onto the line spanned by $\mathbf{e}_3$).

(3) $\operatorname{rank} p = 2 \Leftrightarrow \lambda_1 = \lambda_2 = 1,\ \lambda_3 = 0 \Leftrightarrow p(\mathbf{v}) = (\mathbf{e}_1 \cdot \mathbf{v})\mathbf{e}_1 + (\mathbf{e}_2 \cdot \mathbf{v})\mathbf{e}_2$ (perpendicular projection onto the plane spanned by $\mathbf{e}_1$, $\mathbf{e}_2$). In this case we may also write, since $\mathbf{e}_3 = \mathbf{e}_1 \times \mathbf{e}_2$, that $p(\mathbf{v}) = (\mathbf{e}_3 \times \mathbf{v}) \times \mathbf{e}_3$.

(4) $\operatorname{rank} p = 3 \Leftrightarrow \lambda_1 = \lambda_2 = \lambda_3 = 1 \Leftrightarrow p = 1$ (the identity operator). It also

follows that a symmetric operator with eigenvalues 0 or 1 must be a perpendicular projection.

3. Reflections (Symmetric Involutions)

$$h \text{ reflection} \Leftrightarrow h^* = h = h^{-1}$$

Indeed, $h^* = h = h^{-1} \Leftrightarrow h(\mathbf{v}) = \lambda_1(\mathbf{e}_1 \cdot \mathbf{v})\mathbf{e}_1 + \lambda_2(\mathbf{e}_2 \cdot \mathbf{v})\mathbf{e}_2 + \lambda_3(\mathbf{e}_3 \cdot \mathbf{v})\mathbf{e}_3$, where $\{\mathbf{e}_1, \mathbf{e}_2, \mathbf{e}_3\}$ is an orthonormal basis (because $h^* = h$), and λ_1, λ_2, λ_3 can only be either 1 or –1 (because $h = h^{-1}$). There are four cases:

(1) $\operatorname{trace} h = 3 \Leftrightarrow \lambda_1 = \lambda_2 = \lambda_3 = 1 \Leftrightarrow h = 1$ (identity operator).

(2) $\operatorname{trace} h = 1 \Leftrightarrow \lambda_1 = \lambda_2 = 1, \quad \lambda_3 = -1 \Leftrightarrow h(\mathbf{v}) = (\mathbf{e}_1 \cdot \mathbf{v})\mathbf{e}_1 + (\mathbf{e}_2 \cdot \mathbf{v})\mathbf{e}_2 - (\mathbf{e}_3 \cdot \mathbf{v})\mathbf{e}_3$ (reflection in the plane spanned by $\mathbf{e}_1$, $\mathbf{e}_2$).

(3) $\operatorname{trace} h = -1 \Leftrightarrow \lambda_1 = 1, \lambda_2 = \lambda_3 = -1 \Leftrightarrow h(\mathbf{v}) = (\mathbf{e}_1 \cdot \mathbf{v})\mathbf{e}_1 - (\mathbf{e}_2 \cdot \mathbf{v})\mathbf{e}_2 - (\mathbf{e}_3 \cdot \mathbf{v})\mathbf{e}_3$ (reflection in the line spanned by $\mathbf{e}_1$).

(4) $\operatorname{trace} h = -3 \Leftrightarrow \lambda_1 = \lambda_2 = \lambda_3 = -1 \Leftrightarrow h = -1$ (reflection in the origin).

It also follows that a symmetric operator with eigenvalues 1 or –1 must be a reflection.

It follows from the principal axis theorem that a rotation through 180° is a symmetric operator: it has eigenvalues 1, –1, –1. The identity operator can be thought of as a rotation through 0° (or 360°) and has eigenvalues 1, 1, 1. These are all the exceptional rotations: all other rotations of $\mathbb{R}^3$ have only one eigenvalue (why?) and, therefore, cannot be symmetric (why?).

ANTISYMMETRIC OPERATORS. THE PRINCIPAL TWIST THEOREM

A linear operator a is called *antisymmetric* if $a^* = -a$. This notion is independent of the choice of coordinates, although we determine whether a is antisymmetric or not by inspecting its matrix

$$a = \begin{pmatrix} 0 & a_{12} & a_{13} \\ -a_{12} & 0 & a_{23} \\ -a_{13} & -a_{23} & 0 \end{pmatrix}$$

i.e., the entries of the matrix of a must satisfy $a_{nm} = -a_{mn}$. In particular, the diagonal entries of an antisymmetric matrix must be 0. We also have, on the strength of the star-dot formula, that

$$a^* = -a \Leftrightarrow a(\mathbf{v}) \cdot \mathbf{w} = -\mathbf{v} \cdot a(\mathbf{w})$$

for all $\mathbf{v}$, $\mathbf{w}$.

Antisymmetric operators play a distinguished role in linear algebra and it is important to have their geometric characterization. The principal twist theorem asserts that the antisymmetric operators of $\mathbb{R}^3$ are precisely the cross products by a fixed vector $\mathbf{n}$. As we know, this means that every antisymmetric operator is the product of a perpendicular projection onto the normal plane of $\mathbf{n}$, a rotation around $\mathbf{n}$ through 90°, and a strain in ratio $|\mathbf{n}|$. This brings out the importance of the antisymmetric operators: they are destined to generalize the cross product to higher dimension.

Before we state and prove the principal twist theorem, we shall look at the special case of an antisymmetric twist t. *The following statements about a linear operator t in $\mathbb{R}^3$ are equivalent*:

(1) *t is an antisymmetric twist*: $t^3 = -t = t^*$.
(2) $t = \ell - \ell^*$, *where ℓ is a lift and $\ell \circ \ell^*$, $\ell^* \circ \ell$ are perpendicular projections*.
(3) $t(\mathbf{v}) = (\mathbf{u}_2 \cdot \mathbf{v})\mathbf{u}_1 - (\mathbf{u}_1 \cdot \mathbf{v})\mathbf{u}_2$ for all $\mathbf{v}$, *where $\mathbf{u}_1$, $\mathbf{u}_2$ are fixed perpendicular unit vectors*.
(4) $t(\mathbf{v}) = \mathbf{u} \times \mathbf{v}$ *for all $\mathbf{v}$, where $\mathbf{u}$ is a fixed unit vector*.

The proof of this is left as an exercise.

Examples

1. Let $\mathbf{u} = \cos\alpha\mathbf{i} + \cos\beta\mathbf{j} + \cos\gamma\mathbf{k}$. Then the matrix of the corresponding anti symmetric twist $t(\mathbf{v}) = \mathbf{u} \times \mathbf{v}$ is

$$t = \begin{pmatrix} 0 & -\cos\gamma & \cos\beta \\ \cos\gamma & 0 & -\cos\alpha \\ -\cos\beta & \cos\alpha & 0 \end{pmatrix}$$

2. The linear operator

$$t = \begin{pmatrix} 0 & -1 & 0 \\ 1 & 0 & 0 \\ 0 & 0 & 0 \end{pmatrix}$$

is an antisymmetric twist that can be written in either of the following two forms: $t(\mathbf{v}) = \mathbf{k} \times \mathbf{v} = (\mathbf{i} \cdot \mathbf{v})\mathbf{j} - (\mathbf{j} \cdot \mathbf{v})\mathbf{i}$, for all $\mathbf{v}$, by virtue of the double cross formula.

The Principal Twist Theorem

$$a^* = -a \Leftrightarrow a = \omega t$$

where ω is a fixed scalar and t is an antisymmetric twist: $t^* = -t = t^3$, *both uniquely determined by $a \neq 0$* (up to signature). This theorem, as stated, is valid in $\mathbb{R}^2$ as well as in $\mathbb{R}^3$. In $\mathbb{R}^3$, however, it is equivalent to

$$a^* = -a \Leftrightarrow a(\mathbf{v}) = \omega(\mathbf{u} \times \mathbf{v}) \qquad \text{for all } \mathbf{v}$$

where $\mathbf{u}$ is a fixed unit vector. Observe that every antisymmetric matrix has the form

$$a = \begin{pmatrix} 0 & a_{12} & a_{13} \\ -a_{12} & 0 & a_{23} \\ -a_{13} & -a_{23} & 0 \end{pmatrix}$$

and hence $a(\mathbf{v}) = \mathbf{n} \times \mathbf{v}$ for all $\mathbf{v}$ with $\mathbf{n} = -a_{23}\mathbf{i} + a_{13}\mathbf{j} - a_{12}\mathbf{k}$ and, conversely, the matrix of the linear operator $a(\mathbf{v}) = \mathbf{n} \times \mathbf{v}$ is antisymmetric. Then we can write $\mathbf{n} = \omega\mathbf{u}$, where $\omega = |\mathbf{n}|$ and $\mathbf{u} = \mathbf{n}^0$. However, this proof reveals nothing of the underlying geometry, so we shall give a second proof for $\mathbb{R}^3$ (the proof for $\mathbb{R}^2$ being left as an exercise).

Let $a \neq 0$ be antisymmetric; then

$$a(\mathbf{v}) \cdot \mathbf{v} = -\mathbf{v} \cdot a(\mathbf{v}) \Rightarrow 2\mathbf{v} \cdot a(\mathbf{v}) = 0 \Rightarrow \mathbf{v} \perp a(\mathbf{v}) \qquad \text{for all } \mathbf{v}$$

Suppose that $\mathbf{w}$ is an eigenvector of a; then $a(\mathbf{w}) \parallel \mathbf{w}$ and $a(\mathbf{w}) \perp \mathbf{w}$ as well. This can be true only if $a(\mathbf{w}) = 0$, and we conclude that the one and only eigenvalue of a is 0, and the corresponding eigenvector of a spans $\ker a$. Since $\dim \ker a = 1$, $\dim \operatorname{im} a = 2$ by the law of nullity. Let $\mathbf{u}_1$, $\mathbf{u}_2$ be a pair of perpendicular unit vectors spanning $\operatorname{im} a$. Then $a(\mathbf{u}_1) = \omega\mathbf{u}_2$, hence $\omega = a(\mathbf{u}_1) \cdot \mathbf{u}_2 = -\mathbf{u}_1 \cdot a(\mathbf{u}_2)$ so that $a(\mathbf{u}_2) = -\omega\mathbf{u}_1$. It follows that $a(\mathbf{v}) = \omega((\mathbf{u}_1 \cdot \mathbf{v})\mathbf{u}_2 - (\mathbf{u}_2 \cdot \mathbf{v})\mathbf{u}_1)$ or, with $\mathbf{u} = \mathbf{u}_1 \times \mathbf{u}_2$, $a(\mathbf{v}) = \omega(\mathbf{u} \times \mathbf{v})$ for all $\mathbf{v}$, where $\mathbf{u}$ is the normal unit vector of the plane $\operatorname{im} a$ spanning $\ker a$. Uniqueness of $\mathbf{u}$ is clear.

Conversely, if $a(\mathbf{v}) = \omega(\mathbf{u} \times \mathbf{v}) = \mathbf{n} \times \mathbf{v}$, then

$$\begin{aligned} a(\mathbf{v}) \cdot \mathbf{w} &= (\mathbf{n} \times \mathbf{v}) \cdot \mathbf{w} = -(\mathbf{v} \times \mathbf{n}) \cdot \mathbf{w} = -\mathbf{v} \cdot (\mathbf{n} \times \mathbf{w}) \qquad \text{(cross-dot formula)} \\ &= -\mathbf{v} \cdot a(\mathbf{w}) \Rightarrow a^* = -a \end{aligned}$$

This completes the proof of the principal twist theorem.

The rank of a (nontrivial) antisymmetric operator is always 2. In the higher-dimensional case the rank of an antisymmetric operator can be different from 2, but it must be an even number. A *principal* twist is a twist of rank 2. Then the theorem above can be restated by saying that a linear operator is antisymmetric if, and only if, it is a scalar multiple of a principal antisymmetric twist.

It follows from the principal twist theorem that antisymmetric operators of $\mathbb{R}^3$ are precisely the noninvertible elliptic operators satisfying $\operatorname{im} a \perp \ker a$. Antisymmetric operators have no real eigenvalue other than $\lambda = 0$, and no real eigenvector other than those in $\ker a$. However, the principal antisymmetric twist t of a may be regarded as a substitute for the missing pair of eigenvectors corresponding to the complex conjugate pair of roots of the characteristic equation.

The higher-dimensional version of the principal twist theorem states that

$$a^* = -a \Leftrightarrow a = \omega_1 t_1 + \omega_2 t_2 + \cdots + \omega_k t_k$$

where $\omega_\nu \in \mathbb{R}$, $t_\nu^* = -t_\nu = t_\nu^3$, rank $t_\nu = 2$, $t_\nu \circ t_\mu = 0 = t_\mu \circ t_\nu$, $\nu \neq \mu$; $\mu, \nu = 1, 2, \ldots, k$. It follows that im $t_1, \ldots,$ im t_k are mutually perpendicular planes with the only common point at the origin. Thus a linear operator a is antisymmetric if, and only if, it is a linear combination of mutually perpendicular principal twists. As the principal perpendicular projections are the building blocks of symmetric operators, so the principal antisymmetric twists are the building blocks of antisymmetric operators. This remark explains how the principal twist theorem earns its name.

The antisymmetric operators of $\mathbb{R}^n$ form an $\frac{1}{2}n(n-1)$-dimensional space. This can be seen by counting the independent entries in an antisymmetric $n \times n$ matrix:

$$1 + 2 + \cdots + (n-1) = \frac{n(n-1)}{2}$$

As mentioned in Chap. 4, and discussed further in Chap. 14, the space of antisymmetric operators can be converted into a Lie algebra by the introduction of the product $[f,g] = f \circ g - g \circ f$.

ORTHOGONAL OPERATORS. THE ORTHOGONAL GROUP

A linear operator r is called *orthogonal* if it is invertible and $r^{-1} = r^*$. Thus the inverse of an orthogonal operator is easy to calculate, simply by transposing its matrix.

Since the eigenvalues of r^* are the same as those of r, and since the eigenvalues of r^{-1} are just the reciprocals of the eigenvalues of r, we may conclude that the an orthogonal operator r admits (real) eigenvalues 1 or –1 only.

An orthogonal operator is called *orientation preserving* or *orientation reversing* according as the number of its eigenvalues equal to –1 is even or odd.

Examples

1. Reflections (symmetric involutions) are obviously orthogonal. In $\mathbb{R}^2$, reflection in the origin is orientation preserving, and reflection in a line is orientation reversing. By contrast, in $\mathbb{R}^3$, reflection in the origin is orientaion reversing, and reflection in a line (or, what is the same, rotation through 180° around that line) is orientation preserving.
2. An orthogonal invection, i.e., a linear operator j satisfying $j^* = j^3 = j^{-1}$ has a unique eigenvector belonging to 1 or –1, and is accordingly

orientation preserving or reversing. An orientation-preserving orthogonal invection is just a rotation through 90° (or–90°).

The extraordinary importance of orthogonal operators is due to the fact that they preserve the length and the angle of vectors. In particular, they preserve orthogonality (or perpendicularity). This remark explains how orthogonal operators earn their name. These properties of orthogonal operators follow from the

Theorem An invertible operator r is orthogonal if, and only if, it preserves the dot product of vectors, i.e.,

$$r(\mathbf{v}) \cdot r(\mathbf{w}) = \mathbf{v} \cdot \mathbf{w} \qquad \textit{for all } \mathbf{v}, \mathbf{w}$$

Proof: The proof is based on the star-dot formula. The condition is necessary: $r^{-1} = r^* \Rightarrow r(\mathbf{v}) \cdot r(\mathbf{w}) = \mathbf{v} \cdot r^*(r(\mathbf{w})) = \mathbf{v} \cdot r^{-1}(r(\mathbf{w})) = \mathbf{v} \cdot (r^{-1} \circ r)(\mathbf{w}) = \mathbf{v} \cdot \mathbf{w}$.

The condition is also sufficient:

$$\begin{aligned} r(\mathbf{v}) \cdot r(\mathbf{w}) = \mathbf{v} \cdot \mathbf{w} \quad \text{for all } \mathbf{v}, \mathbf{w} &\Rightarrow \mathbf{v} \cdot (r^* \circ r)(\mathbf{w})) = \mathbf{v} \cdot \mathbf{w} \\ &\Rightarrow \mathbf{v} \cdot (r^* \circ r - 1)(\mathbf{w}) = 0 \Rightarrow \mathbf{v} \perp (r^* \circ r - 1)(\mathbf{w}) \qquad \text{for all } \mathbf{v} \\ &\Rightarrow (r^* \circ r - 1)(\mathbf{w}) = 0 \qquad \text{for all } \mathbf{w} \Rightarrow r^* \circ r - 1 = 0 \\ &\Rightarrow r^* \circ r = 1 \Rightarrow r^* = r \end{aligned}$$

Since the length and angle of vectors can be expressed by the dot product of those vectors, it follows that

$$|r(\mathbf{v})| = |\mathbf{v}|$$

and

$$\measuredangle(r(\mathbf{v}), r(\mathbf{w})) = \measuredangle(\mathbf{v},\mathbf{w})$$

Corollary The orthogonal operators form a group called the orthogonal group.

In Chap. 13 we shall see that for an orthogonal operator r, $\det r = 1$ or -1 according as r is orientation preserving or reversing. We shall also see that $\det(g \circ f) = (\det g)(\det f)$, from which it will follow that *the orientation-preserving orthogonal operators form a group, the special orthogonal group*.

Examples

1. A reflection in a plane is an orientation-reversing orthogonal operator,

because it preserves the length and angle, and hence the dot product of vectors, and its only eigenvector, the normal vector of the plane, belongs to eigenvalue –1.

2. The linear operator f given by the formula

$$f(\mathbf{v}) = \mathbf{k} \times \mathbf{v} - (\mathbf{k} \cdot \mathbf{v})\mathbf{k} = (\mathbf{i} \cdot \mathbf{v})\mathbf{j} - (\mathbf{j} \cdot \mathbf{v})\mathbf{i} - (\mathbf{k} \cdot \mathbf{v})\mathbf{k}$$

has a matrix

$$f = \begin{pmatrix} 0 & -1 & 0 \\ 1 & 0 & 0 \\ 0 & 0 & -1 \end{pmatrix}$$

It is easy to see that f is an orientation-reversing orthogonal operator (in fact, an orthogonal invection). f is clearly orthogonal as

$$f^{-1}(\mathbf{v}) = -(\mathbf{i} \cdot \mathbf{v})\mathbf{j} + (\mathbf{j} \cdot \mathbf{v})\mathbf{i} - (\mathbf{k} \cdot \mathbf{v})\mathbf{k} = f^*(\mathbf{v}) \qquad \text{for all } \mathbf{v}$$

and f is orientation reversing as its only eigenvector $\mathbf{k}$ belongs to eigenvalue –1. f is neither a rotation nor is it a reflection; it is a rotary reflection (product of a rotation and –1.)

There are no rotary reflections in $\mathbb{R}^2$; the product of a rotation through ϕ and a reflection in a line is a rotation through $\phi + 180°$.

In view of the great practical importance of the orthogonal operators, we wish to find a criterion whereby their matrices—the *orthogonal matrices*—can be recognized. As a first step we state and prove another necessary and sufficient condition for orthogonality:

$$r \text{ orthogonal} \Leftrightarrow r(\mathbf{v}) = (\mathbf{i} \cdot \mathbf{v})\mathbf{i}' + (\mathbf{j} \cdot \mathbf{v})\mathbf{j}' + (\mathbf{k} \cdot \mathbf{v})\mathbf{k}'$$

where $\mathbf{i}' = r(\mathbf{i})$, $\mathbf{j}' = r(\mathbf{j})$, $\mathbf{k}' = r(\mathbf{k})$ is a second orthonormal basis (and if it is a right-hand system as well, then r is also orientation preserving). In other words, *r is orthogonal if, and only if, it preserves orthonormal bases*. This is immediate from the fact that orthogonal operators, and only those, preserve the dot product of vectors, because orthonormal bases are defined in terms of the dot product: $\mathbf{i} \cdot \mathbf{i} = \mathbf{j} \cdot \mathbf{j} = \mathbf{k} \cdot \mathbf{k} = 1$ and $\mathbf{i} \cdot \mathbf{j} = \mathbf{j} \cdot \mathbf{k} = \mathbf{k} \cdot \mathbf{i} = 0$.

Since $\mathbf{i}', \mathbf{j}', \mathbf{k}'$ are just the column vectors of the matrix r, we conclude that *a matrix is orthogonal if, and only if, its column vectors are pairwise perpendicular unit vectors* (i.e., *form an orthonormal basis)*. Because the row vectors of the matrix are just the column vectors of the matrix r^{-1}, another orthogonal matrix, we see that the row vectors of an orthogonal matrix are pairwise perpendicular unit vectors, too.

The description of an orthogonal matrix in geometric terms is as follows. All nine entries of the matrix are direction cosines, the columns being the coordinates of the vectors $\mathbf{i}', \mathbf{j}', \mathbf{k}'$ in terms of the basis $\mathbf{i}, \mathbf{j}, \mathbf{k}$ and the rows being the coordinates of the vectors **i**, **j**, **k** in terms of the basis **i**′, **j**′, **k**′:

$$r = \begin{pmatrix} \mathbf{i}\cdot\mathbf{i}' & \mathbf{i}\cdot\mathbf{j}' & \mathbf{i}\cdot\mathbf{k}' \\ \mathbf{j}\cdot\mathbf{i}' & \mathbf{j}\cdot\mathbf{j}' & \mathbf{j}\cdot\mathbf{k}' \\ \mathbf{k}\cdot\mathbf{i}' & \mathbf{k}\cdot\mathbf{j}' & \mathbf{k}\cdot\mathbf{k}' \end{pmatrix} = \begin{pmatrix} \cos\sphericalangle(\mathbf{i},\mathbf{i}') & \cos\sphericalangle(\mathbf{i},\mathbf{j}') & \cos\sphericalangle(\mathbf{i},\mathbf{k}') \\ \cos\sphericalangle(\mathbf{j},\mathbf{i}') & \cos\sphericalangle(\mathbf{j},\mathbf{j}') & \cos\sphericalangle(\mathbf{j},\mathbf{k}') \\ \cos\sphericalangle(\mathbf{k},\mathbf{i}') & \cos\sphericalangle(\mathbf{k},\mathbf{j}') & \cos\sphericalangle(\mathbf{k},\mathbf{k}') \end{pmatrix}$$

Example Show that the matrix

$$r = \begin{pmatrix} -21/29 & 12/29 & 16/29 \\ 12/29 & -11/29 & 24/29 \\ 16/29 & 24/29 & 3/29 \end{pmatrix}$$

is equal to its own inverse.

Answer: Since the matrix is symmetric, it will suffice to show that its column vectors form an orthonormal basis; i.e., they are pairwise perpendicular unit vectors. This will follow from the fact that $21^2 + 12^2 + 16^2 = 12^2 + 11^2 + 24^2 = 16^2 + 24^2 + 3^2 = 29^2$ and $-21(12) - 12(11) + 16(24) = 0$, $12(16) - 11(24) + 24(3) = 0$, and $-21(16) + 12(24) + 16(3) = 0$.

In fact, r is reflection in the line spanned by the unit vector

$$\mathbf{u} = \frac{2}{\sqrt{29}}\mathbf{i} + \frac{3}{\sqrt{29}}\mathbf{j} + \frac{4}{\sqrt{29}}\mathbf{k}$$

APPLICATIONS

4. The Cayley Transform of an Antisymmetric Operator

We shall now show that

$$a^* = -a \Rightarrow r = (1 + a)\circ(1 - a)^{-1} \text{ orthogonal} \qquad \text{i.e.,} \qquad r^* = r^{-1}$$

The orthogonal operator r is called the *Cayley transform* of the antisymmetric operator a. In fact, r is an orientation-preserving orthogonal operator and, as we shall see in Chap. 12, r is a rotation through $\phi \neq 180°$. Conversely, if r is a rotation through $\phi \neq 180°$, then r is the Cayley transform of the antisymmetric operator

$$a = (r + 1)^{-1}\circ(r - 1)$$

and the two Cayley transforms are inverses of one another, setting up a bijective correspondence between the antisymmetric operators and the rotations through $\phi \neq 180°$.

First, observe that if f, g commute and g is bijective, then f, g^{-1} will also commute:

$$g\circ f=f\circ g\Rightarrow g^{-1}\circ f=g^{-1}\circ f\circ(g\circ g^{-1})=g^{-1}\circ(f\circ g)\circ g^{-1}=g^{-1}\circ(g\circ f)\circ g^{-1}=f\circ g^{-1}$$

It is clear that $1+a$, $1-a$ commute; hence $1+a$, $(1-a)^{-1}$ must also commute, provided that $(1-a)^{-1}$ exists. Now

$$\begin{aligned} r^* &= ((1+a)\circ(1-a)^{-1})^* = ((1-a)^{-1}\circ(1+a))^* = (1+a)^*\circ((1-a)^{-1})^* \\ &= (1+a)^*\circ((1-a)^*)^{-1} = (1+a^*)\circ(1-a^*)^{-1} = (1-a)\circ(1+a)^{-1} \\ &= ((1+a)\circ(1-a)^{-1})^{-1} = r^{-1} \end{aligned}$$

and we conclude that r, if it exists, is orthogonal.

In order to show that r^{-1} exists it will suffice to prove that $a^* = -a \Rightarrow (1-a)^{-1}$ exists. By the principal twist theorem, $a=\omega t$, where t is an antisymmetric twist (i.e., $t^* = -t = t^3$) and $\omega \in \mathbb{R}$. We shall be looking for $(1-\omega t)^{-1}$ in the form of $1+At+Bt^2$ (since all the higher powers of t can be reduced to t and t^2). Then

$$\begin{aligned} (1-\omega t)\circ(1+At+Bt^2) &= 1+At+Bt^2-\omega t-A\omega t^2-B\omega t^3 \\ &= 1+[A+(B-1)\omega]t+(B-A\omega)t^2 = 1 \end{aligned}$$

$$\Rightarrow \left.\begin{array}{r} A+B\omega=\omega \\ -A\omega+B=0 \end{array}\right\} \Rightarrow \begin{cases} A=\dfrac{\omega}{1+\omega^2} \\[2ex] B=\dfrac{\omega^2}{1+\omega^2} \end{cases}$$

This means that, if $(1-\omega t)^{-1}$ exists, then

$$(1-\omega t)^{-1} = 1+\frac{\omega}{1+\omega^2}t+\frac{\omega^2}{1+\omega^2}t^2$$

Existence now follows from the calculation

$$(1-\omega t)\circ\left(1+\frac{\omega}{1+\omega^2}t+\frac{\omega^2}{1+\omega^2}t^2\right) = 1-\omega t+\frac{\omega+\omega^3}{1+\omega^2}t = 1$$

We can now write an explicit formula for the Cayley transform r of ωt:

Cayley's Formula

$$r = 1+\frac{2\omega}{1+\omega^2}t+\frac{2\omega^2}{1+\omega^2}t^2$$

which can be verified by direct multiplication: $r=(1+\omega t)\circ(1-\omega t)^{-1}$.

Replacing ω by $-\omega$, we get

$$r^{-1} = (1-\omega t)\circ(1+\omega t)^{-1} = 1-\frac{2\omega}{1+\omega^2}t+\frac{2\omega^2}{1+\omega^2}t^2 = r^*$$

By subtraction,

$$a = \omega t = \frac{1 + \omega^2}{4}(r - r^*)$$

which is another and simpler formula for the Cayley transform of the orientation-preserving orthogonal operator. It is clear that a, given by this formula, is antisymmetric.

Moreover,

$$r = 1 + \frac{2\omega}{1+\omega^2}t + \frac{2\omega^2}{1+\omega^2}t^2 = (1 + t^2) + \frac{1-\omega^2}{1+\omega^2}t^4 + \frac{2\omega}{1+\omega^2}t$$

or (using the rational parametric form of the circle, Chap. 2, Application 12), with $\omega = \tan \tfrac{1}{2}\phi$ ($\phi \neq 180°$), we get the

Standard Form of Orientation-Preserving Orthogonal Operators

$$r = 1 + t^2 + t^4 \cos\phi + t \sin\phi \qquad (t^* = -t = t^3)$$

We shall see in Chap. 12 that orientation-preserving orthogonal operators in $\mathbb{R}^3$ are the rotations around the line $\ker t$ through the angle ϕ. Since t is an antisymmetric twist, $1 + t^2$ is the perpendicular projection onto the line $\ker t$, and $t^4 = -t^2$ is the perpendicular projection onto the plane $\operatorname{im} t$. Therfore, the vector spanning $\ker t$ is the only eigenvector of r, belonging to the eigenvalue $\lambda = 1$.

In the special case when $\alpha = 90°$, $\omega = 1$ and $a = t$; i.e., *the Cayley transform of an antisymmetric twist is an orthogonal invection*:

$$j = 1 + t + t^2$$

satisfying $j^* = j^{-1} = j^3$. If we denote the Cayley transform by the symbol $a \leftrightsquigarrow$ r, we have

$$t \leftrightsquigarrow j$$

$$0 \leftrightsquigarrow 1$$

$$-a \leftrightsquigarrow r^{-1}$$

It is left as an exercise to show that the Cayley transform establishes a bijection between the set of antisymmetric operators and the set of orientation-preserving orthogonal operators.

Examples

1. Find the Cayley transform of the antisymmetric operator

$$a = \begin{pmatrix} 0 & 4 & -8 \\ -4 & 0 & 1 \\ 8 & -1 & 0 \end{pmatrix}$$

Answer: $a = \omega t$, where $\omega = \sqrt{1^2 + 4^2 + 8^2} = \sqrt{81} = 9$ and

$$t = \begin{pmatrix} 0 & 4/9 & -8/9 \\ -4/9 & 0 & 1/9 \\ 8/9 & -1/9 & 0 \end{pmatrix} \qquad t^2 = \begin{pmatrix} -80/81 & 8/81 & 4/81 \\ 8/81 & -17/81 & 32/81 \\ 4/81 & 32/81 & -65/81 \end{pmatrix} \qquad t^3 = -t$$

Then the Cayley transform of a is

$$r = 1 + \frac{2\omega}{1+\omega^2} t + \frac{2\omega^2}{1+\omega^2} t^2 = \begin{pmatrix} -39/41 & 12/41 & -4/41 \\ 4/41 & 24/41 & 33/41 \\ 12/41 & 31/41 & -24/41 \end{pmatrix}$$

which is orthogonal:

$$r^{-1} = 1 - \frac{2\omega}{1+\omega^2} t + \frac{2\omega^2}{1+\omega^2} t^2 = r^*$$

2. The Cayley transform of the antisymmetric operator

$$\begin{pmatrix} 0 & -1 & 0 \\ 1 & 0 & 0 \\ 0 & 0 & 0 \end{pmatrix} \quad \text{is} \quad \begin{pmatrix} 0 & -1 & 0 \\ 1 & 0 & 0 \\ 0 & 0 & 1 \end{pmatrix}$$

3. Find the standard form of the rotation

$$r = \begin{pmatrix} 0.8 & 0.36 & 0.48 \\ -0.36 & 0.928 & -0.096 \\ -0.48 & -0.096 & 0.872 \end{pmatrix}$$

Answer:

$$\frac{1}{2}(r - r^*) = (\sin\alpha) t = \begin{pmatrix} 0 & 0.36 & 0.48 \\ -0.36 & 0 & 0 \\ -0.48 & 0 & 0 \end{pmatrix} = \frac{3}{5}\begin{pmatrix} 0 & 3/5 & 4/5 \\ -3/5 & 0 & 0 \\ -4/5 & 0 & 0 \end{pmatrix}$$

i.e., $\sin\alpha = 3/5$, $\cos\alpha = 4/5$,

$$t = \begin{pmatrix} 0 & 3/5 & 4/5 \\ -3/5 & 0 & 0 \\ -4/5 & 0 & 0 \end{pmatrix}$$

$$r = \begin{pmatrix} 0 & 0 & 0 \\ 0 & 16/25 & -12/25 \\ 0 & -12/25 & 9/25 \end{pmatrix} + \frac{4}{5}\begin{pmatrix} 1 & 0 & 0 \\ 0 & 9/25 & 12/25 \\ 0 & 12/25 & 16/25 \end{pmatrix} + \frac{3}{5}\begin{pmatrix} 0 & 3/5 & 4/5 \\ -3/5 & 0 & 0 \\ -4/5 & 0 & 0 \end{pmatrix}$$

4. The standard form of the rotation

$$r = \begin{pmatrix} 53/117 & 88/117 & -56/117 \\ -56/117 & 77/117 & 68/117 \\ 88/117 & -4/117 & 77/117 \end{pmatrix}$$

is

$$r = \begin{pmatrix} 1/9 & 2/9 & 2/9 \\ 2/9 & 4/9 & 4/9 \\ 2/9 & 4/9 & 4/9 \end{pmatrix} + \frac{5}{13}\begin{pmatrix} 8/9 & -2/9 & -2/9 \\ -2/9 & 5/9 & -4/9 \\ -2/9 & -4/9 & 5/9 \end{pmatrix} + \frac{12}{13}\begin{pmatrix} 0 & 2/3 & -2/3 \\ -2/3 & 0 & 1/3 \\ 2/3 & -1/3 & 0 \end{pmatrix}$$

5. Changing the Orthonormal Basis

We have already discussed the problem of changing the basis and the coordinates (Chap. 10, Application 8) in the most general case: changing from one basis to another, i.e., changing the skew affine coordinate system (axes are not necessarily perpendicular and the units of length along the axes are not necessarily the same).

Here we are interested in a special case: that of changing one orthonormal basis to another, i.e., that of changing the metric rectangular coordinates (axes mutually perpendicular, units of length along the axes are the same). It is clear that in this case the transition operator g must be orthogonal, because both the old basis $\{\mathbf{i},\mathbf{j},\mathbf{k}\}$ and the new basis $\{\mathbf{i}',\mathbf{j}',\mathbf{k}'\}$ are orthonormal. The transition equation $f' = g^{-1} \circ f \circ g = g^* \circ f \circ g$ involves less calculation, as the inverse of the transition operator, g^{-1}, can be quickly obtained by transposition.

Example Find an orthonormal basis in which the symmetric operator

$$s = \begin{pmatrix} 11 & -8 & -2 \\ -8 & 5 & -10 \\ -2 & -10 & 2 \end{pmatrix}$$

will assume diagonal form. Write the transition equation.

Answer: The new basis vectors will be the (normalized) eigenvectors of s and they will also appear as the column vectors of the transition matrix g. $\det(s-\lambda) = 0 \Rightarrow -\lambda^3 + 18\lambda^2 + 81\lambda - 1458 = 0$. We try to find integral solutions via the tabular method, which can have prime factors 2 and 3 only as $1458 = 2(3^6)$.

	-1	18	81	-1458	
9	-1	9	162	0	$\lambda_1 = 9$

Further integral solutions can be found easily if we use the tabular method for the equation from which $\lambda-9$ has been factored out:

	–1	9	162
–9	–1	18	0

$\lambda_2 = -9$ and $\lambda_3 = 18$

The diagonal form of s is

$$\begin{pmatrix} 9 & 0 & 0 \\ 0 & -9 & 0 \\ 0 & 0 & 18 \end{pmatrix}$$

Find the eigenvectors of s:

$\ker(s-9)$:

$$\begin{pmatrix} 2 & -8 & -2 \\ -8 & -4 & -10 \\ -2 & -10 & -7 \end{pmatrix} \Rightarrow \begin{pmatrix} 1 & 0 & 1 \\ 0 & 2 & 1 \end{pmatrix} \Rightarrow \begin{cases} x = 2t \\ y = t \\ z = -2t \end{cases}$$

and the unit vector spanning this line is

$$\mathbf{e}_1 = \frac{2}{3}\mathbf{i} + \frac{1}{3}\mathbf{j} - \frac{2}{3}\mathbf{k}$$

Similar calculation yields

$$\mathbf{e}_2 = \frac{1}{3}\mathbf{i} + \frac{2}{3}\mathbf{j} + \frac{2}{3}\mathbf{k}$$

spanning $\ker(s+9)$ and

$$\mathbf{e}_3 = \frac{2}{3}\mathbf{i} + \frac{2}{3}\mathbf{j} + \frac{1}{3}\mathbf{k}$$

spanning $\ker(s-18)$. Hence

$$g = \begin{pmatrix} 2/3 & 1/3 & 2/3 \\ 1/3 & 2/3 & -2/3 \\ -2/3 & 2/3 & 1/3 \end{pmatrix} \qquad g^{-1} = \begin{pmatrix} 2/3 & 1/3 & -2/3 \\ 1/3 & 2/3 & 2/3 \\ 2/3 & -2/3 & 1/3 \end{pmatrix}$$

because $g^{-1} = g^*$. The transition equation is

$$\begin{pmatrix} 2/3 & 1/3 & -2/3 \\ 1/3 & 2/3 & 2/3 \\ 2/3 & -2/3 & 1/3 \end{pmatrix} \circ \begin{pmatrix} 11 & -8 & -2 \\ -8 & 5 & -10 \\ -2 & -10 & 2 \end{pmatrix} \circ \begin{pmatrix} 2/3 & 1/3 & 2/3 \\ 1/3 & 2/3 & -2/3 \\ -2/3 & 2/3 & 1/3 \end{pmatrix} = \begin{pmatrix} 9 & 0 & 0 \\ 0 & -9 & 0 \\ 0 & 0 & 18 \end{pmatrix}$$

6. Gram-Schmidt Orthonormalization

Examples

1. Given a basis $\{\mathbf{e}_1,\mathbf{e}_2,\mathbf{e}_3\}$, find an *orthonormal* basis $\{\mathbf{u}_1,\mathbf{u}_2,\mathbf{u}_3\}$ such that $\mathbf{e}_1$, $\mathbf{u}_1$ span the same line; and $\{\mathbf{e}_1,\mathbf{e}_2\}$, $\{\mathbf{u}_1,\mathbf{u}_2\}$ span the same plane.

   Answer: Set $\mathbf{u}_1 = \mathbf{e}_1^0 = \mathbf{e}_1/|\mathbf{e}_1|$. Take the perpendicular component of $\mathbf{e}_2$ with respect to $\mathbf{u}_1$ and normalize it to get $\mathbf{u}_2$, i.e.,

$$\mathbf{u}_2 = \frac{\mathbf{e}_2-(\mathbf{u}_1\cdot\mathbf{e}_2)\mathbf{u}_1}{|\mathbf{e}_2-(\mathbf{u}_1\cdot\mathbf{e}_2)\mathbf{u}_1|}$$

   Then clearly $\mathbf{u}_1\cdot\mathbf{u}_1 = \mathbf{u}_2\cdot\mathbf{u}_2 = 1$, $\mathbf{u}_1\cdot\mathbf{u}_2 = 0$. Continuing the pattern, we set

$$\mathbf{u}_3 = \frac{\mathbf{e}_3-(\mathbf{u}_1\cdot\mathbf{e}_3)\mathbf{u}_1-(\mathbf{u}_2\cdot\mathbf{e}_3)\mathbf{u}_2}{|\mathbf{e}_3-(\mathbf{u}_1\cdot\mathbf{e}_3)\mathbf{u}_1-(\mathbf{u}_2\cdot\mathbf{e}_3)\mathbf{u}_2|}$$

   and it is easy to see that $\mathbf{u}_3\cdot\mathbf{u}_3 = 1$ and $\mathbf{u}_1\cdot\mathbf{u}_3 = \mathbf{u}_2\cdot\mathbf{u}_3 = 0$, and that $\{\mathbf{u}_1,\mathbf{u}_2,\mathbf{u}_3\}$ is the required orthonormal basis.

This algorithm works analogously for higher-dimensional metric vector spaces (and in fact is used to prove the existence of orthonormal bases there); it is known as the Gram-Schmidt orthonormalization.

2. Apply Gram-Schmidt orthonormalization to the basis

$$\mathbf{e}_1 = \begin{pmatrix}-1\\4\\-8\end{pmatrix} \qquad \mathbf{e}_2 = \begin{pmatrix}1\\0\\1\end{pmatrix} \qquad \mathbf{e}_3 = \begin{pmatrix}5\\7\\13\end{pmatrix}$$

   Answer: $|\mathbf{e}_1| = 9$,

$$\mathbf{u}_1 = \begin{pmatrix}-1/9\\4/9\\-8/9\end{pmatrix} \qquad \mathbf{u}_1\cdot\mathbf{e}_2 = -1$$

$$\mathbf{e}_2-(\mathbf{u}_1\cdot\mathbf{e}_2)\mathbf{u}_1 = \mathbf{e}_2+\mathbf{u}_1 = \begin{pmatrix}8/9\\4/9\\1/9\end{pmatrix} = \mathbf{u}_2$$

$$\mathbf{u}_1\cdot\mathbf{e}_3 = -9 \qquad \mathbf{u}_2\cdot\mathbf{e}_3 = 9 \qquad \mathbf{e}_3+9\mathbf{u}_1-9\mathbf{u}_2 = \begin{pmatrix}-4\\7\\4\end{pmatrix} \qquad \mathbf{u}_3 = \begin{pmatrix}-4/9\\7/9\\4/9\end{pmatrix}$$

It is easy to check that $\{\mathbf{u}_1,\mathbf{u}_2,\mathbf{u}_3\}$ is the orthonormal basis required.

7. Impact of the Linear Operator on Planes

It would be a mistake to assume that the impact of g on a vector is the same as its impact on the normal vector $\mathbf{n}$ of a plane. If g is an invertible operator mapping the plane π with normal vector $\mathbf{n}$ to the plane $g(\pi)$, then the normal vector of $g(\pi)$ is not $g(\mathbf{n})$ but $g^{*-1}(\mathbf{n})$. Indeed, by virtue of the star-dot formula, for the vectors $g(\mathbf{v})$ in $g(\pi)$,

$$g^{*-1}(\mathbf{n})\cdot g(\mathbf{v})=0 \Leftrightarrow g^{-1*}(\mathbf{n})\cdot g(\mathbf{v})=0 \Leftrightarrow \mathbf{n}\cdot(g^{-1}\circ g)(\mathbf{v})=0 \Leftrightarrow \mathbf{n}\cdot\mathbf{v}=0$$

and we conclude that the equation of the plane $g(\pi)$ is $\mathbf{n}'\cdot\mathbf{v}=0$ with $\mathbf{n}'=g^{*-1}(\mathbf{n})$.

We shall now show that if p is the projection onto the line ℓ along the plane π, then its conjugate $g\circ p\circ g^{-1}$ is the projection onto the line $g(\ell)$ along the plane $g(\pi)$. Indeed,

$$\begin{aligned} p(\mathbf{v})=(\mathbf{n}\cdot\mathbf{v})\mathbf{e},\ \mathbf{n}\cdot\mathbf{e}=1 \Rightarrow (p\circ g^{-1})(\mathbf{v}) &= (\mathbf{n}\cdot g^{-1}(\mathbf{v}))\mathbf{e} \\ \Rightarrow (g\circ p\circ g^{-1})(\mathbf{v}) &= (\mathbf{n}\cdot g^{-1}(\mathbf{v}))g(\mathbf{e}) \\ &= (g^{*-1}(\mathbf{n})\cdot\mathbf{v})g(\mathbf{e})=(\mathbf{n}'\cdot\mathbf{v})\mathbf{e}' \end{aligned}$$

where $\mathbf{n}'\cdot\mathbf{e}'=g^{*-1}(\mathbf{n})\cdot g(\mathbf{e})=\mathbf{n}\cdot\mathbf{e}=1$ by the star-dot formula.

It follows immediately that if f is a dilatation (involution) relative to the plane π along the line ℓ, then its conjugate $g\circ f\circ g^{-1}$ is the dilatation in the ratio λ (involution for $\lambda=-1$) relative to the plane $g(\pi)$ along the line $g(\ell)$.

It is left as an exercise to show that if f is a lift (shear) parallel to the plane π along the line ℓ, then its conjugate $g\circ f\circ g^{-1}$ is the lift (shear) parallel to the plane $g(\pi)$ along the line $g(\ell)$.

EXERCISES

11.1 Without solving the characteristic equation, give a complete geometric description of each of the following linear operators. (*Hint*: Calculate f^2 or f^3 first and observe whether f is symmetric or antisymmetric.)

(i) $\begin{pmatrix} 0 & 0 & 1 \\ 0 & 1 & 0 \\ 1 & 0 & 0 \end{pmatrix}$ (ii) $\begin{pmatrix} 1/3 & 1/3 & -1/3 \\ 1/3 & 1/3 & -1/3 \\ -1/3 & -1/3 & 1/3 \end{pmatrix}$

(iii) $\begin{pmatrix} 1/3 & 2/3 & -2/3 \\ 2/3 & 1/3 & 2/3 \\ -2/3 & 2/3 & 1/3 \end{pmatrix}$ (iv) $\begin{pmatrix} 0 & -2/3 & 2/3 \\ 2/3 & 0 & -1/3 \\ -2/3 & 1/3 & 0 \end{pmatrix}$

(v) $\begin{pmatrix} 2/7 & 3/7 & -6/7 \\ 3/7 & -6/7 & -2/7 \\ -6/7 & -2/7 & -3/7 \end{pmatrix}$ (vi) $\begin{pmatrix} 5/14 & -3/14 & 6/14 \\ -3/14 & 13/14 & 2/14 \\ 6/14 & 2/14 & 10/14 \end{pmatrix}$

(vii) $\begin{pmatrix} 0 & 8/9 & 4/9 \\ -8/9 & 0 & -1/9 \\ -4/9 & 1/9 & 0 \end{pmatrix}$ (viii) $\begin{pmatrix} 8/9 & 2/9 & -2/9 \\ 2/9 & 5/9 & 4/9 \\ -2/9 & 4/9 & 5/9 \end{pmatrix}$

(ix) $\begin{pmatrix} 7/9 & 4/9 & -4/9 \\ 4/9 & 1/9 & 8/9 \\ -4/9 & 8/9 & 1/9 \end{pmatrix}$ (x) $\begin{pmatrix} 1/9 & -2/9 & 2/9 \\ -2/9 & 4/9 & -4/9 \\ 2/9 & -4/9 & 4/9 \end{pmatrix}$

(xi) $\begin{pmatrix} 2/11 & 6/11 & -9/11 \\ 6/11 & 7/11 & 6/11 \\ -9/11 & 6/11 & 2/11 \end{pmatrix}$ (xii) $\begin{pmatrix} -0.8 & 0.6 & 0 \\ 0.6 & 0.8 & 0 \\ 0 & 0 & 1 \end{pmatrix}$

(xiii) $\begin{pmatrix} 0.1 & 0.3 & 0 \\ 0.3 & 0.9 & 0 \\ 0 & 0 & 1 \end{pmatrix}$ (xiv) $\begin{pmatrix} 0 & 0.6 & 0.8 \\ -0.6 & 0 & 0 \\ -0.8 & 0 & 0 \end{pmatrix}$

(xv) $\begin{pmatrix} -0.96 & 0 & 0.28 \\ 0 & 1 & 0 \\ 0.28 & 0 & 0.96 \end{pmatrix}$ (xvi) $\begin{pmatrix} 0.02 & 0 & 0.14 \\ 0 & 1 & 0 \\ 0.14 & 0 & 0.98 \end{pmatrix}$

(xvii) $\begin{pmatrix} 0 & 0.8 & 0.6 \\ 0.8 & -0.36 & 0.48 \\ 0.6 & 0.48 & -0.64 \end{pmatrix}$ (xviii) $\begin{pmatrix} 0.5 & 0.4 & 0.3 \\ 0.4 & 0.32 & 0.24 \\ 0.3 & 0.24 & 0.18 \end{pmatrix}$

(xix) $\begin{pmatrix} 0.5 & -0.4 & -0.3 \\ -0.4 & 0.68 & -0.24 \\ -0.3 & -0.24 & 0.82 \end{pmatrix}$ (xx) $\begin{pmatrix} 0.8 & 0.32 & 0.24 \\ 0.32 & 0.488 & -0.384 \\ 0.24 & -0.384 & 0.712 \end{pmatrix}$

(xxi) $\begin{pmatrix} 0.6 & 0.64 & 0.48 \\ 0.64 & -0.024 & -0.768 \\ 0.48 & -0.768 & 0.424 \end{pmatrix}$ (xxii) $\begin{pmatrix} 0 & 0.168 & -0.576 \\ -0.168 & 0 & 0.8 \\ 0.576 & -0.8 & 0 \end{pmatrix}$

(xxiii) $\begin{pmatrix} 0 & 0.36 & 0.48 \\ -0.36 & 0 & 0.8 \\ 0.48 & -0.8 & 0 \end{pmatrix}$ (xxiv) $\begin{pmatrix} 0 & -0.6 & -0.48 \\ 0.6 & 0 & -0.64 \\ 0.48 & 0.64 & 0 \end{pmatrix}$

(xxv) $\begin{pmatrix} 0.744 & -0.64 & 0.192 \\ -0.64 & -0.6 & 0.48 \\ 0.192 & 0.48 & 0.856 \end{pmatrix}$ (xxvi) $\begin{pmatrix} 0.352 & 0.36 & 0.864 \\ 0.36 & 0.8 & -0.48 \\ 0.864 & -0.48 & -0.152 \end{pmatrix}$

11.2 Find a new orthonormal basis for each antisymmetric twist f below, in terms of which f will take the form

$$\begin{pmatrix} 0 & -1 & 0 \\ 1 & 0 & 0 \\ 0 & 0 & 0 \end{pmatrix}$$

Write the transition equation.

(i) $\begin{pmatrix} 0 & 2/3 & -1/3 \\ -2/3 & 0 & 2/3 \\ 1/3 & -2/3 & 0 \end{pmatrix}$ (ii) $\begin{pmatrix} 0 & 2/7 & -6/7 \\ -2/7 & 0 & 3/7 \\ 6/7 & -3/7 & 0 \end{pmatrix}$

(iii) $\begin{pmatrix} 0 & 4/9 & -8/9 \\ -4/9 & 0 & 1/9 \\ 8/9 & -1/9 & 0 \end{pmatrix}$ (iv) $\begin{pmatrix} 0 & 2/11 & 9/11 \\ -2/11 & 0 & -6/11 \\ -9/11 & 6/11 & 0 \end{pmatrix}$

11.3 Given that

$$f = \begin{pmatrix} 1/3 & 1/3 & 1/3 \\ 1/3 & 1/3 & 1/3 \\ 1/3 & 1/3 & 1/3 \end{pmatrix}$$

is a perpendicular projection onto a line, find a new orthonormal basis in terms of which the matrix of f becomes

$$\begin{pmatrix} 1 & 0 & 0 \\ 0 & 0 & 0 \\ 0 & 0 & 0 \end{pmatrix}$$

Write the transition equation.

11.4 Given that

$$f = \begin{pmatrix} 8/9 & 2/9 & -2/9 \\ 2/9 & 5/9 & 4/9 \\ -2/9 & 4/9 & 5/9 \end{pmatrix}$$

is a perpendicular projection onto a plane, find a new orthonormal basis in terms of which the matrix of f becomes

$$\begin{pmatrix} 1 & 0 & 0 \\ 0 & 1 & 0 \\ 0 & 0 & 0 \end{pmatrix}$$

Write the transition equation.

11.5 Given that

$$f = \begin{pmatrix} 2/7 & 3/7 & -6/7 \\ 3/7 & -6/7 & -2/7 \\ -6/7 & -2/7 & -3/7 \end{pmatrix}$$

is an orientation-preserving symmetric involution (i.e., a rotation through 180°), find a new orthonormal basis in terms of which the matrix of f becomes

$$\begin{pmatrix} -1 & 0 & 0 \\ 0 & -1 & 0 \\ 0 & 0 & 1 \end{pmatrix}$$

Write the transition equation.

11.6 Givên that

$$f = \begin{pmatrix} 4/9 & 7/9 & -4/9 \\ 1/9 & 4/9 & 8/9 \\ 8/9 & -4/9 & 1/9 \end{pmatrix}$$

is an orientation-preserving orthogonal invection (i.e., a rotation through 90°), find a new orthonormal basis in terms of which the matrix of f becomes

$$\begin{pmatrix} 0 & -1 & 0 \\ 1 & 0 & 0 \\ 0 & 0 & 1 \end{pmatrix}$$

Write the transition equation.

11.7 For each of the symmetric operators f listed below, find a new orthonormal basis in terms of which the matrix of f takes the diagonal form. Write the transition equation.

(i) $\begin{pmatrix} 3 & 0 & 6 \\ 0 & -3 & 6 \\ 6 & 6 & 0 \end{pmatrix}$ (ii) $\begin{pmatrix} 3 & -2 & -2 \\ -2 & 6 & 0 \\ -2 & 0 & 2 \end{pmatrix}$ (iii) $\begin{pmatrix} 5 & 2 & 0 \\ 2 & 4 & -2 \\ 0 & -2 & 7 \end{pmatrix}$

(iv) $\begin{pmatrix} -16 & 8 & 20 \\ 8 & 5 & 8 \\ 20 & 8 & 11 \end{pmatrix}$ (v) $\begin{pmatrix} 32 & 6 & 30 \\ 6 & -5 & 24 \\ 30 & 24 & -27 \end{pmatrix}$ (vi) $\begin{pmatrix} 32 & 30 & 54 \\ 30 & 13 & 96 \\ 54 & 96 & -45 \end{pmatrix}$

(vii) $\begin{pmatrix} 71 & -30 & -24 \\ -30 & 130 & 6 \\ -24 & 6 & 93 \end{pmatrix}$

11.8 Find the diagonal form of each of the following symmetric operators (the transition equation is not required).

(i) $\begin{pmatrix} 2 & 1 & -2 \\ 1 & 2 & 2 \\ -2 & 2 & -1 \end{pmatrix}$ (ii) $\begin{pmatrix} 2 & 3 & -6 \\ 3 & 6 & -2 \\ -6 & -2 & -3 \end{pmatrix}$

(iii) $\begin{pmatrix} -8 & 1 & 4 \\ 1 & -8 & 4 \\ 4 & 4 & 7 \end{pmatrix}$ (iv) $\begin{pmatrix} -1 & -2 & 4 \\ -2 & 2 & 2 \\ 4 & 2 & -1 \end{pmatrix}$

(v) $\begin{pmatrix} 1 & 1 & 1 \\ 1 & 1 & -1 \\ 1 & -1 & -1 \end{pmatrix}$ (vi) $\begin{pmatrix} 2 & 6 & -9 \\ 6 & 7 & 6 \\ -9 & 6 & 2 \end{pmatrix}$

(vii) $\begin{pmatrix} -9 & 8 & 12 \\ 8 & -9 & 12 \\ 12 & 12 & 1 \end{pmatrix}$ (viii) $\begin{pmatrix} 1 & 6 & 18 \\ 6 & 17 & -6 \\ 18 & -6 & 1 \end{pmatrix}$

(ix) $\begin{pmatrix} -13 & 16 & -4 \\ 16 & 11 & -8 \\ -4 & -8 & -19 \end{pmatrix}$ (x) $\begin{pmatrix} 23 & 10 & 10 \\ 10 & -25 & 2 \\ 10 & 2 & -25 \end{pmatrix}$

(xi) $\begin{pmatrix} -21 & 12 & 16 \\ 12 & -11 & 24 \\ 16 & 24 & 3 \end{pmatrix}$ (xii) $\begin{pmatrix} 5 & 6 & -30 \\ 6 & -30 & -5 \\ -30 & -5 & -6 \end{pmatrix}$

(xiii) $\begin{pmatrix} -31 & -8 & 8 \\ -8 & 32 & 1 \\ 8 & 1 & -32 \end{pmatrix}$ (xiv) $\begin{pmatrix} 17 & 20 & 20 \\ 20 & 8 & -25 \\ 20 & -25 & -8 \end{pmatrix}$

(xv) $\begin{pmatrix} -17 & 28 & 4 \\ 28 & 16 & 7 \\ 4 & 7 & -32 \end{pmatrix}$

11.9 Find the diagonal form of each of the following symmetric operators, given their eigenvalues (the transition equation is not required).

(i) $\begin{pmatrix} -28 & 24 & 3 \\ 24 & 27 & 8 \\ 3 & 8 & -36 \end{pmatrix}, \quad \lambda = \pm 37$ (ii) $\begin{pmatrix} -39 & 4 & 12 \\ 4 & -33 & 24 \\ 12 & 24 & 31 \end{pmatrix}, \quad \lambda = \pm 41$

(iii) $\begin{pmatrix} -6 & -42 & 7 \\ -42 & 7 & 6 \\ 7 & 6 & 42 \end{pmatrix}, \quad \lambda = \pm 43$ (iv) $\begin{pmatrix} 7 & 30 & 30 \\ 30 & 18 & -25 \\ 30 & -25 & 18 \end{pmatrix}, \quad \lambda = \pm 43$

(v) $\begin{pmatrix} 33 & 4 & 36 \\ 4 & 48 & -9 \\ 36 & -9 & -32 \end{pmatrix}, \quad \lambda = \pm 49$ (vi) $\begin{pmatrix} 1 & 50 & 10 \\ 50 & 1 & -10 \\ 10 & -10 & 49 \end{pmatrix}, \quad \lambda = \pm 51$

11.10 Find orthogonal operators among those listed in Exercise 1. Which of those orthogonal operators are rotations, and which ones are reflections?

11.11 In each of the following, show that f is an orthogonal operator. Calculate f^{-1}. Is there a reflection among the orthogonal operators listed here?

(i) $\begin{pmatrix} 1/3 & -2/3 & 2/3 \\ 2/3 & 2/3 & 1/3 \\ -2/3 & 1/3 & 2/3 \end{pmatrix}$ (ii) $\begin{pmatrix} 3/5 & -4/5 & 0 \\ 4/5 & 3/5 & 0 \\ 0 & 0 & 1 \end{pmatrix}$

(iii) $\begin{pmatrix} 2/7 & -3/7 & 6/7 \\ 3/7 & 6/7 & 2/7 \\ -6/7 & 2/7 & 3/7 \end{pmatrix}$ (iv) $\begin{pmatrix} 2/11 & -9/11 & 6/11 \\ 6/11 & 6/11 & 7/11 \\ -9/11 & 2/11 & 6/11 \end{pmatrix}$

(v) $\begin{pmatrix} 5/13 & 0 & -12/13 \\ 0 & 1 & 0 \\ 12/13 & 0 & 5/13 \end{pmatrix}$ (vi) $\begin{pmatrix} 4/13 & -3/13 & -12/13 \\ 12/13 & 4/13 & 3/13 \\ 3/13 & -12/13 & 4/13 \end{pmatrix}$

(vii) $\begin{pmatrix} 5/15 & 10/15 & 10/15 \\ -14/15 & 5/15 & 2/15 \\ -2/15 & -10/15 & 11/15 \end{pmatrix}$ (viii) $\begin{pmatrix} 1 & 0 & 0 \\ 0 & 8/17 & -15/16 \\ 0 & 15/17 & 8/17 \end{pmatrix}$

(ix) $\begin{pmatrix} 8/17 & 12/17 & -9/17 \\ -9/17 & 12/17 & 8/17 \\ 12/17 & 1/17 & 12/17 \end{pmatrix}$ (x) $\begin{pmatrix} 1/19 & 6/19 & 18/19 \\ 18/19 & -6/19 & 1/19 \\ 6/19 & 17/19 & -6/19 \end{pmatrix}$

(xi) $\begin{pmatrix} 4/21 & 8/21 & 19/21 \\ 16/21 & 11/21 & -8/21 \\ 13/21 & -16/21 & 4/21 \end{pmatrix}$ (xii) $\begin{pmatrix} 5/21 & 4/21 & -20/21 \\ 20/21 & -5/21 & 4/21 \\ 4/21 & 20/21 & 5/21 \end{pmatrix}$

(xiii) $\begin{pmatrix} 3/23 & 6/23 & 22/23 \\ 18/23 & 13/23 & -6/23 \\ -14/23 & 18/23 & -3/23 \end{pmatrix}$ (xiv) $\begin{pmatrix} 7/25 & -24/25 & 0 \\ 24/25 & 7/25 & 0 \\ 0 & 0 & 1 \end{pmatrix}$

(xv) $\begin{pmatrix} 0 & 20/25 & 15/25 \\ 20/25 & -9/25 & 12/25 \\ 15/25 & 12/25 & -16/25 \end{pmatrix}$ (xvi) $\begin{pmatrix} 10/27 & 10/27 & 23/27 \\ 2/27 & -25/27 & 10/27 \\ -25/27 & 2/27 & 10/27 \end{pmatrix}$

(xvii) $\begin{pmatrix} -21/29 & 16/29 & 12/29 \\ 12/29 & 24/29 & -11/29 \\ 16/29 & 3/29 & 24/29 \end{pmatrix}$ (xviii) $\begin{pmatrix} 20/29 & -21/29 & 0 \\ 21/29 & 20/29 & 0 \\ 0 & 0 & 1 \end{pmatrix}$

(xix) $\begin{pmatrix} 30/31 & -6/31 & -5/31 \\ 5/31 & 30/31 & -6/31 \\ 6/31 & 5/31 & 30/31 \end{pmatrix}$ (xx) $\begin{pmatrix} 6/31 & 14/31 & 27/31 \\ 22/31 & -21/31 & 6/31 \\ 21/31 & 18/31 & -14/31 \end{pmatrix}$

(xxi) $\begin{pmatrix} 19/39 & 22/39 & 26/39 \\ -34/39 & 14/39 & 13/39 \\ 2/39 & 29/39 & -26/39 \end{pmatrix}$ (xxii) $\begin{pmatrix} 0.8 & 0.6 & 0 \\ -0.36 & 0.48 & 0.8 \\ 0.48 & -0.64 & 0.6 \end{pmatrix}$

(xxiii) $\begin{pmatrix} 0.64 & 0.48 & 0.6 \\ -0.024 & -0.768 & 0.64 \\ -0.768 & 0.424 & 0.48 \end{pmatrix}$ (xxiv) $\begin{pmatrix} 0.192 & 0.744 & -0.64 \\ 0.48 & -0.64 & -0.6 \\ 0.856 & 0.192 & 0.48 \end{pmatrix}$

(xxv) $\begin{pmatrix} 0.36 & 0.864 & 0.352 \\ 0.8 & -0.48 & 0.36 \\ -0.48 & -0.152 & 0.864 \end{pmatrix}$

11.12 Show that if we permute the row (or column) vectors of an orthogonal matrix, we get another orthogonal matrix. Also, show that if we change the signatures of the entries in a row (or column) of an orthogonal matrix, we get another orthogonal matrix.

11.13 Show that

(i) If p is a perpendicular projection, then $1-2p$ and $2p-1$ are reflections.
(ii) If h is a reflection, then $\frac{1}{2}(1+h)$ and $\frac{1}{2}(1-h)$ are perpendicular projections.

11.14 Show that if t is an antisymmetric twist, then t^4 and $1+t^2$ are perpendicular projections. Describe $\operatorname{im} t^4$, $\ker t^4$, $\operatorname{im}(1+t^2)$, $\ker(1+t^2)$ in terms of $\operatorname{im} t$ and $\ker t$.

11.15 Let $f(\mathbf{v}) = (\mathbf{a}\cdot\mathbf{v})\mathbf{b}$, where $\mathbf{a}$, $\mathbf{b}$ are fixed vectors. Show that

$$f^*(\mathbf{v}) = (\mathbf{b}\cdot\mathbf{v})\mathbf{a}$$

11.16 A lift ℓ is said to be *perpendicular* if $\ell\circ\ell^*$, $\ell^*\circ\ell$ are perpendicular projections. Show that ℓ is perpendicular if, and only if,

$$\ell(\mathbf{v}) = (\mathbf{u}_1\cdot\mathbf{v})\mathbf{u}_2$$

for all $\mathbf{v}$, where $\mathbf{u}_1$, $\mathbf{u}_2$ are perpendicular unit vectors. Let ℓ be a perpendicular lift of $\mathbb{R}^3$; show that

(i) $t=\ell-\ell^*$ is an antisymmetric twist and $\ker t$ is spanned by $\mathbf{u}_1\times\mathbf{u}_2$; $\operatorname{im} t$ is the normal plane of $\mathbf{u}_1\times\mathbf{u}_2$.

(ii) $p=\ell\circ\ell^*$, $q=\ell^*\circ\ell$ are principal perpendicular projections, and $\ker p$, $\ker q$ are the normal planes of $\mathbf{u}_1$,$\mathbf{u}_2$; and $\operatorname{im} p$, $\operatorname{im} q$ are the lines spanned by $\mathbf{u}_2$, $\mathbf{u}_1$, respectively.

(iii) $\ell\circ\ell^*+\ell^*\circ\ell$ is a perpendicular projection onto a plane. Find $\ker(\ell\circ\ell^*+\ell^*\circ\ell)$, $\operatorname{im}(\ell\circ\ell^*+\ell^*\circ\ell)$.

(iv) $h=2(\ell^*\circ\ell)-1$, $k=2(\ell\circ\ell^*)-1$ are rotations through 180° around the axes spanned by the vectors $\mathbf{u}_1$, $\mathbf{u}_2$, respectively.

11.17 Show that if the principal perpendicular projections p, q are mutual zero divisors ($p\circ q=0=q\circ p$), they can be decomposed into the product of a pair of perpendicular lifts: $p=\ell\circ\ell^*$, $q=\ell^*\circ\ell$, $\ell^2=0=\ell^{*2}$.

11.18 Show that every antisymmetric twist can be decomposed into the difference of a pair of perpendicular lifts.

11.19 Find the Cayley transform of the antisymmetric operator

$$a=\begin{pmatrix} 0 & 2 & -1\\ -2 & 0 & 2\\ 1 & -2 & 0\end{pmatrix}$$

and show that it is orientation preserving.

11.20 Find the Cayley transform of the orientation-preserving orthogonal operator

$$r=\begin{pmatrix} 4/5 & 3/5 & 0\\ -3/5 & 4/5 & 0\\ 0 & 0 & 1\end{pmatrix}$$

11.21 Show that the unique eigenvector of an antisymmetric operator $a\neq 0$ and that of its Cayley transform coincide.

11.22 Show that the Cayley transform of an antisymmetric operator is orientation preserving.

11.23 Let $a^* = -a$ and $r = (1 + a) \circ (1 - a)^{-1}$; then $r^* = r^{-1}$ and $a = (r + 1)^{-1} \circ (r - 1)$. Also, prove the converse. (In other words, the Cayley transform between the set of antisymmetric operators and the set of orientation-preserving orthogonal operators is bijective.)

11.24 Prove the principal twist theorem for $\mathbb{R}^2$.

11.25 Show that $t \neq 0$ is an antisymmetric twist if, and only if,

$$t(\mathbf{v}) = \mathbf{u} \times \mathbf{v} \qquad \text{where } |\mathbf{u}| = 1$$

(See Chap. 4, Exercise 12, and Chap. 10, Exercise 9.)

11.26 Show that p is a perpendicular projection onto a line if, and only if,

$$p(\mathbf{v}) = (\mathbf{u} \cdot \mathbf{v})\mathbf{u} \qquad \text{where } |\mathbf{u}| = 1$$

(See Chap. 4, Exercise 14, and Chap. 10, Exercise 2.)

11.27 Show that q is a perpendicular projection onto a plane if, and only if,

$$q(\mathbf{v}) = (\mathbf{u} \times \mathbf{v}) \times \mathbf{u} \qquad \text{where } |\mathbf{u}| = 1$$

(See Chap. 4, Exercise 14, and Chap. 10, Exercise 2.)

11.28 Show that an orthogonal operator r is orientation preserving if, and only if, it preserves the cross product of vectors, i.e.,

$$r(\mathbf{v} \times \mathbf{w}) = r(\mathbf{v}) \times r(\mathbf{w}) \qquad \text{for all } \mathbf{v}, \mathbf{w} \in \mathbb{R}^3$$

Show also that an orientation-reversing orthogonal operator r satisfies

$$r(\mathbf{v} \times \mathbf{w}) = r(\mathbf{w}) \times r(\mathbf{v}) \qquad \text{for all } \mathbf{v}, \mathbf{w} \in \mathbb{R}^3$$

11.29 A linear operator s is defined by the formula

$$s(\mathbf{v}) = (\mathbf{e} \times \mathbf{v}) \times \mathbf{e} + \lambda(\mathbf{e} \cdot \mathbf{v})\mathbf{e}$$

for all $\mathbf{v} \in \mathbb{R}^3$, where $|\mathbf{e}| = 1$. Then s is called a *perpendicular dilatation* in ratio λ along the line spanned by $\mathbf{e}$. Show that s is symmetric; find its eigenvalues and eigenvectors. Show that if $\lambda \neq 0$, then s is invertible and find s^{-1}.

11.30 Show that

$$s = \begin{pmatrix} 13/9 & -4/9 & 2/9 \\ -4/9 & 13/9 & -2/9 \\ 2/9 & -2/9 & 10/9 \end{pmatrix}$$

is a perpendicular dilatation. What is the ratio λ?

11.31 In each of the following, find the invariant plane of the operator f and state the type of f.

(i) $\begin{pmatrix} 3 & -7 & 5 \\ 3 & -6 & 5 \\ 2 & -4 & 3 \end{pmatrix}$ (ii) $\begin{pmatrix} -2 & 4 & 5 \\ -8 & -2 & 2 \\ 2 & 5 & 4 \end{pmatrix}$

11.32 Write the equation of the invariant plane of:
(i) The parabolic operator

$$\begin{pmatrix} \lambda & 1 & 0 \\ 0 & \lambda & 1 \\ 0 & 0 & \lambda \end{pmatrix}$$

(ii) The rotation through 120°

$$\begin{pmatrix} 0 & 0 & 1 \\ 1 & 0 & 0 \\ 0 & 1 & 0 \end{pmatrix}$$

(iii) The rotation through 90°

$$\begin{pmatrix} 4/9 & 7/9 & 4/9 \\ 1/9 & 4/9 & -8/9 \\ -8/9 & 4/9 & 1/9 \end{pmatrix}$$

(iv) The invection

$$\begin{pmatrix} 1 & -2 & 2 \\ 1 & -1 & 1 \\ 0 & 0 & 1 \end{pmatrix}$$

11.33 Let h be a rotation through $\phi \neq 0°$. Show that

$$h^* = h \Leftrightarrow \phi = 180°$$

11.34 Without calculating f^3 show that

$$f = \begin{pmatrix} 0 & 10/111 & -11/111 \\ -10/111 & 0 & 110/111 \\ 11/111 & -110/111 & 0 \end{pmatrix}$$

is a twist.

11.35 Show that the transpose of a projection (involution, twist, invection, respectively) is a projection (involution, twist, invection, respectively).

11.36 Apply Gram-Schmidt orthonormalization to each of the following bases (do not change the order of basis vectors).

(i) $\begin{pmatrix} 2 \\ 1 \\ 2 \end{pmatrix}, \begin{pmatrix} 1 \\ 1 \\ 0 \end{pmatrix}, \begin{pmatrix} 1 \\ 5 \\ 1 \end{pmatrix}$ (ii) $\begin{pmatrix} 6 \\ 2 \\ 3 \end{pmatrix}, \begin{pmatrix} 3 \\ 8 \\ 5 \end{pmatrix}, \begin{pmatrix} 5 \\ 11 \\ -1 \end{pmatrix}$

(iii) $\begin{pmatrix} 2 \\ 6 \\ -9 \end{pmatrix}, \begin{pmatrix} -7 \\ 12 \\ -7 \end{pmatrix}, \begin{pmatrix} -1 \\ 19 \\ -1 \end{pmatrix}$ (iv) $\begin{pmatrix} 4 \\ 12 \\ 3 \end{pmatrix}, \begin{pmatrix} 1 \\ 16 \\ -9 \end{pmatrix}, \begin{pmatrix} -11 \\ 19 \\ -5 \end{pmatrix}$

(v) $\begin{pmatrix} 4 \\ 4 \\ 7 \end{pmatrix}, \begin{pmatrix} -4 \\ 5 \\ 11 \end{pmatrix}, \begin{pmatrix} -3 \\ -3 \\ 15 \end{pmatrix}$

11.37 Find the standard form of each of the following orientation-preserving orthogonal operators (= rotations).

(i) $\begin{pmatrix} 0 & 0 & 1 \\ 1 & 0 & 0 \\ 0 & 1 & 0 \end{pmatrix}$ (ii) $\begin{pmatrix} 4/9 & 7/9 & 4/9 \\ 1/9 & 4/9 & -8/9 \\ -8/9 & 4/9 & 1/9 \end{pmatrix}$

(iii) $\begin{pmatrix} 2/7 & 3/7 & -6/7 \\ 3/7 & -6/7 & -2/7 \\ -6/7 & -2/7 & -3/7 \end{pmatrix}$ (iv) $\begin{pmatrix} -2/3 & -1/3 & -2/3 \\ -1/3 & -2/3 & 2/3 \\ -2/3 & 2/3 & 1/3 \end{pmatrix}$

(v) $\begin{pmatrix} 3/5 & -4/5 & 0 \\ 4/5 & 3/5 & 0 \\ 0 & 0 & 1 \end{pmatrix}$ (vi) $\begin{pmatrix} 29/45 & 28/45 & -20/45 \\ -20/45 & 35/45 & 20/45 \\ 28/45 & -4/45 & 35/45 \end{pmatrix}$

11.38 True or false? In each of the following, furnish a proof if true; furnish a counterexample if false.

(i) Every orthogonal operator is elliptic.
(ii) Every orientation-reversing orthogonal operator is symmetric.
(iii) There exist no antisymmetric orthogonal operators.
(iv) Every antisymmetric operator is elliptic.
(v) The negative of an orthogonal operator is orthogonal.
(vi) If r is an orientation-preserving orthogonal operator, then $-r$ is an orientation-reversing orthogonal operator, and vice versa.
(vii) The inverse of a symmetric operator, if it exists, is symmetric.
(viii) The negative of an antisymmetric operator is antisymmetric.

11.39 True or false? "There are only a finite number of orthogonal 2×2 and 3×3 matrices with nonnegative entries." If true, find these numbers. If false, give a proof.

11.40 True or false? "There are only a finite number of orthogonal 2×2 and 3×3 matrices with integral entries." If true, find these numbers. If false, give a proof.

11.41 Let $\mathbf{e}$ be an eigenvector of f belonging to the eigenvalue λ, and let $\mathbf{e}'$ be an eigenvector of f^* belonging to the eigenvalue λ'. Show that

$$\lambda \neq \lambda' \Rightarrow \mathbf{e} \perp \mathbf{e}'$$

11.42 Show that if f is symmetric or antisymmetric with $\ker f \neq 0$, then $\operatorname{im} f \perp \ker f$. Is the converse statement true? Why?

11.43 Show that transposition makes no sense if the basis fails to be orthonormal. (*Hint*: Change the basis and observe that some symmetric matrix is no longer symmetric in the new basis.)

11.44 Show that if f is a lift parallel to the plane π along the line ℓ, and g is an invertible operator, then the conjugate $g \circ f \circ g^{-1}$ of f is the lift parallel to the plane $g(\pi)$ along the line $g(\ell)$.

Also, show that the conjugate of a shear is a shear and describe $g \circ f \circ g^{-1}$ geometrically.

11.45 Show that the product of two symmetric operators is symmetric if, and only if, they commute.

11.46 A symmetric operator s is called *positive* (resp., *positive definite*) if $s(\mathbf{v}) \cdot \mathbf{w} \geqq 0$ (resp., $s(\mathbf{v}) \cdot \mathbf{w} > 0$) for all $\mathbf{v}$, $\mathbf{w} \neq \mathbf{0}$. Given a positive operator s, define its *square root* $s^{1/2}$ and show that it is positive.

11.47 Show that, for any operator f, $f \circ f^*$ and $f^* \circ f$ are positive. Moreover, $f \circ f^*$ and $f^* \circ f$ are positive definite if, and only if, f is invertible.

11.48 A linear operator f is said to be *normal* if $f \circ f^* = f^* \circ f$. Show that f symmetric or orthogonal $\Rightarrow f$ normal; and f normal $\Rightarrow g \circ f \circ g^{-1}$ normal, for all invertible g. Show that f normal $\Leftrightarrow f$ symmetric or $f = \lambda r$ where r is orthogonal and $\lambda \in \mathbb{R}$.

11.49 Show that any invertible operator f can be decomposed uniquely into the product of a positive definite operator s and an orthogonal operator r, i.e., $f = s \circ r$ or $f = r' \circ s'$.

11.50 Show that the conjugate of a symmetric (antisymmetric) operator by an orthogonal operator is symmetric (resp., antisymmetric). Through examples, show that the statement is false if the requirement *orthogonal* is dropped.

12

Rotations and Reflections: The Classification of Orthogonal Operators†

In this chapter we classify the orthogonal operators of the plane $\mathbb{R}^2$ as well as those of the space $\mathbb{R}^3$. Above all, we want to be able to identify those orthogonal operators which are rotations, and those which are reflections (in a point, in a line, or in a plane). We are also interested in the question of whether these exhaust the set of orthogonal operators. We shall see that the answer in $\mathbb{R}^2$ is "yes," but in $\mathbb{R}^3$ it is "no" because there are orientation-reversing orthogonal operators which are neither a rotation nor a reflection. It turns out that in $\mathbb{R}^n$ $(n \geqq 4)$ there are also orientation-preserving orthogonal operators that fail to be a rotation or a reflection.

CASE OF $\mathbb{R}^2$

An orthogonal operator r is either involutory: $r^* = r = r^{-1}$, or it is noninvolutory, according as its matrix is symmetric or not. We shall first look at the case when r is involutory; then $p = \frac{1}{2}(1 + r)$ is a perpendicular projection: $p^2 = p = p^*$. There are three possibilities, according as $\operatorname{rank} p = \operatorname{trace} p = 2, 1, 0$.

1. $\operatorname{trace} r = 2 \Leftrightarrow \operatorname{rank} p = 2 \Leftrightarrow r = 1$, the identity operator.
2. $\operatorname{trace} r = -2 \Leftrightarrow \operatorname{rank} p = 0 \Leftrightarrow r = -1$, reflection in the origin.
3. $\operatorname{trace} r = 0 \Leftrightarrow \operatorname{rank} p = 1 \Leftrightarrow r$ symmetric involution $\neq 1, -1$
   $\Leftrightarrow r$ reflection in the line $\operatorname{im} \frac{1}{2}(1 + r)$.

In the first two cases, when $\operatorname{trace} r = 2$ or -2, r is orientation preserving, and in the third case, when $\operatorname{trace} r = 0$, r is orientation reversing.

†*Note*: In this chapter all bases are orthonormal and all coordinate systems are rectangular metric.

We shall now consider the important case of the noninvolutory orthogonal operator r. Note that $r+1$ is invertible (otherwise, -1 would be an eigenvalue of r and r would be involutory). The linear operator

$$a=(r+1)^{-1}\circ(r-1)$$

called the *Cayley transform* of r, is antisymmetric. Indeed, $r+1$ and $r-1$ commute; therefore, $(r+1)^{-1}$ and $r-1$ also commute (why?) and

$$\begin{aligned} a^{*} &= ((r+1)^{-1}\circ(r-1))^{*} = ((r+1)^{*})^{-1}\circ(r-1)^{*} = (r^{-1}+1)^{-1}\circ(r^{-1}-1) \\ &= (r^{-1}+r^{-1}\circ r)^{-1}\circ(r^{-1}-1) = (r^{-1}\circ(1+r))^{-1}\circ(r^{-1}-1) \\ &= (1+r)^{-1}\circ r\circ(r^{-1}-1) = (1+r)^{-1}\circ(1-r) = -a \end{aligned}$$

It follows from the principal twist theorem that $a=\omega i$, where $\omega\in\mathbb{R}$ and i is an antisymmetric twist: $i^{*}=-i=i^{3}$. Since i is antisymmetric,

$$i=\begin{pmatrix}0 & x\\ -x & 0\end{pmatrix}$$

Since i is a twist,

$$\begin{pmatrix}0 & x\\ -x & 0\end{pmatrix}^{3}=\begin{pmatrix}0 & -x^{3}\\ x^{3} & 0\end{pmatrix}=\begin{pmatrix}0 & -x\\ x & 0\end{pmatrix}$$

and, therefore, the scalar x must satisfy the equation $x^{3}-x=0$. This cubic equation has three solutions: $x=0, 1, -1$. $x=0$ yields the trivial operator, which is ruled out. The remaining solutions yield

$$i=\begin{pmatrix}0 & -1\\ 1 & 0\end{pmatrix}\qquad\text{and}\qquad -i=\begin{pmatrix}0 & 1\\ -1 & 0\end{pmatrix}$$

which we recognize as the rotations through 90° and –90°. Thus *in the plane there are two, and only two, nontrivial antisymmetric twists, i and –i, which both satisfy*

$$i^{2}=-1\qquad\text{and}\qquad i^{-1}=-i$$

Furthermore, they both are *orthogonal invections*:

$$i^{*}=i^{-1}=i^{3}=-i$$

and there are no other orthogonal invections in the plane. It is a characteristic property of $\mathbb{R}^{2}$ that in it the concept of a (nontrivial) antisymmetric twist coincides with the concept of a (nonidentity) orthogonal invection, and that their number is finite, namely, two. (There are, of course, infinitely many other twists and invections in $\mathbb{R}^{2}$.)

The Cayley transform of $a=(r+1)^{-1}\circ(r-1)$ is the original orthogonal operator r, which also satisfies $r=(1+a)\circ(1-a)^{-1}$ (see Chap. 11,

Application 4). Cayley's Formula yields

$$r = 1 + \frac{2\omega}{1+\omega^2} i + \frac{2\omega^2}{1+\omega^2} i^2 = \frac{1-\omega^2}{1+\omega^2} + \frac{2\omega}{1+\omega^2} i$$

By the rational parametric form of the circle (Chap. 2, Application 12)

$$\omega = \tan \tfrac{1}{2}\alpha \Leftrightarrow \cos\alpha = \frac{1-\omega^2}{1+\omega^2}, \quad \sin\alpha = \frac{2\omega}{1+\omega^2} \qquad (\alpha \neq 180°)$$

Hence we get

Standard Form

$$r = \cos\alpha + i\sin\alpha \qquad (i^2 = -1)$$

Conversely, given a linear operator in this form, we have that

$$r^{-1} = \cos\alpha - i\sin\alpha = r^* \neq r$$

and therefore r is orthogonal and noninvolutory. Thus *the noninvolutory orthogonal operators of the plane are precisely the rotations of the plane around the origin through* $\alpha \neq 180°$. It is clear that they are orientation preserving. The formula $r = \cos\alpha + i\sin\alpha$ actually covers both involutory and noninvolutory rotations: if $\alpha = 180°$, $r = -1$, the reflection in the origin. Let $q = \cos\beta + i\sin\beta$ be another rotation through β; then

$$q \circ r = \cos(\alpha+\beta) + i\sin(\alpha+\beta)$$

is a third rotation through $\alpha + \beta$. Moreover, we have

De Moivre's† Formula

$$r^n = \cos n\alpha + i\sin n\alpha$$

These formulas suggested the exponential notation to Euler:

Euler's‡ Formula

$$e^{\alpha i} = \cos\alpha + i\sin\alpha$$

where $e = 2.71828\ldots$, the basis of natural logarithms, and i is the antisymmetric twist of the plane. Then the formulas above can be summarized as

$$e^{\alpha i} \circ e^{\beta i} = e^{(\alpha+\beta)i}$$

$$(e^{\alpha i})^{-1} = e^{-\alpha i}$$

†Abraham De Moivre (1667-1754), French mathematician working in London.

‡Leonhard Euler (1707-1783) Swiss mathematician working at Berlin and St. Petersburg; the most productive mathematician of all times.

and we see that *rotations around the origin form a one-parameter commutative group.*

Example Find the standard form of the orthogonal operator

$$r = \begin{pmatrix} \cos\alpha & -\sin\alpha \\ \sin\alpha & \cos\alpha \end{pmatrix}$$

Answer:

$$r = \cos\alpha \begin{pmatrix} 1 & 0 \\ 0 & 1 \end{pmatrix} + \sin\alpha \begin{pmatrix} 0 & -1 \\ 1 & 0 \end{pmatrix} = \cos\alpha + i\sin\alpha$$

r is the rotation around the origin through α: $r = e^{\alpha i}$.

There is this remarkable contrast between the two matrices

$$\begin{pmatrix} \cos\alpha & -\sin\alpha \\ \sin\alpha & \cos\alpha \end{pmatrix} \quad \text{and} \quad \begin{pmatrix} \cos\alpha & \sin\alpha \\ \sin\alpha & -\cos\alpha \end{pmatrix}$$

The latter can be brought to the simpler form

$$\begin{pmatrix} 1 & 0 \\ 0 & -1 \end{pmatrix}$$

by changing the orthonormal basis. By contrast, the former cannot be further simplified, no matter how we may change the orthonormal basis. The reason for this is, of course, the principal axis theorem, which applies to the latter (as it is symmetric), but does not apply to the former.

Note that the foregoing discussion has exhausted the set of orthogonal operators in $\mathbb{R}^2$ so that we have the

Classification of Orthogonal Operators in $\mathbb{R}^2$

Orthogonal operators	Involutory	Noninvolutory
Orientation-preserving	trace $r = 2 \Leftrightarrow r = 1$ eigenvalues 1,1 the identity operator	no eigenvalues, no eigenvectors; $r = \cos\alpha + i\sin\alpha = e^{\alpha i}$ *rotation* through $\alpha \neq 180°$
	trace $r = -2 \Leftrightarrow r = -1$ eigenvalues $-1, -1$ the reflection in the origin (180° rot.)	
Orientation-reversing	trace $r = 0$ eigenvalues $1, -1$ $r^* = r = r^{r-1}$, symmetric involution *reflection* in the line im ½$(1 + r)$	———

Thus an orthogonal operator of $\mathbb{R}^2$ is either a rotation or a reflection. There are no noninvolutory orientation-reversing orthogonal operators in $\mathbb{R}^2$. Later in this chapter we shall see that such operators exist in $\mathbb{R}^3$.

Let us eliminate the twist i from Euler's formula:

$$r^2 = \cos^2\alpha + 2i\cos\alpha\sin\alpha - \sin^2\alpha = 2\cos\alpha(\cos\alpha + i\sin\alpha) - 1$$
$$= 2\cos\alpha\, r - 1$$

We conclude that a rotation through α satisfies

$$r^2 - 2r\cos\alpha + 1 = 0$$

This is, of course, the characteristic equation of r which has no (real) root unless $\alpha = 0°$ or $\alpha = 180°$. This abstract algebraic statement has an immediate and concrete geometric significance, which is that *a noninvolutory rotation has no eigenvector.*

APPLICATION

1. Trigonometric Polynomials

We want to find formulas to express the trigonometric polynomials

$$1 + \cos\alpha + \cos 2\alpha + \cos 3\alpha + \cdots + \cos n\alpha$$
$$\sin\alpha + \sin 2\alpha + \sin 3\alpha + \cdots + \sin n\alpha$$

We begin with the obvious equality

$$(r^2 - 1)\circ(1 + r^2 + r^4 + \cdots + r^{2n}) = r^{2n+2} - 1 = r^{n+1}\circ(r^{n+1} - r^{-n-1})$$

Multiplying by r^{-1} on the left, we get

$$(r - r^{-1})\circ(1 + r^2 + r^4 + \cdots + r^{2n}) = r^n\circ(r^{n+1} - r^{-n-1}) = (r^{n+1} - r^{-n-1})\circ r^n$$

From Euler's formula, $r - r^{-1} = 2i\sin\alpha$ and from De Moivre's formula, $r^{n+1} - r^{-n-1} = 2i\sin(n+1)\alpha$, so that we have

$$2i\sin\alpha\circ(1 + r^2 + r^4 + \cdots + r^{2n}) = 2i\sin(n+1)\alpha\circ r^n$$

We may cancel i on both sides since in $\mathbb{R}^2$ the antisymmetric twist is invertible. For $\alpha \neq 0°, 180°$ we have

$$1 + r^2 + r^4 + \cdots + r^{2n} = \frac{\sin(n+1)\alpha}{\sin\alpha} r^n$$

or, putting $\alpha = \tfrac{1}{2}\phi$,

$$1 + r + r^2 + \cdots + r^n = \frac{\sin\tfrac{1}{2}(n+1)\phi}{\sin\tfrac{1}{2}\phi}(\cos\tfrac{1}{2}n\phi + i\sin\tfrac{1}{2}n\phi)$$

By the uniqueness of the representation of the rotation in terms of Euler's formula, for $\phi \neq 0$, we get

$$1+\cos\phi+\cos 2\phi+\cdots+\cos n\phi=\frac{\sin \frac{1}{2}(n+1)\phi}{\sin \frac{1}{2}\phi}\cos \tfrac{1}{2}n\phi$$

$$\sin\phi+\sin 2\phi+\cdots+\sin n\phi=\frac{\sin \frac{1}{2}(n+1)\phi}{\sin \frac{1}{2}\phi}\sin \tfrac{1}{2}n\phi$$

COMPLEX NUMBERS

The vectors of the straight line form a one-dimensional vector space which can be identified with the scalars, or real numbers: $\mathbb{R} = \mathbb{R}^1$. The formal rules of calculation for the real numbers are as follows:

	Addition	*Multiplication*	
	Closure		
(0)	$\lambda, \lambda' \in \mathbb{R} \Rightarrow \lambda + \lambda' \in \mathbb{R}$	$\lambda, \lambda' \in \mathbb{R} \Rightarrow \lambda\lambda' \in \mathbb{R}$	(0*)
	Commutativity		
(1)	$\lambda + \lambda' = \lambda' + \lambda$	$\lambda\lambda' = \lambda'\lambda$	(1*)
	Associativity		
(2)	$\lambda + (\lambda' + \lambda'') = (\lambda + \lambda') + \lambda''$	$\lambda(\lambda'\lambda'') = (\lambda\lambda')\lambda''$	(2*)
	Neutral		
(3)	$\lambda + 0 = \lambda$	$1\lambda = \lambda$	(3*)
	Inverse		
(4)	$\lambda \in \mathbb{R} \Rightarrow -\lambda \in \mathbb{R}$ such that $\lambda + (-\lambda) = 0$	$\lambda \in \mathbb{R}, \lambda \neq 0 \Rightarrow \lambda^{-1} \in \mathbb{R}$ such that $\lambda\lambda^{-1} = 1$	(4*)
	Distributivity		
(5)	$\lambda(\lambda' + \lambda'') = \lambda\lambda' + \lambda\lambda''$		

These rules are summarized in the following ready-reference table:

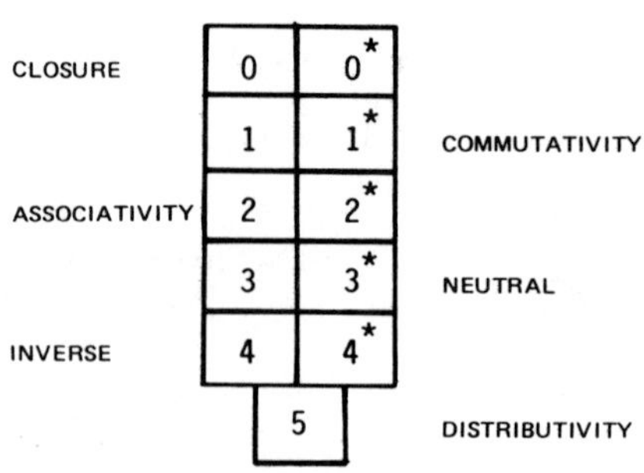

We may now consider these 11 rules as *axioms* and call a set with two operations: addition and multiplication, subject to these eleven axioms a *field*. Examples of fields are: $\mathbb{Q}$, the field of rational numbers, and $\mathbb{R}$, the field of real numbers. $\mathbb{Z}$ is not: Axiom 4^* fails there.

The question now arises whether the vectors of the plane $\mathbb{R}^2$ also form a field. We already have the addition subject to the axioms 0 through 4 defined in $\mathbb{R}^2$, but we lack a product. We shall now see that such a product is being furnished by the orthogonal operators of $\mathbb{R}^2$, thereby converting $\mathbb{R}^2$ into a field, called the field of complex numbers, to be denoted by the symbol $\mathbb{C}$. It will turn out that $\mathbb{R}$ is a subfield of $\mathbb{C}$, and $\mathbb{C}$ has the advantage that the quadratic equations which failed to have a solution in $\mathbb{R}$ can be solved in $\mathbb{C}$. In fact, the fundamental theorem of algebra holds for $\mathbb{C}$: every algebraic equation

$$a_n x^n + a_{n-1}x^{n-1} + \cdots + a_1 x + a_0 = 0$$

where $a_n \neq 0$ and $a_k \in \mathbb{C}\ (k = 0, 1, 2, \ldots, n)$ has a solution in $\mathbb{C}$; moreover, the number of solutions is the same as the degree n of the equation, provided that we count the solutions with their multiplicity. We shall say that the field $\mathbb{C}$ of complex numbers is *algebraically closed*. By contrast, $\mathbb{R}$ and, for the stronger reason, $\mathbb{Q}$ fail to be algebraically closed. (The proof of the fundamental theorem of algebra can be found, for example, in the book *Algebra* by Godement, pp. 517 and 533.)

In order to introduce the product of complex numbers, we need to fix some notation. Throughout this section we shall denote the antisymmetric twist

$$\begin{pmatrix} 0 & -1 \\ 1 & 0 \end{pmatrix}$$

by the symbol i. We should keep in mind the geometric meaning of i: the rotation of the plane about the origin through 90°. As a complex number, i is known as the imaginary unit; since it satisfies the quadratic equation $x^2 + 1 = 0$, it is sometimes written in the form $i = \sqrt{-1}$. Recall Euler's formula:

$$e^{\alpha i} = \cos\alpha + i \sin\alpha$$

Define *complex numbers* $z \in \mathbb{C}$ such that

$$z = \rho e^{\alpha i} = \rho(\cos\alpha + i\sin\alpha) = x + iy$$

with $x = \rho\cos\alpha$, $y = \rho\sin\alpha$, where $\rho \in \mathbb{R}$, $\rho \geqq 0$ is called the *absolute value of the complex number* z, denoted by

$$|z| = \rho = \sqrt{x^2 + y^2}$$

and α is called the *argument of the complex number* z, denoted as

$$\arg z = \alpha \qquad (z \neq 0)$$

The three forms of the complex number z are called the exponential form: $z = \rho e^{\alpha i}$, the trigonometric form: $z = \rho(\cos\alpha + i\sin\alpha)$, and the algebraic form: $z = x + iy$.

We have an isomorphism of vector spaces $\mathbb{C} \to \mathbb{R}^2$ defined by $\rho e^{\alpha i} \rightsquigarrow \rho e^{\alpha i}(\mathbf{v}_0)$, where $\mathbf{v}_0 \in \mathbb{R}^2$ is an arbitrary but fixed vector $\neq \mathbf{0}$. The geometric meaning of this isomorphism is that once the vector $\mathbf{v}_0$ is fixed, there is one, and only one, complex number that will take $\mathbf{v}_0$ into any prescribed vector of $\mathbb{R}^2$. We are constantly using this isomorphism: when we add complex numbers, we think of them as vectors in $\mathbb{R}^2$; when we multiply them, we think of them as linear operators in $\mathbb{C}$. Of course, we could also think of complex numbers as linear operators when we add them but, for one thing, the sum of vectors is easier to visualize than the sum of linear operators; for another, it is not immediately clear that $\mathbb{C}$ is closed under the addition. We need the isomorphism between $\mathbb{C}$ and $\mathbb{R}^2$ to establish the closure axiom (0).

The product of complex numbers is defined to be their product as linear operators:

$$\rho_1 e^{\alpha_1 i} \circ \rho_2 e^{\alpha_2 i} = \rho_1\rho_2 e^{(\alpha_1 + \alpha_2)i}$$

and so is the inverse, or reciprocal, of the complex number:

$$(\rho e^{\alpha i})^{-1} = \rho^{-1} e^{-\alpha i} \qquad (\rho \neq 0)$$

It follows that the product and inverse of complex numbers can also be calculated from the algebraic form, using the formal rules of calculations plus the relation $i^2 = -1$:

$$(x + iy)(x' + iy') = (xx' - yy') + i(xy' + x'y)$$

$$(x + iy)^{-1} = \frac{1}{x + iy} = \frac{x - iy}{x^2 + y^2} \qquad (x^2 + y^2 \neq 0)$$

The *conjugate* of a complex number $z = x + iy \in \mathbb{C}$ is defined as

$$\bar{z} = \overline{x + iy} = x - iy$$

so that $\overline{\rho e^{\alpha i}} = \rho e^{-\alpha i}$. Notice that

$$z\bar{z} = \rho^2 = x^2 + y^2 \in \mathbb{R}$$

is always a nonnegative real number. We may conclude from the foregoing discussion that $\mathbb{C}$ is a field and $\mathbb{R}$ is a subfield of $\mathbb{C}$. The quadratic equation $ax^2 + bx + c = 0$ $(a \neq 0,\ a,\ b,\ c \in \mathbb{R})$ can always be solved in $\mathbb{C}$ and, depending on the discriminant $D = b^2 - 4ac$, we have

Two distinct real roots $\Leftrightarrow D > 0$.

Two identical real roots $\Leftrightarrow D = 0$.

Pair of complex conjugate roots $\Leftrightarrow D < 0$.

This construction of the field of complex numbers presents a geometric picture which is very fruitful in linear algebra. According to it, *positive real numbers are the strains of the plane*, the negative number -1 *is the reflection in the origin*; other negative numbers are the product of a strain and the reflection in the origin. An arbitrary complex number $z \neq 0$ is then seen as the product of a strain and a rotation of the plane, in particular, the ratio of the strain is $\rho = |z|$, and the angle of the rotation, $\alpha = \arg z$. *The complex numbers of absolute value* 1 *are just the rotations of the plane.*

From the definition of the product of complex numbers follows

The Product Rule

$$|z_1 z_2| = |z_1|\,|z_2|$$

$$\arg z_1 z_2 = \arg z_1 + \arg z_2$$

The isomorphism $\mathbb{C} \rightsquigarrow \mathbb{R}^2$ identifies absolute value with length; hence we have the

Triangle Inequality

$$|z_1 + z_2| \leqq |z_1| + |z_2|$$

Moreover, in the triangle inequality the equality sign will hold if, and only if, the arguments of the complex numbers are all equal or, what is the same, the quotient of any pair of the complex numbers is a positive real number:

$$|z_1 + z_2| = |z_1| + |z_2| \Leftrightarrow \arg z_1 = \arg z_2$$

($z_\nu \neq 0$) and, in this case, the three points represented by z_1, z_2, and $z_1 + z_2$ are collinear.

One of the great classical problems of algebra was to find all the n for which the vector space $\mathbb{R}^n$ could be converted into a field, by introducing a product. The answer is $n = 1, 2$: the field of real numbers, and the field of complex numbers. However, if we drop the commutative law of multiplication, we also have $n = 4$: the quaternions; and if we drop the associative law of multiplication, we have $n = 8$: the Cayley numbers. No vector space over $\mathbb{R}$ other than $\mathbb{R}$, $\mathbb{R}^2$, $\mathbb{R}^4$, and $\mathbb{R}^8$ can be given a product having an inverse. These questions are rather deep and must remain outside the scope of this book.

APPLICATION

2. The Theorem of Ptolemy

The product of the diagonals of a quadrilateral is no greater than the sum of the products of the opposite sides. Moreover, equality holds if, and only if, the vertices of the quadrilateral fall on the same circle.

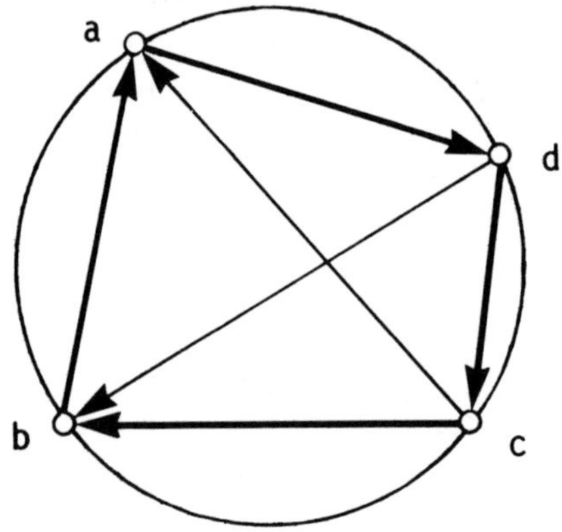

Proof: Let the vertices of the quadrilateral be denoted by a, b, c, d, which can also be interpreted as complex numbers. Then the sides of the qudrilateral are the vectors $a-b$, $b-c$, $c-d$, $d-a$; the diagonals of the quadrilateral are the vectors $a-c$, $b-d$.

A quick calculation (not reproduced here) shows that

$$(a-b)(c-d)+(a-d)(b-c)=(a-c)(b-d)$$

and from the product rule and the triangle inequality,

$$|a-b|\,|c-d|+|a-d|\,|b-c| \geqq |a-c|\,|b-d|$$

follows, proving the first part of Ptolemy's theorem. We must show now that in our relation equality holds if, and only if, a, b, c, d are the points of a circle. First we shall prove that the condition is sufficient. Accordingly, assume that a, b, c, d fall on the same circle. Then the chord $b-c$ can be seen from a at the same angle as from d, i.e.,

$$\arg\frac{a-b}{a-c}=\arg\frac{b-d}{c-d}$$

because a, d are on the Thales circle belonging to the chord $b-c$ (see Chap. 2, Application 3). Also, the chord $c-d$ can be seen from the points a and b at equal angles, i.e.,

$$\arg\frac{a-c}{a-d} = \arg\frac{b-c}{b-d}$$

Let us add these equations:

$$\arg\frac{a-b}{a-c} + \arg\frac{a-c}{a-d} = \arg\frac{b-d}{c-d} + \arg\frac{b-c}{b-d}$$

$$\Rightarrow \arg\frac{a-b}{a-c}\frac{a-c}{a-d} = \arg\frac{b-d}{c-d}\frac{b-c}{b-d} \Rightarrow \arg\frac{a-b}{a-d} = \arg\frac{b-c}{c-d}$$

$$\Rightarrow \left|\frac{a-b}{a-d} + \frac{b-c}{c-d}\right| = \left|\frac{a-b}{a-d}\right| + \left|\frac{b-c}{c-d}\right|$$

$$\Rightarrow |(a-b)(c-d) + (a-d)(b-c)| = |a-b|\,|c-d| + |a-d|\,|b-c|$$

$$\Rightarrow |a-c|\,|b-d| = |a-b|\,|c-d| + |a-d|\,|b-c|$$

This proves that the condition is sufficient. As this chain of implications can be reversed step by step, we can conclude that the condition is also necessary. This completes the proof of Ptolemy's theorem.

All this is usually formalized by the introduction of the *cross ratio* of four points (complex numbers):

$$(z_1,z_2;z_3,z_4) = \frac{z_4-z_1}{z_4-z_3} \div \frac{z_2-z_1}{z_2-z_3}$$

Then we have that the cross ratio of four points is real if, and only if, the points lie on a circle (or a straight line).

CASE OF $\mathbb{R}^3$

The classification of orthogonal operators of $\mathbb{R}^3$ is quite different from the classification of those of $\mathbb{R}^2$. Every orientation-reversing orthogonal operator in $\mathbb{R}^2$ is symmetric; every noninvolutory operator in $\mathbb{R}^2$ is orientation preserving. By contrast, in $\mathbb{R}^3$ there exist orientation-reversing noninvolutory orthogonal operators which, of course, cannot be symmetric.

Let r be an orthogonal operator of $\mathbb{R}^3$; then $r^* = r^{-1}$. Since the eigenvalues of r^* are the same as those of r (why?), and the eigenvalues of r^{-1} are just the reciprocals of those of r (why?), it follows that *an eigenvalue of r must be either* 1 *or* -1.

The orthogonal operator r is either involutory or it is noninvolutory. We shall study the case first when r is involutory: $r^* = r = r^{-1}$. Then $p = ½(1 + r)$ is a perpendicular projection: $p^2 = p = p^*$ and we have four possibilities, according as $\operatorname{rank} p = \operatorname{trace} p = 3, 2, 1$, or 0.

1. trace $r = 3 \Leftrightarrow \operatorname{rank} p = 3 \Leftrightarrow r = 1$, the identity operator.
2. trace $r = 1 \Leftrightarrow \operatorname{rank} p = 2 \Leftrightarrow$ eigenvalues $1, 1, -1$
   $\Leftrightarrow r$ is the reflection in the plane $\operatorname{im} \frac{1}{2}(1 + r)$.
3. trace $r = -1 \Leftrightarrow \operatorname{rank} p = 1 \Leftrightarrow$ eigenvalues $1, -1, -1$
   $\Leftrightarrow r$ is the reflection in the line $\operatorname{im} \frac{1}{2}(1 + r)$
   [or, what is the same, r is the rotation around the line $\operatorname{im} \frac{1}{2}(1 + r)$ through 180°].
4. trace $r = -3 \Leftrightarrow \operatorname{rank} p = 0 \Leftrightarrow r = -1$, the reflection in the origin.

Note that the involutory orthogonal operator r is orientation preserving if, and only if, trace $r = 3$ or -1; i.e., the number of eigenvalues equal to -1 is *even*.

Now we shall study the case when r is noninvolutory. Then r has a unique eigenvector belonging to the eigenvalue 1 or -1. First suppose that the unique eigenvalue of the orthogonal operator r is 1. We shall see now that in this case r is a rotation through $\alpha \neq 0$, 180°; in fact, r is orientation preserving.

In Chap. 11, Application 4, we have seen that an orientation-preserving orthogonal operator r of $\mathbb{R}^3$ determines (to within signature) an antisymmetric twist t, which has led to the standard form

$$r = 1 + t^2 + t^4 \cos \alpha + t \sin \alpha \qquad (t^* = -t = t^3)$$

From this we can see that r is a rotation around the axis $\ker t$ through the angle α. Indeed, $1 + t^2$ is the perpendicular projection onto the normal plane of the line $\ker t$, i.e., the plane $\operatorname{im} t$. The vector spanning $\ker t$ is the only eigenvector of r, belonging to the eigenvalue $\lambda = 1$.

Conversely, let a linear operator be given by $r = 1 + t^2 + t^4 \cos \alpha + t \sin \alpha$, where $t^* = -t = t^3$ and $\alpha \neq 180°$. Then

$$r^{-1} = 1 + t^2 + t^4 \cos \alpha - t \sin \alpha = r^*$$

Therefore r is a noninvolutory orientation-preserving orthogonal operator (having a unique eigenvector belonging to $\lambda = 1$) if, and only if, r is a rotation through the angle $\alpha \neq 180°$ around the line $\ker t$.

The standard form actually covers both the involutory and noninvolutory rotations. If $\alpha = 180°$, we have $r = 1 + t^2 - t^4 = 1 + 2t^2$ as a special case. Other notable special cases:

$$\alpha = 90° \Leftrightarrow r = 1 + t + t^2$$

$$\alpha = -90° \Leftrightarrow r = 1 - t + t^2$$

(orthogonal invections).

Let $q = 1 + t^2 + t^4 \cos \beta + t \sin \beta$ be another rotation around the same axis through the angle β; then

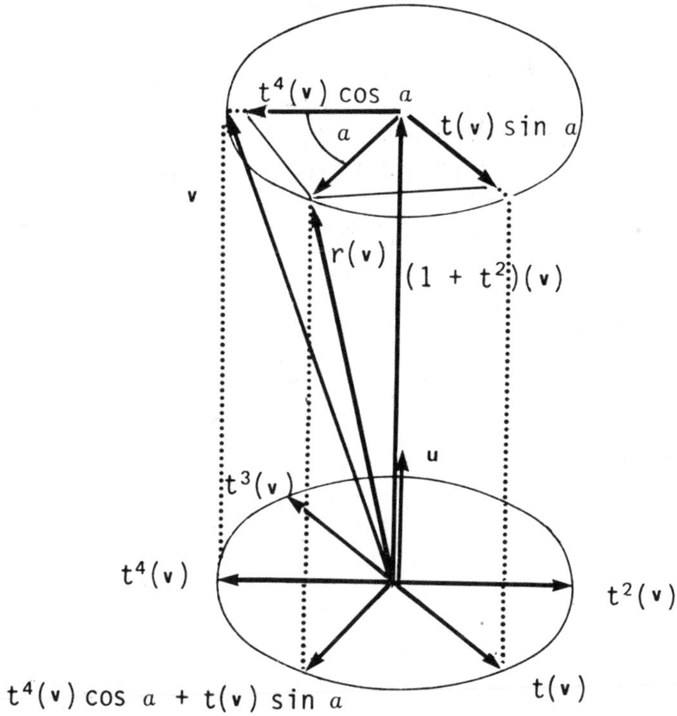

$$q \circ r = 1 + t^2 + t^4 \cos(\alpha + \beta) + t \sin(\alpha + \beta)$$

is a third rotation around the same axis, through the angle $\alpha + \beta$. Moreover,

$$r^n = 1 + t^2 + t^4 \cos n\alpha + t \sin n\alpha \qquad (n \in \mathbb{Z})$$

These formulas suggest the exponential notation for rotations:

Steenrod's Formula

$$e^{\alpha t} = 1 + t^2 + t^4 \cos \alpha + t \sin \alpha \qquad (t^* = -t = t^3)$$

This, of course, is a generalization of Euler's formula to $\mathbb{R}^3$. Indeed, $t = i \Rightarrow 1 + t^2 = 0$ and $t^4 = 1$. From Steenrod's formula it follows that

$$e^{\alpha t} \circ e^{\beta t} = e^{(\alpha + \beta)t}$$

$$(e^{\alpha t})^{-1} = e^{-\alpha t}$$

We conclude that *rotations around the same axis form a one-parameter commutative group*.

In Chap. 11 we have seen that the Cayley transform assigns the noninvolutory rotation $r = 1 + t^2 + t^4 \cos\alpha + t\sin\alpha$ to the antisymmetric operator $a = \omega t$, where $\omega = \tan \frac{1}{2}\alpha$. This assignment is clearly injective; we shall now show that it is, in fact, bijective. We have

$$(r+1)^{-1} = (e^{\alpha t}+1)^{-1} = (e^{(1/2)\alpha t} + e^{-(1/2)\alpha t})^{-1} \circ e^{-(1/2)\alpha t}$$
$$= \tfrac{1}{2}(1 + t^2 + t^4 \cos \tfrac{1}{2}\alpha)^{-1} \circ e^{-(1/2)\alpha t}$$

and

$$r - 1 = e^{\alpha t} - 1 = (e^{(1/2)\alpha t} - e^{-(1/2)\alpha t}) \circ e^{(1/2)\alpha t} = 2t \sin \tfrac{1}{2}\alpha \circ e^{(1/2)\alpha t}$$

Therefore,

$$a = (r+1)^{-1} \circ (r-1) = [1 + (1 - \cos \tfrac{1}{2}\alpha)t^2]^{-1} \circ e^{-(1/2)\alpha t} \circ t \circ e^{(1/2)\alpha t} \sin \tfrac{1}{2}$$
$$= \sin \tfrac{1}{2}\alpha \left[1 + \left(1 - \frac{1}{\cos \frac{1}{2}\alpha}\right) t^2 \right) \circ t = \sin \tfrac{1}{2}\alpha \left(- \frac{1}{\cos \frac{1}{2}\alpha} t^3 \right)$$
$$= -t^3 \tan \tfrac{1}{2}\alpha = t \tan \tfrac{1}{2}\alpha = \omega t$$

because

$$[1 + (1 - \cos \tfrac{1}{2}\alpha)t^2]^{-1} = 1 + \left(1 - \frac{1}{\cos \frac{1}{2}\alpha}\right) t^2$$

Moreover, a is antisymmetric and, clearly, the Cayley transform of a is r. We conclude that the Cayley transform is surjective and, hence, bijective.

We have seen that the following conditions on a linear operator r are equivalent in $\mathbb{R}^3$:

1. r is a noninvolutory rotation.
2. r is a nonsymmetric, orientation-preserving orthogonal operator.
3. r is orthogonal and -1 is not an eigenvalue of r.

The fourth and last type of orthogonal operators are the noninvolutory orientation-reversing orthogonal operators. They have a unique eigenvector belonging to the eigenvalue $\lambda = -1$. They can be described geometrically as a rotation $e^{\alpha t}$ through $\alpha \neq 180°$ around the axis $\ker t$ followed by reflection in the origin, i.e., $r = -e^{\alpha t}$. We call an orthogonal operator of this type a rotary reflection through α.

We have now exhausted the set of orthogonal operators in $\mathbb{R}^3$.

Classification of Orthogonal Operators in $\mathbb{R}^3$

<table>
<tr><th>Orthogonal operators</th><th>Involutory</th><th>Noninvolutory</th></tr>
<tr><td rowspan="2">Orientation-preserving</td><td>trace $r = 3 \Leftrightarrow r = 1$
the identity operator
eigenvalues 1, 1, 1</td><td rowspan="2">eigenvalue 1
$r = 1 + t^2 + t^4\cos\alpha + t\sin\alpha = e^{\alpha t}$
rotation through $\alpha \neq 180$ around ker t</td></tr>
<tr><td>trace $r = -1$
r reflection in the line im ½$(1 + r)$, i.e., rotation through 180 around same line; eigenvalues $1, -1, -1$</td></tr>
<tr><td rowspan="2">Orientation-reversing</td><td>trace $r = 1$
r reflection in the plane im½$(1 + r)$
eigenvalues $1, 1, -1$</td><td rowspan="2">eigenvalue -1
$r = -e^{\alpha t}$
rotary reflection through $\alpha \neq 180$ around ker t</td></tr>
<tr><td>trace $r = -3 \Leftrightarrow r = -1$
the reflection in the origin
eigenvalues $-1, -1, -1$</td></tr>
</table>

It is left as an exercise to show that the characteristic equation of the orthogonal operator r is

$$(r-1)\circ(r^2-2r\cos\alpha+1)=0$$

or

$$(r+1)\circ(r^2-2r\cos\alpha+1)=0$$

according as r is orientation preserving or orientation reversing.

APPLICATIONS

3. The Ergodic Theorem

For every rotation through $\alpha \neq 0$ we have (see Application 1 above)

$$1+r+r^2+\cdots+r^n=(n+1)(1+t^2)+\frac{\sin \tfrac{1}{2}(n+1)\alpha}{\sin \tfrac{1}{2}\alpha}(t^4\cos \tfrac{1}{2}n\alpha+t\sin \tfrac{1}{2}n\alpha)$$

If we divide both sides by $n + 1$ and let n tend to infinity, then we obtain the

Ergodic Theorem

$$\lim_{n\to\infty} \frac{1 + r + r^2 + \cdots + r^n}{n+1} = 1 + t^2$$

i.e., the arithmetic mean of the successive powers of a rotation converges to the perpendicular projection onto the axis of the rotation. To justify this it will suffice to observe that the length of the vector

$$\frac{\sin \tfrac{1}{2}(n+1)\alpha}{\sin \tfrac{1}{2}\alpha}(t^4 \cos \tfrac{1}{2}n\alpha + t \sin \tfrac{1}{2}n\alpha)(\mathbf{v}) \qquad \text{for any fixed } \mathbf{v}$$

is bounded so that, when divided by $n+1$, it will converge to $\mathbf{0}$ as n approaches infinity.

4. Axis and Angle of a Rotation

The ergodic theorem provides us with a method which, in principle, can be used to determine the axis of a rotation. In practice, however, we have a quicker way. $r - r^{-1} = r - r^* = 2t \sin \alpha$ is an antisymmetric operator the matrix of which is easy to get. By the principal twist theorem, $(r - r^*)(\mathbf{v}) = \mathbf{n} \times \mathbf{v}$ for all $\mathbf{v}$, where $\mathbf{n}$ spans the axis of the rotation r.

The coordinates of the vector $\mathbf{n}$ can be readily obtained from the matrix $r - r^*$. The case $\mathbf{n} = \mathbf{0}$ may occur nontrivially, when $r = r^*$, i.e., when r is the rotation through 180°. In this case the method breaks down. However, if r is an involutory rotation, $p = \tfrac{1}{2}(1 + r)$ is the perpendicular projection onto the axis of the rotation; the axis can be determined by calculating $\operatorname{im} p$.

A third method of finding the axis of a rotation r would be to calculate the unique eigenvector $\mathbf{e} \in \ker(r - 1)$ spanning the axis of r.

The formula $r - r^* = 2t \sin \alpha$ can also be used to calculate the angle of a rotation. In Chap. 13 we shall see a simpler formula:

$$\cos \alpha = \tfrac{1}{2}(\operatorname{trace} r - 1)$$

Examples

1. Show that

$$f = \begin{pmatrix} 0 & 20/25 & 15/25 \\ 20/25 & -9/25 & 12/25 \\ 15/25 & 12/25 & -16/25 \end{pmatrix}$$

is a rotation. Find the axis and the angle of the rotation.

Answer: f is an orthogonal matrix; it is also symmetric.

Therefore, f is a rotation through 180° provided that -1 is an eigenvalue with multiplicity 2. Calculating $\ker(f+1)$, we have

$$\begin{pmatrix} 25 & 20 & 15 \\ 20 & 16 & 12 \\ 15 & 12 & 9 \end{pmatrix} \Rightarrow (5 \quad 4 \quad 3)$$

which shows that all the eigenvectors belonging to -1 lie in the plane $5x + 4y + 3z = 0$. Thus the normal vector of that plane, $5\mathbf{i} + 4\mathbf{j} + 3\mathbf{k}$, is the axis of rotation, and the angle of rotation is 180°.

2. Show that

$$f = \begin{pmatrix} -39/41 & 12/41 & -4/41 \\ 4/41 & 24/41 & 33/41 \\ 12/41 & 31/41 & -24/41 \end{pmatrix}$$

is a rotation. Find the axis and the angle of rotation.

Answer: f is an orthogonal matrix; if it is a rotation, then 1 is an eigenvalue with multiplicity 1 and $\ker(f-1)$ is the axis of rotation. Calculating $\ker(f-1)$, we have

$$\begin{pmatrix} -39 & 12 & -4 \\ 4 & 24 & 33 \\ 12 & 31 & -24 \end{pmatrix} \Rightarrow \begin{pmatrix} -4 & 0 & 1 \\ -8 & 1 & 0 \end{pmatrix} \Rightarrow \begin{cases} x = t \\ y = 8t \\ z = 4t \end{cases}$$

We conclude that f is a rotation around the axis $\mathbf{i} + 8\mathbf{j} + 4\mathbf{k}$. The angle of rotation α is such that

$$\cos\alpha = \tfrac{1}{2}(\operatorname{trace} f - 1) = \tfrac{1}{2}\left(-\frac{39}{41} - 1\right) = -\frac{40}{41}$$

5. Rotations of the Cube

A beautiful application of orthogonal matrices is the problem of finding the rotations of the cube, as well as the products of these rotations. The cube has exactly 24 rotations, belonging to 13 axes.

If we have a coordinate system fitted to the cube in such a way that the coordinate axes are perpendicular to the faces of the cube, while the origin is at its geometric center, these 13 axes can be described as follows:

The three coordinate axes: the E & W, N & S, and U & L axes

The four equiangular lines: the UNE & LSW, LSE & UNW, USW & LNE, and LNW & USE equiangular lines

The six bisector axes: the NE & SW, NW & SE, UE & LW, UN & LS, UW & LE, and US & LN bisector axes

Each coordinate axis has three rotations around it: through 90°, 180°, and 270° (= –90°). This accounts for nine rotations.

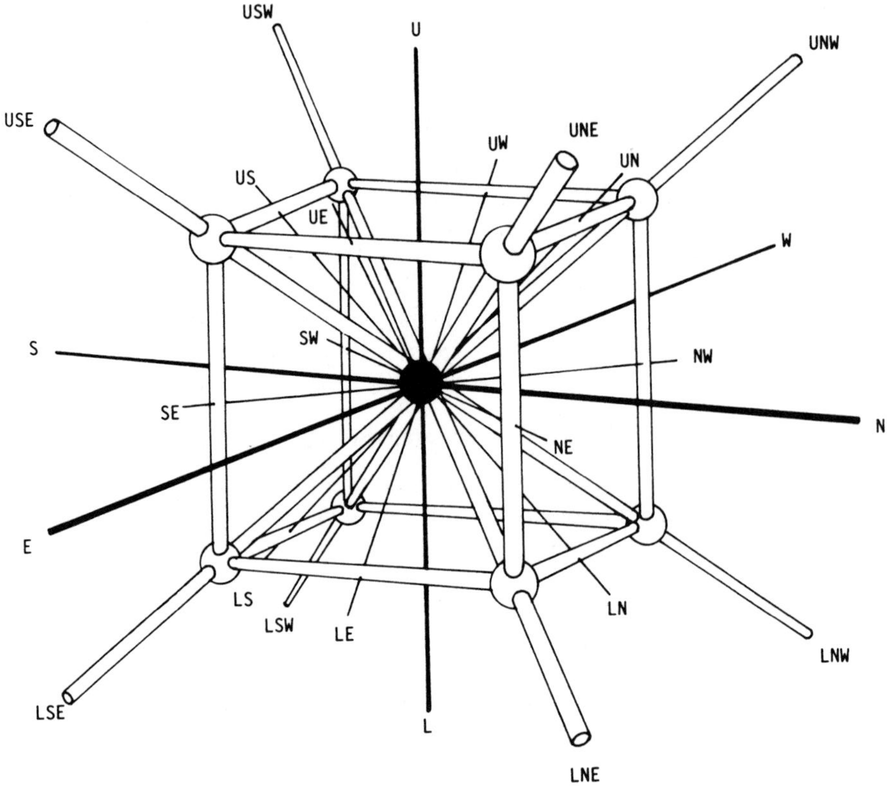

The thirteen axes of rotation of the cube

Each equiangular line has two rotations around it: through 120° and 240° (= –120°). This accounts for eight rotations.

Each bisector axis has one rotation around it: through 180°. This accounts for six rotations.

So far we have accounted for $9 + 8 + 6 = 23$ rotations. The twenty-fourth and last rotation is the identity operator. The 24 rotations of the cube form a group, as we shall see on the pages that follow.

Complete the table describing the 24 rotations of the cube

Symbol	Axis of rotation	Angle	Matrix	Symbol	Axis of rotation	Angle	Matrix
g_1	E & W coordinate axis	90°	$\begin{pmatrix} 1 & 0 & 0 \\ 0 & 0 & -1 \\ 0 & 0 & 0 \end{pmatrix}$	f_2^{-1}			$\begin{pmatrix} 0 & -1 & 0 \\ 0 & 0 & 1 \\ -1 & 0 & 0 \end{pmatrix}$
g_1^{-1}	E & W coordinate axis	270°	$\begin{pmatrix} 1 & 0 & 0 \\ 0 & 0 & 1 \\ 0 & -1 & 0 \end{pmatrix}$	f_3		120°	$\begin{pmatrix} 0 & 0 & -1 \\ 1 & 0 & 0 \\ 0 & -1 & 0 \end{pmatrix}$
k_1	E & W coordinate axis	180°	$\begin{pmatrix} 1 & 0 & 0 \\ 0 & -1 & 0 \\ 0 & 0 & -1 \end{pmatrix}$	f_3^{-1}			$\begin{pmatrix} 0 & 1 & 0 \\ 0 & 0 & -1 \\ -1 & 0 & 0 \end{pmatrix}$
g_2		90°	$\begin{pmatrix} 0 & 0 & -1 \\ 0 & 1 & 0 \\ 1 & 0 & 0 \end{pmatrix}$	f_4	UNE & LSW equiangular line	120°	$\begin{pmatrix} 0 & 0 & 1 \\ 1 & 0 & 0 \\ 0 & 1 & 0 \end{pmatrix}$
g_2^{-1}			$\begin{pmatrix} 0 & 0 & 1 \\ 0 & 1 & 0 \\ -1 & 0 & 0 \end{pmatrix}$	f_4^{-1}			$\begin{pmatrix} 0 & 1 & 0 \\ 0 & 0 & 1 \\ 1 & 0 & 0 \end{pmatrix}$
k_2		180°	$\begin{pmatrix} -1 & 0 & 0 \\ 0 & 1 & 0 \\ 0 & 0 & -1 \end{pmatrix}$	p_1		180°	$\begin{pmatrix} -1 & 0 & 0 \\ 0 & 0 & -1 \\ 0 & -1 & 0 \end{pmatrix}$

(continued)

Symbol	Axis of rotation	Angle	Matrix	Symbol	Axis of rotation	Angle	Matrix
g_3		90°	$\begin{pmatrix} 0 & -1 & 0 \\ 1 & 0 & 0 \\ 0 & 0 & 1 \end{pmatrix}$	q_1		180°	$\begin{pmatrix} -1 & 0 & 0 \\ 0 & 0 & 1 \\ 0 & 1 & 0 \end{pmatrix}$
g_3^{-1}			$\begin{pmatrix} 0 & 1 & 0 \\ -1 & 0 & 0 \\ 0 & 0 & 1 \end{pmatrix}$	p_2		180°	$\begin{pmatrix} 0 & 0 & 1 \\ 0 & -1 & 0 \\ 1 & 0 & 0 \end{pmatrix}$
k_3		180°	$\begin{pmatrix} -1 & 0 & 0 \\ 0 & -1 & 0 \\ 0 & 0 & 1 \end{pmatrix}$	q_2		180°	$\begin{pmatrix} 0 & 0 & -1 \\ 0 & -1 & 0 \\ -1 & 0 & 0 \end{pmatrix}$
f_1		120°	$\begin{pmatrix} 0 & 0 & 1 \\ -1 & 0 & 0 \\ 0 & -1 & 0 \end{pmatrix}$	p_3		180°	$\begin{pmatrix} 0 & -1 & 0 \\ -1 & 0 & 0 \\ 0 & 0 & -1 \end{pmatrix}$
f_1^{-1}			$\begin{pmatrix} 0 & -1 & 0 \\ 0 & 0 & -1 \\ 1 & 0 & 0 \end{pmatrix}$	q_3		180°	$\begin{pmatrix} 0 & 1 & 0 \\ 1 & 0 & 0 \\ 0 & 0 & -1 \end{pmatrix}$
f_2		120°	$\begin{pmatrix} 0 & 0 & -1 \\ -1 & 0 & 0 \\ 0 & 1 & 0 \end{pmatrix}$	1	the identity	0°	$\begin{pmatrix} 1 & 0 & 0 \\ 0 & 1 & 0 \\ 0 & 0 & 1 \end{pmatrix}$

Complete the table listing all the products of the rotations of the cube .

	1	k_1	k_2	k_3	f_1	f_2	f_3	f_4	f_1^{-1}	f_2^{-1}	f_3^{-1}	f_4^{-1}	g_1	g_1^{-1}	p_1	q_1	g_2	g_2^{-1}	p_2	q_2	g_3	g_3^{-1}	p_3	q_3
1																								
k_1																								
k_2								f_3																
k_3																								
f_1																								
f_2																								
f_3																								
f_4																								
f_1^{-1}																								
f_2^{-1}																								
f_3^{-1}																								
f_4^{-1}																								
g_1																								
g_1^{-1}																								
p_1																								
q_1																								
g_2																								
g_2^{-1}																								
p_2																								
q_2																								
g_3																								
g_3^{-1}																								
p_3																								
q_3																								

For example,

$$k_2 \circ f_4 = \begin{pmatrix} -1 & 0 & 0 \\ 0 & 1 & 0 \\ 0 & 0 & -1 \end{pmatrix} \circ \begin{pmatrix} 0 & 0 & 1 \\ 1 & 0 & 0 \\ 0 & 1 & 0 \end{pmatrix} = \begin{pmatrix} 0 & 0 & -1 \\ 1 & 0 & 0 \\ 0 & -1 & 0 \end{pmatrix} = f_3$$

6. Symmetries of the Cube

The 24 rotations of the cube are only the orientation-preserving symmetries of the cube; in addition, there are also the orientation-reversing symmetries. The number of these is also 24, so that the total number of the symmetries of the cube is 48. The orientation-reversing symmetries of the cube we can get from the orientation-preserving ones simply by attaching the negative signature.

The 48 symmetries of the cube, just as the 24 rotations, form a group. These is no point in compiling a multiplication table for the 48 symmetries if we already have one for the 24 rotations of the cube. In the larger group, the arithmetic rules governing the multiplication of negative numbers will apply. That is, the product of two orientation-reversing symmetries is an orientation-preserving symmetry, and the product of an orientation-preserving symmetry with an orientation-reversing one is orientation-reversing.

In more detail, the cube has nine planes of symmetry, as follows:

The three coordinate planes: the x, y, and z planes

The six bisector planes: the NE & SW, NW & SE, UE & LW, UW & LE, UN & LS, and US & LN bisector planes.

These planes account for 15 symmetries of the cube, namely, nine reflections in planes: $-k_1, -k_2, -k_3, -p_1, -q_1, -p_2, -q_2, -p_3, -q_3$, and the six rotary reflections through 90° or –90° around the coordinate axes: $-g_1, -g_1^{-1}, -g_2, -g_2^{-1}, -g_3, -g_3^{-1}$. Next, there are the eight rotary reflections through 120° and –120° around the four equiangular lines: $-f_1, -f_1^{-1}, -f_2, -f_2^{-1}, -f_3, -f_3^{-1}, -f_4, -f_4^{-1}$. Finally, there is the reflection in the origin, –1, accounting for the $15 + 8 + 1 = 24$ additional symmetries. (See table on p.303.)

7. Decomposition of an Orthogonal Operator into the Product of Reflections

We shall now show that every orthogonal operator r of $\mathbb{R}^3$ can be decomposed into the product of at most three reflections. If r is a reflection, there is nothing to prove. We may therefore assume that r is elliptic; it may be orientation preserving or orientation reversing. In the first case we have

$$r = e^{\phi t} = 1 + t^2 + t^4 \cos \phi + t \sin \phi \qquad (t^* = -t = t^3)$$

Complete the table describing the 24 orientation-reversing symmetries of the cube.

Symbol	Description of symmetry	Matrix	Symbol	Description of symmetry	Matrix
$-g_1$	Rotary reflection through 90° around the x-axis	$\begin{pmatrix} -1 & 0 & 0 \\ 0 & 0 & 1 \\ 0 & -1 & 0 \end{pmatrix}$	$-f_2^{-1}$		
$-g_1^{-1}$			$-f_3$		
$-k_1$	Reflection in yz plane	$\begin{pmatrix} -1 & 0 & 0 \\ 0 & 1 & 0 \\ 0 & 0 & 1 \end{pmatrix}$	$-f_3^{-1}$		
$-g_2$			$-f_4$	Rotary reflection through 120° around UNE & LSW equiangular line	$\begin{pmatrix} 0 & 0 & -1 \\ -1 & 0 & 0 \\ 0 & -1 & 0 \end{pmatrix}$
$-g_2^{-1}$			$-f_4^{-1}$		
$-k_2$			$-p_1$	Reflection in UN & LS bisector plane	$\begin{pmatrix} 1 & 0 & 0 \\ 0 & 0 & 1 \\ 0 & 1 & 0 \end{pmatrix}$

(continued)

Symbol	Description of symmetry	Matrix	Symbol	Description of symmetry	Matrix
$-g_3$			$-q_1$	Reflection in US & LN bisector plane	$\begin{pmatrix} 1 & 0 & 0 \\ 0 & 0 & 1 \\ 0 & 1 & 0 \end{pmatrix}$
$-g_3^{-1}$			$-p_2$		
$-k_3$			$-q_2$		
$-f_4$			$-p_3$		
$-f_4^{-1}$			$-q_3$		
$-f_2$			-1	Reflection in the origin	$\begin{pmatrix} -1 & 0 & 0 \\ 0 & -1 & 0 \\ 0 & 0 & -1 \end{pmatrix}$

Let $t = \ell - \ell^*$ be the decomposition of the antisymmetric twist t into the difference of a pair of perpendicular lifts (see Exercises 11.16 and 11.18). Then $\ell^2 = 0 = \ell^{*2}$, $(\ell \circ \ell^*)^2 = \ell \circ \ell^*$, $(\ell^* \circ \ell)^2 = \ell^* \circ \ell$. Set

$$h = 1 - \ell \circ \ell^* - \ell^* \circ \ell + (\ell \circ \ell^* - \ell^* \circ \ell)\cos \tfrac{1}{2}\phi + (\ell + \ell^*)\sin \tfrac{1}{2}\phi$$
$$k = 1 - \ell \circ \ell^* - \ell^* \circ \ell + (\ell \circ \ell^* - \ell^* \circ \ell)\cos \tfrac{1}{2}\phi - (\ell + \ell^*)\sin \tfrac{1}{2}\phi$$

Clearly, $h^* = h$, $k^* = k$. A simple calculation (not reproduced here) shows that $h^2 = 1 = k^2$ and $k \circ h = r$, $h \circ k = r^{-1} = r^*$. In particular, r is the product of two reflections h, k.

Finally, if r is orientation reversing, then $r = -e^{\alpha t} = (-1) \circ k \circ h$, r is the product of three reflections.

Compare this decomposition with that of 10.29.

CASE OF $\mathbb{R}^n$

It is not the task of this book to study the classification of orthogonal operators in $\mathbb{R}^n$ for $n \geqq 4$. It may, however, be helpful for the understanding of $\mathbb{R}^2$ and $\mathbb{R}^3$ if we mention in passing the highlights of the general case.

In $\mathbb{R}^n$, as in the lower-dimensional cases, an orthogonal operator r can have eigenvalues 1 and -1 only. r is orientation preserving or reversing according as the number of eigenvalues equal to -1 (counted with the appropriate multiplicity) is even or odd.

The simplest orientation-preserving orthogonal operators are the *rotations* defined via Steenrod's formula,

$$e^{\phi t} = 1 + t^2 + t^4 \cos \phi + t \sin \phi$$

where t is an antisymmetric twist, i.e., $t^* = -t = t^3$, and where ϕ is called the *angle* of rotation. We define the *degree* of the rotation by the formula

$$\deg e^{\phi t} = \operatorname{rank} t$$

and the *axis* of rotation by

$$\operatorname{ax} e^{\phi t} = \ker t$$

The degree of rotation is necessarily an even number and, by the law of nullity, the dimension of the axis is

$$n - \deg e^{\phi t}$$

It is clear that rotations about the same axis form a one-parameter commutative group, where the angle ϕ is the parameter. The rotation $e^{\phi t}$ is involutory if, and only if, $\phi = 180°$ in which case there are n real eigenvalues, namely, an even number, $\deg e^{\phi t}$ of them equal to -1, and the rest, $n - \deg e^{\phi t}$ of them equal to 1. Let t, t' be two antisymmetric twists; we shall say that they are *perpendicular* to one another if $t \circ t' = 0 = t' \circ t$. If this is the case, then the rotations $e^{\phi t}$, $e^{\phi' t'}$ are also said to be perpendicular. It is clear that perpendicular rotations commute:

$$e^{\phi t} \circ e^{\phi' t'} = e^{\phi' t'} \circ e^{\phi t}$$

and that the product is an orientation-preserving orthogonal operator. Conversely, every orientation-preserving orthogonal operator can be decomposed into the product of pairwise perpendicular and commuting rotations (at most one of which may be involutory):

$$r = e^{\phi_1 t_1} \circ \cdots \circ e^{\phi_\nu t_\nu}$$

where the angles $\phi_1, \ldots, \phi_\nu$ are distinct. This decomposition is, moreover, unique. The degree of r is the sum of degrees of the factors.

A *reflection* (or symmetric involution) is defined by the formula

$$r = 1 - 2p$$

where p is a perpendicular projection, i.e., $p^* = p = p^2$, and consequently r satisfies $r^* = r = r^{-1}$. We shall say that r is a reflection in $\ker p$. r is said to be a *principal reflection* if $\operatorname{rank} p = 1$. A principal reflection can also be described as a reflection in the hyperplane $\ker p$ of $\mathbb{R}^n$. A principal reflection is necessarily orientation reversing: it has exactly one eigenvalue equal to -1, and $n-1$ eigenvalues equal to 1. It follows that r is a principal reflection only if $\operatorname{trace} r = n - 2$.

Let p, p' be two perpendicular projections; we shall say the p, p' are *perpendicular* to one another if $p \circ p' = 0 = p' \circ p$. If this is the case, the reflections $u = 1 - 2p$, $u' = 1 - 2p'$ are also said to be perpendicular and they will in fact commute:

$$u \circ u' = 1 - 2p - 2p' = u' \circ u$$

From Steenrod's formula it is clear that the orientation-preserving reflections are precisely the rotations through 180°, because $\phi = 180°$ yields $r = 1 + t^2 - t^4 = 1 - 2t^4 = 1 - 2p$, with $p = t^4$ satisfying $p^* = p = p^2$.

On the other hand, an orientation-reversing reflection is the product of a principal reflection and a perpendicular and commuting rotation through 180°. This decomposition, however, is not unique. The positive integer

$$\deg r = \tfrac{1}{2}(n - \operatorname{trace} r)$$

is called the *degree* of the reflection r; it is the same as the number of eigenvalues of r equal to -1. We have $1 \leqq \deg r \leqq n$, and

$$\deg r = 1 \Leftrightarrow r \text{ is a principal reflection}$$

$$\deg r = n \Leftrightarrow r = -1 \text{ the reflection in the origin}$$

Moreover, the reflection r is orientation preserving if, and only if, $\deg r$ is even. Thus the reflection in the origin is orientation preserving in an even-dimensional vector space, whereas it is orientation reversing in an odd-dimensional vector space.

We say that the principal perpendicular projection p and the antisymmetric twist t are *perpendicular* to one another if $p \circ t = 0 = t \circ p$. In this case the principal reflection $u = 1 - 2p$ and the rotation $e^{\phi t}$ are also said to be perpendicular, and they will commute:

$$u \circ e^{\phi t} = e^{\phi t} \circ u$$

Every orientation-reversing orthogonal operator r can be decomposed into the product of pairwise perpendicular and commuting rotations (one of which may be involutory) and a principal reflection which is perpendicular to and commutes with each rotation. This decomposition is not necessarily unique, but the sum of the degrees of the factors, called the degree of r, is unique. The degree of an orientation-reversing orthogonal operator is odd.

Classification of Orthogonal Operators in $\mathbb{R}^n$ $(n \geq 4)$

Orthogonal operators	Involutory	Noninvolutory
Orientation-preserving (even degree)	Product of an even number of perpendicular and commuting principal reflections: Reflections of even degree	Rotations: $e^{\phi t} = 1 + t^2 + t^4 \cos\phi + t \sin\phi$, $t^* = -t = t^3$; $\phi \neq 180°$
		Product of perpendicular and commuting rotations through distinct angles
Orientation-reversing (odd degree)	Product of an odd number of perpendicular and commuting principal reflections: Reflections of odd degree	Product of an orientation-preserving orthogonal operator and a perpendicular and commuting principal reflection

In $\mathbb{R}^n$ a rotation r can be decomposed into the product of $\deg r$ principal reflections (proof along the lines of Application 7 above). This gives us the result that every orthogonal operator $r \neq 1$ of $\mathbb{R}^n$ can be decomposed into the product of exactly $\deg r$ principal reflections, $1 \leqq \deg r \leqq n$. This decomposition is not necessarily unique.

EXERCISES

12.1 In each of the following, show that r is a rotation, and find its axis and angle.

(i) $\begin{pmatrix} 2/3 & 1/3 & -2/3 \\ 1/3 & 2/3 & 2/3 \\ 2/3 & -2/3 & 1/3 \end{pmatrix}$ (ii) $\begin{pmatrix} 2/7 & -3/7 & 6/7 \\ 3/7 & 6/7 & 2/7 \\ -6/7 & 2/7 & 3/7 \end{pmatrix}$

(iii) $\begin{pmatrix} 4/9 & -8/9 & 1/9 \\ 4/9 & 1/9 & -8/9 \\ 7/9 & 4/9 & 4/9 \end{pmatrix}$ (iv) $\begin{pmatrix} 2/11 & -9/11 & 6/11 \\ 6/11 & 6/11 & 7/11 \\ -9/11 & 2/11 & 6/11 \end{pmatrix}$

(v) $\begin{pmatrix} 4/13 & -3/13 & -12/13 \\ 12/13 & 4/13 & 3/13 \\ 3/13 & -12/13 & 4/13 \end{pmatrix}$ (vi) $\begin{pmatrix} 5/15 & 10/15 & 10/15 \\ -14/15 & 5/15 & 2/15 \\ -2/15 & -10/15 & 11/15 \end{pmatrix}$

(vii) $\begin{pmatrix} -9/17 & 8/17 & 12/17 \\ 8/17 & -9/17 & 12/17 \\ 12/17 & 12/17 & 1/17 \end{pmatrix}$ (viii) $\begin{pmatrix} 1/19 & 18/19 & 6/19 \\ 6/19 & -6/19 & 17/19 \\ 18/19 & 1/19 & -6/19 \end{pmatrix}$

(ix) $\begin{pmatrix} 4/21 & 8/21 & -19/21 \\ 16/21 & 11/21 & 8/21 \\ 13/21 & -16/21 & -4/21 \end{pmatrix}$ (x) $\begin{pmatrix} 5/21 & 4/21 & 20/21 \\ 20/21 & -5/21 & -4/21 \\ 4/21 & 20/21 & -5/21 \end{pmatrix}$

(xi) $\begin{pmatrix} 3/23 & 6/23 & 22/23 \\ 18/23 & 13/23 & -6/23 \\ -14/23 & 18/23 & -3/23 \end{pmatrix}$ (xii) $\begin{pmatrix} 10/27 & -10/27 & 23/27 \\ 2/27 & 25/27 & 10/27 \\ -25/27 & -2/27 & 10/27 \end{pmatrix}$

(xiii) $\begin{pmatrix} -21/29 & 12/29 & 16/29 \\ 12/29 & -11/29 & 24/29 \\ 16/29 & 24/29 & 3/29 \end{pmatrix}$ (xiv) $\begin{pmatrix} 30/31 & -6/31 & -5/31 \\ 5/31 & 30/31 & -6/31 \\ 6/31 & 5/31 & 30/31 \end{pmatrix}$

(xv) $\begin{pmatrix} 6/31 & 14/31 & 27/31 \\ 22/31 & -21/31 & 6/31 \\ 21/31 & 18/31 & -14/31 \end{pmatrix}$ (xvi) $\begin{pmatrix} 8/33 & 8/33 & 31/33 \\ 1/33 & -32/33 & 8/33 \\ 32/33 & -1/33 & -8/33 \end{pmatrix}$

(xvii) $\begin{pmatrix} -17/33 & 28/33 & 4/33 \\ 28/33 & 16/33 & 7/33 \\ 4/33 & 7/33 & -32/33 \end{pmatrix}$ (xviii) $\begin{pmatrix} -17/33 & -20/33 & 20/33 \\ -20/33 & -8/33 & -25/33 \\ 20/33 & -25/33 & 8/33 \end{pmatrix}$

(xix) $\begin{pmatrix} 30/35 & -17/35 & 6/35 \\ 15/35 & 30/35 & 10/35 \\ -10/35 & -6/35 & 33/35 \end{pmatrix}$ (xx) $\begin{pmatrix} 3/37 & 24/37 & 28/37 \\ 8/37 & 27/37 & -24/37 \\ -36/37 & 8/37 & -3/37 \end{pmatrix}$

(xxi) $\begin{pmatrix} 2/39 & 29/39 & -26/39 \\ 19/39 & 22/39 & 26/39 \\ 34/39 & -14/39 & -13/39 \end{pmatrix}$ (xxii) $\begin{pmatrix} 7/43 & -42/43 & -6/43 \\ 6/43 & 7/43 & -42/43 \\ 42/43 & 6/43 & 7/43 \end{pmatrix}$

(xxiii) $\begin{pmatrix} -7/43 & -30/43 & -30/43 \\ -30/43 & -18/43 & 25/43 \\ -30/43 & 25/43 & -18/43 \end{pmatrix}$ (xxiv) $\begin{pmatrix} 4/49 & -36/49 & 33/49 \\ 48/49 & 9/49 & 4/49 \\ -9/49 & 32/49 & 36/49 \end{pmatrix}$

(xxv) $\begin{pmatrix} 1/51 & 10/51 & 50/51 \\ 50/51 & -10/51 & 1/51 \\ 10/51 & 49/51 & -10/51 \end{pmatrix}$ (xxvi) $\begin{pmatrix} 1/51 & -22/51 & 46/51 \\ 38/51 & 31/51 & 14/51 \\ -34/51 & 34/51 & 17/51 \end{pmatrix}$

(xxvii) $\begin{pmatrix} -7/57 & -40/57 & -40/57 \\ -40/57 & -25/57 & 32/57 \\ -40/57 & 32/57 & -25/57 \end{pmatrix}$ (xxviii) $\begin{pmatrix} 9/59 & 30/59 & 50/59 \\ 50/59 & -30/59 & 9/59 \\ 30/59 & 41/59 & -30/59 \end{pmatrix}$

(xxix) $\begin{pmatrix} 39/65 & 0 & 52/65 \\ 20/65 & 60/65 & -15/65 \\ -48/65 & 25/65 & 36/65 \end{pmatrix}$ (xxx) $\begin{pmatrix} -21/79 & -30/79 & -70/79 \\ -70/79 & -21/79 & 30/79 \\ -30/79 & 70/79 & -21/79 \end{pmatrix}$

(xxxi) $\begin{pmatrix} 1/81 & 44/81 & 68/81 \\ 76/81 & 23/81 & -16/81 \\ -28/81 & 64/81 & -41/81 \end{pmatrix}$ (xxxii) $\begin{pmatrix} 16/81 & -47/81 & 64/81 \\ 79/81 & 16/81 & -8/81 \\ -8/81 & 64/81 & 49/81 \end{pmatrix}$

(xxxiii) $\begin{pmatrix} -1/99 & -70/99 & -70/99 \\ 70/99 & 49/99 & -50/99 \\ 70/99 & -50/99 & 49/99 \end{pmatrix}$ (xxxiv) $\begin{pmatrix} -10/111 & -11/111 & -110/111 \\ -11/111 & 110/111 & -10/111 \\ -110/111 & -10/111 & 11/111 \end{pmatrix}$

(xxxv) $\begin{pmatrix} 0 & 0.6 & -0.8 \\ 0.8 & 0.48 & 0.36 \\ 0.6 & -0.64 & -0.48 \end{pmatrix}$ (xxxvi) $\begin{pmatrix} 0.6 & 0.48 & 0.64 \\ 0.64 & -0.768 & -0.024 \\ 0.48 & 0.424 & -0.768 \end{pmatrix}$

(xxxvii) $\begin{pmatrix} 0.744 & 0.64 & 0.192 \\ -0.64 & 0.6 & 0.48 \\ 0.192 & -0.48 & 0.856 \end{pmatrix}$ (xxxviii) $\begin{pmatrix} -0.352 & -0.36 & -0.864 \\ -0.36 & -0.8 & 0.48 \\ -0.864 & 0.48 & 0.152 \end{pmatrix}$

12.2 Find the geometric meaning of the orthogonal operator

$$f = \begin{pmatrix} \cos\phi & \sin\phi \\ \sin\phi & -\cos\phi \end{pmatrix}$$

Find a new orthonormal basis in which the matrix of f assumes its simplest form. Write the transition equation.

12.3 Show that the rotation through 90° of the plane can be decomposed into the product of reflections in lines making an angle 45°. More generally, show that a rotation through ϕ of the plane can be decomposed into the product of reflections in lines making an angle $\frac{1}{2}\phi$.

12.4 Complete the table giving all the products of the 24 rotations of the cube (see Application 5 above).

12.5 Decompose the rotation

$$\begin{pmatrix} 15/25 & 20/25 & 0 \\ -12/25 & 9/25 & 20/25 \\ 16/25 & -12/25 & 15/25 \end{pmatrix}$$

into the product of reflections in two planes; show that the angle of rotation is twice the angle made by the planes.

12.6 Show that the characteristic equation of an orthogonal operator in $\mathbb{R}^3$ is

$$(r-1)\circ(r^2-2r\cos\alpha+1)=0 \qquad \text{or} \qquad (r+1)\circ(r^2-2r\cos\alpha+1)=0$$

according as r is orientation preserving or reversing.

12.7 Let the axis of the rotation r in $\mathbb{R}^3$ be spanned by the unit vector $\mathbf{u}$. Show that then

$$r(\mathbf{v}) = (\cos\alpha)\,\mathbf{v} + (1-\cos\alpha)(\mathbf{u}\cdot\mathbf{v})\mathbf{u} + \sin\alpha(\mathbf{u}\times\mathbf{v})$$

for all $\mathbf{v}\in\mathbb{R}^3$, where α is the angle of rotation.

12.8 Let r be a rotation of $\mathbb{R}^3$. Show that the antisymmetric operators

$$a=(r+1)^{-1}\circ(r-1) \qquad a' = \tfrac{1}{2}(r-r^*)$$

are related by the equation $a'=(1+\cos\alpha)a$, where α is the angle of rotation.

12.9 Show that the following conditions on the rotations $e^{\alpha t}$; $e^{\beta t'}$ are equivalent:

(1) $e^{\alpha t}\circ e^{\beta t'} = e^{\beta t'}\circ e^{\alpha t}$.

(2) $[t,t']=0$.

(3) $t\circ t'$ is a symmetric operator.

12.10 Check the ergodic formula through the example of the rotation

$$r=\begin{pmatrix} 2/3 & -2/3 & -1/3 \\ 1/3 & 2/3 & -2/3 \\ 2/3 & 1/3 & 2/3 \end{pmatrix}$$

13

The Determinant and the Trace of a Linear Operator†

DETERMINANT OF A LINEAR OPERATOR

Informally, we have been using the determinant of a linear operator all along. Let us now formalize this important idea. We begin with the

Star-Box Formula

$$[f(\mathbf{a}), f(\mathbf{b}), f(\mathbf{c})]\,[\mathbf{a}'\mathbf{b}'\mathbf{c}'] = [\mathbf{abc}]\,[f^*(\mathbf{a}'), f^*(\mathbf{b}'), f^*(\mathbf{c}')]$$

for all linear operators f and vectors $\mathbf{a}$, $\mathbf{b}$, $\mathbf{c}$, $\mathbf{a}'$, $\mathbf{b}'$, $\mathbf{c}'$. It reminds one of the star-dot formula. The proof is based on just that, plus the double box formula of Chapter 4:

$$[f(\mathbf{a}), f(\mathbf{b}), f(\mathbf{c})]\,[\mathbf{a}'\mathbf{b}'\mathbf{c}'] = \begin{vmatrix} f(\mathbf{a})\cdot\mathbf{a}' & f(\mathbf{b})\cdot\mathbf{a}' & f(\mathbf{c})\cdot\mathbf{a}' \\ f(\mathbf{a})\cdot\mathbf{b}' & f(\mathbf{b})\cdot\mathbf{b}' & f(\mathbf{c})\cdot\mathbf{b}' \\ f(\mathbf{a})\cdot\mathbf{c}' & f(\mathbf{b})\cdot\mathbf{c}' & f(\mathbf{c})\cdot\mathbf{c}' \end{vmatrix}$$

$$= \begin{vmatrix} \mathbf{a}\cdot f^*(\mathbf{a}') & \mathbf{b}\cdot f^*(\mathbf{a}') & \mathbf{c}\cdot f^*(\mathbf{a}') \\ \mathbf{a}\cdot f^*(b') & \mathbf{b}\cdot f^*(\mathbf{b}') & \mathbf{c}\cdot f^*(\mathbf{b}') \\ \mathbf{a}\cdot f^*(c') & \mathbf{b}\cdot f^*(\mathbf{c}') & \mathbf{c}\cdot f^*(\mathbf{c}') \end{vmatrix}$$

$$= [\mathbf{abc}]\,[f^*(\mathbf{a}'), f^*(\mathbf{b}'), f^*(\mathbf{c}')]$$

Notice that $[f^*(\mathbf{a}'), f^*(\mathbf{b}'), f^*(\mathbf{c}')] = [f(\mathbf{a}'), f(\mathbf{b}'), f(\mathbf{c}')]$ on the strength of the reflection property of determinants, so that the stars may simply be omitted from the star-box formula. Moreover, if we assume that $\{\mathbf{a},\mathbf{b},\mathbf{c}\}$ and $\{\mathbf{a}',\mathbf{b}',\mathbf{c}'\}$ are linearly independent sets, then $[\mathbf{abc}] \neq 0$, $[\mathbf{a}'\mathbf{b}'\mathbf{c}'] \neq 0$ and

†*Note*: The bases in this chapter are not necessarily orthonormal, and hence the coordinates are skew affine, except where stipulated otherwise.

we may divide by their product to get the

Box Invariance Formula

$$\frac{[f(\mathbf{a}'),f(\mathbf{b}'),f(\mathbf{c}')]}{[\mathbf{a}'\mathbf{b}'\mathbf{c}']}=\frac{[f(\mathbf{a}),f(\mathbf{b}),f(\mathbf{c})]}{[\mathbf{abc}]}$$

It asserts that a linear operator f changes the volume of parallelepipeds (and, hence, all volumes) in the same way: It multiplies volume by a factor of proportionality that depends only on f. The box invariance formula gives us an invariant of linear operators, calling for a definition:

The *determinant* of the linear operator f, in symbols $\det f$, is defined to be the scalar such that, for all vectors $\mathbf{a}$, $\mathbf{b}$, $\mathbf{c}$

$$[f(\mathbf{a}),f(\mathbf{b}),f(\mathbf{c})]=\det f[\mathbf{abc}]$$

It is the contents of the box invariance formula that such a scalar does, in fact, exist. If the matrix of f is

$$f=\begin{pmatrix} a_{11} & a_{12} & a_{13} \\ a_{21} & a_{22} & a_{23} \\ a_{31} & a_{32} & a_{33} \end{pmatrix}$$

then we have that

$$\det f=\begin{vmatrix} a_{11} & a_{12} & a_{13} \\ a_{21} & a_{22} & a_{23} \\ a_{31} & a_{32} & a_{33} \end{vmatrix}$$

but it is clear that, although the matrix of f generally changes when the basis is changed, $\det f$ remains unchanged. If we use rectangular metric coordinates, i.e., if $\{\mathbf{i},\mathbf{j},\mathbf{k}\}$ is an orthonormal basis, then

$$\det f=[f(\mathbf{i}),f(\mathbf{j}),f(\mathbf{k})]$$

because $[\mathbf{i},\mathbf{j},\mathbf{k}]=1$ is the volume of the unit cube. Thus $\det f$ can be described geometrically as the signed volume of the parallelepiped into which f deforms the unit cube.

We note that if f has three distinct eigenvalues λ_1, λ_2, λ_3, then

$$\det f=\lambda_1\lambda_2\lambda_3$$

FORMAL PROPERTIES OF THE DETERMINANT

Product Rule

$$\det(g\circ f)=(\det g)(\det f)$$

$$\det 1=1$$

$$\det f^{-1} = (\det f)^{-1} \qquad \det f \neq 0$$
$$\det f^{*} = \det f$$

Proof:

$$\begin{aligned}\det(g\circ f)[\mathbf{abc}] &= [(g\circ f)(\mathbf{a}),(g\circ f)(\mathbf{b}),(g\circ f)(\mathbf{c})]\\ &= [g(f(\mathbf{a})),g(f(\mathbf{b})),g(f(\mathbf{c}))]\\ &= (\det g)[f(\mathbf{a}),f(\mathbf{b}),f(\mathbf{c})] = (\det g)(\det f)[\mathbf{abc}]\end{aligned}$$

for all $\mathbf{a}$, $\mathbf{b}$, $\mathbf{c} \in \mathbb{R}^3$, and the product rule follows.

The formula for $\det f^{-1}$ is a special case of this with $g = f^{-1}$. Another consequence of the product rule is

$$\det(g\circ f\circ g^{-1}) = \det f$$

i.e., the determinant of f remains unchanged under a change of coordinates. Furthermore,

$$\det(f\circ g\circ f^{-1}\circ g^{-1}) = 1$$

i.e., the determinant of the commutator $f\circ g\circ f^{-1}\circ g^{-1}$ of the operators f, g is 1.

In order to extend these results to $\mathbb{R}^n$, we may start by defining the *box product of n vectors* $\mathbf{a}_1, \ldots, \mathbf{a}_n \in \mathbb{R}^n$

$$[\mathbf{a}_1,\ldots,\mathbf{a}_n] \in \mathbb{R}$$

as follows. $[\mathbf{a}_1,\ldots,\mathbf{a}_n]$ is a skew-symmetric function of n vector variables, which is linear in each of the variables. Skew symmetry means that

$$[\mathbf{a}_{\pi(1)},\ldots,\mathbf{a}_{\pi(n)}] = (\operatorname{sgn}\pi)\,[\mathbf{a}_1,\ldots,\mathbf{a}_n]$$

where π is a permutation of the set $\{1,2,\ldots,n\}$ and $\operatorname{sgn}\pi$ is the signature of the permutation π. In other words,

$$\operatorname{sgn}\pi = (-1)^{\operatorname{par}\pi}$$

where $\operatorname{par}\pi$ is the parity of the permutation π; i.e., $\operatorname{par}\pi$ is even or odd according as the permutation π is even or odd. Thus the signature of even permutations, including 1, is $+1$, and the signature of odd permutations is -1.

It can be shown that a nontrivial box product in $\mathbb{R}^n$ exists, and any two such box products $[\ ,\ldots,\ ]_1, [\ ,\ldots,\ ]_2$ are related by the equation

$$[\mathbf{a}_1,\ldots,\mathbf{a}_n]_1 = \lambda[\mathbf{a}_1,\ldots,\mathbf{a}_n]_2$$

where $\lambda \in \mathbb{R}$ is fixed. We shall assume that a nontrivial box product in $\mathbb{R}^n$ has been selected. Then it can be shown that

$$[\mathbf{a}_1,\ldots,\mathbf{a}_n] \neq 0 \Leftrightarrow \{\mathbf{a}_1,\ldots,\mathbf{a}_n\} \text{ linearly independent}$$

The choice of a nontrivial box product in $\mathbb{R}^n$ is tantamount to defining the signed volume of the parallelepiped $\triangle$ spanned by the vectors $\mathbf{a}_1,\ldots,\mathbf{a}_n$:

$$\text{vol}(\triangle) = [\mathbf{a}_1,\ldots,\mathbf{a}_n]$$

To extend our results on the determinant of a linear operator to $\mathbb{R}^n$, we may now establish the star-box formula, the box invariance formula, and adopt the definition

$$[f(\mathbf{a}_1),\ldots,f(\mathbf{a}_n)] = (\det f)[\mathbf{a}_1,\ldots,\mathbf{a}_n]$$

All the formal properties of $\det f$ carry over to $\mathbb{R}^n$. The details are left as an exercise. These remarks also show that $\det f$ does not depend on the choice of the dot product, only on the box product. It is not a metric idea, but an affine idea.

ORIENTATION IN $\mathbb{R}^n$

The discriminating reader may ask the question how orientation is defined in $\mathbb{R}^n$ ($n \geqq 4$), where the distinction between clockwise and counterclockwise rotation, and the distinction between left hand and right hand, are lost. The following remarks are designed to answer this question. We shall see that the set of bases splits, in a natural way, into two disjoint classes. Two bases belong to the same class if they have the same sense. To give $\mathbb{R}^n$ an orientation means to distinguish, through an arbitrary choice, one of the two classes by calling its elements positively oriented. Once it has been given an orientation in this way, $\mathbb{R}^n$ is called an oriented vector space. Here are the details.

Let us pick a basis $\{\mathbf{e}_1,\ldots,\mathbf{e}_n\}$ and distinguish it by the name *preferred basis* of $\mathbb{R}^n$. We then fix the choice of the box product of $\mathbb{R}^n$ such that

$$[\mathbf{e}_1,\ldots,\mathbf{e}_n] > 0$$

We shall say that two bases have the same *sense* if the box product of their elements, in their proper order, have the same signature. The important fact to notice is that this splitting of the set of bases into two classes is natural; i.e., it is independent of the choice of the preferred basis. Under any other choice precisely the same bases would belong to the same class, although, clearly, the new preferred basis may now belong to the other class. The bases belonging to the same class as the preferred basis are called *positively oriented*. An invertible operator f is called *orientation preserving* or *orientation reversing* according as $\det f > 0$ or $\det f < 0$. The former leaves the classes unchanged, and the latter switches them around.

Again, the distinction between these two sets of invertible linear operators is independent of the choice of orientation. The orientation-preserving operators of $\mathbb{R}^n$ form a group denoted $GL^+(\mathbb{R}^n)$.

VOLUME IN $\mathbb{R}^n$

In the same order of ideas, we want to discuss how volume arises in $\mathbb{R}^n$. We shall see that the set of all bases of $\mathbb{R}^n$ splits, in a natural way, into mutually disjoint classes (as many as there are real numbers). Two bases fall into the same class if, and only if, one can be transformed into the other by an operator h satisfying $\det h = 1$. Defining volume in $\mathbb{R}^n$ means to distinguish one of the classes by stipulating that the volume of the parallelepipeds spanned by its elements be $+1$. Then all parallelepipeds will automatically get their volume, too, and we say that $\mathbb{R}^n$ has been given volume. The details of the construction are as follows.

Let us pick a basis $\{\mathbf{e}_1,\ldots,\mathbf{e}_n\}$, called the preferred basis of $\mathbb{R}^n$. We fix the choice of box product such that

$$[\mathbf{e}_1,\ldots,\mathbf{e}_n] = 1$$

Then the signed volume of the parallelepiped spanned by any other basis $\{f(\mathbf{e}_1),\ldots,f(\mathbf{e}_n)\}$ is, by definition, $\det f$. This definition splits the set of bases into mutually disjoint classes such that two bases belong to the same class if, and only if, the parallelepipeds spanned by them have the same volume:

$$\text{vol}(\triangle) = \text{vol}(\triangle') \Leftrightarrow \triangle' = h(\triangle) \qquad \text{with } \det h = 1$$

Moreover, this splitting is natural; i.e., it is independent of the choice of the preferred basis. In other words, a different choice of the preferred basis will leave the classes undisturbed: it will merely reassign the real numbers to the classes in a different way. This is a nontrivial result that we shall now proceed to establish.

An invertible operator h is called *volume preserving* if $\det h = 1$. It is clear that the volume-preserving operators of $\mathbb{R}^n$ form a group, called the *special linear group* $SL(\mathbb{R}^n)$, a subgroup of the *general linear group* $GL(\mathbb{R}^n)$ consisting of the invertible operators of $\mathbb{R}^n$.

Theorem Let h be a volume-preserving operator. Then for any invertible operator f there exists another volume-preserving operator h' such that

$$f \circ h = h' \circ f$$

In symbols:

$$h \in SL(\mathbb{R}^n), f \in GL(\mathbb{R}^n) \Rightarrow f \circ h = h' \circ f \qquad \text{with } h' \in SL(\mathbb{R}^n)$$

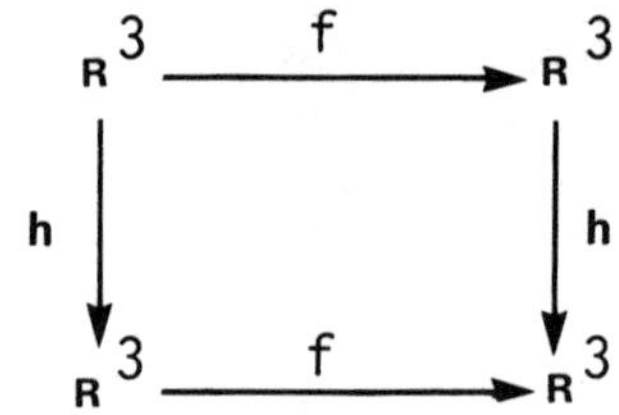

In fact, h' is uniquely determined by the formula $h' = f \circ h \circ f^{-1}$. The proof of this theorem is based on the product rule of determinants:

$$\det h' = \det(f \circ h \circ f^{-1}) = (\det f)(\det h)(\det f^{-1}) = (\det f)1(\det f)^{-1} = 1$$

The following corollary will bring out the geometric contents of the theorem.

Corollary

$$\text{vol}(\triangle) = \text{vol}(\triangle') \Leftrightarrow \text{vol}(f(\triangle)) = \text{vol}(f(\triangle'))$$

where f is an arbitrary invertible operator. In other words, f takes equal volumes into equal volumes. We may also express this fact by saying that invertible operators *respect volume*. Note that respecting volume is not the same as preserving it: only operators h with $\det h = 1$ preserve it.

Proof of the Corollary:

$$\begin{aligned} \text{vol}(\triangle) = \text{vol}(\triangle') &\Rightarrow \triangle' = h(\triangle) \quad \text{with } \det h = 1 \\ &\Rightarrow f(\triangle') = f(h(\triangle)) \\ &\qquad = h'(f(\triangle)) \quad \text{with } \det h' = 1 \text{ by the theorem} \\ &\Rightarrow \text{vol}(f(\triangle)) = \text{vol}(f(\triangle')) \end{aligned}$$

The converse follows if we apply this implication with the operator f^{-1}.

Now the bases $\{\mathbf{e}_1,\ldots,\mathbf{e}_n\}$ and $\{\mathbf{e}'_1,\ldots,\mathbf{e}'_n\}$ belong to the same class if, and only if, $\text{vol}(\triangle) = \text{vol}(\triangle')$ for the parallelepipeds $\triangle$, $\triangle'$ spanned by them. But changing the preferred basis means that $\{\mathbf{e}_1,\ldots,\mathbf{e}_n\}$ and $\{\mathbf{e}'_1,\ldots,\mathbf{e}'_n\}$ are transformed into $\{f(\mathbf{e}_1),\ldots,f(\mathbf{e}_n)\}$, $\{f(\mathbf{e}'_1),\ldots,f(\mathbf{e}'_n)\}$, respectively. The latter belong to the same class as well, by the corollary. Therefore, changing the preferred basis leaves the classes undisturbed, as claimed above. The splitting of the set of bases is natural.

The important result stated in the theorem is expressed, in the language of group theory, by saying that the special linear group $SL(\mathbb{R}^n)$ is a *normal subgroup* of the general linear group $GL(\mathbb{R}^n)$. By contrast, the length-preserving operators, i.e., the orthogonal operators, while they also form a subgroup of $GL(\mathbb{R}^n)$, fail to be a normal subgroup. This fact can be established by furnishing a simple two-dimensional counterexample, as follows.

It is not true that, given an orthogonal operator r of $\mathbb{R}^2$, for every invertible operator f there exists another orthogonal operator r' satisfying $f \circ r = r' \circ f$ because $r' = f \circ r \circ f^{-1}$ is not, in general, an orthogonal operator. If it were, every invertible operator f would *respect* length, which is clearly not the case:

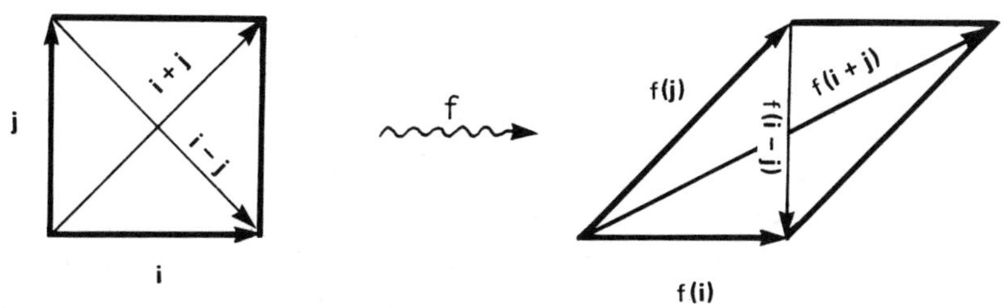

Here the unit square spanned by $\{\mathbf{i},\mathbf{j}\}$ is transformed by a shear $f(\mathbf{v}) = \mathbf{v} + (\mathbf{j} \cdot \mathbf{v})\mathbf{i}$ into a parallelogram spanned by $f(\mathbf{i}) = \mathbf{i}$, $f(\mathbf{j}) = \mathbf{i} + \mathbf{j}$. Whereas the diagonals of the square have the same length, those of the parallelogram no longer do: $\text{length}(\mathbf{i}+\mathbf{j}) = \text{length}(\mathbf{i}-\mathbf{j})$, but $\text{length}(f(\mathbf{i}+\mathbf{j})) \neq \text{length}(f(\mathbf{i}-\mathbf{j}))$. Therefore, not all invertible operators respect length, as they respect volume. Hence the group of orthogonal operators $O(\mathbb{R}^n)$ is not a normal subgroup of $GL(\mathbb{R}^n)$.

This fine distinction is reflected by the phrase that "the concept of volume is natural in linear algebra, while the concept of length is not." Volume can be introduced without the benefit of the dot product, but length and angle cannot. If we discard the dot product and replace it by another, volume will not notice the change, but length and angle will. This can also be described as a distinction between affine and metric concepts. Orientation, volume, and parallelism are affine ideas which are inherent in the concept of a vector space. By contrast, length, angle, and perpendicularity are not inherent in the concept of a vector space; they are a matter of endowment, converting the vector space into a *metric* vector space.

It is worth noting that $SL(\mathbb{R}^n)$ is also a normal subgroup of $GL^+(\mathbb{R}^n)$, the group of orientation-preserving operators, which, in turn, is a normal subgroup of $GL(\mathbb{R}^n)$.

Examples

1. A shear $f = 1 + \ell$, where ℓ is a lift, i.e., $\ell^2 = 0$, has a matrix, in a suitable basis,

$$f=\begin{pmatrix} 1 & 0 & \cdots & 0 & 1 \\ 0 & 1 & \cdots & 0 & 0 \\ \cdots & \cdots & \cdots & \cdots & \cdots \\ 0 & 0 & \cdots & 0 & 1 \end{pmatrix}$$

and hence $\det f = 1$. We conclude that every shear is volume preserving.

2. An involution h of $\mathbb{R}^n$ such that trace $h = n-2k$ ($k = 0, 1, \ldots, n$) has a matrix, in a suitable basis,

$$h=\begin{pmatrix} -1 & 0 & 0 & \cdots & 0 \\ 0 & -1 & 0 & \cdots & 0 \\ 0 & 0 & -1 & \cdots & 0 \\ \cdots & \cdots & \cdots & \cdots & \cdots \\ 0 & 0 & 0 & \cdots & 1 \end{pmatrix}$$

(k diagonal elements equal to -1) satisfies $\det h = 1 \Leftrightarrow k = \tfrac{1}{2}(n - \text{trace } h)$ is *even*. We conclude that an involution h is volume preserving if, and only if, $\tfrac{1}{2}(n-\text{trace } h)$ is even. Otherwise, h is orientation reversing. In particular, the reflection in the origin, -1, is volume preserving in an even-dimensional vector space, but not in an odd-dimensional one. Reflection in a hyperplane is never volume preserving.

3. Let t be a twist, not necessarily antisymmetric. The linear operator

$$f = 1 + t^2 + t^4 \cos\alpha + t \sin\alpha$$

has a matrix, in a suitable basis,

$$f=\begin{pmatrix} \cos\alpha & -\sin\alpha & & & \\ \sin\alpha & \cos\alpha & & & \\ & & \cos\alpha & -\sin\alpha & \\ & & \sin\alpha & \cos\alpha & \\ & & & & \ddots \\ & & & & & 1 \end{pmatrix}$$

and hence $\det f = 1$. We conclude that f is volume preserving. Notice that f is not orthogonal (rotation) unless $t^* = -t$.

4. An invection j such that $j = 1 + t + t^2$, where t is a twist, is volume preserving. This is a special case of preceding example with $\alpha = 90°$. By contrast, the invection $-j$ has $\det(-j) = -1$ and hence is not volume preserving.

APPLICATION

1. Orientation-Preserving Orthogonal Operators

Let r be an orthogonal operator; then $\det r = 1$ or -1. To see this, we write

$$\begin{aligned} r^* = r^{-1} \Rightarrow \quad & r \circ r^* = 1 \\ & \Rightarrow \det(r \circ r^*) = 1 \\ & \Rightarrow (\det r)(\det r^*) = 1 \\ & \Rightarrow (\det r)^2 = 1 \Rightarrow \det r = \pm 1 \end{aligned}$$

In other words, orthogonal operators preserve the unsigned volume, but they may change the orientation. An orthogonal operator is volume preserving if, and only if, it is orientation preserving. We can give further necessary and sufficient conditions.

Any two of the following conditions on an orthogonal operator r are equivalent:

1. $\det r = 1$.
2. r is volume preserving.
3. r is orientation preserving.
4. $[r(\mathbf{a}_1),\ldots,r(\mathbf{a}_n)] = [\mathbf{a}_1,\ldots,\mathbf{a}_n]$ for all $\mathbf{a}_1, \ldots, \mathbf{a}_n \in \mathbb{R}^n$.

The orientation-preserving orthogonal operators form a group, called the *special orthogonal group* $SO(\mathbb{R}^n)$.

In $\mathbb{R}^3$ we also have the result that an orthogonal operator r is orientation preserving if, and only if,

$$r(\mathbf{a} \times \mathbf{b}) = r(\mathbf{a}) \times r(\mathbf{b}) \qquad \text{for all } \mathbf{a}, \mathbf{b}$$

The proof of this is left as an exercise.

TRACE OF A LINEAR OPERATOR

In addition to det f, we also had occasion to use informally another invariant, trace f, of the linear operator f. We shall now formalize this concept.

Trace Invariance Formula

$$\frac{[f(\mathbf{a}),\mathbf{b},\mathbf{c}] + [\mathbf{a},f(\mathbf{b}),\mathbf{c}] + [\mathbf{a},\mathbf{b},f(\mathbf{c})]}{[\mathbf{a},\mathbf{b},\mathbf{c}]}$$

$$= \frac{[f(\mathbf{a}'),\mathbf{b}',\mathbf{c}'] + [\mathbf{a}',f(\mathbf{b}'),\mathbf{c}')] + [\mathbf{a}',\mathbf{b}',f(\mathbf{c}')]}{[\mathbf{a}',\mathbf{b}',\mathbf{c}']}.$$

for all linear operators f, and for all linearly independent sets $\{\mathbf{a},\mathbf{b},\mathbf{c}\}$ and $\{\mathbf{a}',\mathbf{b}',\mathbf{c}'\}$. The proof of this formula is similar to that of the box invariance formula, and is left as an exercise. The trace invariance formula asserts that the left-hand side depends only on the operator f, not on the choice of the linearly independent set $\{\mathbf{a},\mathbf{b},\mathbf{c}\}$. Hence it exhibits an invariant of f, called the *trace* of f. Accordingly, we define trace f such that

$$[f(\mathbf{a}),\mathbf{b},\mathbf{c}] + [\mathbf{a},f(\mathbf{b}),\mathbf{c}] + [\mathbf{a},\mathbf{b},f(\mathbf{c})] = (\text{trace } f)[\mathbf{a},\mathbf{b},\mathbf{c}]$$

for all vectors **a**, **b**, **c**. If the matrix of f is

$$f = \begin{pmatrix} a_{11} & a_{12} & a_{13} \\ a_{21} & a_{22} & a_{23} \\ a_{31} & a_{32} & a_{33} \end{pmatrix}$$

then

$$\text{trace } f = a_{11} + a_{22} + a_{33}$$

This is immediate if $\{\mathbf{i},\mathbf{j},\mathbf{k}\}$ is an orthornormal basis. In this case $[\mathbf{i},\mathbf{j},\mathbf{k}] = 1$ and trace $f = f(\mathbf{i}) \cdot \mathbf{i} + f(\mathbf{j}) \cdot \mathbf{j} + f(\mathbf{k}) \cdot \mathbf{k}$. The proof for an arbitrary basis is left as an exercise. Note that if f has three (real) eigenvalues, then

$$\text{trace } f = \lambda_1 + \lambda_2 + \lambda_3$$

FORMAL PROPERTIES OF THE TRACE

$$\text{trace}(f + g) = \text{trace } f + \text{trace } g$$

$$\text{trace}(\lambda f) = \lambda \text{ trace } f$$

$$\text{trace } f^* = \text{trace } f$$

$$\text{trace } 0 = 0$$

$$\text{trace } 1 = n \qquad 1 \in GL(\mathbb{R}^n)$$

$$\text{Vanishing Trace Property} \qquad \text{trace}[f,g] = 0$$

All these properties, except the last, are obvious. To prove the last property, we first observe that it is equivalent to

$$\text{trace}(g \circ f) = \text{trace}(f \circ g)$$

which follows immediately from the matrix product rule (telling us that the diagonal elements of the matrices of $f \circ g$ and $g \circ f$ are the same).

In particular,

$$\text{trace}(g^{-1} \circ f \circ g) = \text{trace } f$$

i.e., the trace of f remains unchanged under a change of coordinates.

The foregoing properties are also valid in $\mathbb{R}^n$ where trace f is defined in terms of the box product of n vectors by the formula

$$[f(\mathbf{a}_1),\mathbf{a}_2,\ldots,\mathbf{a}_n] + [\mathbf{a}_1,f(\mathbf{a}_2),\ldots\mathbf{a}_n] + \cdots + [\mathbf{a}_1,\mathbf{a}_2,\ldots,f(\mathbf{a}_n)]$$
$$= (\text{trace } f)[\mathbf{a}_1,\mathbf{a}_2,\ldots,\mathbf{a}_n]$$

for arbitrary vectors $\mathbf{a}_1$, $\mathbf{a}_2$, ..., $\mathbf{a}_n$.

APPLICATION

2. Angle and Trace of a Rotation

From Steenrod's formula, in $\mathbb{R}^3$,

$$\begin{aligned}\text{trace } e^{\alpha t} &= \text{trace } (1 + t^2 + \cos\,\alpha\, t^4 + \sin\,\alpha\, t)\\ &= \text{trace } (1 + t^2) + \cos\,\alpha \text{ trace } t^4 + \sin\,\alpha \text{ trace } t\\ &= 1 + 2\cos\,\alpha\end{aligned}$$

because $1 + t^2$ is a projection onto a line, t^4 is a projection onto a plane, and their traces are equal to their ranks, which are 1 and 2, respectively. On the other hand, trace $t = 0$ because t is antisymmetric. This establishes the formula: If r is a rotation of $\mathbb{R}^3$ through α, then

$$\cos\,\alpha = \tfrac{1}{2}\,(\text{trace } r - 1)$$

Special Cases For a rotation r in $\mathbb{R}^3$: $-1 \leqq \text{trace}\, r \leqq 3$, in particular:

$$\begin{aligned}\text{trace } r = 3 &\Leftrightarrow \alpha = 0°\\ \text{trace } r = 2 &\Leftrightarrow \alpha = \pm 60°\\ \text{trace } r = 1 &\Leftrightarrow \alpha = \pm 90°\\ \text{trace } r = 0 &\Leftrightarrow \alpha = \pm 120°\\ \text{trace } r = -1 &\Leftrightarrow \alpha = 180°\end{aligned}$$

$$\text{trace } r > 1 \Leftrightarrow \alpha \text{ acute} \qquad -1 < \text{trace } r < 1 \Leftrightarrow \alpha \text{ obtuse}$$

The angle of rotation r is determined to within signature; if the axis is given an orientation, the angle of rotation will be uniquely determined as the signed angle of rotation, observed from the vantage point of the positive half of the axis.

It is left as an exercise to show that the angle α of the rotation r in $\mathbb{R}^n$ can be calculated from

$$\cos\,\alpha = \frac{\text{trace}\, r - n}{\deg r} + 1$$

The trace of a rotation in $\mathbb{R}^n$ satisfies the inequality

$$n - 2\deg r \leqq \text{trace } r \leqq n$$

where deg r is the degree of the rotation r.

EXERCISES

13.1 Show that the dilatations $d = 1 + (\lambda - 1)p$ along the same line $\operatorname{im} p\,(p^2 = p)$ form a commutative group which is isomorphic to $\mathbb{R}^*$, the multiplicative group of real numbers $\neq 0$. Show that the isomorphism is furnished by $\det d$.

13.2 Show that the shears of $\mathbb{R}^3$ with a common eigenspace form a commutative group which is isomorphic to the additive group of vectors in the plane.

13.3 Show that the shears of $\mathbb{R}^3$ with a common double eigenvector form a commutative group which is isomorphic to the additive group of vectors in the plane.

13.4 Show that two shears of $\mathbb{R}^3$ commute if, and only if, either they have a double eigenvector, or they have an eigenspace of dimension 2 in common.

13.5 Show that the group $GL^+(\mathbb{R}^n)$ of orientation-preserving operators is a normal subgroup of the general linear group $GL(\mathbb{R}^n)$.

13.6 Show that the group $SL(\mathbb{R}^n)$ of volume-preserving operators is a normal subgroup of $GL^+(\mathbb{R}^n)$.

13.7 Prove the formula

$$([f(\mathbf{a}),\mathbf{b},\mathbf{c}] + [\mathbf{a},f(\mathbf{b}),\mathbf{c}] + [\mathbf{a},\mathbf{b},f(\mathbf{c})])[\mathbf{a}',\mathbf{b}',\mathbf{c}']$$
$$= [\mathbf{a},\mathbf{b},\mathbf{c}]([f^*(\mathbf{a}'),\mathbf{b}',\mathbf{c}'] + [\mathbf{a}',f^*(\mathbf{b}'),\mathbf{c}'] + [\mathbf{a}',\mathbf{b}',f^*(\mathbf{c}')])$$

where $f \in L(\mathbb{R}^3, \mathbb{R}^3)$ and $\mathbf{a}$, $\mathbf{b}$, $\mathbf{c}$, $\mathbf{a}'$, $\mathbf{b}'$, $\mathbf{c}'$, $\in \mathbb{R}^3$.

13.8 Using the previous result, prove the trace invariance formula.

13.9 Show that

$$\text{trace } f = a_{11} + a_{22} + a_{33}$$

i.e., the trace of f can be calculated as the sum of the diagonal elements of the matrix of f.

13.10 Prove the formula

$$([f(\mathbf{a}),f(\mathbf{b}),\mathbf{c}] + [f(\mathbf{a}),\mathbf{b},f(\mathbf{c})] + [\mathbf{a},f(\mathbf{b}),f(\mathbf{c})])[\mathbf{a}',\mathbf{b}',\mathbf{c}']$$
$$= [\mathbf{a},\mathbf{b},\mathbf{c}]([f^*(\mathbf{a}'),f^*(\mathbf{b}'),\mathbf{c}'] + [f^*(\mathbf{a}'),\mathbf{b}',f^*(\mathbf{c}')] + [\mathbf{a}',f^*(\mathbf{b}'),f^*(\mathbf{c}')])$$

where f is any linear operator, and $\mathbf{a}$, $\mathbf{b}$, $\mathbf{c}$, $\mathbf{a}'$, $\mathbf{b}'$, $\mathbf{c}'$ are arbitrary vectors of $\mathbb{R}^3$.

13.11 Using the previous result show that

$$\text{inv } f = \frac{[f(\mathbf{a}),f(\mathbf{b}),\mathbf{c}] + [f(\mathbf{a}),\mathbf{b},f(\mathbf{c})] + [\mathbf{a},f(\mathbf{b}),f(\mathbf{c})]}{[\mathbf{a},\mathbf{b},\mathbf{c}]}$$

does not depend on the choice of the linearly independent set of vectors $\{\mathbf{a},\mathbf{b},\mathbf{c}\}$, but only on the linear operator f. Thus inv f is another invariant of f, in addition to det f and trace f.

13.12 Show that if the matrix of the linear operator f is

$$f = \begin{pmatrix} a_{11} & a_{12} & a_{13} \\ a_{21} & a_{22} & a_{23} \\ a_{31} & a_{32} & a_{33} \end{pmatrix}$$

then

$$\text{inv } f = \begin{vmatrix} a_{11} & a_{12} \\ a_{21} & a_{22} \end{vmatrix} + \begin{vmatrix} a_{11} & a_{13} \\ a_{31} & a_{33} \end{vmatrix} + \begin{vmatrix} a_{22} & a_{23} \\ a_{32} & a_{33} \end{vmatrix}$$

13.13 Show that the characteristic equation of $f \in L(\mathbb{R}^3, \mathbb{R}^3)$ is

$$\lambda^3 - (\operatorname{trace} f)\lambda^2 + (\operatorname{inv} f)\lambda - \det f = 0$$

13.14 Show that if f has three distinct eigenvalues λ_1, λ_2, λ_3 then

$$\operatorname{inv} f = \lambda_1\lambda_2 + \lambda_2\lambda_3 + \lambda_3\lambda_1$$

13.15 Show that $\operatorname{inv} f = (\operatorname{trace} f^{-1}) \det f$ provided that f is bijective.

13.16 Show that for the angle α of a rotation r of $\mathbb{R}^3$,

$$\cos\alpha = \tfrac{1}{4}\det(r+1) - 1$$

13.17 Show that for the angle α of a rotation r of $\mathbb{R}^n$

$$\cos\alpha = \frac{\operatorname{trace} r - n}{\deg r} + 1$$

13.18 Show that for a rotation of r of $\mathbb{R}^n$

$$n - 2\deg r \leqq \operatorname{trace} r \leqq n$$

13.19 f is called *nilpotent* if $f^n = 0$ $(f \neq 0)$. Show that f is nilpotent if, and only if, trace $f^k = 0$ for $k = 1, 2, \ldots$

13.20 Show that $p^2 = p \Rightarrow \operatorname{rank} p = \operatorname{trace} p$.

13.21 Show that $t^3 = -t \Rightarrow \operatorname{rank} t = -\operatorname{trace} t^2$.

13.22 Show that for an involution h of $\mathbb{R}^n$,

$$h \text{ orientation preserving} \Leftrightarrow \operatorname{trace} h \equiv n (\operatorname{mod} 4)$$

13.23 Let $f \in L(\mathbb{R}^3, \mathbb{R}^3)$, $\det f = 0$. Show that the characteristic equation of f is

$$\lambda^3 - (\operatorname{trace} f)\lambda^2 + \tfrac{1}{2}(\operatorname{trace}^2 f - \operatorname{trace} f^2)\lambda = 0$$

13.24 Let $f \in L(\mathbb{R}^3, \mathbb{R}^3)$, $\det f \neq 0$. Show that the characteristic equation of f is

$$\lambda^3 - (\operatorname{trace} f)\lambda^2 + (\operatorname{trace} f^{-1})(\det f)\lambda - \det f = 0$$

Hence conclude that

$$f \text{ elliptic} \Leftrightarrow \operatorname{trace}^3 f \det f^{-1} + \operatorname{trace}^3 f^{-1} \det f + 27 > (\tfrac{1}{2}\operatorname{trace} f \operatorname{trace} f^{-1} + 9)^2$$

while the opposite inequality is a necessary and sufficient condition for f to be hyperbolic. Show also that f has a multiple eigenvalue if, and only if,

$$\operatorname{trace}^3 f \det f^{-1} + \operatorname{trace}^3 f^{-1} \det f + 27 = (\tfrac{1}{2}\operatorname{trace} f \operatorname{trace} f^{-1} + 9)^2$$

13.25 Let $f \in L(\mathbb{R}^3, \mathbb{R}^3)$, $\det f \neq 0$. Show that

$$f^{-1} \text{ elliptic} \Leftrightarrow f \text{ elliptic}$$

$$f^{-1} \text{ hyperbolic} \Leftrightarrow f \text{ hyperbolic}$$

$$f^{-1} \text{ parabolic} \Leftrightarrow f \text{ parabolic}$$

13.26 Show that the characteristic equation of $f \in L(\mathbb{R}^2, \mathbb{R}^2)$ is

$$\lambda^2 - (\text{trace } f)\lambda + \det f = 0$$

Hence deduce that

$$f \text{ elliptic} \Leftrightarrow \text{trace}^2 f < 4 \det f$$

$$f \text{ hyperbolic} \Leftrightarrow \text{trace}^2 f > 4 \det f$$

Moreover, $\text{trace}^2 f = 4 \det f$ is a necessary and sufficient condition for f to have a double eigenvalue (then f is either parabolic or is a strain).

13.27 The linear operator f of $\mathbb{R}^3$ has three invariants: trace f, inv f, and det f (see Exercise 13.11). How many invariants does f of $\mathbb{R}^n$ have, and what are they?

13.28 Show that a subgroup N of G, with half as many elements as G has, is always normal in G.

13.29 Without any further calculation, simply by inspecting the multiplication table of the group G of rotations of the cube (see Exercise 12.4), find *two* normal subgroups of G different from G and the trivial group.

13.30 gN is called a (*left*) *coset* of the group G (modulo N). Show that the cosets of G, under the rule

$$(gN) \circ (g'N) = (g \circ g')N$$
$$(gN)^{-1} = (g^{-1})N$$

form a group, called the *quotient group of* G (modulo N) [denoted G/N], provided that N is normal in G.

Use the multiplication table 12.4 to illustrate the cosets and to demonstrate multiplication in the quotient group.

13.31 Give an example of a nontrivial group homomorphism that is not injective.

13.32 Show that trace $f = 0 \Leftrightarrow f = [g,h]$ in $\mathbb{R}^2$.

14
The Exponential Functor†

We want to tie up the loose ends. In this final chapter we shall see that the Cayley transform, Steenrod's formula, the relationship between the group of orientation-preserving orthogonal operators and the Lie algebra of antisymmetric operators, and the relationship between det f and trace f are far from being isolated facts. They are all special cases of the same phenomenon, the exponential functor, one of the great unifying ideas in modern mathematics.

THE EXPONENTIAL OF A LINEAR OPERATOR

We start by recalling the following

Examples

1. *The group of rotations induced by an antisymmetric operator*: If f is an antisymmetric operator ($f^* = -f$) of $\mathbb{R}^3$, then $f = \omega t$ where t is an antisymetric twist ($t^3 = -t = t^*$) and $\omega \in \mathbb{R}$ by the principal twist theorem. Therefore $e^{\alpha f} = 1 + t^2 + t^4 \cos \omega\alpha + t \sin \omega\alpha$ is a rotation around the line ker f with angular velocity ω. These rotations form a commutative group parametrized by α:

$$e^{\alpha f} \circ e^{\beta f} = 1 + t^2 + t^4 \cos \omega(\alpha + \beta) + \sin \omega(\alpha + \beta) = e^{(\alpha + \beta)f}$$
$$(e^{\alpha f})^{-1} = 1 + t^2 + t^4 \cos \omega\alpha - t \sin \omega\alpha = e^{-\alpha f}$$

Note that trace $f = 0$ and det $e^{\alpha f} = 1$ for all $\alpha \in \mathbb{R}$.

†*Note*: The bases in this chapter are not necessarily orthonormal, and hence the coordinates are skew affine, except where noted otherwise.

2. *The group of shears induced by a lift:* If ℓ is a lift ($\ell^2=0$), then $e^{\alpha\ell}=1+\alpha\ell$ is a shear for all $\alpha \in \mathbb{R}$. These shears form a commutative group parametrized by α:

$$e^{\alpha\ell}\circ e^{\beta\ell}=(1+\alpha\ell)\circ(1+\beta\ell)=1+(\alpha+\beta)\ell=e^{(\alpha+\beta)\ell}$$
$$(e^{\alpha\ell})^{-1}=(1+\alpha\ell)^{-1}=1-\alpha\ell=e^{-\alpha\ell}$$

Note that trace $\ell=0$ and det $e^{\alpha\ell}=1$ for all $\alpha \in \mathbb{R}$.

3. *The group of dilatations induced by a principal projection:* Let p be a principal projection ($p^2=p$, rank $p=1$). Then $e^{\alpha p}=1+(e^{\alpha}-1)p$ is a dilatation in ratio e^{α} along the line im p relative to the plane ker p. These dilatations form a commutative group parametrized by $\alpha \in \mathbb{R}$:

$$e^{\alpha p}\circ e^{\beta p}=(1+(e^{\alpha}-1)p)\circ(1+(e^{\beta}-1)p)=1+e^{(\alpha+\beta)p}$$
$$(e^{\alpha p})^{-1}=(1+(e^{\alpha}-1)p)^{-1}=1+(e^{-\alpha}-1)p=e^{-\alpha p}$$

Note that trace $p=1$ and det $e^{\alpha p}=e^{\alpha}$ for all $\alpha \in \mathbb{R}$.

4. *The group of positive definite operators induced by a symmetric operator:* If f is a symmetric operator ($f^*=f$) of $\mathbb{R}^n$, then by the principal axis theorem, $f=\lambda_1p_1+\lambda_2p_2+\cdots+\lambda_np_n$ where $p_1, p_2, \ldots, p_n$ are symmetric principal projections that are pairwise perpendicular (i.e., $p_\nu^2=p_\nu=p_\nu^*$; rank $p_\nu=1$; $p_\nu\circ p_\mu=0=p_\mu\circ p_\nu$; $\nu, \mu=1, 2, \ldots, n$; $\nu\neq\mu$; $p_1+p_2+\cdots+p_n=1$), therefore

$$\begin{aligned}e^{\alpha f}&=e^{\alpha(\lambda_1p_1+\lambda_2p_2+\cdots+\lambda_np_n)}=e^{\alpha\lambda_1p_1}\circ e^{\alpha\lambda_2p_2}\circ\cdots\circ e^{\alpha\lambda_np_n}\\&=(1+(e^{\alpha\lambda_1}-1)p_1)\circ(1+(e^{\alpha\lambda_2}-1)p_2)\circ\cdots\circ(1+(e^{\alpha\lambda_n}-1)p_n)\\&=1-(p_1+p_2+\cdots+p_n)+e^{\alpha\lambda_1}p_1+e^{\alpha\lambda_2}p_2+\cdots+e^{\alpha\lambda_n}p_n\\&=e^{\alpha\lambda_1}p_1+e^{\alpha\lambda_2}p_2+\cdots+e^{\alpha\lambda_n}p_n\end{aligned}$$

Obviously, the eigenvectors of $e^{\alpha f}$ are the same as those of f, while its eigenvalues are all positive: $e^{\alpha\lambda_\nu}>0$, $\nu=1, 2, \ldots, n$. Moreover, $e^{\alpha f}$ is symmetric:

$$\begin{aligned}(e^{\alpha f})^*&=(e^{\alpha\lambda_1}p_1+e^{\alpha\lambda_2}p_2+\cdots+e^{\alpha\lambda_n}p_n)^*\\&=e^{\alpha\lambda_1}p_1^*+e^{\alpha\lambda_2}p_2^*+\cdots+e^{\alpha\lambda_n}p_n^*\\&=e^{\alpha\lambda_1}p_1+e^{\alpha\lambda_2}p_2+\cdots+e^{\alpha\lambda_n}p_n=e^{\alpha f}\end{aligned}$$

We conclude that if f is symmetric, then $e^{\alpha f}$ is positive definite (see Exercise 11.46, 11.47). These positive definite operators form a commutative group parametrized by $\alpha \in \mathbb{R}$:

$$e^{\alpha f}\circ e^{\alpha' f}=e^{(\alpha+\alpha')\lambda_1}p_1+e^{(\alpha+\alpha')\lambda_2}p_2+\cdots+e^{(\alpha+\alpha')\lambda_n}p_n=e^{(\alpha+\alpha')f}$$

Note that trace $f=\lambda_1+\lambda_1+\ldots+\lambda_n$ and det $e^{\alpha f}=e^{\alpha\lambda_1}e^{\alpha\lambda_2}\cdots e^{\alpha\lambda_n}=e^{\alpha(\lambda_1+\lambda_2+\cdots+\lambda_n)}=e^{\alpha\,\text{trace}\,f}$ for all $\alpha \in \mathbb{R}$.

The foregoing examples motivate the following definition: for an arbitrary linear operator f we define the *exponential* of f, in symbols $e^{\alpha f}$, by the formula

$$e^{\alpha f} = 1 + \frac{\alpha f}{1!} + \frac{\alpha^2 f^2}{2!} + \frac{\alpha^3 f^3}{3!} + \ldots \qquad (\alpha \in \mathbb{R})$$

It is true, but we shall not prove it in this book, that the infinite series defining $e^{\alpha f}$ is convergent. It is left as an exercise to show that $e^{\alpha f}$ is independent of the choice of coordinates. The definition of $e^{\alpha f}$ is not suitable for calculation except when f is a lift, projection, twist, involution, invection, nilpotent (see Exercise 13.19) or a scalar multiple thereof, in which cases it reduces to the formulas given in the Examples. Not every operator occurs as an exponential: for example,

$$\begin{pmatrix} -1 & \alpha \\ 0 & -1 \end{pmatrix}$$

can be an exponential only for $\alpha = 0$. Moreover, if an invertible operator occurs as an exponential, then it may do so in infinitely many different ways. Let $\mathbf{w}$ be an eigenvector of f belonging to the eigenvalue λ. Then $\mathbf{w}$ is also an eigenvector of $e^{\alpha f}$ belonging to the eigenvalue of $e^{\alpha\lambda}$:

$$\begin{aligned} f(\mathbf{w}) = \lambda\mathbf{w} \Rightarrow e^{\alpha f}(\mathbf{w}) &= \mathbf{w} + \frac{\alpha}{1!} f(\mathbf{w}) + \frac{\alpha^2}{2!} f^2(\mathbf{w}) + \cdots \\ &= \mathbf{w} + \frac{\alpha}{1!}\lambda\mathbf{w} + \frac{\alpha^2}{2!}\lambda^2\mathbf{w} + \cdots \\ &= \left(1 + \frac{\alpha\lambda}{1!} + \frac{\alpha^2\lambda^2}{2!} + \cdots\right)\mathbf{w} = e^{\alpha\lambda}\mathbf{w} \end{aligned}$$

We leave it as an exercise to show that $e^{\alpha f}$ has no other eigenvectors.

FORMAL PROPERTIES OF THE EXPONENTIAL

Product Rule

$$e^{\alpha f} \circ e^{\alpha' f} = e^{(\alpha + \alpha')f}$$
$$(e^{\alpha f})^{-1} = e^{-\alpha f}$$
$$e^{0f} = 1$$
$$(e^{\alpha f})^* = e^{\alpha f^*}$$

Det-Trace Formula

$$\det e^{\alpha f} = e^{\alpha \operatorname{trace} f}$$

Exponential Property

$$[f,g]=0 \Rightarrow e^{\alpha(f+g)} = e^{\alpha f} \circ e^{\alpha g}$$

A consequence of the first three properties is that the exponentials of f form a one-parameter commutative group. We can deduce from the fourth property that $e^{\alpha f}$ positive definite $\Leftrightarrow f$ symmetric; $e^{\alpha f}$ orthogonal $\Leftrightarrow$ f antisymmetric. It follows from the det-trace formula that det $e^{\alpha f} > 0$ so that the exponential of a linear operator is orientation preserving for all $\alpha \in \mathbb{R}$. We leave the proofs of the formal properties as an exercise. The det-trace formula is obvious for operators $f \in L(\mathbb{R}^3, \mathbb{R}^3)$ with three distinct eigenvalues λ_1, λ_2, λ_3:

$$e^{\alpha \text{ trace } f} = e^{\alpha(\lambda_1+\lambda_2+\lambda_3)} = e^{\alpha\lambda_1}e^{\alpha\lambda_2}e^{\alpha\lambda_3} = \det e^{\alpha f}.$$

GEOMETRIC MEANING OF THE TRACE

The geometric meaning of trace f is buried under what may be called the infinitesimal structure of the linear operator f. It is brought out by the det-trace formula.

From now on we shall regard α as time. We shall be talking about the *flow* induced by the linear operator f. This is the steady flow that causes a point with location vector $\mathbf{v}_0$ at time 0 to move to the point with location vector $e^{\alpha f}(\mathbf{v}_0)$ at time α. The moving points trace out *streamlines*: through every point in the space there passes exactly one such streamline. The equation of the streamline passing through the point $\mathbf{v}_0$ is the vector-valued function of the scalar variable α:

$$\mathbf{v}_\alpha = e^{\alpha f}(\mathbf{v}_0)$$

Moreover, the velocity vector of the flow at the point $\mathbf{v}_0$ is precisely $f(\mathbf{v}_0)$. The flow is called *steady* as the velocity vector does not depend on the time α, but only on the position $\mathbf{v}_0$. In particular, the streamlines do not change as α varies.

In general, however, the flow will not be incompressible. A ball is slightly distorted under the flow into an ellipsoid. The volume of the ball will change by a factor equal to det $e^{\alpha f} = e^{\alpha \text{ trace } f}$. Indeed, this factor varies with the time α. In more detail, the flow induced by f is

$$\text{Incompressible} \Leftrightarrow \text{trace } f = 0$$

$$\text{Expansive} \Leftrightarrow \text{trace } f > 0$$

$$\text{Compressive} \Leftrightarrow \text{trace } f < 0$$

In either case, trace f *is the time rate of change of the volume factor of the flow induced by the linear operator f at $\alpha = 0$*. Indeed, the det-trace formula implies that, whenever α is very small, $y = \det e^{\alpha f}$ can be represented in a coordinate system by a straight line with y intercept 1 and slope trace f. In other words, the volume changes proportionally with time, and the factor of proportionality is just trace f.

Examples

1. The flow induced by the lift ℓ has streamlines $\mathbf{v}_\alpha = \mathbf{v}_0 + \alpha\ell(\mathbf{v}_0)$, which is a family of parallel lines. Since the double eigenvector of ℓ belongs to $\lambda = 0$, the particles in the plane im ℓ are actually not moving, and the streamlines in that plane are stationary. The time rate of change of the volume factor of the flow is trace $\ell = 0$; the flow is incompressible.
2. The flow induced by the principal projection p has streamlines $\mathbf{v}_\alpha = \mathbf{v}_0 + (e^\alpha - 1)p(\mathbf{v}_0)$, which is a family of parallel lines. The time rate of change of the volume factor of the flow is trace $p = 1$; this is an example of an expansive flow. Notice the contrast between this and the preceding example: although the streamlines of the two flows are identical, the flows are very different. This is so because the velocity of the particles are different, in fact, one flow changes the volume factor and the other does not.
3. The flow induced by the strain λ has streamlines $\mathbf{v}_\alpha = e^{\lambda\alpha}\mathbf{v}_0$ which is a family of straight lines (more precisely, rays) emanating from the origin. The origin as a single point is considered as a separate streamline. It is customary to say that, for this flow pattern, the origin is a *source* ($\lambda > 0$) or a *sink* ($\lambda < 0$). The time rate of change of the volume factor is 3.

   The geometric picture that is emerging is this: Every linear operator f induces a steady flow in specifying the velocity vector of a particle at the point $\mathbf{v}_0$ to be $f(\mathbf{v}_0)$, for all $\mathbf{v}_0 \in \mathbb{R}^3$. The steady flow can be visualized as a family of streamlines $\mathbf{v}_\alpha = e^{\alpha f}(\mathbf{v}_0)$, filling the space $\mathbb{R}^3$. Another way of describing all this is to say that a curve C followed by particle P is also followed by any other particle P' also in C; thus C flows (slides) along itself. The particle P moving along C will do so with varying velocity, but when it reaches the place $\mathbf{v}_0$, its velocity will be the prescribed $f(\mathbf{v}_0)$. It is this picture of streamlines, filling the space and not changing shape as time passes, that we call a steady flow. The eigenvectors of f (if any) will determine streamlines which are actually straight lines, passing through the origin. The eigenvalues, in turn, will determine the nature of the streamlines in a neighborhood of the eigenvectors.

   Then trace f measures the *divergence* of the flow (or, if trace $f < 0$, its convergence) at the time moment $\alpha = 0$. Indeed, it is customary in the literature on mathematical physics to call the trace of a linear operator f the divergence of the vector field $f(\mathbf{v})$.

APPLICATIONS

1. Flows Induced by Hyperbolic, Elliptic, and Parabolic Operators

Examples

1. Find the streamlines of the flow induced by the hyperbolic operator

$$f = \begin{pmatrix} 1 & 0 \\ 0 & 2 \end{pmatrix}$$

Answer: We have

$$e^{\alpha f} = \begin{pmatrix} e^{\alpha} & 0 \\ 0 & e^{2\alpha} \end{pmatrix}$$

The streamline passing through $\mathbf{v}_0 = x_0\mathbf{i} + y_0\mathbf{j}$ $(x_0 \neq 0,\ y_0 \neq 0)$ is $\mathbf{v}_\alpha = e^{\alpha f}(\mathbf{v}_0) = x_0 e^{\alpha}\mathbf{i} + y_0 e^{2\alpha}\mathbf{j}$, i.e.,

$$\left.\begin{aligned} x &= x_0 e^{\alpha} \\ y &= y_0 e^{2\alpha} \end{aligned}\right\} \Rightarrow \frac{x^2}{y} = \frac{x_0^2}{y_0} = \text{const} = C \Rightarrow y = Cx^2$$

a family of parabolas tangent to the x axis at the origin.

2. Find the streamlines of the flow induced by the hyperbolic operator

$$f = \begin{pmatrix} 1 & 0 \\ 0 & -1 \end{pmatrix}$$

Answer: We have

$$e^{\alpha f} = \begin{pmatrix} e^{\alpha} & 0 \\ 0 & e^{-\alpha} \end{pmatrix}$$

The equation of the streamline passing through the point $\mathbf{v}_0 = x_0\mathbf{i} + y_0\mathbf{j}$ is

$$\mathbf{v}_\alpha = e^{\alpha f}(\mathbf{v}_0) = x_0 e^{\alpha}\mathbf{i} + y_0 e^{-\alpha}\mathbf{j} \Rightarrow \begin{cases} x = x_0 e^{\alpha} \\ y = y_0 e^{-\alpha} \end{cases} \Rightarrow xy = x_0 y_0 = \text{const}$$

which is a family of hyperbolas with the coordinate axes as their common asymptotes.

3. Find the streamlines of the flow induced by the elliptic operator

$$f = \begin{pmatrix} -3/4 & -5/4 \\ 5/4 & 3/4 \end{pmatrix}$$

Answer:

$$f^3 = -f \Rightarrow f \text{ is a twist} \Rightarrow e^{\alpha f} = \cos\alpha + f\sin\alpha$$

$$= \begin{pmatrix} \tfrac{1}{4}(4\cos\alpha - 3\sin\alpha) & -\tfrac{5}{4}\sin\alpha \\ \tfrac{5}{4}\sin\alpha & \tfrac{1}{4}(4\cos\alpha + 3\sin\alpha) \end{pmatrix}$$

The equation of the streamlines is

$$\mathbf{v}_\alpha = e^{\alpha f}(\mathbf{v}_0) \Rightarrow \begin{cases} x = \tfrac{1}{4}(4\cos\alpha - 3\sin\alpha)x_0 - \tfrac{5}{4}(\sin\alpha)y_0 \\ y = \tfrac{5}{4}(\sin\alpha)x_0 + \tfrac{1}{4}(\cos\alpha + 3\sin\alpha)y_0 \end{cases}$$

Calculating x^2, xy, y^2 and eliminating $\cos\alpha$, $\sin\alpha$ from the equations we get

$$5x^2 + 6xy + 5y^2 = 5x_0^2 + 6x_0y_0 + 5y_0^2 = \text{const} = C$$
$$\Rightarrow (x-y)^2 + 4(x+y)^2 = C$$
$$\Rightarrow \frac{(x-y)^2}{4} + \frac{(x+y)^2}{1} = C$$
$$\Rightarrow \frac{(x/\sqrt{2} - y/\sqrt{2})^2}{2^2} + \frac{(x/\sqrt{2} - y/\sqrt{2})^2}{1^2} = C$$

We conclude that the streamlines are ellipses centered at the origin, with the semimajor axis twice as long as the semiminor axis, and making an angle 45° with the x axis.

4. Find the streamlines of the flow induced by the parabolic operator

$$f = \begin{pmatrix} \lambda & 0 \\ 1 & \lambda \end{pmatrix} \qquad f = \lambda + \ell,\ \ell^2 = 0$$

Answer:

$$e^{\alpha f} = e^{\alpha(\lambda+\ell)} = e^{\alpha\lambda} \circ e^{\alpha\ell} = e^{\alpha\lambda}(1 + \alpha\ell) = \begin{pmatrix} e^{\alpha\lambda} & 0 \\ \alpha e^{\alpha\lambda} & e^{\alpha\lambda} \end{pmatrix}$$

The equation of the streamlines:

$$\mathbf{v}_\alpha = e^{\alpha f}(\mathbf{v}_0) \Rightarrow \begin{cases} x = e^{\alpha\lambda}x_0 \\ y = \alpha e^{\alpha\lambda}x_0 + e^{\alpha\lambda}y_0 \end{cases} \Rightarrow \frac{y}{x} = \alpha + \frac{y_0}{x_0}$$

If $\lambda = 0, f = \ell$ is a lift and the streamlines are vertical lines. Suppose now that $\lambda \neq 0$, $x_0 \neq 0$. Then

$$\frac{y}{x} = \frac{1}{\lambda}\log\frac{x}{x_0} + \frac{y_0}{x_0}$$

or, for $y_0 = 0$, $x_0 = 1$ the streamline is

$$y = \frac{1}{\lambda}\log x$$

Moreover, for $y_0 = 0$ and arbitrary $x_0 > 0$, the streamlines are

$$y' = \frac{1}{\lambda}\log x'$$

where

$$\begin{cases} x' = \dfrac{x}{x_0} \\ y' = \dfrac{y}{x_0} \end{cases}$$

are obtained by a strain. If $x_0 > 0$, the streamlines are defined for $x > 0$; if $x_0 < 0$, the streamlines are obtained by a reflection in the y axis. If $x_0 = 0$, the corresponding streamline is the (positive) y axis.

STREAMLINES OF OPERATORS OF $\mathbb{R}^2$ BY TYPE

Elliptic

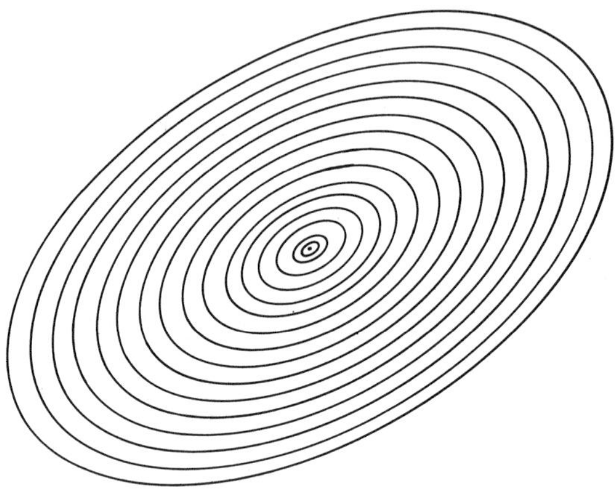

Hyperbolic

1. λ_1, λ_2 have the same signature: $\lambda_1\lambda_2 > 0$

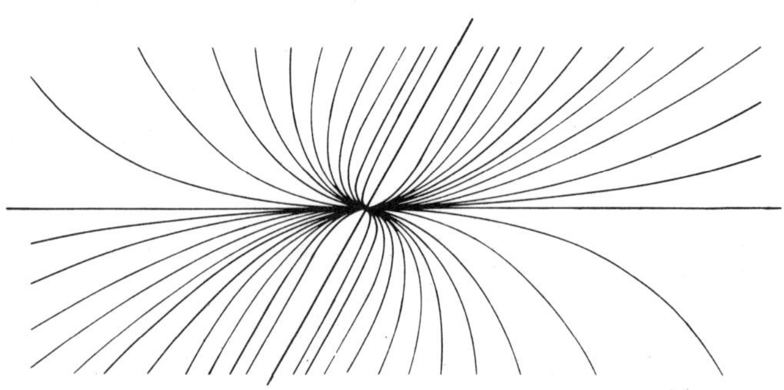

2. λ_1, λ_2 have opposite signature: $\lambda_1\lambda_2 < 0$

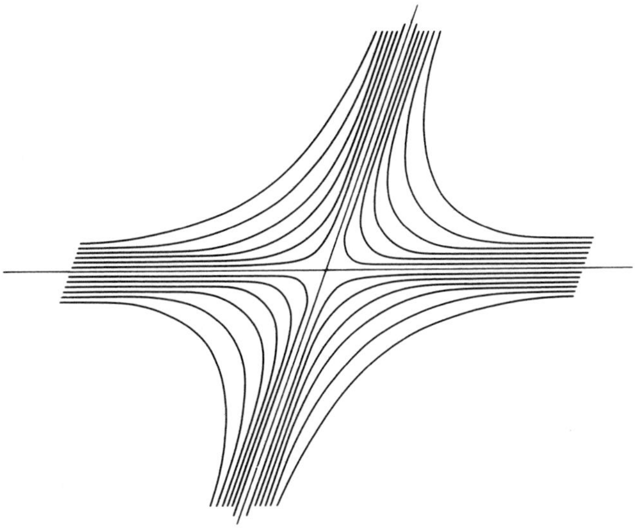

3. One of λ_1, λ_2 is zero: $\lambda_1\lambda_2 = 0$

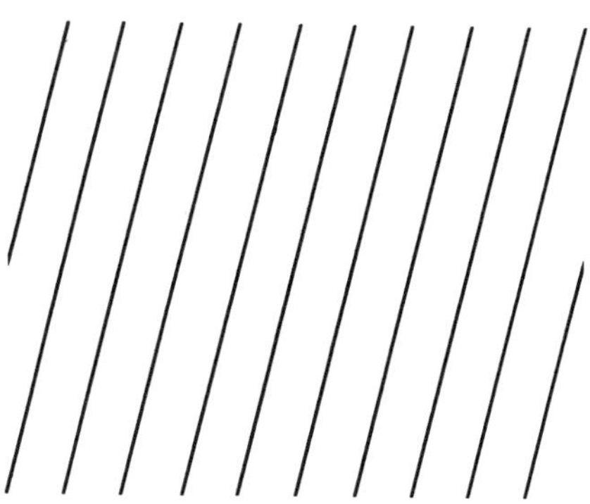

Parabolic

1. $\lambda \neq 0$

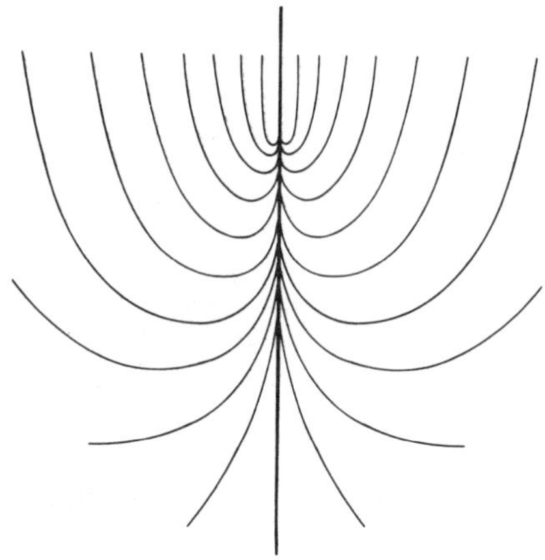

2. $\lambda = 0$

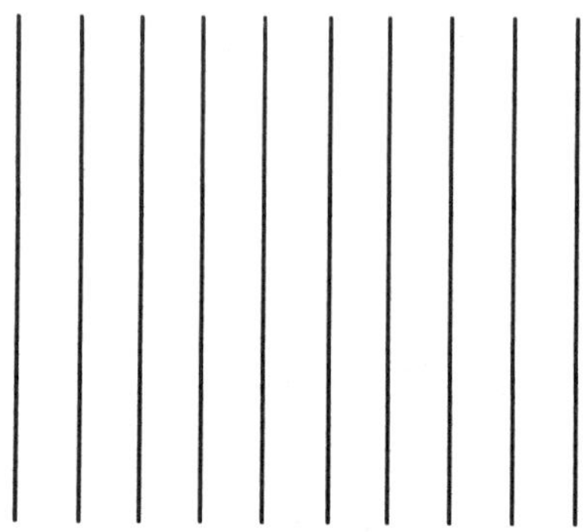

Strain

$\lambda \neq 0$

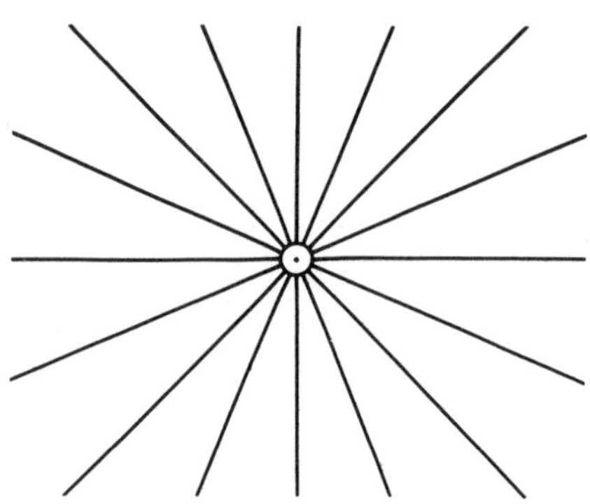

APPLICATIONS

2. Flows Induced by Symmetric and Antisymmetric Operators†

Example Find the streamlines of the flow induced by the anti-symmetric operator

$$f = \begin{pmatrix} 0 & -1 \\ 1 & 0 \end{pmatrix}$$

Answer:

$$e^{\alpha f} = \cos\alpha + f\sin\alpha = \begin{pmatrix} \cos\alpha & -\sin\alpha \\ \sin\alpha & \cos\alpha \end{pmatrix}$$

The streamline passing through $\mathbf{v}_0 = x_0\mathbf{i} + y_0\mathbf{j}$ is

$$\mathbf{v}_\alpha = e^{\alpha f}(\mathbf{v}_0) = (x_0\cos\alpha - y_0\sin\alpha)\mathbf{i} + (x_0\sin\alpha + y_0\cos\alpha)\mathbf{j}$$

$$\Rightarrow \begin{cases} x = x_0\cos\alpha - y_0\sin\alpha \\ y = x_0\sin\alpha + y_0\cos\alpha \end{cases} \Rightarrow x^2 + y^2 = x_0^2 + y_0^2 = \text{const}$$

This is a family of concentric circles centered at the origin.

Case of $\mathbb{R}^3$ If f is *symmetric*, then $f = \lambda_1 p_1 + \lambda_2 p_2 + \lambda_3 p_3$ by the principal axis theorem, where $\lambda_1, \lambda_2, \lambda_3$ are the eigenvalues of f and p_1, p_2, p_3, are perpendicular projections onto the (pairwise perpendicular) principal axes of f. The streamlines of the flow induced by f are

$$\mathbf{v}_\alpha = e^{\alpha\lambda_1 f}(\mathbf{v}_0) + e^{\alpha\lambda_2 f}(\mathbf{v}_0) + e^{\alpha\lambda_3 f}(\mathbf{v}_0)$$

If there is a multiple eigenvalue, all the streamlines in the plane of the corresponding eigenvectors will be straight lines. If all three eigenvalues are distinct, the only streamlines which are actually straight lines will be the three principal axes (or, rather, the six semiaxis determined by them). The shape of the remaining streamlines will depend on the signature of the eigenvalues.

If f is *antisymmetric*, then $f = \omega t$ where $\omega \in \mathbb{R}$ and $t^* = -t = t^3$, by the principal twist theorem. According to Steenrod's formula, the streamlines of the flow induced by f are

$$\mathbf{v}_\alpha = e^{\alpha t}(\mathbf{v}_0) = \mathbf{v}_0 + t^2(\mathbf{v}_0) + \cos\omega\alpha\, t^4(\mathbf{v}_0) + \sin\omega\alpha\, t(\mathbf{v}_0)$$

which is a family of circles lying in planes parallel to im t and centered on the line ker t. Let $\mathbf{u}$ denote the unit vector spanning ker t. Then we have

†*Note*: In this application the bases are assumed to be orthonormal and all coordinates are assumed to be rectangular metric.

$f(\mathbf{v}) = \omega\mathbf{u} \times \mathbf{v}$, for all $\mathbf{v} \in \mathbb{R}^3$. In the literature of mathematical physics the vector $\omega\mathbf{u}$ is called the *curl* of this vector field. The flow is a whirlpool (vortex) around the curl vector and ω is the (uniform) angular velocity of the particles in the flow.

3. Divergence and Curl of a Linear Operator in $\mathbb{R}^3$

Having analyzed so successfully the flows induced by a symmetric and an antisymmetric operator, it would be reasonable to expect an analysis of an arbitrary linear operator f. In fact, every linear operator f can be uniquely decomposed into the sum of its symmetric and anti-symmetric part s, a, respectively, namely, $s = \frac{1}{2}(f + f^*)$, $a = \frac{1}{2}(f - f^*)$ satisfy $s^* = s$, $a^* = -a$, $s + a = f$ and, conversely, if s and a satisfy these three conditions, they cannot be anything but those given above.

However, the problem of finding the flow of an arbitrary linear operator in terms of the flows of its symmetric and antisymmetric parts is not as easy as it might appear at the first glance. If $e^{\alpha s}$, $e^{\alpha a}$ are the exponentials of s, a, respectively, one would hope (in view of the det-trace formula) that the product $e^{\alpha s} \circ e^{\alpha a}$ might furnish the exponential of $s + a$; but this is not always the case. In fact, the product $e^{\alpha s} \circ e^{\alpha a}$ furnishes the exponential of the linear operator $g = s + e^{\alpha s} \circ a \circ e^{-\alpha s}$. Therefore, if a and $e^{\alpha s}$ commute, then the second term reduces to a and $e^{\alpha s} \circ e^{\alpha a}$ is indeed the exponential of $f = s + a$. But, in general, $e^{\alpha s}$ and a do not commute. However, when α is small, $e^{\alpha s}$ is near $e^{0s} = 1$ which commutes with all linear operators. Thus, for small values of α , $e^{\alpha s}$ very nearly commutes with a and $\mathbf{v}_\alpha = e^{\alpha s} \circ e^{\alpha a}(\mathbf{v}_0)$ will approximate the streamline through the point v_0 of the flow induced by f.

Accordingly, there will be a rotational effect measured by curl and a divergence effect measured by trace s under the flow induced by $f = s + a$ ($s^* = s$, $a^* = -a$). It is customary to call trace s = trace f the *divergence* of f; and the vector $\omega\mathbf{u}$ (such that $a(\mathbf{v}) = \omega\mathbf{u} \times \mathbf{v}$ for all $\mathbf{v}$) the *curl* of f.

4. Hyperbolic and Trigonometric Functions of Higher Order

The hyperbolic functions

$$\cosh \alpha = 1 + \frac{\alpha^2}{2!} + \frac{\alpha^4}{4!} + \cdots$$

$$\sinh \alpha = \frac{\alpha}{1!} + \frac{\alpha^3}{3!} + \frac{\alpha^5}{5!} + \cdots$$

can be introduced via the exponential of the involution in $\mathbb{R}^2$

$$f = \begin{pmatrix} 0 & 1 \\ 1 & 0 \end{pmatrix}$$

satisfying $f^2 = 1$ and trace $f = 0$ as follows.

$$e^{\alpha f} = 1 + \frac{\alpha f}{1!} + \frac{\alpha^2 f^2}{2!} + \cdots = \cosh\,\alpha + f\sinh\,\alpha$$

$$= \begin{pmatrix} \cosh\alpha & \sinh\alpha \\ \sinh\alpha & \cosh\alpha \end{pmatrix}$$

Since trace $f = 0$, we have det $e^{\alpha f} = 1$, i.e.,

$$\cosh^2\alpha - \sinh^2\alpha = 1$$

Moreover, on the strength of $e^{\alpha f} \circ e^{\beta f} = e^{(\alpha+\beta)f}$, we have the

Addition Formulas

$$\cosh(\alpha+\beta) = \cosh\,\alpha\cosh\,\beta + \sinh\,\alpha\sinh\,\beta$$
$$\sinh(\alpha+\beta) = \cosh\,\alpha\sinh\,\beta + \sinh\,\alpha\cosh\,\beta$$

Now we shall introduce the *hyperbolic functions of order* 3: $h(\alpha)$, $h'(\alpha)$, $h''(\alpha)$ via the exponential of the operator of $\mathbb{R}^3$

$$f = \begin{pmatrix} 0 & 0 & 1 \\ 1 & 0 & 0 \\ 0 & 1 & 0 \end{pmatrix}$$

satisfying $f^3 = 1$ and trace $f = 0$.

$$e^{\alpha f} = 1 + \frac{\alpha f}{1!} + \frac{\alpha^2 f^2}{2!} + \cdots$$

$$= h(\alpha)f^2 + h'(\alpha)f + h''(\alpha) = \begin{pmatrix} h''(\alpha) & h(\alpha) & h'(\alpha) \\ h'(\alpha) & h''(\alpha) & h(\alpha) \\ h(\alpha) & h'(\alpha) & h''(\alpha) \end{pmatrix}$$

where

$$h(\alpha) = \frac{\alpha^2}{2!} + \frac{\alpha^5}{5!} + \frac{\alpha^8}{8!} + \cdots$$
$$h'(\alpha) = \frac{\alpha}{1!} + \frac{\alpha^4}{4!} + \frac{\alpha^7}{7!} + \cdots$$
$$h''(\alpha) = 1 + \frac{\alpha^3}{3!} + \frac{\alpha^6}{6!} + \cdots$$

Since trace $f = 0$, we have det $e^{\alpha f} = 1$, i.e.,

$$\begin{vmatrix} h''(\alpha) & h(\alpha) & h'(\alpha) \\ h'(\alpha) & h''(\alpha) & h(\alpha) \\ h(\alpha) & h'(\alpha) & h''(\alpha) \end{vmatrix} = 1 \qquad \text{for all } \alpha \in \mathbb{R}$$

Moreover, on the strength of $e^{\alpha f} \circ e^{\beta f} = e^{(\alpha+\beta)f}$, we have the

Addition Formulas

$$h(\alpha+\beta) = h(\alpha)h''(\beta) + h'(\alpha)h'(\beta) + h''(\alpha)h(\beta)$$
$$h'(\alpha+\beta) = h'(\alpha)h''(\beta) + h''(\alpha)h'(\beta) + h(\alpha)h(\beta)$$
$$h''(\alpha+\beta) = h''(\alpha)h''(\beta) + h(\alpha)h'(\beta) + h'(\alpha)h(\beta)$$

Let us now calculate the hyperbolic functions of order 3 in terms of the elementary functions, by using the third root of unity

$$\varepsilon = -\frac{1}{2} + \frac{\sqrt{3}}{2}i$$

satisfying $1 + \varepsilon + \varepsilon^2 = 0$ and $\varepsilon^3 = 1$. We have

$$e^{\alpha} = h''(\alpha) + h'(\alpha) + h(\alpha)$$
$$e^{\varepsilon\alpha} = h''(\alpha) + \varepsilon h'(\alpha) + \varepsilon^2 h(\alpha)$$
$$e^{\varepsilon^2\alpha} = h''(\alpha) + \varepsilon^2 h'(\alpha) + \varepsilon h(\alpha)$$

Hence $3h''(\alpha) = e^{\alpha} + e^{\varepsilon\alpha} + e^{\varepsilon^2\alpha}$ and

$$h''(\alpha) = \frac{1}{3} e^{\alpha} + \frac{2}{3} e^{-(1/2)\alpha} \cos \frac{\sqrt{3}}{2} \alpha$$

Similar calculations yield

$$h'(\alpha) = \frac{1}{3} e^{\alpha} + \frac{2}{3} e^{-(1/2)\alpha} \cos(\frac{\sqrt{3}}{2}\alpha - 120°)$$
$$h(\alpha) = \frac{1}{3} e^{\alpha} + \frac{2}{3} e^{-(1/2)\alpha} \cos(\frac{\sqrt{3}}{2}\alpha - 240°)$$

The *hyperbolic functions* $h, h', \ldots, h^{(n-1)}$ *of order* n are defined via the exponential of the operator of $\mathbb{R}^n$

$$f = \begin{pmatrix} 0 & 0 & 0 & \cdots & 1 \\ 1 & 0 & 0 & \cdots & 0 \\ 0 & 1 & 0 & \cdots & 0 \\ \cdots & \cdots & \cdots & \cdots & \cdots \end{pmatrix}$$

satisfying $f^n = 1$ and trace $f = 0$.

$$e^{\alpha f}=h(\alpha)f^{n-1}+h'(\alpha)f^{n-2}+\cdots+h^{(n-2)}(\alpha)f+h^{(n-1)}(\alpha)$$

$$=\begin{pmatrix} h^{(n-1)}(\alpha) & h(\alpha) & h'(\alpha) & \cdots & h^{(n-2)}(\alpha) \\ h^{(n-2)}(\alpha) & h^{(n-1)}(\alpha) & h(\alpha) & \cdots & h^{(n-3)}(\alpha) \\ h^{(n-3)}(\alpha) & h^{(n-2)}(\alpha) & h^{(n-1)}(\alpha) & \cdots & h^{(n-4)}(\alpha) \\ \cdots\cdots & \cdots\cdots & \cdots\cdots & \cdots\cdots & \cdots\cdots \end{pmatrix}$$

Since trace $f=0$, det $e^{\alpha f}=1$, i.e.,

$$\begin{vmatrix} h^{(n-1)}(\alpha) & h(\alpha) & \cdots & h^{(n-2)}(\alpha) \\ h^{(n-2)}(\alpha) & h^{(n-1)}(\alpha) & \cdots & h^{(n-3)}(\alpha) \\ h^{(n-3)}(\alpha) & h^{(n-2)}(\alpha) & \cdots & h^{(n-4)}(\alpha) \\ \cdots\cdots & \cdots\cdots & \cdots\cdots & \cdots\cdots \end{vmatrix}=1 \qquad \text{for all } \alpha\in\mathbb{R}$$

The addition formulas can be obtained from $e^{\alpha f}\circ e^{\beta f}=e^{(\alpha+\beta)f}$. It is left as an exercise to find the power series expansions of the hyperbolic functions of order n.

The *trigonometric function* $s, s', \ldots, s^{(n-1)}$ *of order* n are defined as

$$s(\alpha)=\frac{\alpha^{n-1}}{(n-1)!}-\frac{\alpha^{2n-1}}{(2n-1)!}+\frac{\alpha^{3n-1}}{(3n-1)!}-+\cdots$$

$$s'(\alpha)=\frac{\alpha^{n-2}}{(n-2)!}-\frac{\alpha^{2n-2}}{(2n-2)!}+\frac{\alpha^{3n-2}}{(3n-2)!}-+\cdots$$

$$\vdots$$

$$s^{(n-1)}(\alpha)=1-\frac{\alpha^{n}}{n!}+\frac{\alpha^{2n}}{(2n)!}-\frac{\alpha^{3n}}{(3n)!}+-\cdots$$

These functions, like the hyperbolic functions, can also be introduced via the exponential of a linear operator f of $\mathbb{R}^n$ such that $f^n=-1$ and trace $f=0$. It is left as an exercise to find f, $e^{\alpha f}$, det $e^{\alpha f}$, and the addition formulas for the trigonometric functions of order n.

The ordinary trigonometric functions sin α, cos α are of order 2. It is left as an exercise to show that $e^{\alpha f}$ is orthogonal for all $\alpha\in\mathbb{R}$ if, and only if, $n=2$. This fact explains the extraordinary utility of the trigonometric functions of order 2, not shared by the trigonometric functions of order $n\neq 2$, nor by the hyperbolic functions.

LIE ALGEBRAS OF LINEAR OPERATORS

The concept of a Lie algebra was introduced in Chap. 4.

Examples

1. $L=L(\mathbb{R}^n,\mathbb{R}^n)$ is an n^2-dimensional Lie algebra under the Lie product

$$[f,g] = f \circ g - g \circ f$$

This is to say, the vector space L satisfies the following five axioms:

(0^x) Closure

$$f,\ g \in L \Rightarrow [f,g] \in L$$

(1^x) Anticommutativity

$$[g,f] = -[f,g]$$

(2^x) Scalar Associativity

$$[\lambda f,g] = \lambda[f,g] = [f,\lambda g] \qquad \lambda \in \mathbb{R}$$

(3^x) Jacobi Identity

$$[f,[g,h]] + [g,[h,f]] + [h,[f,g]] = 0$$

(5^x) Distributivity

$$[f,\ g+g'] = [f,g] + [f,g']$$
$$[f+f',\ g] = [f,g] + [f',g]$$

We prove 3^x; the proofs of the others are left as an exercise.

$$\begin{aligned}
&[f,[g,h]] + [g,[h,f]] + [h,[f,g]] = f \circ [g,h] - [g,h] \circ f + \cdots \\
&= f \circ (g \circ h - h \circ g) - (g \circ h - h \circ g) \circ f + \cdots \\
&= f \circ g \circ h - f \circ h \circ g - g \circ h \circ f + h \circ g \circ f + g \circ h \circ f - g \circ f \circ h - h \circ f \circ g + f \circ h \circ g \\
&\qquad + h \circ f \circ g - h \circ g \circ f - f \circ g \circ h + g \circ f \circ h = 0
\end{aligned}$$

In order to show that $\dim L = n^2$, we shall produce a set consisting of n^2 elements $\ell_{\mu\nu} \in L$ $(\mu,\nu = 1, \ldots, n)$ satisfying the following conditions:

(1)
$$\ell_{\mu\nu}^2 = \begin{cases} \ell_{\mu\mu} \Leftrightarrow \mu = \nu \\ 0 \quad \Leftrightarrow \mu \neq \nu \end{cases}$$

i.e., n elements of the set are projections, and the rest of them are lifts;

(2) $$[\ell_{\mu\nu},\ell_{\pi\rho}] = \begin{cases} \ell_{\mu\rho} \Leftrightarrow \nu = \pi \text{ and } \mu \neq \rho \\ 0 \quad \text{otherwise} \end{cases}$$

(3) The set is linearly independent in L.
(4) The set spans L.

To prove this, start with an arbitrary basis $\{\mathbf{e}_1,\dots,\mathbf{e}_n\}$ of $\mathbb{R}^n$. Construct the dual basis $\{\mathbf{m}_1,\dots,\mathbf{m}_n\}$ satisfying

$$\mathbf{m}_\nu \cdot \mathbf{e}_\mu = \begin{cases} 1 \Leftrightarrow \nu = \mu \\ 0 \Leftrightarrow \nu \neq \mu \end{cases}$$

The details of constructing the dual basis are left as an exercise. Define $\ell_{\mu\nu}$ by setting

$$\ell_{\mu\nu}(\mathbf{v}) = (\mathbf{m}_\nu \cdot \mathbf{v})\mathbf{e}_\mu \qquad (\mathbf{v} \in \mathbb{R}^n)$$

It is easy to check that the set of $\ell_{\mu\nu} \in L$ satisfies the conditions (1) through (4) and is therefore a basis for L. It follows that $\dim L = n^2$.

If the basis $\{\mathbf{e}_1, \dots, \mathbf{e}_n\}$ is orthornormal, it coincides with its dual: $\mathbf{m}_\nu = \mathbf{e}_\nu$, $\nu = 1, \dots, n$. In this case $\ell_{\mu\mu}$ are perpendicular projections:

$$\ell^*_{\mu\mu} = \ell_{\mu\mu} = \ell^2_{\mu\mu}$$

and $\ell_{\mu\nu}$, $\mu \neq \nu$, are perpendicular lifts (see Exercise 11.16):

$$\ell_{\mu\nu} \circ \ell^*_{\mu\nu} = \ell_{\mu\mu}, \qquad \ell^*_{\mu\nu} \circ \ell_{\mu\nu} = \ell_{\nu\nu}$$

In fact, we have

$$\ell^*{}_{\mu\nu} = \ell_{\nu\mu}, \qquad \mu, \nu = 1, \dots, n$$

We may convert L into a metric vector space if we define a dot product $\langle\ ,\ \rangle$ in L by putting

$$\langle \ell_{\mu\nu}, \ell_{\pi\rho} \rangle = \text{trace}(\ell_{\mu\nu} \circ \ell^*_{\pi\rho})$$

and extend it by linearity to the whole of L. It is left as an exercise to show that this definition converts L into a metric vector space, and that in terms of this metric, the basis $\{\ell_{\mu\nu} \mid \mu,\nu = 1,\dots,n\}$ is orthornormal. (The metric in general is not positive definite.)

2. Let $A = \{f \in L \mid \text{trace} f = 0\}$. By the vanishing trace property, $\text{trace}[f,g] = 0$ for all f, $g \in L$, so, for the stronger reason, A is closed under the Lie product. It is also closed under the sum and scalar multiple of linear operators, on the strength of the bilinearity property of the trace. We conclude that A is a Lie subalgebra of L, called the *special Lie algebra* of $\mathbb{R}^n$, in symbols, $sL = sL(\mathbb{R}^n, \mathbb{R}^n)$. Actually, we have more: the special Lie algebra sL is an *ideal* of the general Lie algebra L, i.e.,

$$g \in sL, f \in L \Rightarrow [f,g] \in sL$$

This immediate from the vanishing trace property.

As $\dim L = n^2$, and as sL is given by the linear equation $a_{11} + \cdots + a_{nn} = 0$, the latter is a hyperplane of the former and, hence, $\dim sL = n^2 - 1$.

3. Let A be the vector space of antisymmetric operators:

$$A = \{f \in L \mid f^* = -f\}$$

A is closed under the Lie product:

$$\begin{aligned} f, g \in A &\Rightarrow f^* = -f,\ g^* = -g \\ &\Rightarrow [f,g]^* = (f \circ g - g \circ f)^* = (f \circ g)^* - (g \circ f)^* \\ &\qquad = g^* \circ f^* - f^* \circ g^* = (-g) \circ (-f) - (-f) \circ (-g) \\ &\qquad = g \circ f - f \circ g = -[f,g] \Rightarrow [f,g] \in A \end{aligned}$$

A is also closed under the sum and scalar multiple of linear operators, on the strength of the bilinearity property of transposition. Therefore A is a Lie subalgebra of L, and since $f^* = -f \Rightarrow \text{trace } f = 0$, it is a Lie subalgebra of the ideal sL of linear operators with trace zero as well. In order to clarify the structure of A we shall now construct a basis for it. Set

$$t_{\mu\nu} = (-1)^{\mu+\nu}(\ell_{\mu\nu} - \ell^*_{\mu\nu})$$

where $\{\ell_{\mu\nu} \mid \mu,\nu = 1,\ldots,n\}$ is the orthonormal basis for L introduced in the first example above. In other words,

$$t_{\mu\nu}(\mathbf{v}) = (-1)^{\mu+\nu}((\mathbf{e}_\nu \cdot \mathbf{v})\mathbf{e}_\mu - (\mathbf{e}_\mu \cdot \mathbf{v})\mathbf{e}_\nu) \qquad (\mathbf{v} \in \mathbb{R}^n)$$

Then the set $\{t_{\mu\nu} \mid 1 \leq \mu < \nu \leq n\}$ satisfies the following conditions:

(a) Each of the $\frac{1}{2}n(n-1)$ elements is an antisymmetric twist, i.e.,

$$t^*_{\mu\nu} = -t_{\mu\nu} = t^3_{\mu\nu}$$

(b)

$$[t_{\mu\nu}, t_{\pi\rho}] = \begin{cases} t_{\mu\rho} \Leftrightarrow \nu = \pi \\ 0 \quad \Leftrightarrow \nu \neq \pi \end{cases}$$

(c) The set is linearly independent in A.

(d) The set spans A.

(e) In the metric of L introduced above, the set is orthogonal, i.e.,

$$\langle t_{\mu\nu}, t_{\pi\rho} \rangle = \begin{cases} 2 & \text{if } \mu = \pi \text{ and } \nu = \rho \\ 0 & \text{otherwise} \end{cases}$$

The details are left as an exercise. We conclude that A is a $\frac{1}{2}n(n-1)$-dimensional Lie subalgebra of L.

4. We shall now generalize the preceding example in constructing other $\frac{1}{2}n(n-1)$-dimensional isomorphic Lie subalgebras of L with a basis consisting of twists. Unlike the construction in the preceding example, this one is *not* using the metric of the vector space $\mathbb{R}^n$ in an essential way.

Let $\{\ell_{\mu\nu} \mid \mu, \nu = 1,\ldots,n\}$ be the basis of L of the first example above, i.e.,

$$\ell_{\mu\nu}(\mathbf{v}) = (\mathbf{m}_\nu \cdot \mathbf{v})\mathbf{e}_\mu \qquad (\mathbf{v} \in \mathbb{R}^n)$$

where $\{\mathbf{e}_1, \ldots, \mathbf{e}_n\}$ and $\{\mathbf{m}_1, \ldots, \mathbf{m}_n\}$ are dual bases of $\mathbb{R}^n$. Put

$$t_{\mu\nu} = (-1)^{\mu+\nu}(\ell_{\mu\nu} - \ell_{\nu\mu})$$

Then the set

$$\{t_{\mu\nu} \mid 1 \leq \mu < \nu \leq n\}$$

satisfies the following conditions:

(a) Each of the $n(n-1)/2$ elements is a twist:

$$t_{\mu\nu}^3 = -t_{\mu\nu}$$

(b)

$$[t_{\mu\nu}, t_{\pi\rho}] = \begin{cases} t_{\mu\rho} \Leftrightarrow \nu = \pi \\ 0 \quad \Leftrightarrow \nu \neq \pi \end{cases}$$

(c) The set is linearly independent in L.

The proof is left as an exercise.

Now consider the vector space T spanned by this set. Clearly, T is a Lie subalgebra of L, in view of (b). We shall call T a *twist algebra* of $\mathbb{R}^n$. If T' is another twist algebra of $\mathbb{R}^n$, then T and T' are isomorphic as Lie algebras. The isomorphism is furnished by the transition operator changing the basis used in the construction of T into that used in the construction of T'. Since trace $t_{\mu\nu} = 0$, the twist algebras are Lie subalgebras of the ideal sL also. The Lie algebra A of antisymmetric operators is just one of the twist algebras of $\mathbb{R}^n$. In particular, A is isomorphic to T. In order to survey the twist algebras of $\mathbb{R}^n$, we shall now introduce the idea of conjugation.

CONJUGATION

The *conjugate* of $f \in L$ by $g \in GL(\mathbb{R}^n)$ is $f^g = g \circ f \circ g^{-1} \in L$. The function $L \to L$ such that $f \rightsquigarrow f^g$ is called conjugation by g (cf. see Chapter 8, Application 4). By the cancellation law, conjugation is bijective.

Formal Properties of Conjugation

$$(f+f')^g = f^g + f'^g$$
$$(\alpha f)^g = \alpha(f^g) \qquad \alpha \in \mathbb{R}$$
$$0^g = 0$$

Product Rule

$$(f \circ f')^g = f^g \circ f'^g$$
$$(f^{-1})^g = (f^g)^{-1}$$
$$1^g = 1$$

Conjugate of Exp

$$(e^{\alpha f})^g = e^{\alpha f^g}$$

Conjugate of Lie Bracket

$$[f, f']^g = [f^g, f'^g]$$

i.e., conjugation preserves the sum, scalar multiple, product, inverse of operators. Moreover, the conjugate of the exponential is the exponential of the conjugate, and the conjugate of the Lie product is the Lie product of the conjugates. Let us prove the product rule:

$$(f \circ f')^g = g \circ f \circ f' \circ g^{-1} = (g \circ f \circ g^{-1}) \circ (g \circ f' \circ g^{-1}) = f^g \circ f'^g$$

We also calculate the conjugate of the Lie bracket, leaving the proofs of the remaining formal rules as an exercise.

$$[f,f']^g = (f \circ f' - f' \circ f)^g = (f \circ f')^g - (f' \circ f)^g = f^g \circ f'^g - f'^g \circ f^g = [f^g, f'^g]$$

It follows from the formal properties that conjugation preserves twists. However, in general, conjugation does not preserve antisymmetric operators.

The conjugate of a Lie subalgebra A of L by g is defined as

$$A^g = \{f^g \mid f \in A\}$$

It is clear that A^g is also a Lie subalgebra. However, the conjugate of the Lie algebra of antisymmetric operators is no longer antisymmetric. At any rate, the twist algebras of $\mathbb{R}^n$ can now be described as the conjugates of the Lie algebra of antisymmetric operators.

The conjugate of a subgroup H of $GL(\mathbb{R}^n)$ by g is defined as

$$H^g = \{h^g \mid h \in H\}$$

It is clear that H^g is also a subgroup. Note that H is a normal subgroup of $GL(\mathbb{R}^n)$ if, and only if, all the conjugates of H coincide, i.e., $H^g = H$ for all $g \in GL(\mathbb{R}^n)$. Normal subgroups remain invariant under conjugation.

We shall in particular be interested in the following

Example The conjugates of the special orthogonal group $SO(\mathbb{R}^n)$ in $SL(\mathbb{R}^n)$ are called the *invection groups* of $\mathbb{R}^n$. They consist of products of skew rotations. The conjugates of rotations through 90° are just the orientation-preserving invections, revealing how the invection groups earn their name. This also explains why the conjugates of the rotations are called *skew rotations*, and the invections themselves *skew rotations through 90°* (see Chap. 10, Applications 6 and 9). Conjugation preserves the angle of skew rotations. Moreover, the conjugate of a reflection is a skew reflection or involution (see Chap. 7, Application 2). We are going to see the important relationship between the twist algebras and the invection groups of $\mathbb{R}^n$ next.

THE CONCEPT OF A FUNCTOR

The *exponential functor* exp is the bijective correspondence

$$L = L(\mathbb{R}^n, \mathbb{R}^n) \overset{\exp}{\rightsquigarrow} GL^+(\mathbb{R}^n) = G$$

from the Lie algebra of linear operators to the Lie group of orientation-preserving operators of $\mathbb{R}^n$, assigning a Lie subgroup H of G to every Lie subalgebra A of L, in symbols: $H = \exp A$. The terms *functor* and *Lie group* will be explained in some detail below. If A is a k-dimensional Lie subalgebra, then $\exp A$ is a k-parameter Lie group. If A is a commutative Lie subalgebra, then $\exp A$ is a commutative Lie group. If A is an ideal of L, then $\exp A$ is a normal subgroup of G, and vice versa.

Examples

1. L is an n^2-dimensional Lie algebra, and G is an n^2-parameter Lie group. Let $f \in L$, $f \neq 0$. Then f spans a one-dimensional commutative Lie subalgebra of L, $A = \{\lambda f \mid \lambda \in \mathbb{R}\}$. We have $\exp A = \{e^{\alpha f} \mid \alpha \in \mathbb{R}\}$, a one-parameter commutative Lie subgroup of G.
2. Let $f, g \in L$ be linearly independent operators such that $[f,g] = \xi f + \eta g$, ξ, $\eta \in \mathbb{R}$. Then f, g generate a two-dimensional Lie algebra

$$A = \{\lambda f + \mu g \mid \lambda, \mu \in \mathbb{R}\}$$

and $\exp A$ is the group generated by $e^{\alpha f}$, $e^{\beta g}$ for all α, $\beta \in \mathbb{R}$: it is a two-parameter Lie group.

If f, g commute, i.e., $[f,g]=0$, then A is a commutative Lie algebra and exp A is a commutative Lie group. This follows from the formula

$$e^{\alpha(f,g)} = e^{\beta f} \circ e^{\beta g} \circ e^{-\beta f} \circ e^{-\beta g}$$

(the proof of which is left as an exercise), because

$$[f,g]=0 \Rightarrow e^{\beta f} \circ e^{\beta g} \circ e^{-\beta f} \circ e^{-\beta g} = 1 \Rightarrow e^{\beta f} \circ e^{\beta g} = e^{\beta g} \circ e^{\beta f}$$

More generally, if the Lie algebra A has a basis $\{f_1, \ldots, f_k\}$ such that $[f_\mu, f_\nu]=0$ for all $\mu, \nu = 1, \ldots, k$, then A is a commutative Lie algebra and exp A is a commutative Lie group.

3. By the det-trace formula, $e^{\alpha f}$ is volume preserving for all $\alpha \in \mathbb{R}$ if, and only if, f has zero trace. Therefore,

$$\exp sL = SL(\mathbb{R}^n)$$

the special linear group. Since $\dim sL = n^2-1$, $SL(\mathbb{R}^n)$ is an (n^2-1)-parameter Lie group.

We draw attention to the facts (already observed) that $SL(\mathbb{R}^n)$ is a normal subgroup of $G = GL^+(\mathbb{R}^n)$, and sL is an ideal of the Lie algebra L. This is no accident as it is true in general that exp A is a normal subgroup of G if, and only if, A is an ideal of the Lie algebra L.

4. Let A be the Lie algebra of antisymmetric operators. Since

$$f^* = -f \Leftrightarrow (e^{\alpha f})^* = e^{\alpha f^*} = e^{-\alpha f} = (e^{\alpha f})^{-1}$$
$$\Leftrightarrow e^{\alpha f} \text{ is orthogonal for all } \alpha \in \mathbb{R}$$

we conclude that

$$\exp A = SO(\mathbb{R}^n)$$

the group of orientation-preserving orthogonal operators, a $\frac{1}{2}n(n-1)$-parameter Lie group.

We draw attention to the special case $n=3$, already discussed in Chap. 4. This case is remarkable because $n=3 \Rightarrow \frac{1}{2}n(n-1)=3$, so that the Lie algebra belonging to the group $SO(\mathbb{R}^3)$ of rotations is just the Lie algebra $\mathbb{R}^3$ of vectors endowed with the cross product. As observed in Chap. 4, $\mathbb{R}^3$ is the only vector space that can be converted into a non-commutative Lie algebra in a way that is compatible with the metric. $\mathbb{R}^n$ fails to have a Lie algebra structure compatible with the metric for $n \geqq 4$.

In spite of this failure, the cross product has a natural generalization: the Lie algebra of antisymmetric operators of $\mathbb{R}^n$, and $\mathbf{i} \times \mathbf{j} = \mathbf{k}$ generalizes to the Lie product of twists $[t_{\mu\nu}, t_{\nu\rho}] = t_{\mu\rho}$.

5. Let T be a twist algebra, then exp T is an invection group. Invection groups have been defined as the conjugates of the special orthogonal

group. They could also be defined as the exponentials of the twist algebras: the two definitions are equivalent.

Indeed, Steenrod's formula can be generalized for an arbitrary twist $f \in T$:

$$f^3 = -f \Rightarrow e^{\alpha f} = 1 + \frac{\alpha f}{1!} + \frac{\alpha^2 f^2}{2!} + \cdots$$

$$= 1 + (\frac{\alpha}{1!} - \frac{\alpha^3}{3!} + - \cdots)f + (-\frac{\alpha^2}{2!} + \frac{\alpha^4}{4!} - + \cdots)f^4$$

$$= 1 + f \sin \alpha + f^4 \cos \alpha - f^4 = (1 + f^2) + f^4 \cos \alpha + f \sin \alpha$$

which is a volume-preserving elliptic operator (skew rotation) for $\alpha \neq 0°$, 180°; an orientation-preserving involution (skew reflection in a line) for $\alpha = 180°$. If $\alpha = 90°$, we have the Cayley transform of f (see Chap. 10, Application 5), an orientation-preserving invection (skew rotation through 90°).

From the conjugate of exp property it follows that if T^g, $g \in SL(\mathbb{R}^n)$, is another twist algebra, then $\exp(T^g) = (\exp T)^g$. In particular, for the Lie algebra A of antisymmetric operators, $\exp(A^g) = (\exp A)^g$, showing the equivalence of the two definitions of an invection group.

The Cayley transform now appears as a bijective function between the elements of twist algebras and the elements of invection groups which do not have -1 as an eigenvalue. For $T = A$ we recapture the Cayley transform between anti-symmetric operators and orientation-preserving orthogonal operators (see Chap. 11, Application 4).

The invection group $\exp T$ is a $\frac{1}{2}n(n-1)$-parameter Lie group which is isomorphic to $SO(\mathbb{R}^n)$ under conjugation.

It is not the task of this book to give the precise definition of a Lie group. Such a definition falls within the domain of differential geometry. Here we content ourselves with an intuitive geometric description of this concept. A Lie group with k parameters can be visualized as a k-dimensional curved surface coordinatized by the parameters $\alpha_1, \ldots, \alpha_k$. The origin is the identity operator 1 belonging to parameter values $(0, \ldots, 0)$. For any fixed parameter values $\alpha_\nu \in \mathbb{R}$ we get a certain group element as a point of the curved surface, and every element of the Lie group can be uniquely represented in this way. If we fix all but one of the parameter values, we get a curve on the surface called a coordinate line. Through every point there are exactly k such coordinate lines. The k coordinate lines passing through the origin 1 are the coordinate axes. Each coordinate axis is a one-parameter Lie group, with the origin as identity and the corresponding coordinate as the parameter. The coordinate lines of the Lie group fall into k families labeled by the parameters. Coordinate lines belonging to the same family have no point in common. Thus a Lie group appears as a cobweb of coordinate lines belong to k families; we shall say that the dimension of the Lie group is k. A one-dimensional Lie group is necessarily commutative, as it is coordinatized by a single parameter α, and it is subject to $e^{\alpha f} \circ e^{\alpha' f} = e^{(\alpha + \alpha')f}$. Higher dimensional Lie groups may or may not be commutative, and elements that do not lie on the same coordinate axis may or may not commute.

The intuitive picture of the exponential functor is as follows. Given the vector space $\mathbb{R}^n$, the n^2-dimensional Lie group $GL^+(\mathbb{R}^n)$ is visualized as a curved surface; the Lie algebra $L = L(\mathbb{R}^n, \mathbb{R}^n)$ then appears as the tangent space to that surface at the origin 1 (= the identity operator) such that the point of L in contact with 1 is 0 (= the zero operator). Every smooth curve passing through the origin 1 has a uniquely determined tangent line in the tangent space passing through 0. The converse is not true though: given a line in the tangent space through 0, there will be infinitely many smooth curves on the surface through 1 whose common tangent is the given line.

However, not every smooth curve through 1 is a Lie subgroup, although every line through 0 is a Lie subalgebra. The remarkable fact is that the exponential functor establishes a bijective correspondence between the Lie subgroups of $GL^+(\mathbb{R}^n)$ and the Lie subalgebras of L, so that if A is a Lie subalgebra represented as a subspace of the tangent space, then $\exp A$ is a Lie subgroup represented as a unique subsurface tangent to A. Thus the exponential functor can be described intuitively as a projection from the tangent space L to the curved surface $GL^+(\mathbb{R}^n)$ which respects the tangency of subspaces and subsurfaces.

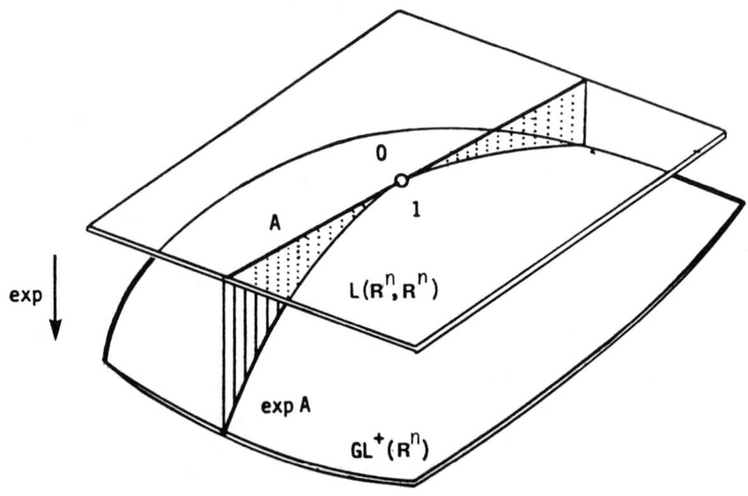

The exponential functor does not depend on the metric of $\mathbb{R}^n$: it is an affine concept which is available in the case of a vector space without dot product. The exponential functor earns its name by specializing to the *exponential function* $x \rightsquigarrow e^x$ taking the sum of real numbers into the product of the exponentials of the summands: $x+y \rightsquigarrow e^x e^y$. Indeed, this special case arises for $n=1$. It is left as an exercise to show that $L(\mathbb{R},\mathbb{R}) = \mathbb{R}$, the commutative Lie algebra of the real numbers; and $GL^+(\mathbb{R}) = \mathbb{R}^*$, the multiplicative group of positive real numbers.

It is not the task of this book to give the precise definition of a functor. Such a definition falls within the domain of category theory. Here we may content ourselves with pointing out that, in general, a functor fr is an assignment whereby to every set X with a certain type of mathematical structure another set fr X with a certain other type of mathematical structure is assigned. Moreover, if $f:X \to Y$ is a map preserving the first structure, then the functor induces a map fr$f$: fr $X \to$ fr Y preserving the second structure such that

$$\mathrm{fr}(g \circ f) = (\mathrm{fr}\ g) \circ (\mathrm{fr}\ f) \qquad \mathrm{fr}(1_X) = 1_{\mathrm{fr}X}$$

In words, the induced map of the product is the product of the induced maps, and the induced map of the identity is the identity. These formulas are called the *functorial properties*. The discussion of the functorial properties of exp, i.e., the fact that a Lie algebra homomorphism $f:A \to B$ induces a group homomorphism $\exp f : \exp A \to \exp B$ such that $\exp(g \circ f) = (\exp g) \circ (\exp f)$ and $\exp(1_A) = 1_{\exp A}$ is left as an exercise (see Exercise 14.20).

Examples

1. A *group homomorphism* $h : G \to G'$ is defined as a map such that for x, $y \in G$, $(x \circ y)^h = x^h \circ y^h$ and $1^h = 1$. Therefore h satisfies the functorial properties and is a functor. Conjugation of subgroups H of G by $g \in G$ is a functor because it is a group homomorphism $g : H \to H^g$, as $(x \circ y)^g = x^g \circ y^g$ for all x, $y \in H$.
2. A *Lie algebra homomorphism* $f : A \to B$ is defined as a linear transformation that also preserves the Lie product: $f([x,y]) = [f(x),f(y)]$ for $x,y \in A$. It is left as an exercise to show that

$$\mathrm{trace} : L(\mathbb{R}^n,\mathbb{R}^n) \to L(\mathbb{R},\mathbb{R}) = \mathbb{R}$$

is a Lie algebra homomorphism. In fact, the induced homomorphism of trace via the exponential functor is

$$\det : GL^+(\mathbb{R}^n) \to GL^+(\mathbb{R}) = \mathbb{R}^*$$

in symbols exp(trace) = det (see Exercise 14.21).

These are examples of a *covariant* functor. Another variety is called a *contravariant* functor cfr where the induced maps satisfy

$$\mathrm{cfr}(g \circ f) = (\mathrm{cfr}\, f) \circ (\mathrm{cfr}\, g)$$

(note the change in the order of the maps!) and

$$\mathrm{cfr}(1_X) = 1_{\mathrm{cfr} X}$$

Examples

1. Transposition of linear operators is a contravariant functor:

$$(g \circ f)^* = f^* \circ g^* \qquad \text{and} \qquad 1^* = 1$$

2. Taking the inverse of bijective operators is a contravariant functor:

$$(g \circ f)^{-1} = f^{-1} \circ g^{-1} \qquad \text{and} \qquad 1^{-1} = 1$$

CHANGING THE METRIC

A vector space per se is a model for affine ideas. Metric ideas arise only after additional structure has been imposed on a vector space.

The metric of the vector space $\mathbb{R}^3$ has been derived from a priori geometric principles such as those governing length, and perpendicularity, and therefore it can be considered "natural." However, the same is no longer true for $\mathbb{R}^n$, $n \geqq 4$, where the metric is a matter of endowment and is not part of the paraphernalia inherent in the vector space structure. In order to have the correct perspective on this fine distinction we shall now discuss the problem of changing the metric.

This problem is usually described in terms of changing the dot product. While that approach is quicker, it has the disadvantage of obscuring the geometry: it is not immediately clear what effect the changing of the dot product will have on rotations, reflections, etc. Therefore, we follow another route. We shall see that changing the dot product is not the cause but the effect of changing the metric. The true cause is changing the cross product or, (since the Lie algebra of antisymmetric operators has superseded the cross product), *changing the Lie algebra of antisymmetric operators*. More generally, if $\mathbb{R}^n$ has not had a metric in the first place, the introduction of a metric is effected not by the endowment of a dot product, but by the endowment of a cross product or, more precisely, by *the endowment of a Lie algebra of antisymmetric operators*. This will then introduce the group of orthogonal operators via the exponential functor, as well as all the other metric ideas. The details are as follows.

Pick a basis $\{\mathbf{e}_1, \ldots, \mathbf{e}_n\}$ of $\mathbb{R}^n$ and distinguish it by the name *preferred basis*. As we have seen, this choice fixes the orientation and volume in $\mathbb{R}^n$; we shall now see that it also fixes a twist algebra T and a metric in such a way that $\{\mathbf{e}_1, \ldots, \mathbf{e}_n\}$ will be an orthonormal basis in terms of that metric. Consider the basis for L,

$$\{\ell_{\mu\nu} \mid \mu, \nu = 1, \ldots, n\}$$

to which $\{\mathbf{e}_1, \ldots, \mathbf{e}_n\}$ gives rise through the construction described earlier. Recall that $\ell_{\mu\mu}$, $\mu = 1, \ldots, n$, are principal projections such that $\mathbf{e}_\mu$ is the eigenvector of $\ell_{\mu\mu}$ belonging to the eigenvalue $\lambda = 1$.

The *preferred twist algebra T* is spanned by the set of twists

$$\{t_{\mu\nu} = (-1)^{\mu+\nu}(\ell_{\mu\nu} - \ell_{\nu\mu}) \mid 1 \leq \mu < \nu \leq n\}$$

Introduce the *adjoint* * for the basis elements of L by the formula

$$\ell^*_{\mu\nu} = \ell_{\nu\mu}$$

and extend it to the whole of L by linearity. It is easy to see that the adjoint has all the formal properties of the transpose which it supersedes (see Chap. 11) and that

$$f^* = -f \Leftrightarrow f \in T$$

i.e., the *preferred twist algebra has now become the algebra of antisymmetric operators*. Using the properties of the exponential, it can also be seen that for a linear operator f with $\det f = 1$,

$$f^* = f^{-1} \Leftrightarrow f \in \exp T$$

This recaptures the concept of orthogonal operators and that of the orthogonal group:

$$\exp T = SO(\mathbb{R}^n)$$

In order to recapture the dot product in $\mathbb{R}^n$, first observe that there is an isomorphism Φ between $\mathbb{R}^n$ and the commutative Lie algebra spanned by the set of principal perpendicular projections $\{\ell_{\mu\mu} \mid \mu = 1, 2, \ldots, n\}$ such that $\lambda_1\mathbf{e}_1 + \lambda_2\mathbf{e}_2 + \cdots + \lambda_n\mathbf{e}_n \overset{\Phi}{\leftrightsquigarrow} \lambda_1\ell_{11} + \lambda_2\ell_{22} + \cdots + \lambda_n\ell_{nn}$. Then the new dot product $\langle\ ,\ \rangle$ is defined by the formula

$$\langle \mathbf{v}, \mathbf{w} \rangle = \operatorname{trace}(\Phi(\mathbf{v}) \circ (\Phi(\mathbf{w})^*)$$

It is easy to see that this dot product satisfies the formal rules (see Chap. 2) and thus converts $\mathbb{R}^n$ into a metric vector space. Moreover, the preferred basis is orthonormal in terms of the new metric.

To summarize, introducing a new metric in $\mathbb{R}^n$ means distinguishing one of the twist algebras T, by promoting it to the status of the Lie algebra of antisymmetric operators. This choice will also fix the group of rotations to be one of the invection groups, namely, exp T. Changing the metric in $\mathbb{R}^n$ means discarding T as the Lie algebra of antisymmetric operators and replacing it by another twist algebra T'. This will also change the group of rotations from exp T to exp T', another invection group, since exp T and exp T' are conjugate subgroups of $SL(\mathbb{R}^n)$.

The metric is "natural" in $\mathbb{R}^3$ by virtue of the natural isomorphism between the Lie algebra of antisymmetric operators (with Lie product $[\ ,\ ]$) and the Lie algebra $\mathbb{R}^3$ (with the cross product $\times$). The metric is also natural in $\mathbb{R}^2$ because the Lie algebra of antisymmetric operators is a one-dimensional commutative Lie algebra which is naturally isomorphic to the Lie algebra $\mathbb{R}$. However, the metric in $\mathbb{R}^n$ is no longer natural for $n \geq 4$. The absence of a natural isomorphism makes the metric a matter of choice, subject to change.

EXERCISES

14.1 Calculate $e^{\alpha f}$ for each of the following operators:

(i) $\begin{pmatrix} 0 & 0 & 1 \\ 0 & 0 & 0 \\ 0 & 0 & 0 \end{pmatrix}$ (ii) $\begin{pmatrix} 0 & 1 & 0 \\ 0 & 0 & 1 \\ 0 & 0 & 0 \end{pmatrix}$ (iii) $\begin{pmatrix} 1 & 0 & 0 \\ 4 & 5 & 4 \\ -6 & -6 & -5 \end{pmatrix}$

(iv) $\begin{pmatrix} 4 & 8 & 12 \\ -3 & -6 & 9 \\ -1 & -2 & 3 \end{pmatrix}$ (v) $\begin{pmatrix} -15 & 12 & 21 \\ -8 & 6 & 10 \\ -6 & 5 & 9 \end{pmatrix}$ (vi) $\begin{pmatrix} 4 & -7 & 5 \\ 3 & -6 & 5 \\ 2 & -4 & 3 \end{pmatrix}$

(vii) $\begin{pmatrix} 4 & -8 & 12 \\ -1 & 2 & -3 \\ -2 & 4 & -6 \end{pmatrix}$ (viii) $\begin{pmatrix} 2 & 1 & 0 \\ 0 & 2 & 1 \\ 0 & 0 & 2 \end{pmatrix}$ (ix) $\begin{pmatrix} 2 & 1 & 1 \\ 1 & 2 & 1 \\ 1 & 1 & 2 \end{pmatrix}$

(x) $\begin{pmatrix} 2 & -4 & 5 \\ 1 & -3 & 5 \\ 1 & -3 & 4 \end{pmatrix}$ (xi) $\begin{pmatrix} 1 & 1 \\ -2 & -1 \end{pmatrix}$ (xii) $\begin{pmatrix} 3 & -4 \\ 2 & -3 \end{pmatrix}$

14.2 Show that $\mathbf{w}$ is an eigenvector of $e^{\alpha f}$ if, and only if, $\mathbf{w}$ is an eigenvector of f.

14.3 Prove the det-trace formula

14.4 Prove the conjugate of exp property.

14.5 Prove that the exponential of f does not depend on the choice of coordinates and is, in fact, a linear operator.

14.6 Show that the streamlines of

$$f = \begin{pmatrix} 0 & 1 & 0 \\ 0 & 0 & 1 \\ 0 & 0 & 0 \end{pmatrix}$$

are plane curves.

14.7 Find the streamlines of

$$f = \begin{pmatrix} 5 & 13 \\ -2 & -5 \end{pmatrix}$$

14.8 Calculate the hyperbolic functions of order 4 in terms of the elementary functions.

14.9 Write the power series expansion of the hyperbolic functions of order n.

14.10 Let $s, s', s'', \ldots, s^{(n-1)}$ be the trigonometric functions of order n ($n \geq 2$). Find a linear operator f of $\mathbb{R}^n$ such that $f^n = -1$, trace $f = 0$ and

$$e^{\alpha f} = s(\alpha)f^{n-1} + s'(\alpha)f^{n-2} + s''(\alpha)f^{n-3} + \cdots + s^{(n-2)}(\alpha)f + s^{(n-1)}(\alpha)$$

14.11 Show that $e^{\alpha f}$ in the previous exercise is orthogonal if, and only if, $n = 2$.

14.12 Calculate the trigonometric functions s, s', s'' of order 3 in terms of the elementary functions. Calculate $e^{\alpha f}$ for the linear operator f of $\mathbb{R}^3$ such that $f^3 = -1$ and trace $f = 0$. Write the addition formulas for the trigonometric functions of order 3.

14.13 Calculate the trigonometric functions of order 4 in terms of the elementary functions.

14.14 Show that $e^{\alpha[f,g]} = e^{\beta f} \circ e^{\beta g} \circ e^{-\beta f} \circ e^{-\beta g}$, i.e., the exponential of the Lie product is the commutator of the exponentials.

14.15 Given an arbitrary basis $\{\mathbf{e}_1, \ldots, \mathbf{e}_n\}$ of a metric vector space $\mathbb{R}^n$, show how to construct its dual basis $\{\mathbf{m}_1, \ldots, \mathbf{m}_n\}$ such that

$$\mathbf{e}_\mu \cdot \mathbf{m}_\nu = \begin{cases} 1 \Leftrightarrow \mu = \nu \\ 0 \Leftrightarrow \mu \neq \nu \end{cases}$$

14.16 Show that a basis is orthonormal if, and only if, it is self-dual.

14.17 Show that there exists no linear operators f, g in $\mathbb{R}^n$ satisfying

$$f \circ g - g \circ f = 1$$

14.18 Show that $L(\mathbb{R}, \mathbb{R}) = \mathbb{R}$, the (commutative) Lie algebra of real numbers.

14.19 Show that $GL^+(\mathbb{R}) = \mathbb{R}^*$, the multiplicative group of positive real numbers.

14.20 Show that under the exponential functor every Lie algebra homomorphism induces a Lie group homomorphism $\exp f$: $\exp A \to \exp B$ such that

$$\exp(g \circ f) = (\exp g) \circ (\exp f)$$

and

$$\exp(1_A) = 1_{\exp A}$$

14.21 Show that trace : $L(\mathbb{R}^n, \mathbb{R}^n) \to L(\mathbb{R}, \mathbb{R}) = \mathbb{R}$ is a Lie algebra homomorphism with exp(trace) = det, i.e., its induced Lie group homomorphism is det : $GL^+(\mathbb{R}^n) \to GL^+(\mathbb{R}) = \mathbb{R}^*$.

14.22 Show that every skew rotation r of $\mathbb{R}^3$ is the Cayley transform of $f = \omega t$ where $\omega \in \mathbb{R}$ and t is a twist, i.e., $r = (1 + f) \circ (1 - f)^{-1}$.

14.23 Show that every rotation is a commutator $r_1 \circ r_2 \circ r_1^{-1} \circ r_2^{-1}$ with $r_1, r_2, \in O(\mathbb{R}^n)$.

14.24 Show that every volume-preserving linear operator of $\mathbb{R}^2$, other than -1, can be decomposed into the product of not more than two shears. Show that -1 is the product of three, but not two, shears. Conclude that every element of $SL(\mathbb{R}^2)$ can be decomposed into the product of not more than two involutions (see Chap. 8, Application 5; also Exercises 10.29 and 10.42).

14.25 Show that every volume-preserving linear operator of $\mathbb{R}^3$ can be decomposed into the product of not more than three shears. Give examples of a volume-preserving operator that are the product of three, but not two, shears. Conclude that every element of $SL(\mathbb{R}^3)$ can be decomposed into the product of two or four involutions (see Chap. 8, Application 5; also Exercises 10.29 and 10.43).

14.26 Let s be a fixed positive definite operator (see Exercise 11.46). Show that the set of operators $G = \{f \mid f^* \circ s \circ f = s\}$ is a subgroup of $GL(\mathbb{R}^n)$. In particular, G is the orthogonal group for $s = 1$.

14.27 Show that the set of operators $A = \{f \mid f^* \circ s + s \circ f = 0\}$ is a Lie subalgebra of $L(\mathbb{R}^n, \mathbb{R}^n)$ (s is as in Exercise 14.26). In particular, A is the Lie algebra of antisymmetric operators for $s = 1$.

14.28 Show that every element $r \in G$ which does not have -1 as an eigenvalue is the Cayley transform of an $a \in A$ which does not have -1 as an eigenvalue, i.e., $r = (1-a) \circ (1+a)^{-1}$. ($A$ and G are as in Exercises 14.26 and 14.27).

14.29 Prove the exponential property.

14.30 Calculate $e^{\alpha f}$ for

$$f = \begin{pmatrix} \lambda & 1 & 0 & 0 & \cdots & 0 \\ 0 & \lambda & 1 & 0 & \cdots & 0 \\ 0 & 0 & \lambda & 1 & \cdots & 0 \\ \cdots & \cdots & \cdots & \cdots & \cdots & \cdots \\ 0 & 0 & 0 & 0 & \cdots & 1 \\ 0 & 0 & 0 & 0 & \cdots & \lambda \end{pmatrix}$$

14.31 A group is said to be *simple* if it has no normal subgroups other than the trivial group and itself. Show that an invection group is simple.

14.32 Let H be a subgroup of G. H is said to be a *maximal* subgroup of G if for every subgroup K of G, $H \subseteq K \subseteq G \Rightarrow K = H$ or $K = G$. Show that an invection group is maximal in $SL(\mathbb{R}^n)$.

14.33 Show that $SL(\mathbb{R}^n)$ is simple.

APPENDIX 1
Numerical Methods of Linear Algebra: Determinants. Gaussian Elimination

DEFINITION OF AN $n \times n$ DETERMINANT

For $n \geqq 2$, we define an $n \times n$ determinant by the formula

$$\begin{vmatrix} a_{11} & a_{12} & \cdots & a_{1n} \\ a_{12} & a_{22} & \cdots & a_{2n} \\ \cdots & \cdots & \cdots & \cdots \\ a_{n1} & a_{n2} & \cdots & a_{nn} \end{vmatrix} = \sum_{\pi} (-1)^{\text{par}\,\pi} a_{1\pi(1)} a_{2\pi(2)} \cdots a_{n\pi(n)}$$

where π is a permutation of the set $\{1, 2, \ldots, n\}$ and Σ_π means that the summation is extended over all the $n! = 1 \cdot 2 \cdots n$ permutations of that set. The symbol par π stands for the *parity* of π, so that half of the terms on the right-hand side keep their signature (for par π = even), and the other other half change their signature (for par π = odd). Note that in each term we have a product of elements exactly one of which coming from each row and exactly one from each column. The *main diagonal* of the determinant corresponds to the permutation $\pi = 1$, the identity, yielding the term $a_{11}a_{22} \cdots a_{nn}$. The *value* of the determinant is the number obtained on the right-hand side.

The definition of the determinant is not suitable for calculation, except in the cases $n = 2$ and 3:

Examples

1. $$\begin{vmatrix} a_{11} & a_{12} \\ a_{21} & a_{22} \end{vmatrix} = a_{11}a_{22} - a_{12}a_{21}$$

2. $$\begin{vmatrix} a_{11} & a_{12} & a_{13} \\ a_{21} & a_{22} & a_{23} \\ a_{31} & a_{32} & a_{33} \end{vmatrix} = a_{11}a_{22}a_{33} + a_{12}a_{23}a_{31} + a_{13}a_{21}a_{32} - a_{13}a_{22}a_{31} - a_{12}a_{21}a_{33} - a_{11}a_{23}a_{32}$$

For $n \geqq 4$ other methods of evaluation are available, which are based on one or more of the 11 properties of determinants which we shall now discuss.

PROPERTIES OF $n \times n$ DETERMINANTS

(1) **Reflection Property** The value of a determinant remains unchanged under the reflection of its elements in the main diagonal.

In other words, the rows and columns of a determinant may be interchanged. In particular, every property of the determinant involving rows is automatically valid for columns as well. This property is a direct consequence of the definition as par $\pi^{-1} = \text{par } \pi$ for every permutation π.

Example

$$\begin{vmatrix} 1 & \sqrt{2}+1 \\ \sqrt{2}-1 & 1 \end{vmatrix} = \begin{vmatrix} 1 & \sqrt{2}-1 \\ \sqrt{2}+1 & 1 \end{vmatrix} = 1 - 1 = 0$$

(2) **Switching Property** The value of a determinant changes its signature if we switch around any two rows (or any two columns).

An even number of switches will leave the value of the determinant unchanged. This property is a direct consequence of the definition, as the switching of two columns has the effect of changing the parity of permutations.

Example

$$\begin{vmatrix} \cos\alpha & -\sin\alpha \\ \sin\alpha & \cos\alpha \end{vmatrix} = 1 \qquad \begin{vmatrix} \sin\alpha & \cos\alpha \\ \cos\alpha & -\sin\alpha \end{vmatrix} = -1$$

(3) **Repetition Property** If a determinant has a repeating row (column), then its value is zero.

This is clear in view of the switching property, because we may switch around the repeating rows (columns) without changing the determinant, yet at the same time changing the signature. This contradiction can be resolved only by assuming that the value of the determinant is zero.

Example

$$\begin{vmatrix} 99 & 98 & 97 \\ 96 & 95 & 94 \\ 96 & 95 & 94 \end{vmatrix} = 99(95 \cdot 94 - 94 \cdot 95) + \cdots = 99 \cdot 0 + \cdots = 0$$

(4) Scalar Multiple Property In multiplying the elements of a row (column) of a determinant by a scalar λ, the value of the determinant gets multiplied by λ.

This follows directly from the definition, on the strength of the distributive law.

The scalar multiple property is often used "backwards": *We may remove a common factor from a row (column)*, thereby simplifying the determinant.

(5) All-Zero Property If all the elements of a row (column) are equal to zero, then the value of the determinant is zero.

This is a special case of the scalar multiple property with $\lambda = 0$.

(6) Proportionality Property If two rows (columns) are proportional, then the value of the determinant is equal to zero.

This is a consequence of the scalar multiple property and the repetition property.

Examples

1.
$$\begin{vmatrix} 91 & 39 & 65 \\ 56 & 24 & 40 \\ 49 & 64 & 81 \end{vmatrix} = 0$$

because the first two rows are proportional (in ratio 8/13).

2.
$$\begin{vmatrix} \sqrt{3}+\sqrt{2} & 1 \\ 1 & \sqrt{3}-\sqrt{2} \end{vmatrix} = 0$$

because the second row is $(\sqrt{3}-\sqrt{2})$ times the first.

3.
$$\begin{vmatrix} 1 & 2 & 4 \\ 8 & 16 & 32 \\ 64 & 128 & 256 \end{vmatrix} = 8$$

More generally, if the entries of a determinant are the consecutive members of a geometric progression, then the value of the determinant is zero.

(7) Summation Property The sum of two determinants which are identical except for the elements of one pair of corresponding rows (columns) can be calculated as a single determinant by adding the corresponding rows (columns) elementwise, while leaving the other rows (columns) unchanged.

This property follows directly from the definition.

Example

$$\begin{vmatrix} a_1 & b_1 & c_1 \\ a_2 & b_2 & c_2 \\ a_3 & b_3 & c_3 \end{vmatrix} + \begin{vmatrix} a_1' & b_1' & c_1' \\ a_2 & b_2 & c_2 \\ a_3 & b_3 & c_3 \end{vmatrix} = \begin{vmatrix} a_1 + a_1' & b_1 + b_1' & c_1 + c_1' \\ a_2 & b_2 & c_2 \\ a_3 & b_3 & c_3 \end{vmatrix}$$

(8) Invariance Property We may add a scalar multiple of a row to another row (or of a column to another column); the value of the determinant will remain invariant.

This is an immediate consequence of the proportionality property combined with the summation property applied "backwards."

The invariance property can be used to simplify determinants, by making entries smaller, or by making them zero.

Examples

1. $$\begin{vmatrix} 1 & 2 & 3 \\ 4 & 5 & 6 \\ 7 & 8 & 9 \end{vmatrix} = \begin{vmatrix} 1 & 2 & 3 \\ 3 & 3 & 3 \\ 3 & 3 & 3 \end{vmatrix} = 0 \quad \text{by the repetition property}$$

   More generally, if the entries of an $n \times n$ determinant, $n \geqq 3$, are the consecutive members of an arithmetic progression, then the value of the determinant is zero.

2. $$\begin{vmatrix} 1 & 4 & 9 & 16 \\ 25 & 36 & 49 & 64 \\ 81 & 100 & 121 & 144 \\ 169 & 196 & 225 & 256 \end{vmatrix} = \begin{vmatrix} 1 & 4 & 9 & 16 \\ 24 & 32 & 40 & 48 \\ 56 & 64 & 72 & 80 \\ 88 & 96 & 104 & 112 \end{vmatrix} = \begin{vmatrix} 1 & 4 & 9 & 16 \\ 24 & 32 & 40 & 48 \\ 32 & 32 & 32 & 32 \\ 32 & 32 & 32 & 32 \end{vmatrix} = 0$$

   More generally, if the entries of an $n \times n$ determinant, $n \geqq 4$, are the consecutive members of an arithmetic progression of order 2, then the value of the determinant is zero.

Further properties of determinants have to do with subdeterminants and cofactors. If we delete the μth row and the νth column of an $n \times n$ determinant, then we obtain an $(n-1) \times (n-1)$ determinant called the *subdeterminant* belonging to the element $a_{\mu\nu}$. If we prefix this subdeterminant by the signature $(-1)^{\mu+\nu}$, we get the *cofactor* $A_{\mu\nu}$ belonging to the element $a_{\mu\nu}$:

$$A_{\mu\nu} = (-1)^{\mu+\nu} \begin{vmatrix} a_{11} & \cdots & \not{a}_{1\nu} & \cdots & a_{1n} \\ \cdots & \cdots & \cdots & \cdots & \cdots \\ \not{a}_{\mu 1} & \cdots & \not{a}_{\mu\nu} & \cdots & \not{a}_{\mu n} \\ \cdots & \cdots & \cdots & \cdots & \cdots \\ a_{n1} & \cdots & \not{a}_{n\nu} & \cdots & a_{nn} \end{vmatrix}$$

In practice, the signature is determined via the chessboard rule:

$$\begin{matrix} + & - & + & - & + & - & \cdots \\ - & + & - & + & - & + & \cdots \\ + & - & + & - & + & - & \cdots \\ \cdots & \cdots & \cdots & \cdots & \cdots & \cdots & \cdots \end{matrix}$$

(9) Expansion Property The sum of elements of a certain row (column) multiplied by the cofactors of the same elements is equal to the value of the determinant.

This form of calculation is called the *expansion* of the determinant with respect to a row (column).

Examples

1. $$\begin{vmatrix} a_1 & b_1 & c_1 \\ a_2 & b_2 & c_2 \\ a_3 & b_3 & c_3 \end{vmatrix} = a_1A_1 + b_1B_1 + c_1C_1 \qquad \text{where } A_1 = \begin{vmatrix} b_2 & c_2 \\ b_3 & c_3 \end{vmatrix},$$

   etc., is expansion with respect to the first row;

2. $$\begin{vmatrix} a_1 & b_1 & c_1 \\ a_2 & b_2 & c_2 \\ a_3 & b_3 & c_3 \end{vmatrix} = a_1A_1 + a_2A_2 + a_3A_3$$

   is expansion with respect to the first column. The 3×3 determinant has six such expansions.

The expansion property can be proved by grouping the terms of the sum

$$\sum_{\pi} (-1)^{\operatorname{par}\pi} a_{1\pi(1)} a_{2\pi(2)} \cdots a_{n\pi(n)}$$

into n summands according as they contain one of the n elements of the row (column) with respect to which the expansion is made. The details of the proof are left to the reader.

(10) Vanishing Property The sum of elements of a certain row (column) multiplied by the corresponding cofactors of elements of another row (column) of the determinant is equal to 0, e.g.,

$$a_{\mu 1}A_{\nu 1} + a_{\mu 2}A_{\nu 2} + \cdots + a_{\mu n}A_{\nu n} = 0$$
$$(a_{1\mu}A_{1\nu} + a_{2\mu}A_{2\nu} + \cdots + a_{n\mu}A_{n\nu} = 0)$$

for $\mu \neq \nu$; $\mu, \nu = 1, 2, \ldots, n$.

This is a consequence of the expansion property and the repetition property of determinants.

(11) Triangle Property A *triangular determinant* is one with all zero entries below (above) the main diagonal. The value of a triangular determinant is the product of its diagonal elements:

$$\begin{vmatrix} a_{11} & a_{12} & \cdots & a_{1n} \\ 0 & a_{22} & \cdots & a_{2n} \\ \cdots & \cdots & \cdots & \cdots \\ 0 & 0 & \cdots & a_{nn} \end{vmatrix} = a_{11}a_{22}\cdots a_{nn}$$

This is a simple consequence of the expansion property.

With the aid of the invariance property and the scalar multiple property, we can reduce any determinant to a triangular determinant, the value of which can be easily calculated.

Examples

1. $$\begin{vmatrix} 1 & 2 & 3 & 4 \\ 2 & 1 & -2 & -1 \\ 3 & 0 & 1 & 2 \\ 0 & 1 & 0 & 2 \end{vmatrix} = \begin{vmatrix} 1 & 2 & 3 & 4 \\ 0 & -3 & -8 & -9 \\ 0 & -6 & -8 & -10 \\ 0 & 1 & 0 & 2 \end{vmatrix} = 2\begin{vmatrix} 1 & * & * & * \\ 0 & 3 & 8 & 9 \\ 0 & 3 & 4 & 5 \\ 0 & 1 & 0 & 2 \end{vmatrix} = -2\begin{vmatrix} 1 & * & * & * \\ 0 & 1 & 0 & 2 \\ 0 & 3 & 4 & 5 \\ 0 & 3 & 8 & 9 \end{vmatrix}$$

$$= -2\begin{vmatrix} 1 & * & * & * \\ 0 & 1 & * & * \\ 0 & 0 & 4 & -1 \\ 0 & 0 & 8 & 3 \end{vmatrix} = -2\begin{vmatrix} 1 & * & * & * \\ 0 & 1 & * & * \\ 0 & 0 & 4 & * \\ 0 & 0 & 0 & 5 \end{vmatrix} = -40$$

2. $$\begin{vmatrix} 1 & 2 & 3 & 4 \\ -2 & 1 & -4 & 3 \\ 3 & -4 & -1 & 2 \\ 4 & 3 & -2 & -1 \end{vmatrix} = \begin{vmatrix} 1 & * & * & * \\ 0 & 5 & 2 & 11 \\ 0 & -10 & -10 & -10 \\ 0 & -5 & -14 & -17 \end{vmatrix} = 10\begin{vmatrix} 1 & * & * & * \\ 0 & 5 & 2 & 11 \\ 0 & 1 & 1 & 1 \\ 0 & 0 & 9 & 12 \end{vmatrix}$$

$$= 30\begin{vmatrix} 1 & * & * & * \\ 0 & 0 & -3 & 6 \\ 0 & 1 & 1 & 1 \\ 0 & 0 & 3 & 4 \end{vmatrix} = -90\begin{vmatrix} 1 & * & * & * \\ 0 & 1 & * & * \\ 0 & 0 & -1 & * \\ 0 & 0 & 0 & 10 \end{vmatrix} = 900$$

3. $$\begin{vmatrix} 1 & 2 & 3 & 4 & 1 \\ 2 & 1 & 0 & 1 & 2 \\ 3 & 1 & 2 & 0 & 3 \\ 1 & 0 & 0 & 2 & 3 \\ 5 & 1 & 0 & 1 & 0 \end{vmatrix} = \begin{vmatrix} 0 & 2 & 3 & 2 & -2 \\ 0 & 1 & 0 & -3 & -4 \\ 0 & 1 & 2 & -6 & -6 \\ 1 & 0 & 0 & 2 & 3 \\ 0 & 1 & 0 & -9 & -15 \end{vmatrix} = - \begin{vmatrix} 1 & * & * & * & * \\ 0 & 1 & 0 & -3 & -4 \\ 0 & 1 & 2 & -6 & -6 \\ 0 & 2 & 3 & 2 & -1 \\ 0 & 1 & 0 & -9 & -15 \end{vmatrix}$$

$$= - \begin{vmatrix} 1 & * & * & * & * \\ 0 & 1 & * & * & * \\ 0 & 0 & 2 & -3 & -2 \\ 0 & 0 & 3 & 8 & 6 \\ 0 & 0 & 0 & -6 & -11 \end{vmatrix} = - \begin{vmatrix} 1 & * & * & * & * \\ 0 & 1 & * & * & * \\ 0 & 0 & 1 & * & * \\ 0 & 0 & 0 & -1 & 26 \\ 0 & 0 & 0 & -6 & -11 \end{vmatrix} = -167$$

4. $$\begin{vmatrix} 1 & 2 & 3 & 0 & 4 & 0 \\ 5 & 6 & 7 & 1 & 8 & 2 \\ 9 & 1 & 2 & 0 & 3 & 0 \\ 4 & 5 & 6 & 1 & 7 & 2 \\ 1 & 2 & 3 & 4 & 4 & 1 \\ 1 & 1 & 1 & 0 & 2 & 0 \end{vmatrix} = \begin{vmatrix} 0 & 1 & 2 & 0 & 2 & 0 \\ 0 & 1 & 2 & 1 & -2 & 2 \\ 0 & -8 & -7 & 0 & -15 & 0 \\ 0 & 1 & 2 & 1 & -1 & 2 \\ 0 & 1 & 2 & 4 & 2 & 1 \\ 1 & 1 & 1 & 0 & 2 & 0 \end{vmatrix}$$

$$= - \begin{vmatrix} 1 & * & * & * & * & * \\ 0 & 0 & 0 & 1 & -4 & 2 \\ 0 & 0 & 9 & 0 & 1 & 0 \\ 0 & 0 & 0 & 1 & -3 & 2 \\ 0 & 0 & 0 & 4 & 0 & 1 \\ 0 & 1 & 2 & 0 & 2 & 0 \end{vmatrix} = \begin{vmatrix} 1 & * & * & * & * & * \\ 0 & 1 & * & * & * & * \\ 0 & 0 & 9 & 0 & 1 & 0 \\ 0 & 0 & 0 & 1 & -3 & 2 \\ 0 & 0 & 0 & 0 & 12 & -7 \\ 0 & 0 & 0 & 0 & -1 & 0 \end{vmatrix}$$

$$= - \begin{vmatrix} 1 & * & * & * & * & * \\ 0 & 1 & * & * & * & * \\ 0 & 0 & 9 & * & * & * \\ 0 & 0 & 0 & 1 & * & * \\ 0 & 0 & 0 & 0 & -1 & * \\ 0 & 0 & 0 & 0 & 0 & -7 \end{vmatrix} = -63$$

APPLICATIONS

1. Signed Area of the Triangle in $\mathbb{R}^2$

Let the vertices of a triangle $\triangle$ be given as $P_1(x_1,y_1)$, $P_2(x_2,y_2)$, $P_3(x_3,y_3)$. Then the signed area of $\triangle$ is

$$\text{Area}(\triangle) = \tfrac{1}{2} \begin{vmatrix} x_1 & y_1 & 1 \\ x_2 & y_2 & 1 \\ x_3 & y_3 & 1 \end{vmatrix}$$

We shall see below (Application 4) that

Area $(\triangle) > 0 \Leftrightarrow P_1, P_2, P_3$ conform to counterclockwise rotation

2. Necessary and Sufficient Condition for Collinearity

As a special case of the previous application, we have that $P_1(x_1,y_1)$, $P_2(x_2,y_2)$, $P_3(x_3,y_3)$ are collinear if, and only if,

$$\begin{vmatrix} x_1 & y_1 & 1 \\ x_2 & y_2 & 1 \\ x_3 & y_3 & 1 \end{vmatrix} = 0$$

3. Equation of the Line Through Two Points

As a further consequence of Application 1, we get the equation of the line passing through the points $P_1(x_1,y_1)$, $P_2(x_2,y_2)$ in determinant form:

$$\begin{vmatrix} x & y & 1 \\ x_1 & y_1 & 1 \\ x_2 & y_2 & 1 \end{vmatrix} = 0$$

4. Clockwise and Counterclockwise Rotation

If the three points $P_1(x_1,y_1)$, $P_2(x_2,y_2)$, $P_3(x_3,y_3)$ are not collinear, then they fall on the same circle. The question arises whether P_1, P_2, P_3, follow one another on the circle in the clockwise or in the counterclockwise sense. We have that

$$\overleftarrow{P_1 \to P_2 \to P_3} \text{ counterclockwise rotation} \Leftrightarrow \begin{vmatrix} x_1 & y_1 & 1 \\ x_2 & y_2 & 1 \\ x_3 & y_3 & 1 \end{vmatrix} > 0$$

5. Necessary and Sufficient Condition for Two Lines to Be Parallel

Let the equation of the lines be ℓ_1: $A_1x + B_1y + C_1 = 0$, ℓ_2: $A_2x + B_2y + C_2 = 0$. Then

$$\ell_1 \parallel \ell_2 \Leftrightarrow \begin{vmatrix} A_1 & B_1 \\ A_2 & B_2 \end{vmatrix} = 0$$

6. Necessary Condition for Three Lines to Have a Point in Common

Let the equations of the lines be ℓ_1: $A_1x + B_1y + C_1 = 0$, ℓ_2: $A_2x + B_2y + C_2 = 0$, ℓ_3: $A_3x + B_3y + C_3 = 0$. Then

$$\ell_1, \ell_2, \ell_3 \text{ have a point in common} \Rightarrow \begin{vmatrix} A_1 & B_1 & C_1 \\ A_2 & B_2 & C_2 \\ A_3 & B_3 & C_3 \end{vmatrix} = 0$$

The condition that the determinant formed of the coefficients in the equations of the lines be zero is necessary but not sufficient. Indeed, if the three lines are pairwise parallel, then they have no point in common yet the determinant will vanish. However, if the determinant vanishes and there are two nonparallel lines among the three, then we may conclude that the three lines have a point in common.

Examples

1.
$$\begin{aligned} 2x+7y-8&=0 \\ 3x+2y+5&=0 \\ x-5y+13&=0 \end{aligned} \qquad \begin{vmatrix} 2 & 7 & -8 \\ 3 & 2 & 5 \\ 1 & -5 & 13 \end{vmatrix} = 0$$

Hence the lines are either parallel or they have a point in common. But the first two lines have slopes $-2/7$ and $-3/2$ and hence are not parallel. We conclude that the three lines have a point in common.

2.
$$\begin{aligned} (3+\sqrt{5})x-(2+2\sqrt{5})y+1&=0 \\ (1+\sqrt{5})x-4y+2&=0 \\ x+(1-\sqrt{5})y+3&=0 \end{aligned} \qquad \begin{vmatrix} 3+\sqrt{5} & -2-2\sqrt{5} & 1 \\ 1+\sqrt{5} & -4 & 2 \\ 1 & 1-\sqrt{5} & 3 \end{vmatrix} = 0$$

However,

$$\begin{vmatrix} 3+\sqrt{5} & -2-2\sqrt{5} \\ 1+\sqrt{5} & -4 \end{vmatrix} = 0 \qquad \begin{vmatrix} 1+\sqrt{5} & -4 \\ 1 & 1-\sqrt{5} \end{vmatrix} = 0$$

Therefore, the three lines are pairwise parallel.

7. Necessary and Sufficient Condition for Two Pairs of Direction Numbers to Determine Parallel Lines

Let $\mathbf{v}_1 = a_1\mathbf{i} + b_1\mathbf{j}$, $\mathbf{v}_2 = a_2\mathbf{i} + b_2\mathbf{j}$. Then

$$\mathbf{v}_1 \| \mathbf{v}_2 \Leftrightarrow \begin{vmatrix} a_1 & b_1 \\ a_2 & b_2 \end{vmatrix} = 0$$

8. Equation of a Line Through A Given Point, Given the Direction Numbers

Given a point $P_0(x_0,y_0)$ and direction numbers a, b; the point $P(x,y)$ falls on the line through P_0 and parallel to the given direction if, and only if,

$$\begin{vmatrix} x-x_0 & y-y_0 \\ a & b \end{vmatrix} = 0$$

or, what is the same,

Symmetric Form

$$\frac{x-x_0}{a}=\frac{y-y_0}{b}$$

9. Necessary Condition for Four Points to Fall on a Circle

The four points $P_1(x_1,y_1)$, $P_2(x_2,y_2)$, $P_3(x_3,y_3)$, $P_4(x_4,y_4)$ may fall on a circle only if

$$\begin{vmatrix} x_1^2+y_1^2 & x_1 & y_1 & 1 \\ x_2^2+y_2^2 & x_2 & y_2 & 1 \\ x_3^2+y_3^2 & x_3 & y_3 & 1 \\ x_4^2+y_4^2 & x_4 & y_4 & 1 \end{vmatrix}=0$$

Proof: If there is a circle passing through P_1, P_2, P_3, P_4, then the homogeneous system

$$\begin{aligned} A(x_1^2+y_1^2)+Bx_1+Cy_1+D&=0\\ A(x_2^2+y_2^2)+Bx_2+Cy_2+D&=0\\ A(x_3^2+y_3^2)+Bx_3+Cy_3+D&=0\\ A(x_4^2+y_4^2)+Bx_4+Cy_4+D&=0 \end{aligned}$$

of linear equations in the unknowns A, B, C, D has a nontrivial solution; i.e., the determinant of the system must be zero (otherwise a contradiction to Cramer's rule would arise).

Example Show that the four points $P_1(4,5)$, $P_2(-3,4)$, $P_3(-2,-3)$, $P_4(5,-2)$ fall on a circle.

Answer

$$\begin{vmatrix} 41 & 4 & 5 & 1 \\ 25 & -3 & 4 & 1 \\ 13 & -2 & -3 & 1 \\ 29 & 5 & -2 & 1 \end{vmatrix}=\begin{vmatrix} 12 & -1 & 7 & 0 \\ -4 & -8 & 6 & 0 \\ -16 & -7 & -1 & 0 \\ 29 & 5 & -2 & 1 \end{vmatrix}$$

$$=\begin{vmatrix} 100 & 50 & 0 & 0 \\ 100 & 50 & 0 & 0 \\ 13 & -2 & -1 & 0 \\ * & * & * & 1 \end{vmatrix}=0 \quad \text{by the repetition property}$$

We conclude that the four points are either collinear, or they fall on a circle. By Application 2, the first three points are not collinear:

$$\begin{vmatrix} 4 & 5 & 1 \\ -3 & 4 & 1 \\ -2 & -3 & 1 \end{vmatrix} = \begin{vmatrix} 6 & 8 & 0 \\ -1 & 7 & 0 \\ -2 & -3 & 1 \end{vmatrix} = \begin{vmatrix} 6 & 8 & 0 \\ -1 & 7 & 0 \\ -2 & -3 & 1 \end{vmatrix} = \begin{vmatrix} 6 & 8 \\ -1 & 7 \end{vmatrix} = 50 \neq 0$$

Therefore, the given four points fall on a circle.

10. Equation of the Circle Passing Through Three Given Points

Three points $P_1(x_1,y_1)$, $P_2(x_2,y_2)$, $P_3(x_3,y_3)$ are given; the equation of the circle passing through them is

$$\begin{vmatrix} x^2+y^2 & x & y & 1 \\ x_1^2+y_1^2 & x_1 & y_1 & 1 \\ x_2^2+y_2^2 & x_2 & y_2 & 1 \\ x_3^2+y_3^2 & x_3 & y_3 & 1 \end{vmatrix} = 0$$

If P_1, P_2, P_3 are collinear, there will be no circle passing through them. In this case no quadratic terms will enter the equation above and it will reduce to a linear equation representing the line containing the three points.

Examples Find the equation of the circle passing through the points $P_1(-3,4)$, $P_2(-2,-3)$, $P_3(5,-2)$.

Answer

$$\begin{vmatrix} x^2+y^2 & x & y & 1 \\ 25 & -3 & 4 & 1 \\ 13 & -2 & -3 & 1 \\ 29 & 5 & -2 & 1 \end{vmatrix} = \begin{vmatrix} x^2+y^2-29 & x-5 & y+2 \\ -4 & -8 & 6 \\ -16 & -7 & -1 \end{vmatrix}$$

$$= 50(x^2+y^2-29) - 100(x-5) - 100(y+2) = 0$$

or

$$x^2+y^2-2x-2y-23 = 0$$

CRAMER'S RULE†

A system of linear equations is called *regular* if (*a*) it has the same number of equations as unknowns and (*b*) it has a unique set of solutions. Notice that

†Compare Cramer's formula, Chap. 4.

(a) does not in itself guarantee (b): the system

$$\begin{aligned} x + y \quad\quad &= 0 \\ x \quad\quad + z &= 0 \\ y - z &= 1 \end{aligned}$$

is not regular since it has no solution (both y and z should be equal to the negative of x according to the first equations, but they are not equal according to the third, and this is a contradiction). On the other hand, the homogeneous system

$$\begin{aligned} x + y \quad\quad &= 0 \\ x \quad\quad + z &= 0 \\ y - z &= 0 \end{aligned}$$

fails to be regular for another reason: it has no unique set of solutions (x can be any number, then $y = -x = z$ will satisfy all three equations).

We want to find a formula for the solution of the system

$$\begin{aligned} a_{11}x_1 + a_{11}x_2 + \cdots + a_{1n}x_n &= r_1 \\ a_{21}x_1 + a_{22}x_2 + \cdots + a_{2n}x_n &= r_2 \\ &\vdots \\ a_{n1}x_1 + a_{n2}x_2 + \cdots + a_{nn}x_n &= r_n \end{aligned} \tag{*}$$

First, we calculate x_1. Let us multiply the first equation by A_{11}, the second equation by $A_{21}, \ldots$, the nth equation by A_{n1} and add them:

$$(a_{11}A_{11} + a_{21}A_{21} + \cdots + a_{n1}A_{n1})x_1 + (a_{12}A_{11} + a_{22}A_{21} + \cdots + a_{n2}A_{n1})x_2 + \cdots$$
$$= r_1A_{11} + r_2A_{21} + \cdots + r_nA_{n1}$$

But by the expansion property of determinants,

$$a_{11}A_{11} + a_{21}A_{21} + \cdots + a_{n1}A_{n1} = \begin{vmatrix} a_{11} & a_{12} & \cdots & a_{1n} \\ a_{21} & a_{22} & \cdots & a_{2n} \\ \cdots & \cdots & \cdots & \cdots \\ a_{n1} & a_{n2} & \cdots & a_{nn} \end{vmatrix} = D$$

and

$$r_1A_{11} + r_2A_{21} + \cdots + r_nA_{n1} = \begin{vmatrix} r_1 & a_{12} & \cdots & a_{1n} \\ r_2 & a_{22} & \cdots & a_{2n} \\ \cdots & \cdots & \cdots & \cdots \\ r_n & a_{n2} & \cdots & a_{nn} \end{vmatrix}$$

Furthermore, by the vanishing property of determinants,

$$a_{1\nu}A_{11} + a_{2\nu}A_{21} + \cdots + a_{n\nu}A_{n1} = 0 \qquad \nu = 2, 3, \ldots, n$$

Therefore,

$$x_1 = \frac{\begin{vmatrix} r_1 & a_{12} & \cdots & a_{1n} \\ r_2 & a_{22} & \cdots & a_{2n} \\ \cdots & \cdots & \cdots & \cdots \\ r_n & a_{n2} & \cdots & a_{nn} \end{vmatrix}}{D}$$

provided that $D \neq 0$. Similar calculations yield

$$x_2 = \frac{\begin{vmatrix} a_{11} & r_1 & \cdots & a_{1n} \\ a_{21} & r_2 & \cdots & a_{2n} \\ \cdots & \cdots & \cdots & \cdots \\ a_{n1} & r_n & \cdots & a_{nn} \end{vmatrix}}{D}, \quad \ldots$$

$$x_n = \frac{\begin{vmatrix} a_{11} & a_{12} & \cdots & r_1 \\ a_{21} & a_{22} & \cdots & r_2 \\ \cdots & \cdots & \cdots & \cdots \\ a_{n1} & a_{n2} & \cdots & r_n \end{vmatrix}}{D}$$

These formulas, furnishing the unique solution to the system (*), are known as Cramer's rule. We see that $D \neq 0$ is a necessary and sufficient condition for the system (*) to be regular.

Example

1. $\begin{aligned} 3x + 5y &= 8 \\ 4x - 2y &= 1 \end{aligned} \qquad D = \begin{vmatrix} 3 & 5 \\ 4 & -2 \end{vmatrix} = -26 \qquad x = \dfrac{\begin{vmatrix} 8 & 5 \\ 1 & -2 \end{vmatrix}}{-26} = \dfrac{21}{26}$

$$y = \frac{\begin{vmatrix} 3 & 8 \\ 4 & 1 \end{vmatrix}}{-26} = \frac{29}{26}$$

2. $\begin{aligned} 2x + y + z &= 2 \\ x + y + 5z &= -7 \\ 2x + 3y - 3z &= 14 \end{aligned} \qquad D = \begin{vmatrix} 2 & 1 & 1 \\ 1 & 1 & 5 \\ 2 & 3 & -3 \end{vmatrix} = -22$

$$x = \frac{\begin{vmatrix} 2 & 1 & 1 \\ -7 & 1 & 5 \\ 14 & 3 & -3 \end{vmatrix}}{-22} = \frac{-22}{-22} = 1 \qquad y = 2 \qquad z = -2$$

METHOD OF GAUSSIAN ELIMINATION

The method of gaussian elimination also furnishes the solutions to the system of linear equations, and it is applicable in any case, whether $D \neq 0$ or not. In fact, it is applicable even if the number of equations differs from the number of unknowns.

The idea of gaussian elimination is to eliminate the unknown x_1 from all but the first equation, then to eliminate x_2 from all but the second equation, ..., finally, to eliminate x_n from all but the nth equation. For example, x_1 can be eliminated by subtracting a suitable multiple of the first equation from the others. If at the end we are left with more unknowns than equations, then the solution will not be unique, but will depend on one or more free parameters.

In order to cut down on the amount of writing, we shall work with the coefficients put in a matrix form, and the numbers on the right-hand side will be separated in the matrix from the coefficients by a vertical bar (augmented matrix). It should be kept in mind that each row of the matrix stands for an equation of the system; thus two identical rows indicate that the system is redundant and one of the corresponding equations may be dropped without loss of information; two rows which are identical apart from the element following the vertical bar indicate that the system is inconsistent, and has no solution.

Examples Find the solutions, if any, of the following systems of linear equations:

1.
$$\begin{aligned} x + y + z &= 1 \\ 3x + 4y - z &= 2 \\ -2x + y - z &= 8 \end{aligned}$$

Answer

$$\left(\begin{array}{rrr|r} 1 & 1 & 1 & 1 \\ 3 & 4 & -1 & 2 \\ -2 & 1 & -1 & 8 \end{array}\right) \Rightarrow \left(\begin{array}{rrr|r} 1 & 1 & 1 & 1 \\ 0 & 1 & -4 & -1 \\ 0 & 3 & 1 & 10 \end{array}\right) \Rightarrow \left(\begin{array}{rrr|r} 1 & 0 & 5 & 2 \\ 0 & 1 & -4 & -1 \\ 0 & 0 & 13 & 13 \end{array}\right)$$

$$\Rightarrow \left(\begin{array}{rrr|r} 1 & 0 & 0 & -3 \\ 0 & 1 & 0 & 3 \\ 0 & 0 & 1 & 1 \end{array}\right) \Rightarrow \begin{cases} x = -3 \\ y = 3 \\ z = 1 \end{cases}$$

2.
$$\left.\begin{aligned} 56x - 14y + 63z &= 49 \\ -24x + 6y - 27z &= -21 \\ 72x - 18y + 81z &= 63 \end{aligned}\right\} \Rightarrow \left(\begin{array}{rrr|r} 56 & -14 & 63 & 49 \\ -24 & 6 & -27 & -21 \\ 72 & -18 & 81 & 63 \end{array}\right) \Rightarrow (8 \quad -2 \quad 9|7)$$

$$\Rightarrow \begin{cases} x = 2s \\ z = 2t \\ y = -7/2 + 8s + 9t \end{cases}$$

In this example we had to drop the second and third equations as redundant, and hence we have two free parameters s, t.

3. $$\begin{array}{l} 3x + 5y + 2z = 11 \\ 6x + 15y + 4z = 26 \\ 9x + 5y + 6z = 25 \end{array} \Rightarrow \left(\begin{array}{ccc|c} 3 & 5 & 2 & 11 \\ 6 & 15 & 4 & 26 \\ 9 & 5 & 6 & 25 \end{array}\right) \Rightarrow \left(\begin{array}{ccc|c} 3 & 5 & 2 & 11 \\ 0 & 5 & 0 & 4 \end{array}\right)$$

$$\Rightarrow \left(\begin{array}{ccc|c} 3 & 0 & 2 & 7 \\ 0 & 5 & 0 & 4 \end{array}\right) \Rightarrow \begin{cases} x = 2t \\ y = 4/5 \\ z = 7/2 - 3t \end{cases}$$

In this example we had to drop the third equation as redundant, hence the free parameter t.

4. $$\begin{array}{l} 2x + 3y + 6z = 2 \\ 5x + 7y + 7z = 4 \\ 3x - 5y + 13z = 10 \\ 3x + 4y + z = 2 \end{array} \Rightarrow \begin{pmatrix} 2 & 3 & 6 & 2 \\ 5 & 7 & 7 & 4 \\ 3 & -5 & 13 & 20 \\ 3 & 4 & 2 & 2 \end{pmatrix} \Rightarrow \left(\begin{array}{ccc|c} -1 & -1 & 5 & 0 \\ 5 & 7 & 7 & 4 \\ 3 & -5 & 13 & 10 \\ 3 & 4 & 1 & 2 \end{array}\right)$$

$$\Rightarrow \left(\begin{array}{ccc|c} -1 & -1 & 5 & 0 \\ 0 & 2 & 32 & 4 \\ 0 & -8 & 28 & 10 \\ 0 & 1 & 16 & 2 \end{array}\right) \Rightarrow \left(\begin{array}{ccc|c} 1 & 0 & -21 & 2 \\ 0 & 1 & 16 & 2 \\ 0 & 0 & 156 & 26 \end{array}\right) \Rightarrow \left(\begin{array}{ccc|c} 1 & 0 & 0 & 3/2 \\ 0 & 1 & 0 & -2/3 \\ 0 & 0 & 1 & 1/6 \end{array}\right)$$

$$\Rightarrow \begin{cases} x = 3/2 \\ y = -2/3 \\ z = 1/6 \end{cases}$$

5. $$\left.\begin{array}{l} x - y + z - u = 1 \\ 2x - y - 3u = 2 \\ 3x - z + u = -3 \end{array}\right\} \Rightarrow \left(\begin{array}{cccc|c} 1 & -1 & 1 & -1 & 1 \\ 2 & -1 & 0 & -3 & 2 \\ 3 & 0 & -1 & 1 & -3 \end{array}\right) \Rightarrow \left(\begin{array}{cccc|c} 1 & -1 & 1 & -1 & 1 \\ 0 & 1 & -2 & -1 & 0 \\ 0 & 3 & -4 & 4 & -6 \end{array}\right)$$

$$\left(\begin{array}{cccc|c} 1 & 0 & -1 & -2 & 1 \\ 0 & 1 & -2 & -1 & 0 \\ 0 & 0 & 2 & 7 & -6 \end{array}\right) \Rightarrow \left(\begin{array}{cccc|c} 1 & 0 & 0 & 3/2 & -2 \\ 0 & 1 & 0 & 6 & -6 \\ 0 & 0 & 1 & 7/2 & -3 \end{array}\right) \Rightarrow \begin{cases} x = -2 - 3t \\ y = -6 - 12t \\ z = -3 - 7t \\ u = 2t \end{cases}$$

If the coefficients are integers, it is advisable to postpone divison of a row to the last. In this way we can avoid unnecessary calculations with fractions. More often than not, we can make the coefficient of x_k in the kth row 1 by successive subtractions of rows, rather than by division.

It may happen that we "lose the 1 from a column" in the course of gaussian elimination. This means that, in trying to eliminate x_k from all but the kth equation, we may have inadvertently eliminated x_{k+1} from the $(k+1)$st and all subsequent rows, and therefore we have no way of making $a_{k+1,k+1}$ equal to 1. This means that x_{k+1} is a free parameter (or scalar multiple of it).

Example

$$1.\quad \left.\begin{aligned} 2x+3y+z+4u&=7\\ 2x+3y+z+5u&=8\\ 2x+3y+z+6u&=9 \end{aligned}\right\} \Rightarrow \left(\begin{array}{cccc|c} 2&3&1&4&7\\ 2&3&1&5&8\\ 2&3&1&6&9 \end{array}\right) \Rightarrow \left(\begin{array}{cccc|c} 2&3&1&4&7\\ 0&0&0&1&1 \end{array}\right)$$

$$\Rightarrow \left(\begin{array}{cccc|c} 2&3&1&0&3\\ 0&0&0&1&1 \end{array}\right) \Rightarrow \begin{cases} x = 3/2-3s-t\\ y=2s\\ z=2t\\ u=1 \end{cases}$$

Here we have lost the 1 in columns 2 and 3.

$$2.\quad \left.\begin{aligned} x+3y-4z&=1\\ 2x+\ y-\ z&=2\\ x-2y+3z&=3 \end{aligned}\right\} \Rightarrow \left(\begin{array}{ccc|c} 1&3&-4&1\\ 2&1&-1&2\\ 1&-2&3&3 \end{array}\right) \Rightarrow \left(\begin{array}{ccc|c} 1&3&-4&1\\ 0&-5&7&0\\ 0&-5&7&2 \end{array}\right)$$

No solution, because the last two equations are inconsistent.

In the homogeneous case, we proceed exactly as in the non-homogeneous one. We simplify notation by writing

$$\begin{pmatrix} * & * & * & \cdots\\ * & * & * & \cdots\\ \cdots&\cdots&\cdots&\cdots \end{pmatrix} \quad \text{instead of} \quad \left(\begin{array}{cccc|c} * & * & * & \cdots & 0\\ * & * & * & \cdots & 0\\ \cdots&\cdots&\cdots&\cdots & 0 \end{array}\right)$$

Examples

$$1.\quad \left.\begin{aligned} 0.8\ x+0.32\ y+0.24\ z&=0\\ 0.32x+0.488y-0.384z&=0\\ 0.24x-0.384y+0.712z&=0 \end{aligned}\right\} \Rightarrow \begin{pmatrix} 0.8 & 0.32 & 0.24\\ 0.32 & 0.488 & -0.384\\ 0.24 & -0.384 & 0.712 \end{pmatrix}$$

$$\Rightarrow \begin{pmatrix} 0.8 & 0.32 & 0.24\\ 0 & 0.36 & -0.48 \end{pmatrix} \Rightarrow \begin{pmatrix} 0.8 & 0.32 & 0.24\\ 0 & 0.18 & -0.24 \end{pmatrix} \Rightarrow \begin{pmatrix} 0.8 & 0.5 & 0\\ 0 & 0.18 & -0.24 \end{pmatrix}$$

$$\Rightarrow \begin{pmatrix} 8 & 5 & 0\\ 0 & 3 & -4 \end{pmatrix} \Rightarrow \begin{cases} x=5t\\ y=-8t\\ z=-6t \end{cases}$$

$$2.\quad \left.\begin{aligned} 0.5x+0.4\ y+0.3\ z&=0\\ 0.4x+0.32y+0.24z&=0\\ 0.3x+0.24y=0.18z&=0 \end{aligned}\right\} \Rightarrow \begin{pmatrix} 0.5 & 0.4 & 0.3\\ 0.4 & 0.32 & 0.24\\ 0.3 & 0.24 & 0.18 \end{pmatrix} \Rightarrow (5\quad 4\quad 3)$$

$$\Rightarrow \begin{cases} x = -4s - 3t \\ y = 5s \\ z = 5t \end{cases}$$

The theory and practice of solving systems of linear equations is treated more thoroughly in Chap. 9.

APPENDIX 2
The Field of Real Numbers

CONCEPT OF A COMPLETE ORDERED FIELD

"Real linear algebra" means the study of (finite-dimensional) vector spaces over the filed $\mathbb{R}$ of real numbers as scalars. The discriminating reader may rightfully expect that this book will give him a definition of $\mathbb{R}$. Without going into the question of existence and uniqueness, we shall now give such a definition. $\mathbb{R}$ is defined as an *archimedean field which is complete with respect to the order*. A field was defined in Chap. 12 in terms of 11 axioms:

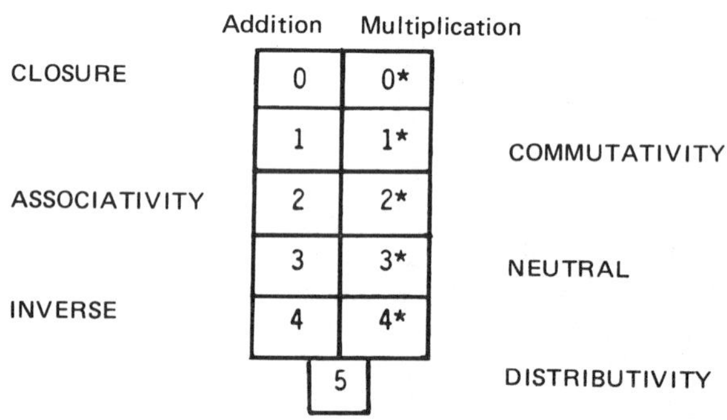

A complete ordered field is subject to eight additional axioms concerning the properties of the order $\leqq$ and its relationship to the field operations, addition and multiplication. The additional axioms are summarized in the following ready-reference table

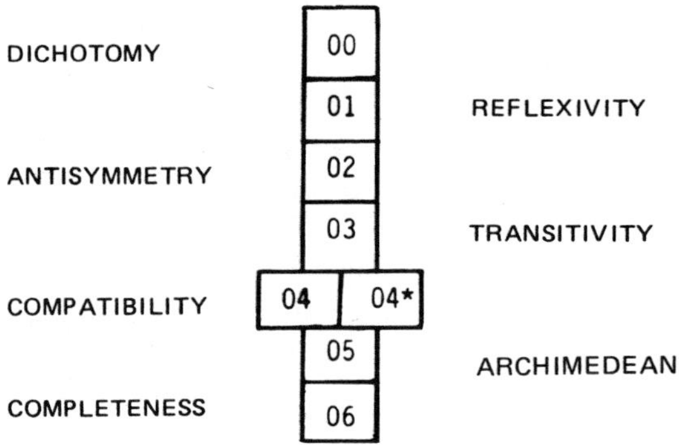

FIELD OF RATIONAL NUMBERS

The order axioms will be spelled out in due course; let us first look at the order in the field $\mathbb{Q}$ of rational numbers:

$$\mathbb{Q} = \{m/n \mid m, n \in \mathbb{Z}, n > 0\}$$

(note that we insist on *positive* denominators). Equality and order in $\mathbb{Q}$ are defined such that

$$\frac{m}{n} = \frac{m'}{n'} \Leftrightarrow mn' = m'n$$

$$\frac{m}{n} \geqq \frac{m'}{n'} \Leftrightarrow mn' \geqq m'n$$

Thus the order in $\mathbb{Q}$ is defined in terms of the order in $\mathbb{Z}$, the number system of positive and negative integers. It is clear that $\mathbb{Q}$ inherits the following properties from $\mathbb{Z}$:

(00) **Dichotomy** $\quad \alpha, \beta \in \mathbb{Q} \Rightarrow$ either $\alpha \leqq \beta$ or $\alpha \geqq \beta$

(01) **Reflexivity** $\quad \alpha \leqq \alpha$

(02) **Antisymmetry** $\quad \alpha \leqq \beta$ and $\beta \leqq \alpha \Rightarrow \alpha = \beta$

(03) **Transitivity** $\quad \alpha \leqq b$ and $\beta \leqq \gamma \Rightarrow \alpha \leqq \gamma$

(04) $\quad \alpha \leqq \beta \Rightarrow \alpha + \gamma \leqq \beta + \gamma$

Compatibility

(04*) $\quad \alpha \leqq \beta, \gamma > 0 \Rightarrow \alpha\gamma \leqq \beta\gamma$

A field satisfying these axioms is called an *ordered field*. Thus $\mathbb{Q}$ is an example of an ordered field. $\mathbb{Q}$ is in fact more; it is an archimedean field, as its order satisfies the following

(O5) **Archimedean Property** Given $\varepsilon > 0$, however small, and $K > 0$, however large, there is a positive integer n such that $n\varepsilon > K$.

A subset X of $\mathbb{Q}$ may or may not have a maximal element $\max X \in X$ (such that $x \in X \Rightarrow x \leqq \max X$) or a minimal element $\min X \in X$ (such that $x \in X \Rightarrow \min X \leqq x$). At any rate, if X is finite, $\max X$ and $\min X$ exist. It is not surprising that $\max X$ may not exist for an infinite subset X, because X may not be bounded.

As a consequence of the archimedean property, $\max X$ may not exist even if X is bounded above, and $\min X$ may not exist even if X is bounded below, e.g.,

$$X = \{1/n \mid n = 1,2,\ldots\}$$

has no $\min X$, even though it is bounded below (since all of its elements are positive). Indeed, on the strength of the archimedean property, given any $\varepsilon > 0$, however small, $1/n < \varepsilon$ for a large enough positive integer n. Since there are elements in X smaller than any positive number, $\min X$ does not exist.

LEAST UPPER BOUND. GREATEST LOWER BOUND

A promising substitute for $\max X$ is $\operatorname{lub} X$, the *least upper bound of X*, defined such that

$$x \in X \Rightarrow x \leqq \operatorname{lub} X$$
$$(x \leqq K \text{ for all } x \in X) \Rightarrow \operatorname{lub} X \leqq K$$

Of course, if $\max X$ exists, then $\operatorname{lub} X = \max X$; but the interesting case is where $\max X$ does not exist. Similarly, a substitute for $\min X$ is $\operatorname{glb} X$, the *greatest lower bound of X*. It is left as an exercise for the reader to give the definition of $\operatorname{glb} X$.

The really disturbing, even demoralizing, discovery for the mathematicians of the ancient world was the fact that $\operatorname{lub} X$ may also fail to exist for a *bounded* subset X of $\mathbb{Q}$. The pupils of Pythagoras were studying the set X of rational numbers

$$X = \{x \in \mathbb{Q} \mid x^2 < 2\}$$

clearly bounded above, and they were horrified to find that $\operatorname{lub} X$ did not exist. They realized that, if $\operatorname{lub} X$ existed, it could be written as

$$\text{lub } X = \frac{p}{q}$$

where p, q were relatively prime integers, and $p^2/q^2 = 2$. However,

$$\begin{aligned} p^2/q^2 = 2 \Rightarrow p^2 = 2q^2 \Rightarrow p^2 \text{ even } &\Rightarrow p \text{ even, hence } q \textit{ odd} \\ &\Rightarrow p = 2r \Rightarrow p^2 = 4r^2 \Rightarrow 2q^2 = p^2 = 4r^2 \\ &\Rightarrow q^2 = 2r^2 \Rightarrow q^2 \text{ even } \Rightarrow q \text{ even} \Rightarrow q \text{ both } \textit{even} \text{ and } \textit{odd} \end{aligned}$$

which is *impossible*. Conclusion: lub X *does not exist.*

Geometrically, this means that there are "holes" in the line represented by rational points, in spite of the fact that rational points are everywhere dense on the line, i.e., between any two rational points, no matter how close to one another they may be, there are still infinitely many rational points. As a consequence, the circle centered at the origin and passing through the point $P(1,1)$ does not have a point in common with either one of the coordinate axes, which violates one's geometric intuition.

IRRATIONAL NUMBERS

This discovery of the mathematicians of the ancient world presented them with a problem that they failed to solve. In fact, a complete solution had to wait until the nineteenth century, when the first successful completion of the field of rational numbers was given by the introduction of the irrational numbers, thereby constructing the real number system $\mathbb{R}$. Thus $\mathbb{R}$ must satisfy the axiom:

(O6) **Completeness** Every bounded subset X of $\mathbb{R}$ has a least upper bound, lub X, which is such that

$$x \in X \Rightarrow x \leqq \text{lub } X$$

$$(x \leqq K \text{ for all } x \in X) \Rightarrow \text{lub } X \leqq K$$

An archimedean ordered field which also satisfies the completeness axiom, is called a *complete ordered field*. Note that there is no need to postulate the existence of glb X, because it already follows from axioms 4, 04, 06.

It is far from obvious that $\mathbb{Q}$ can be made into a complete ordered field by the introduction of the irrational numbers, nor is it obvious that this completion of $\mathbb{Q}$ is unique. An account of the construction of $\mathbb{R}$ and the proof of its uniqueness is given in the book *Set Theory and Abstract Algebra*, by T. S. Blyth (1975).

ALGEBRAIC CLOSURE

$\mathbb{R}$ fails to be algebraically closed; e.g., the equation $x^2 + 1 = 0$ has no solution $x \in \mathbb{R}$. As we have seen in Chap. 12, this failure can be remedied by a further extension of the number system from $\mathbb{R}$ to $\mathbb{C}$, by the inclusion of the complex numbers.

However, something is irretrievably lost in this process; the field $\mathbb{C}$ of complex numbers cannot be ordered so as to satisfy axioms 00 through 06 (or even axioms 00 through 04). For example, there is no way of defining what is meant by a "positive complex number." Yet the order structure of $\mathbb{R}$ is the vehicle through which geometric ideas (especially the idea of points populating a line) are conveyed. Losing the order structure amounts to losing much of the geometry.

On the other hand, no geometric gain is made as a result of achieving algebraic closure. As we know, the algebraic fact that the equation

$$x^2 + 1 = 0$$

has no real solution is tantamount to the geometric fact that *no two lines can be parallel and perpendicular to each other at the same time*. Thus, in a very "real" sense of the word, the demand that $\mathbb{R}$ be made algebraically closed through field extension is "ungeometric"; it can be met only by sacrificing more of the geometry than one can find an immediate justification for. Thus, in a vector space over the complex numbers $\mathbb{C}$, a vector $\mathbf{v} \neq 0$ may be perpendicular to itself, or it may have zero length, e.g.,

$$\mathbf{v} = \begin{pmatrix} 1 \\ i \end{pmatrix}$$

This remark reveals the dilemma that teachers of mathematics face in the choice of scalars. We are hereby warned that a generalization of the vector space structure to complex scalars and beyond is justified only after real linear algebra has been thoroughly understood and mastered.

Solutions to Exercises

Chapter 0

1. For $g \circ f$ to be injective it is necessary and sufficient that (a) f be injective and (b) f have a left inverse f' which factorizes through g, i.e.,

$$g \circ f \text{ injective} \Leftrightarrow f' \text{ exists satisfying } f' \circ f = 1,\ f' = h \circ g$$

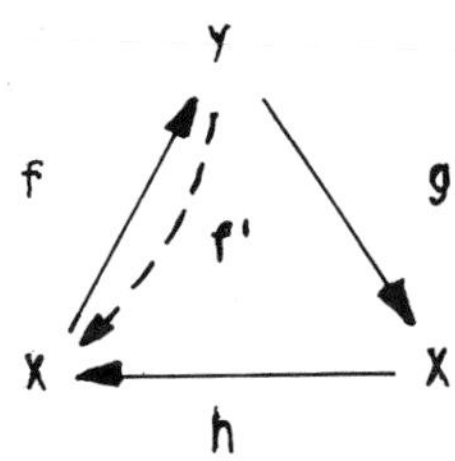

Proof: The condition is sufficient:

$$f' \circ f = 1,\ f' = h \circ g \Rightarrow h \circ (g \circ f) = (h \circ g) \circ f = f' \circ f = 1$$
$$\Rightarrow g \circ f \text{ injective}$$

The condition is necessary:

$$g \circ f \text{ injective} \Rightarrow g \circ f \text{ left invertible}$$
$$\Rightarrow h \circ (g \circ f) = 1 \text{ for some } h$$
$$\Rightarrow f' \circ f = 1 \text{ with } f' = h \circ g$$

We see that the necessary condition can be strengthened by requiring that f have a left inverse f' factorizing through g.

2. The necessary condition can be strengthened to a necessary and sufficient condition by requiring that g have a right inverse g' factorizing through f. Note

that the proof of this involves the axiom of choice. The equivalent condition derived in Exercise 0.4 manages to avoid this involvement.

3. $g \circ f$ injective $\Leftrightarrow f$ injective and the restriction of g to $\operatorname{im} f$ is injective
4. $g \circ f$ surjective $\Leftrightarrow$ the restriction of g to $\operatorname{im} f$ is surjective
5. For any country x, location(capital(x)) $= x$ because there is no country whose capital city is located in a foreign country. The statement now follows from the condition given in Exercise 6.
6. $g \circ f = 1 \Rightarrow (f \circ g)^2 = f \circ (g \circ f) \circ g = f \circ 1 \circ g = f \circ g$

   $\Rightarrow f \circ g$ projection
7. For any married person x, spouse(spouse(x)) $= x$, so that spouse$^2 = 1$.
8. $i^{-1} = i \Rightarrow i \circ i = 1$ is immediate from the definition of the inverse. Conversely, $i \circ i = 1$ implies that i is both injective and surjective, hence i^{-1} exists. But then

$$i \circ i = 1 = i \circ i^{-1} \Rightarrow i = i^{-1} \qquad \text{(by cancellation)}$$

9. $(i \circ j)^2 = 1 \Rightarrow i \circ j \circ i \circ j = 1 \Rightarrow i^2 \circ j \circ i \circ j^2 = i \circ 1 \circ j \Rightarrow j \circ i = i \circ j$ because $i^2 = 1, j^2 = 1$. Conversely, $(i \circ j)^{-1} = j^{-1} \circ i^{-1} = j \circ i = i \circ j$ since $j^{-1} = j, i^{-1} = i$, so $i \circ j$ is an involution.
10. The condition is sufficient by Exercise 6. To show that it is also necessary, consider the decomposition of $p{:}X \to X$ into the product $p = g \circ f$

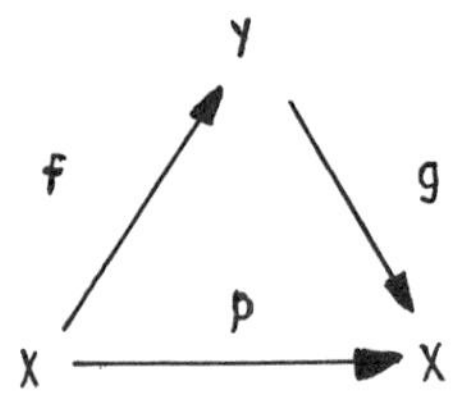

where Y is the image of p, and f, g are defined by the formulas

$$f(x) = p(x) \quad \text{for all } x \in X \qquad g(y) = y \quad \text{for all } y \in Y$$

The argument can now be completed by calculating $f \circ g$ and $g \circ f$.

Chapter 1

1. $$\alpha\mathbf{a} + \beta\mathbf{b} + \gamma\mathbf{c} = 0 \Rightarrow \left.\begin{matrix} \alpha + 2\beta - \gamma = 0 \\ 3\alpha - 3\beta + 2\gamma = 0 \\ -\alpha - 11\beta + 6\gamma = 0 \end{matrix}\right\} \Rightarrow \begin{pmatrix} 1 & 2 & -1 \\ 3 & -3 & 2 \\ -1 & -11 & 6 \end{pmatrix} \Rightarrow \begin{pmatrix} 1 & 2 & -1 \\ 0 & 9 & -5 \end{pmatrix}$$

$$\Rightarrow \begin{pmatrix} 1 & 0 & 1/9 \\ 0 & 1 & -5/9 \end{pmatrix} \Rightarrow \begin{cases} \alpha = -t \\ \beta = 5t \\ \gamma = 9t \end{cases}$$

For $t = 1$ we have $-\mathbf{a} + 5\mathbf{b} + 9\mathbf{c} = \mathbf{0}$. Hence $\mathbf{a}$, $\mathbf{b}$, $\mathbf{c}$ are linearly dependent.

2. $\alpha\mathbf{a}+\beta\mathbf{b}+\gamma\mathbf{c}=\mathbf{0}\Rightarrow\begin{pmatrix}1&2&0\\-2&-2&1\\3&0&7\end{pmatrix}\Rightarrow\begin{pmatrix}1&0&0\\0&1&0\\0&0&1\end{pmatrix}\Rightarrow\alpha=\beta=\gamma=0$

3. Linearly dependent, because the sum of the first two is equal to the third vector.

4. $\begin{pmatrix}1&-1&-1\\2&2&-2\\3&2&3\end{pmatrix}\Rightarrow\begin{pmatrix}1&0&0\\0&1&0\\0&0&1\end{pmatrix}\Rightarrow\alpha=\beta=\gamma=0$. The vectors are linearly independent.

5. $\alpha\mathbf{a}+\beta\mathbf{b}+\gamma\mathbf{c}=\mathbf{d}\Rightarrow\left(\begin{array}{ccc|c}1&1&1&6\\2&-1&1&3\\4&1&3&15\end{array}\right)\Rightarrow\left(\begin{array}{ccc|c}1&0&2/3&3\\0&1&1/3&3\end{array}\right)\Rightarrow\begin{cases}\alpha=3-2t\\\beta=3-t\\\gamma=3t\end{cases}$

   There are infinitely many solutions, e.g., for $t=1$, $\mathbf{a}+2\mathbf{b}+3\mathbf{c}=\mathbf{d}$.

6. No, it cannot.

Chapter 2

1. $|\mathbf{i}+\mathbf{j}+\mathbf{k}|=\sqrt{3}$
2. $|\overrightarrow{AB}|=|\mathbf{i}-7\mathbf{j}+6\mathbf{k}|=\sqrt{86}$
3. $\mathbf{a}^0=\dfrac{\mathbf{a}}{|\mathbf{a}|}=\dfrac{5\mathbf{i}-6\mathbf{j}+30\mathbf{k}}{\sqrt{961}}=(5/31)\mathbf{i}-(6/31)\mathbf{j}+(30/31)\mathbf{k}$
4. $\cos\sphericalangle(\mathbf{a},\mathbf{b})=\mathbf{a}^0\cdot\mathbf{b}^0=-2/\sqrt{7}$. The angle is obtuse.
5. $|\overrightarrow{CA}|=|\overrightarrow{CB}|=\sqrt{46}$
6. $|\overrightarrow{AB}|=|4\mathbf{i}+8\mathbf{j}+8\mathbf{k}|=12$
   $|\overrightarrow{BC}|=|2\mathbf{i}-5\mathbf{j}-14\mathbf{k}|=15$
   $|\overrightarrow{AC}|=|6\mathbf{i}+3\mathbf{j}-6\mathbf{k}|=9$. $9^2+12^2=15^2$, hence $\sphericalangle A$ is a right angle. Indeed, $\overrightarrow{AB}\perp\overrightarrow{AC}$ because $\overrightarrow{AB}\cdot\overrightarrow{AC}=0$.
7. The angle at B is obtuse, because $\overrightarrow{BA}\cdot\overrightarrow{BC}=-64<0$.
8. (i) 2/7, −3/7, −6/7. (ii) 4/21, 5/21, −20/21.
9. (i) $\mathbf{u}^2=\mathbf{v}^2=\mathbf{w}^2=\dfrac{1+4+4}{9}=$; $\mathbf{u}\cdot\mathbf{v}=\mathbf{v}\cdot\mathbf{w}=\mathbf{w}\cdot\mathbf{u}=\dfrac{2+2-4}{9}=0$
10. The diagonal of the unit cube is $\mathbf{i}+\mathbf{j}+\mathbf{k}$; $\cos\alpha=\cos\sphericalangle(\mathbf{i},\mathbf{i}+\mathbf{j}+\mathbf{k})=\mathbf{i}\cdot(\mathbf{i}+\mathbf{j}+\mathbf{k})^0=1/\sqrt{3}$. Hence $\alpha=\arccos(1/\sqrt{3})=54°45'$ (approximately, from trigonometric tables).
11. The direction cosines of UNE are $1/\sqrt{3}$, $1/\sqrt{3}$, $1/\sqrt{3}$. Hence $\alpha=\beta=\gamma=\arccos(1/\sqrt{3})=54°45'$.
12. The direction cosines of LSW are $-1/\sqrt{3}$, $-1/\sqrt{3}$, $-1/\sqrt{3}$. Hence $\alpha=\beta=\gamma=\arccos(-1/\sqrt{3})=180°-\arccos(-1/\sqrt{3})=125°15'$ (approximately).

13. (i) $|\overrightarrow{AB}| = |\overrightarrow{BC}| = |\overrightarrow{CA}| = 3\sqrt{2}$. This is an equilateral triangle, hence $\sphericalangle A = \sphericalangle B = \sphericalangle C = 60°$. Indeed,

$$\cos \sphericalangle A = \frac{\overrightarrow{AB} \cdot \overrightarrow{AC}}{|\overrightarrow{AB}|\,|\overrightarrow{AC}|} = \frac{1}{2} \Rightarrow \sphericalangle A = 60°, \text{ etc.}$$

(ii) This is an isosceles right triangle: $\sphericalangle A = 90°$, $\sphericalangle B = \sphericalangle C = 45°$.
(iii) This is a right triangle with $\sphericalangle A = 30°$, $\sphericalangle B = 60°$, $\sphericalangle C = 90°$.
(iv) This is an isosceles obtuse triangle with $\sphericalangle A = \sphericalangle C = 30°$, $\sphericalangle B = 120°$.

14 The direction vectors are $\mathbf{i} + \mathbf{j} + \mathbf{k}$, $\mathbf{i} + \mathbf{j}$. Hence

$$\cos \alpha = \frac{(\mathbf{i}+\mathbf{j}+\mathbf{k}) \cdot (\mathbf{i}+\mathbf{j})}{|\mathbf{i}+\mathbf{j}+\mathbf{k}|\,|\mathbf{i}+\mathbf{j}|} = \frac{2}{\sqrt{3}\sqrt{2}} = \sqrt{2/3}$$

$\alpha = \arccos(\sqrt{2/3}) = 35°16'$(approximately).

15. The direction vectors are $\mathbf{i} + \mathbf{k}$, $\mathbf{i} + \mathbf{j}$; $\cos \alpha = ½$, $\alpha = \arccos ½ = 60°$.

16. $\alpha = 45°$

17. $\alpha = \arccos(1/3)$

18. (i) Outside. (ii) On. (iii) Inside. (iv) On. (v) Inside.

19. $C(P) = 25$; the length of the tangent is $\sqrt{25} = 5$.

20. $(\alpha_1 - \alpha_2)^2 + (\beta_1 - \beta_2)^2 = \rho_1^2 + \rho_2^2$

21. The radical axis of the circles passes through the points of intersection. The equation of the radical axis is just the difference of the normal equations of the circles, i.e., $7x - 4y = 0$.

22. The radical axis $7x + 3y - 16 = 0$ is the common tangent to the circles.

23. P is on the circle because its coordinates satisfy the equation of the circle. The equation of the tangent to the circle at P can be obtained as the radical axis of the given circle and P considered as a null circle, i.e., $(x-5)^2 + (y-5)^2 = 0$. The radical axis is $3x + 4y + 15 = 0$.

24. If there is such a point, it must be the radical center of the three circles, i.e., the point where the three radical axes meet: $P(3,5)$. Indeed, the length of the tangents is $\sqrt{32}$.

25. The required points are the points of intersection of the radical axis with the coordinate axis. The radical axis in intercept form is $x/(4/3) + y/4 = 1$; the required points are $P(4/3,0)$, $Q(0,4)$. The lengths of the tangents are 8/3 and 4.

Chapter 3

1. $$\mathbf{a} \times \mathbf{b} = \begin{vmatrix} \mathbf{i} & \mathbf{j} & \mathbf{k} \\ 1 & -2 & 3 \\ 0 & 4 & -5 \end{vmatrix} = -2\mathbf{i} + 5\mathbf{j} + 4\mathbf{k} \perp \mathbf{a}, \mathbf{b}$$

2. $\mathbf{w} = \mathbf{u} \times \mathbf{v} = -(2/3)\mathbf{i} + (2/3)\mathbf{j} - (1/3)\mathbf{k}$ and $-\mathbf{w}$ are the only solutions.

3. $\mathbf{a} \times \mathbf{b} = -52\mathbf{i} + 156\mathbf{j} - 39\mathbf{k} = -13(4\mathbf{i} - 12\mathbf{j} + 3\mathbf{k})$, $|\mathbf{a} \times \mathbf{b}| = 13\sqrt{16 + 144 + 9} = 13^2$

is the area of the parallelogram. Notice that $|\mathbf{a}| = |\mathbf{b}| = 13$, $\mathbf{a} \perp \mathbf{b}$, so, in fact, the parallelogram is a square.

4. *First method*: $\mathbf{a} \times \mathbf{b} = \begin{vmatrix} \mathbf{i} & \mathbf{j} & \mathbf{k} \\ 1 & 2 & 2 \\ -2 & -10 & 11 \end{vmatrix} = 42\mathbf{i} - 15\mathbf{j} - 6\mathbf{k} = 3(14\mathbf{i} - 5\mathbf{j} - 2\mathbf{k});$

$|\mathbf{a} \times \mathbf{b}| = 3\sqrt{196 + 25 + 4} = 3\sqrt{225} = 45$

Second method: $|\mathbf{a} \times \mathbf{b}| = \sqrt{|\mathbf{a}|^2|\mathbf{b}|^2 - (\mathbf{a}\cdot\mathbf{b})^2} = \sqrt{9(225) - 0} = 45$ and, incidentally, we learn that the parallelogram is, in fact, a rectangle.

5. (i) $\frac{1}{2}|\overrightarrow{CA} \times \overrightarrow{CB}| = \frac{1}{2}(49)$. (ii) 14. (iii) 14.

6. (i) $\dfrac{|\overrightarrow{CA} \times \overrightarrow{CB}|}{|\overrightarrow{CB}|} = \dfrac{4\sqrt{2}}{2\sqrt{2}} = 2$. (ii) $9\sqrt{3/2}$.

7. (i) $\mathbf{v} = \left(\dfrac{17}{3}\mathbf{i} - \dfrac{10}{3}\mathbf{j} - \dfrac{14}{3}\mathbf{k}\right) + (-7)\left(\dfrac{2}{3}\mathbf{i} + \dfrac{2}{3}\mathbf{j} + \dfrac{1}{3}\mathbf{k}\right)$

(ii) $\mathbf{v} = (2\mathbf{i} - 3\mathbf{j}) + 4\mathbf{k}$

8. $[\mathbf{a},\mathbf{b},\mathbf{c}] = 0$ in each case.

9. (i) $[\mathbf{a},\mathbf{b},\mathbf{c}] = 729 = 9^3$. We may notice that $|\mathbf{a}| = |\mathbf{b}| = |\mathbf{c}| = 9$, so that, in fact, $\mathbf{a},\mathbf{b},\mathbf{c}$ span a cube.

(ii) 17. (iii) 1.

10. $[\overrightarrow{DA},\overrightarrow{DB},\overrightarrow{DC}] = 0$.

11. (i) $\dfrac{1}{6}[\overrightarrow{DA},\overrightarrow{DB},\overrightarrow{DC}] = 9/2$. (ii) 3. (iii) 11/3.

12. (i) $\dfrac{|[\overrightarrow{DA}, \overrightarrow{DB}, \overrightarrow{DC}]|}{|\overrightarrow{DB} \times \overrightarrow{DC}|} = 7$. Notice that $|\overrightarrow{AB}| = 7$ as well, so that the vector $\overrightarrow{AB}$ furnishes the perpendicular distance.

(ii) $11/\sqrt{21}$

13. $[\overrightarrow{AB}, \overrightarrow{AC}, \overrightarrow{CD}\,] = 0$

14. (i) $\dfrac{|[\overrightarrow{CA}, \overrightarrow{AB}, \overrightarrow{CD}]|}{|\overrightarrow{AB} \times \overrightarrow{CD}|} = 3$. Notice that $|\overrightarrow{AC}| = 3$ as well, so that the vector $\overrightarrow{AC}$ furnishes the orthogonal transversal.

(ii) 13.

15. (i) $\dfrac{12}{17}\mathbf{i} + \dfrac{12}{17}\mathbf{j} + \dfrac{1}{17}\mathbf{k}$ (ii) $\dfrac{6}{19}\mathbf{i} + \dfrac{17}{19}\mathbf{j} - \dfrac{6}{19}\mathbf{k}$

(iii) $-\dfrac{19}{21}\mathbf{i} + \dfrac{8}{21}\mathbf{j} - \dfrac{4}{21}\mathbf{k}$ (iv) $\dfrac{20}{21}\mathbf{i} - \dfrac{4}{21}\mathbf{j} - \dfrac{5}{21}\mathbf{k}$

(v) $\frac{22}{23}\mathbf{i}-\frac{6}{23}\mathbf{j}-\frac{3}{23}\mathbf{k}$ (vi) $\frac{23}{27}\mathbf{i}+\frac{10}{27}\mathbf{j}+\frac{10}{27}\mathbf{k}$

(vii) $\frac{16}{29}\mathbf{i}+\frac{24}{29}\mathbf{j}+\frac{3}{29}\mathbf{k}$ (viii) $-\frac{5}{31}\mathbf{i}-\frac{6}{31}\mathbf{j}+\frac{30}{31}\mathbf{k}$

(ix) $\frac{27}{31}\mathbf{i}+\frac{6}{31}\mathbf{j}-\frac{14}{31}\mathbf{k}$ (x) $\frac{31}{33}\mathbf{i}+\frac{8}{33}\mathbf{j}-\frac{8}{33}\mathbf{k}$

(xi) $\frac{4}{33}\mathbf{i}+\frac{7}{33}\mathbf{j}-\frac{32}{33}\mathbf{k}$ (xii) $\frac{20}{33}\mathbf{i}-\frac{25}{33}\mathbf{j}-\frac{8}{33}\mathbf{k}$

(xiii) $\frac{6}{35}\mathbf{i}+\frac{10}{35}\mathbf{j}+\frac{33}{35}\mathbf{k}$ (xiv) $\frac{24}{37}\mathbf{i}-\frac{24}{37}\mathbf{j}-\frac{3}{37}\mathbf{k}$

(xv) $-\frac{26}{39}\mathbf{i}+\frac{26}{39}\mathbf{j}-\frac{13}{39}\mathbf{k}$ (xvi) $-\frac{4}{41}\mathbf{i}+\frac{33}{41}\mathbf{j}-\frac{24}{41}\mathbf{k}$

(xvii) $-\frac{6}{43}\mathbf{i}-\frac{42}{43}\mathbf{j}+\frac{7}{43}\mathbf{k}$ (xviii) $-\frac{30}{43}\mathbf{i}+\frac{25}{43}\mathbf{j}-\frac{18}{43}\mathbf{k}$

(xix) $\frac{33}{49}\mathbf{i}+\frac{4}{49}\mathbf{j}+\frac{36}{49}\mathbf{k}$ (xx) $\frac{50}{51}\mathbf{i}+\frac{1}{51}\mathbf{j}-\frac{10}{51}\mathbf{k}$

(xxi) $\frac{46}{51}\mathbf{i}+\frac{14}{51}\mathbf{j}+\frac{17}{51}\mathbf{k}$ (xxii) $\mathbf{k}$

(xxiii) $-\frac{40}{57}\mathbf{i}+\frac{32}{57}\mathbf{j}-\frac{25}{57}\mathbf{k}$ (xxiv) $\frac{50}{59}\mathbf{i}+\frac{9}{59}\mathbf{j}-\frac{30}{59}\mathbf{k}$

(xxv) $\frac{52}{65}\mathbf{i}-\frac{15}{65}\mathbf{j}+\frac{36}{65}\mathbf{k}$ (xxvi) $-\frac{70}{79}\mathbf{i}+\frac{30}{79}\mathbf{j}-\frac{21}{79}\mathbf{k}$

(xxvii) $\frac{68}{81}\mathbf{i}-\frac{16}{81}\mathbf{j}-\frac{41}{81}\mathbf{k}$ (xxviii) $\frac{64}{81}\mathbf{i}-\frac{8}{81}\mathbf{j}+\frac{49}{81}\mathbf{k}$

(xxix) $-\frac{70}{99}\mathbf{i}-\frac{50}{99}\mathbf{j}+\frac{49}{99}\mathbf{k}$ (xxx) $-\frac{110}{111}\mathbf{i}-\frac{10}{111}\mathbf{j}+\frac{11}{111}\mathbf{k}$

(xxxi) $-0.8\mathbf{i}+0.36\mathbf{j}-0.48\mathbf{k}$ (xxxii) $0.64\mathbf{i}-0.024\mathbf{j}-0.768\mathbf{k}$

(xxxiii) $0.192\mathbf{i}+0.48\mathbf{j}+0.856\mathbf{k}$ (xxxiv) $-0.864\mathbf{i}+0.48\mathbf{j}+0.152\mathbf{k}$

Chapter 4

1. A, B, C, P coplanar $\Leftrightarrow \mathbf{a}-\mathbf{p}, \mathbf{b}-\mathbf{p}, \mathbf{c}-\mathbf{p}$ coplanar (see Chap. 1)

$$\Leftrightarrow [\mathbf{a}-\mathbf{p},\ \mathbf{b}-\mathbf{p},\ \mathbf{c}-\mathbf{p}] = 0$$

$$\Leftrightarrow [\mathbf{a},\mathbf{b},\mathbf{c}] - [\mathbf{a},\mathbf{b},\mathbf{p}] - [\mathbf{a},\mathbf{p},\mathbf{c}] - [\mathbf{p},\mathbf{b},\mathbf{c}] = 0$$

2. $[\mathbf{a}\times\mathbf{b},\ \mathbf{b}\times\mathbf{c},\ \mathbf{c}\times\mathbf{a}] = ((\mathbf{a}\times\mathbf{b})\times(\mathbf{b}\times\mathbf{c}))\cdot(\mathbf{c}\times\mathbf{a})$

$$= ([\mathbf{a},\mathbf{b},\mathbf{c}]\mathbf{b})\cdot(\mathbf{c}\times\mathbf{a}) \qquad \text{(by the triple-cross formula)}$$

$$= [\mathbf{a},\mathbf{b},\mathbf{c}][\mathbf{b},\mathbf{c},\mathbf{a}] = [\mathbf{a},\mathbf{b},\mathbf{c}]^2$$

3. $(\mathbf{b}\times\mathbf{c})\cdot(\mathbf{a}\times\mathbf{d}) + (\mathbf{c}\times\mathbf{a})\cdot(\mathbf{b}\times\mathbf{d}) + (\mathbf{a}\times\mathbf{b})\cdot(\mathbf{c}\times\mathbf{d})$

$$= \begin{vmatrix} \mathbf{b}\cdot\mathbf{a} & \mathbf{c}\cdot\mathbf{a} \\ \mathbf{b}\cdot\mathbf{d} & \mathbf{c}\cdot\mathbf{d} \end{vmatrix} + \begin{vmatrix} \mathbf{c}\cdot\mathbf{b} & \mathbf{a}\cdot\mathbf{b} \\ \mathbf{c}\cdot\mathbf{d} & \mathbf{a}\cdot\mathbf{d} \end{vmatrix} + \begin{vmatrix} \mathbf{a}\cdot\mathbf{c} & \mathbf{b}\cdot\mathbf{c} \\ \mathbf{a}\cdot\mathbf{d} & \mathbf{b}\cdot\mathbf{d} \end{vmatrix} \qquad \text{(by the cross-dot-cross formula)}$$

$$= (\mathbf{b}\cdot\mathbf{a})(\mathbf{c}\cdot\mathbf{d}) - (\mathbf{c}\cdot\mathbf{a})(\mathbf{b}\cdot\mathbf{d}) + (\mathbf{c}\cdot\mathbf{b})(\mathbf{a}\cdot\mathbf{d}) - (\mathbf{a}\cdot\mathbf{b})(\mathbf{c}\cdot\mathbf{d}) + (\mathbf{a}\cdot\mathbf{c})(\mathbf{b}\cdot\mathbf{d}) - (\mathbf{b}\cdot\mathbf{c})(\mathbf{a}\cdot\mathbf{d})$$

$$= 0$$

4. *Hint*: Use the triple cross formula.

5. and 6. Let $*$ stand for either $\cdot$ or $\times$. We have
$(\mathbf{a}-\mathbf{d}) * (\mathbf{b}-\mathbf{c}) + (\mathbf{b}-\mathbf{d}) * (\mathbf{c}-\mathbf{a}) + (\mathbf{c}-\mathbf{d}) * (\mathbf{a}-\mathbf{b})$

$$= \mathbf{a} * (\mathbf{b}-\mathbf{c}) + \mathbf{b} * (\mathbf{c}-\mathbf{a}) + \mathbf{c} * (\mathbf{a}-\mathbf{b}) - \mathbf{d} * (\mathbf{b}-\mathbf{c}+\mathbf{c}-\mathbf{a}+\mathbf{a}-\mathbf{b})$$

$$= \mathbf{a}*\mathbf{b} - \mathbf{a}*\mathbf{c} + \mathbf{b}*\mathbf{c} - \mathbf{b}*\mathbf{a} + \mathbf{c}*\mathbf{a} - \mathbf{c}*\mathbf{b}$$

$$= (\mathbf{a}*\mathbf{b} - \mathbf{b}*\mathbf{a}) + (\mathbf{b}*\mathbf{c} - \mathbf{c}*\mathbf{b}) + (\mathbf{c}*\mathbf{a} - \mathbf{a}*\mathbf{c})$$

$$= \begin{cases} 0 & \text{if } * = \cdot \\ 2(\mathbf{a}\times\mathbf{b} + \mathbf{b}\times\mathbf{c} + \mathbf{c}\times\mathbf{a}) & \text{if } * = \times \end{cases}$$

7. Set $S(\mathbf{p}) = x^2 + y^2 + z^2 + \alpha x + \beta y + \gamma z + \delta = (\mathbf{p}-\mathbf{c})^2 - \rho^2$

$$= \mathbf{d}^2 - \rho^2 = (\mathbf{d}+\rho)(\mathbf{d}-\rho) = PA\cdot PB$$

where $\mathbf{d} = |\mathbf{p}-\mathbf{c}|$ is the distance between the point P (with location vector $\mathbf{p}$) and the center C (with location vector $\mathbf{c}$) of the sphere.

8. $S(\mathbf{p}) = S'(\mathbf{p}) \Leftrightarrow (\alpha-\alpha')x + (\beta-\beta')y + (\gamma-\gamma')z + (\delta-\delta') = 0$
is the equation of a plane, namely, the radical plane of the nonconcentric spheres S, S'.

9. $\mathbf{n} = (\alpha-\alpha')\mathbf{i} + (\beta-\beta')\mathbf{j} + (\gamma-\gamma')\mathbf{k}$

$\mathbf{n}' = (\alpha'-\alpha'')\mathbf{i} + (\beta'-\beta'')\mathbf{j} + (\gamma'-\gamma'')\mathbf{k}$

$\mathbf{n}'' = (\alpha''-\alpha)\mathbf{i} + (\beta''-\beta)\mathbf{j} + (\gamma''-\gamma)\mathbf{k}$

are the normal vectors of the three radical planes. Clearly, $\mathbf{n} + \mathbf{n}' + \mathbf{n}'' = \mathbf{0}$, so the normal vectors are linearly dependent, i.e., coplanar. Since the centers of the spheres are not collinear, the three radical planes cannot be parallel, and the line of intersection of two is also contained in the third radical plane.

Let $\mathbf{a}$, $\mathbf{b}$, $\mathbf{c}$ denote the centers of the spheres; then $\mathbf{a}-\mathbf{b}$, $\mathbf{b}-\mathbf{c}$, $\mathbf{c}-\mathbf{a}$ are the normal vectors of the radical planes, and

$$(\mathbf{a}-\mathbf{b})\times(\mathbf{b}-\mathbf{c}) = \mathbf{a}\times\mathbf{b} - \mathbf{a}\times\mathbf{c} + \mathbf{b}\times\mathbf{c} = \mathbf{a}\times\mathbf{b} + \mathbf{b}\times\mathbf{c} + \mathbf{c}\times\mathbf{a}$$

is the direction vector of the radical axis of the three spheres.

11. (See Chap. 10, Exercise 9, and Chap. 11, Exercise 24.)

$$t(\mathbf{v}) = (\mathbf{n}'\cdot\mathbf{v})\mathbf{e}' - (\mathbf{n}\cdot\mathbf{v})\mathbf{e} \qquad \text{(by the double cross formula)}$$
$$t^2(\mathbf{v}) = (\mathbf{n}'\cdot t(\mathbf{v}))\mathbf{e}' - (\mathbf{n}\cdot t(\mathbf{v}))\mathbf{e} = (\mathbf{n}\cdot\mathbf{v})\mathbf{e}' - (\mathbf{n}'\cdot\mathbf{v})\mathbf{e}$$
$$t^3(\mathbf{v}) = t(t^2(\mathbf{v})) = \cdots = -t(\mathbf{v})$$

12. (See Chap. 10, Exercise 9, and Chap. 11, Exercise 24.)
$$\begin{aligned} t(\mathbf{v}) &= (\mathbf{n}\times\mathbf{v})\times\mathbf{e} - (\mathbf{e}\times\mathbf{v})\times\mathbf{n} \\ &= (\mathbf{e}\cdot\mathbf{v})\mathbf{n} - (\mathbf{n}\cdot\mathbf{v})\mathbf{e} \qquad \text{(by the double-cross formula)} \\ &= (\mathbf{e}\times\mathbf{n})\times\mathbf{v} = \mathbf{u}\times\mathbf{v} \qquad \text{(by the double-cross formula)} \end{aligned}$$
where $\mathbf{u} = \mathbf{e}\times\mathbf{n}$ is a unit vector, and $t^*(\mathbf{v}) = \mathbf{v}\times\mathbf{u} = -t(\mathbf{v})$.

Conversely, if the twist t is antisymmetric, then (by the principal twist theorem, see Chap. 11) $t(\mathbf{v}) = \mathbf{u}\times\mathbf{v}$, where $|\mathbf{u}| = 1$. We now pick unit vectors $\mathbf{n}$, $\mathbf{e}\perp\mathbf{u}$. Then $\mathbf{u} = \mathbf{e}\times\mathbf{n}$ and
$$t(\mathbf{v}) = (\mathbf{n}\times\mathbf{v})\times\mathbf{e} - (\mathbf{e}\times\mathbf{v})\times\mathbf{n}$$

13. $p(\mathbf{v}) = \mathbf{v} - (\mathbf{n}\cdot\mathbf{v})\mathbf{e}$ (by the double-cross formula); $q(\mathbf{v}) = (\mathbf{n}\cdot\mathbf{v})\mathbf{e}$ is the projection onto the line spanned by $\mathbf{e}$ along the normal plane of $\mathbf{n}$. Therefore $p = 1 - q$ is the complementary projection, i.e., the projection onto the normal plane of $\mathbf{n}$ along the line spanned by $\mathbf{e}$ (see Chap. 10, Exercise 2 and Chap. 11, Exercise 27).
14. (See Chap. 10, Exercise 2, and Chap. 11, Exercise 27.) $\mathbf{n} = \mathbf{e}$, $\mathbf{n}\cdot\mathbf{e} = 1 \Rightarrow \mathbf{n}$ is a unit vector and $p(\mathbf{v}) = (\mathbf{n}\times\mathbf{v})\times\mathbf{n}$ is the perpendicular projection onto the normal plane of $\mathbf{n}$. Conversely, it will follow from the principal axis theorem (see Chap. 11) that every perpendicular projection onto a plane is of this form.

Chapter 5

1. $\dfrac{x-1}{19} = \dfrac{y+2}{11} = \dfrac{z-3}{5}$
2. $x/3 = -(y+1) = -(z-3)/3$
3. $(x-1)/(-3) = y = (z+1)/(-2)$
4. $(x-1)/2 = (y-1)/5 = (z+3)/11$
5. $(x-2/5)/(-7) = (y+7/5)/2 = z/5$
6. $(x-1)/(-5) = (y+2)/(-2) = (z-3)/7$
7. $-x = y/8 = z/7$
8. $x = y = -z/2$
9. The UNE & LSW equiangular line
10. $(x-2)/2 = (y+1)7 = z/4$
11. $P(2,-1,1)$
12. $x + 4y + 7z + 16 = 0$
13. $\begin{vmatrix} x-1 & y+2 & z-5 \\ 2 & -3 & 5 \\ 3 & 2 & -2 \end{vmatrix} = 0$ or $4x - 19y - 13z + 23 = 0$

14. $9x - y + 7z - 40 = 0$
15. $15x + 14y - 18z + 1 = 0$
16. $6x - 20y - 11z + 1 = 0$
17. $x - 8y - 13z + 8 = 0$
18. $x/4 = y + 9 = (z - 9)/(-8)$
19. $2x - 2y - z - 9 = 0$
20. $7x - y - 5z = 0$
21. $4x - y - 2z - 9 = 0$
22. $2\sqrt{2}$
23. The origin, $A(2,0,0)$, $B(0,3,0)$, $C(0,0,5)$
24. $x + 4y + 7z + 16 = 0$
25. 90
26. The perpendicular distance of the first plane from the origin is 1, and that of the second is 2. The planes fall on the opposite sides of the origin, because the constant terms in their equations have the opposite signature. Therefore, the perpendicular distance between the planes is $1 + 2 = 3$.
27. $(x - 2)/2 = y/(-3) = (z + 3)/5$ or $x = 2t + 2$, $y = -3t$, $z = 5t - 3$
28. $$\begin{vmatrix} x_1 - x_2 & y_1 - y_2 & z_1 - z_2 \\ 2 & 3 & -4 \\ 1 & -4 & 1 \end{vmatrix} = \begin{vmatrix} -8 & -1 & 10 \\ 2 & 3 & -4 \\ 1 & -4 & 1 \end{vmatrix} = 0$$
29. $\sqrt{2}$
30. 6
31. 3
32. 7
33. $P(21/9,-10/9,-17/9)$, $Q(13/9,-6/9,-25/9)$; $d = 4/3$

Chapter 6

1. $$\begin{pmatrix} \cos\phi & -\sin\phi & 0 \\ \sin\phi & \cos\phi & 0 \\ 0 & 0 & 1 \end{pmatrix}, \quad \begin{pmatrix} \cos\phi & 0 & -\sin\phi \\ 0 & 1 & 0 \\ \sin\phi & 0 & \cos\phi \end{pmatrix}$$
2. $P \rightsquigarrow P'(47/25,-196/25,-3)$
3. $$\begin{pmatrix} 13/14 & 2/14 & -3/14 \\ 2/14 & 10/14 & 6/14 \\ -3/14 & 6/14 & 5/14 \end{pmatrix}$$
4. q is the perpendicular projection onto the normal plane of $\mathbf{i} + 2\mathbf{j} + 2\mathbf{k}$.
5. $$\begin{pmatrix} 9/25 & 12/25 & 20/25 \\ 12/25 & 16/25 & -15/25 \\ -20/25 & 15/25 & 0 \end{pmatrix}$$

6. $x - y + 3z = 0$
7. *Hint*: Find the matrix of $r(\mathbf{v}) = (\mathbf{u}\cdot\mathbf{v})\mathbf{u} - \mathbf{u} \times \mathbf{v}$.
8. $\begin{pmatrix} 1/2 & 1/2 & 0 \\ 1/2 & 1/2 & 0 \\ 0 & 0 & 1 \end{pmatrix}$
9. Perpendicular projection onto the NE & SW bisector axis
10. $\begin{pmatrix} 1/3 & 1/3 & 1/3 \\ 1/3 & 1/3 & 1/3 \\ 1/3 & 1/3 & 1/3 \end{pmatrix}$
11. Reflection in the line $x = y = z/4$
12. Perpendicular projection onto the plane $x - y - 2z = 0$
13. $\begin{pmatrix} -1/3 & 2/3 & 2/3 \\ 2/3 & -1/3 & 2/3 \\ 2/3 & 2/3 & -1/3 \end{pmatrix}$
14. Reflection in the plane $x + y + z = 0$
15. The axis is the NE & SW bisector; the angle is 90°.
16. *Hint*: Use the rational parametric form of the circle (see Chap. 2, Application 12).
17. ℓ is the $\pm\mathbf{e}$ lift of $\mathbf{n}$, with $\mathbf{e} = a\mathbf{i} + b\mathbf{j}$, $\mathbf{n} = A\mathbf{i} + B\mathbf{j}$.
18. f is the shear parallel to the line $Ax + By = 0$ along the line $bx - ay = 0$.

Chapter 7

1. $f^2 = f$
2. $f^2 = 1$
3. $f^2 = 0$
4. $f^3 = -f$
5. $f^4 = 1$ $(f^2 \neq 1)$
6. $t^3 = -t \Rightarrow (1 + t + t^2)^2 = 1 + 2t^2$
   $$\Rightarrow (1 + t + t^2)^4 = (1 + 2t^2)^2 = 1 + 4t^2 + 4t^4 = 1$$
   Also, $(1 + t + t^2)^3 = (1 + 2t^2)\circ(1 + t + t^2) = 1 - t + t^2 = (1 + t + t^2)^{-1}$.
7. $t^3 = -t \Rightarrow (1 + t^2)^2 = 1 + 2t^2 + t^4 = 1 + 2t^2 - t^2 = 1 + t^2$
8. $j^4 = 1 \Rightarrow [\tfrac{1}{2}(j - j^3)]^3 = (1/8)(j^3 - 3j + 3j^3 - j) = \tfrac{1}{2}(j^3 - j) = -\tfrac{1}{2}(j - j^3)$
9. p is the projection onto the line spanned by $\mathbf{e} = a\mathbf{i} + b\mathbf{j}$ along the line with normal vector $\mathbf{n} = A\mathbf{i} + B\mathbf{j}$
10. $q = 1 - p$ is the projection onto the line with normal vector $\mathbf{n} = A\mathbf{i} + B\mathbf{j}$ along the line spanned by $\mathbf{e} = a\mathbf{i} + b\mathbf{j}$
11. *Hint*: proceed as in $\mathbb{R}^3$ (see Application 9)

12. $j^4 = 1 \Rightarrow (\tfrac{1}{4}(1 + j + j^2 + j^3))^2 = (1/8)(2 + 2j + 2j^2 + 2j^3)$

13. (i) $\begin{pmatrix} -10 & -4 & -7 \\ 6 & 14 & 4 \\ -6 & 7 & -1 \end{pmatrix}$ (ii) 0

14. (i) $\begin{pmatrix} \cos n\phi & -\sin n\phi \\ \sin n\phi & \cos n\phi \end{pmatrix}$ (ii) $\begin{pmatrix} 1 & n \\ 0 & 1 \end{pmatrix}$

(iii) 0 for $n \geq 3$

(iv) $\begin{pmatrix} 1 & 0 & n \\ 0 & 1 & 0 \\ 0 & 0 & 1 \end{pmatrix}$ (v) $f^n = \begin{cases} 1 & \text{for } n \text{ even} \\ f & \text{for } n \text{ odd} \end{cases}$

(vi) $f^n = \begin{cases} f & \text{for } n = 4k+1 \\ f^2 & \text{for } n = 4k+2 \\ -f & \text{for } n = 4k+3 \\ -f^2 & \text{for } n = 4k \end{cases}$ (vii) $f^n = \begin{cases} f & \text{for } n = 4k+1 \\ f^2 & \text{for } n = 4k+2 \\ f^3 & \text{for } n = 4k+3 \\ 1 & \text{for } n = 4k \end{cases}$

(viii) $f^n = f$ for $n \geqq 2$ (ix) $f^n = 0$ for $n \geqq 2$

(x) $\begin{pmatrix} \lambda^n & 0 & 0 \\ 0 & 1 & 0 \\ 0 & 0 & 1 \end{pmatrix}$ (xi) $\begin{pmatrix} \lambda^n & 0 & 0 \\ 0 & 0 & 0 \\ 0 & 0 & 0 \end{pmatrix}$

(xii) $\begin{pmatrix} \kappa^n & 0 & 0 \\ 0 & \lambda^n & 0 \\ 0 & 0 & \mu^n \end{pmatrix}$

15. $\ell(v) = (\mathbf{k}\cdot\mathbf{v})\mathbf{i}$; put $p(\mathbf{v}) = (\mathbf{k}\cdot\mathbf{v})(\mathbf{k} + \mathbf{i})$, $q(\mathbf{v}) = (\mathbf{k}\cdot\mathbf{v})\mathbf{k}$ and show that $p^2 = p$, $q^2 = q$. We have

$$\ell = p - q = \begin{pmatrix} 0 & 0 & 1 \\ 0 & 0 & 0 \\ 0 & 0 & 1 \end{pmatrix} - \begin{pmatrix} 0 & 0 & 0 \\ 0 & 0 & 0 \\ 0 & 0 & 1 \end{pmatrix}$$

Chapter 8

1. (i) $\begin{pmatrix} 1 & 0 & 0 \\ 0 & 0 & -1 \\ 0 & 1 & 0 \end{pmatrix}^{-1} = \begin{pmatrix} 1 & 0 & 0 \\ 0 & 0 & 1 \\ 0 & -1 & 0 \end{pmatrix}$ (ii) $\begin{pmatrix} 1 & 0 & n \\ 0 & 1 & 0 \\ 0 & 0 & 1 \end{pmatrix}^{-1} = \begin{pmatrix} 1 & 0 & -n \\ 0 & 1 & 0 \\ 0 & 0 & 1 \end{pmatrix}$

(iii) $\begin{pmatrix} \kappa & 0 & 0 \\ 0 & \lambda & 0 \\ 0 & 0 & \mu \end{pmatrix}^{-1} = \begin{pmatrix} \kappa^{-1} & 0 & 0 \\ 0 & \lambda^{-1} & 0 \\ 0 & 0 & \mu^{-1} \end{pmatrix}$ (iv) $f^{-1} = f$; f is an involution.

(v) $\begin{pmatrix} -3 & -6 & 7 \\ 3 & 5 & -6 \\ 7 & 12 & -14 \end{pmatrix}$ (vi) $\begin{pmatrix} -4 & 2 & 1 \\ -5 & 1 & 1 \\ -13 & 5 & 3 \end{pmatrix}$

(vii) $\begin{pmatrix} -1 & 1 & 2 \\ -5 & 4 & 7 \\ -11 & 7 & -3 \end{pmatrix}$ (viii) $\begin{pmatrix} 2 & 0 & 1 \\ 0 & -7 & 3 \\ 1 & -6 & 3 \end{pmatrix}$

(ix) $\begin{pmatrix} 2 & -1 & 1 \\ 8 & -5 & 2 \\ -11 & 7 & -3 \end{pmatrix}$ (x) $\begin{pmatrix} 1 & -1/2 & -1/12 \\ 0 & 1/4 & -5/24 \\ 0 & 0 & 1/6 \end{pmatrix}$

(xi) $\begin{pmatrix} 1 & -1 & 0 \\ \frac{1}{2} & \frac{1}{2} & -\frac{1}{2} \\ -\frac{1}{2} & \frac{1}{2} & \frac{1}{2} \end{pmatrix}$ (xii) $\begin{pmatrix} 1/7 & 5/7 & -3/7 \\ 13/49 & 16/49 & -11/49 \\ -2/49 & -10/49 & 13/49 \end{pmatrix}$

2. *Hint*: Call the axis of rotations the z axis.

3. The group of $f = \begin{pmatrix} \lambda & 0 \\ 0 & 1 \end{pmatrix}$ for all $\lambda \in \mathbb{R}$, $\lambda \neq 0$.

4. The group of $f = \begin{pmatrix} 1 & \alpha \\ 0 & 1 \end{pmatrix}$ for all $\alpha \in \mathbb{R}$.

5. *Hint*: Apply the condition of Chap. 0, Exercise 9.

6. Follows from the product matrix and the inverse matrix rules.

7. $f^3 = 0 \Rightarrow (1+f+f^2)\circ(1-f) = 1$, $(1-f+f^2)\circ(1+f) = 1$

8. $f^{-1} = \begin{pmatrix} \cos\phi & \sin\phi \\ -\sin\phi & \cos\phi \end{pmatrix}$

9. (i) $\begin{pmatrix} 5 & -2 \\ -2 & 1 \end{pmatrix}$ (ii) $\begin{pmatrix} 1 & -2 & 7 \\ 0 & 1 & -2 \\ 0 & 0 & 1 \end{pmatrix}$

(iii) $\begin{pmatrix} 1 & -3 & 11 & -38 \\ 0 & 1 & -2 & 5 \\ 0 & 0 & 0 & -2 \\ 0 & 0 & 0 & 1 \end{pmatrix}$ (iv) $\begin{pmatrix} 1 & -4 & -3 \\ 1 & -5 & -3 \\ -1 & 6 & 4 \end{pmatrix}$

(v) $\begin{pmatrix} \frac{1}{4} & \frac{1}{4} & \frac{1}{4} & \frac{1}{4} \\ \frac{1}{4} & \frac{1}{4} & -\frac{1}{4} & -\frac{1}{4} \\ \frac{1}{4} & -\frac{1}{4} & \frac{1}{4} & \frac{1}{4} \\ \frac{1}{4} & -\frac{1}{4} & \frac{1}{4} & \frac{1}{4} \end{pmatrix}$ (vi) $\begin{pmatrix} 2 & -1 & 0 & 0 \\ -3 & 2 & 0 & 0 \\ 31 & -19 & 3 & -4 \\ -23 & 14 & -2 & 3 \end{pmatrix}$

(vii) $\begin{pmatrix} -2/3 & 1/3 & 1/3 & 1/3 \\ 1/3 & -2/3 & 1/3 & 1/3 \\ 1/3 & 1/3 & -2/3 & 2/3 \\ 1/3 & 1/3 & 1/3 & -2/3 \end{pmatrix}$

10. (i) $f=\begin{pmatrix} 2 & -23 \\ 0 & 8 \end{pmatrix}$ (unique solution)

(ii) $f=\begin{pmatrix} -3 & 2 & 0 \\ -4 & 5 & -2 \\ -5 & 3 & 0 \end{pmatrix}$ (unique solution)

(iii) There are infinitely many solutions:

$$f=\begin{pmatrix} \alpha & \beta \\ 2-2\alpha & 1-2\beta \end{pmatrix} \quad \text{for all } \alpha, \beta \in \mathbb{R}$$

(iv) No solution, because the right-hand side is bijective, whereas the left-hand side is not.

11. $\ell = \tfrac{1}{2}(f-1) = \begin{pmatrix} 0 & 0 & 1 \\ 0 & 0 & 0 \\ 0 & 0 & 0 \end{pmatrix}$ is a lift;

$$\ell = p - q = \begin{pmatrix} 0 & 0 & 1 \\ 0 & 0 & 0 \\ 0 & 0 & 1 \end{pmatrix} - \begin{pmatrix} 0 & 0 & 0 \\ 0 & 0 & 0 \\ 0 & 0 & 1 \end{pmatrix}$$

is the decomposition of ℓ into the difference of principal projections (see Exercise 7.15). Put

$$h = 1-2p = \begin{pmatrix} 1 & 0 & -2 \\ 0 & 1 & 0 \\ 0 & 0 & -1 \end{pmatrix} \qquad k = 1-2q = \begin{pmatrix} 1 & 0 & 0 \\ 0 & 1 & 0 \\ 0 & 0 & -1 \end{pmatrix}$$

and check that $h \circ k = f$.

13. *Hint*: If $d = 1 + (\lambda-1)p$ with $\lambda \neq 1$, then look at $(d-1)/(\lambda-1)$.

14. Given a shear $1 + \ell$ ($\ell^2 = 0$), its *square root* $1 + \tfrac{1}{2}\ell$ is also a shear which can be factorized as a product of involutions (see Application 5): $1 + \tfrac{1}{2}\ell = h \circ k$, $h^{-1} = h$, $k^{-1} = k$. Then

$$1 + \ell = (h \circ k)^2 = h \circ k \circ h \circ k = h \circ k \circ h^{-1} \circ k^{-1}$$

16. $g^{-1} \circ (g \circ f \circ g^{-1} \circ f^{-1}) \circ g = f \circ g^{-1} \circ f^{-1} \circ (g^{-1})^{-1}$.

Chapter 9

1. $\ker f$ is the line $-x = y/2 = z$.
2. 2
3. $\ker f$ is the plane $x + 2y - 3z = 0$; $\operatorname{im} f$ is the line $-x/4 = y/3 = z$.

4. (i) $\begin{pmatrix} 2 \\ -1 \\ 1 \end{pmatrix}$. (ii) $\begin{pmatrix} 1 \\ 2 \\ 3 \end{pmatrix}$.

(iii) $\begin{pmatrix} 2 \\ 1 \\ 3 \end{pmatrix}$. (iv) $\begin{pmatrix} 1 \\ 2 \\ 3 \end{pmatrix}$.

(v) $\begin{pmatrix} 1 \\ -3 \\ -2 \end{pmatrix}$. (vi) $\begin{pmatrix} -3 \\ 0 \\ 0 \end{pmatrix} + t\begin{pmatrix} 2 \\ 8 \\ 9 \end{pmatrix}$.

(vii) $\begin{pmatrix} -5 \\ 2 \\ 0 \end{pmatrix} + t\begin{pmatrix} -1 \\ -2 \\ 1 \end{pmatrix}$. (viii) $\begin{pmatrix} 2 \\ 2 \\ 0 \end{pmatrix} + t\begin{pmatrix} -1 \\ 2 \\ 1 \end{pmatrix}$.

(ix) No solution.

(x) $\begin{pmatrix} -19 \\ -14 \end{pmatrix}$.

5. (i) $2\mathbf{a} - \mathbf{b} + \mathbf{c} = \mathbf{0}$, linearly dependent. (ii) Linearly independent.
(iii) $\mathbf{a} + \mathbf{b} - \mathbf{c} - \mathbf{d} = \mathbf{0}$, linearly dependent.
(iv) $\mathbf{a} - 2\mathbf{b} + \mathbf{c} = \mathbf{0}$, linearly dependent.
(v) $\mathbf{b} + 2\mathbf{c} = \mathbf{0}$, linearly dependent. (vi) Linearly independent.

6. (i) No solution. (ii) $\mathbf{v} = \mathbf{a} + 2\mathbf{b} + \mathbf{c} - \mathbf{d}$ (unique).
(iii) $\mathbf{v} = 3\mathbf{a} + 4\mathbf{b} + 5\mathbf{c}$ (unique). (iv) $\mathbf{v} = 4.1\mathbf{a} + \mathbf{b} - 2.3\mathbf{c} + 0.1\mathbf{d}$ (unique).
(v) $\mathbf{v} = 2\mathbf{a} + (5/3)\mathbf{b} - (4/3)\mathbf{c}$ (unique).
(vi) $\mathbf{v} = (5/4)\mathbf{a} - (1/4)\mathbf{b}$ and infinitely many other solutions.

7. $x = ½(a+b)$, $y = ½(a+c)$, $z = ½(b+c)$

8. $5a - 2b - c = 0$ because

$$\left(\begin{array}{ccc|c} 1 & 2 & -3 & a \\ 2 & 6 & -11 & b \\ 1 & -2 & 7 & c \end{array}\right) \Rightarrow \left(\begin{array}{ccc|c} 1 & 2 & -3 & a \\ 0 & 2 & -5 & b-2a \\ 0 & 0 & 0 & 2b+c-5a \end{array}\right)$$

9. $2a - b - c = 0$

10. $$\left(\begin{array}{ccc|c} k & 1 & 1 & 1 \\ 1 & k & 1 & 1 \\ 1 & 1 & k & 1 \end{array}\right) \Rightarrow \left(\begin{array}{ccc|c} 1 & 1 & k & 1 \\ 0 & k-1 & 1-k & 0 \\ 0 & 0 & (1-k)(2+k) & 1-k \end{array}\right)$$

Hence, if $k = 1$, there are infinitely many solutions. If $k \neq 1$ and $k = -2$, there are no solutions. If $k \neq 1$ and $k \neq -2$, then

$$\begin{pmatrix} k & 1 & 1 & | & 1 \\ 1 & k & 1 & | & 1 \\ 1 & 1 & k & | & 1 \end{pmatrix} \Rightarrow \begin{pmatrix} 1 & 0 & 0 & | & 1/(k+2) \\ 0 & 1 & 0 & | & 1/(k+2) \\ 0 & 0 & 1 & | & 1/(k+2) \end{pmatrix} \Rightarrow x = y = z = \frac{1}{k+2}$$

is the unique solution.

11. $\left.\begin{matrix} x + y + z = 14 \\ x + 5y + 10z = 89 \end{matrix}\right\} \Rightarrow \left.\begin{matrix} x + y + z = 75 \\ 4y + 9z = 75 \end{matrix}\right\}$. We are looking for an *integral* parameter n.

$$4y + 9z = 75 \Rightarrow y = \frac{75-9z}{4} = 18-2z + \frac{3-z}{4}$$

$$\Rightarrow 3-z = 4n \Rightarrow \begin{cases} x = -1-5n \\ y = 12+9n \\ z = \;\;3-4n \end{cases}$$

The only value of $n \in \mathbb{Z}$ to yield a positive solution is $n = -1$: $x = 4$, $y = 3$, $z = 7$.

12. (i) $\operatorname{im} p = \ker(1-p)$ is the plane $x + y + z = 0$; $\ker p$ is the line $x = 0$, $y = 2t$, $z = 3t$.
    (ii) $\operatorname{im} p$ is the line $-x/4 = y/3 = z$; $\ker p$ is the plane $x + 2y - 3z = 0$.
    (iii) $\operatorname{im} p$ is the plane $2x - 3y + 5z = 0$; $\ker p$ is the line $-x/3 = y = z/2$.
    (iv) $\operatorname{im} p$ is the plane $3x + y + 2z = 0$; $\ker p$ is the line $x = y/2 = -z/3$.
    (v) $\operatorname{im} p$ is the plane $3x + 2y + 3z = 0$; $\ker p$ is the line $-x/9 = y/7 = z/4$.

13. $\mathbf{v}, \mathbf{v}' \in \ker f \Rightarrow f(\mathbf{v}) = \mathbf{0}, f(\mathbf{v}') = \mathbf{0}$

$$\Rightarrow f(\mathbf{v} + \mathbf{v}') = f(\mathbf{v}) + f(\mathbf{v}') = \mathbf{0} \Rightarrow \mathbf{v} + \mathbf{v}' \in \ker f$$

$\lambda \in \mathbb{R}, \mathbf{v} \in \ker f \Rightarrow f(\mathbf{v}) = \mathbf{0}$

$$\Rightarrow f(\lambda\mathbf{v}) = \lambda f(\mathbf{v}) = \mathbf{0} \Rightarrow \lambda\mathbf{v} \in \ker f$$

14. $\mathbf{w}, \mathbf{w}' \in \operatorname{im} f \Rightarrow \mathbf{w} = f(\mathbf{v}), \mathbf{w}' = f(\mathbf{v}')$

$$\Rightarrow \mathbf{w} + \mathbf{w}' = f(\mathbf{v}) + f(\mathbf{v}') = f(\mathbf{v} + \mathbf{v}') \Rightarrow \mathbf{w} + \mathbf{w}' \in \operatorname{im} f$$

$\lambda \in \mathbb{R}, \mathbf{w} \in \operatorname{im} f \Rightarrow \mathbf{w} = f(\mathbf{v})$

$$\Rightarrow \lambda\mathbf{w} = \lambda f(\mathbf{v}) = f(\lambda\mathbf{v}) \Rightarrow \lambda\mathbf{w} \in \operatorname{im} f$$

15. Every principal projection p is of the form $p(\mathbf{v}) = (\mathbf{n}\cdot\mathbf{v})\mathbf{e}$ $(\mathbf{n}\cdot\mathbf{e} = 1)$, where $\mathbf{n}$ is the normal vector of the plane $\ker p$ and $\mathbf{e}$ is the vector spanning the line $\operatorname{im} p$. Take a plane containing $\operatorname{im} p$ with normal vector $\mathbf{n}'$ and a line in $\ker p$, other than the line of intersection, spanned by $\mathbf{e}'$. The $\mathbf{e}\cdot\mathbf{n}' = 0 = \mathbf{e}'\cdot\mathbf{n}$ and $\ell(\mathbf{v}) = (\mathbf{n}\cdot\mathbf{v})\mathbf{e}'$, $\ell'(\mathbf{v}) = (\mathbf{n}'\cdot\mathbf{v})\mathbf{e}$ are lifts satisfying $\ell' \circ \ell = p$. (*Note*: This decomposition is not unique.)

16. Let f be injective; then

$$\mathbf{v} \in \ker f \Rightarrow f(\mathbf{v}) = \mathbf{0} = f(\mathbf{0}) \Rightarrow \mathbf{v} = \mathbf{0}$$

and conclude that $\ker f = \mathbf{0}$. Conversely, let $\ker f = \mathbf{0}$; then

$$f(\mathbf{v}) = f(\mathbf{v}') \Rightarrow f(\mathbf{v}) - f(\mathbf{v}) = f(\mathbf{v} - \mathbf{v}') = \mathbf{0}$$
$$\Rightarrow \mathbf{v} - \mathbf{v}' \in \ker f = 0 \Rightarrow \mathbf{v} = \mathbf{v}'$$

and conclude that f is injective.

17. If p, q are principal projections along the same plane with normal vector $\mathbf{n}$, and $\operatorname{im} p$, $\operatorname{im} q$ are spanned by $\mathbf{a}$, $\mathbf{b}$, respectively, then for all $\mathbf{v}$, $p(\mathbf{v}) = (\mathbf{n}\cdot\mathbf{v})\mathbf{a}$, $q(\mathbf{v}) = (\mathbf{n}\cdot\mathbf{v})\mathbf{b}$, where the lengths of $\mathbf{a}$, $\mathbf{b}$ have been so fixed that $\mathbf{n}\cdot\mathbf{a} = 1 = \mathbf{n}\cdot\mathbf{b}$. Then

$$\frac{\alpha p + \beta q}{\alpha + \beta}(\mathbf{v}) = (\mathbf{n}\cdot\mathbf{v})\frac{\alpha\mathbf{a} + \beta\mathbf{b}}{\alpha + \beta}$$

and, obviously,

$$\mathbf{n}\cdot\frac{\alpha\mathbf{a} + \beta\mathbf{b}}{\alpha + \beta} = 1$$

18. One way is to find the line $\operatorname{im} f$ and the plane $\ker f$, and show that the direction vector of the line is perpendicular to the normal vector of the plane. The other way is to show that $f^2 = 0$.
19. $\ker g = \operatorname{im} f$, both being spanned by $2\mathbf{i} + \mathbf{j} - \mathbf{k}$.
20. (i) $5x - 35y - 23z = 0$. (ii) $x + y - z = 0$.
21. $x/4 = -y/7 = z/3$
22. See Dieudonné p. 35, p. 190.
23. See Dieudonné p. 35, p. 190.
24. See Greub p. 34.
25. See Dieudonné p. 35, p. 190.

Chapter 10

1. (i) $f^2 = 1 \Rightarrow \lambda_1 = 1$, $\lambda_2 = -1$, $\mathbf{e}_1 = \mathbf{j} + \mathbf{k}$; $\mathbf{e}_2$, $\mathbf{e}_2'$ are linearly independent vectors in the plane $3x + 2y - z = 0$, e.g., $\mathbf{e}_2 = \mathbf{i} + 3\mathbf{k}$, $\mathbf{e}_2' = \mathbf{j} + 2\mathbf{k}$.
   (ii) $f^2 = f \Rightarrow \lambda_1 = 0$, $\lambda_2 = 1$; $\mathbf{e}_1 = 2\mathbf{j} - 3\mathbf{k}$; $\mathbf{e}_2$, $\mathbf{e}_2'$ are linearly independent in the plane $x + y + z = 0$, e.g., $\mathbf{e}_2 = \mathbf{i} - \mathbf{j}$, $\mathbf{e}_2' = \mathbf{j} - \mathbf{k}$.
   (iii) $f^2 = f$. (iv) $f^2 = 1$. (v) $f^2 = f$. (vi) $f^2 = 1$. (vii) $f^2 = f$.
   (viii) $f^3 = -f$, $\lambda = 0$, $\mathbf{e} = 3\mathbf{i} + 2\mathbf{j} + \mathbf{k}$. (ix) $f^2 = 1$. (x) $f^2 = f$.
   (xi) $f^2 = 0$, $\lambda = 0$; $\mathbf{e}$, $\mathbf{e}'$ are linearly independent in the plane $x - 2y + 3z = 0$, e.g., $\mathbf{e} = 2\mathbf{i} + \mathbf{j}$, $\mathbf{e}' = 3\mathbf{i} - \mathbf{k}$.
   (xii) $f^2 = 1$. (xiii) $f^2 = -1$, no eigenvalue or eigenvector.
   (xiv) $f^2 = -1$. (xv) $f^2 = 1$. (xvi) $f^2 = -1$.
2. A projection $p \neq 1, 0$ in $\mathbb{R}^3$ is either a principal projection or the complementary projection of a principal projection. In the former case, $p(\mathbf{v}) = (\mathbf{n}\cdot\mathbf{v})\mathbf{e}$, with $\mathbf{n}\cdot\mathbf{e} = 1$, $\mathbf{n}$ the normal vector of $\ker p$, $\mathbf{e}$ the direction vector of $\operatorname{im} p$. In the latter, $p(\mathbf{v}) = \mathbf{v} - (\mathbf{n}\cdot\mathbf{v})\mathbf{e} = (\mathbf{e} \times \mathbf{v}) \times \mathbf{n}$ by the double cross formula, where $\mathbf{n}$ is the normal vector of $\operatorname{im} p$ and $\mathbf{e}$ is the direction vector of $\ker p$.

   p is a perpendicular projection $\Rightarrow \mathbf{e} = \mathbf{n} = \mathbf{u}$, in which case $\mathbf{u}$ must be a unit vector by virtue of $\mathbf{n}\cdot\mathbf{e} = 1$.

3. A twist f in $\mathbb{R}^3$ has a unique eigenvector belonging to $\lambda = 0$, which spans $\ker f$, namely, $\mathbf{e} = \mathbf{i} + 3\mathbf{j} - \mathbf{k}$.
4. An invection f in $\mathbb{R}^3$ has a unique eigenvector which belongs to $\lambda = 1$ if f is orientation preserving, and $\lambda = -1$ otherwise. $\ker(f-1)$ is spanned by $\mathbf{e} = 3\mathbf{i} + 2\mathbf{j} + \mathbf{k}$; conclude that f is orientation preserving.
5. *Hint*: Use the multinomial formula.
6. $t(\mathbf{e}) = \mathbf{0} \Rightarrow j(\mathbf{e}) = \mathbf{e}$ is obviously true if $j = 1 + t + t^2$.
7. $j^3(\mathbf{v}) = (\mathbf{u}\cdot\mathbf{v})\mathbf{u} - \mathbf{u} \times \mathbf{v} = j^{-1}(\mathbf{v})$, $t(\mathbf{v}) = \mathbf{u} \times \mathbf{v}$. Indeed, j is the rotation through 90° around $\mathbf{u}$. $1 + t + t^2 = j$.
8. $j = 1 + \ell - \ell' - \ell \circ \ell' - \ell' \circ \ell$
9. *Hint*: Use the decomposition $t = \ell - \ell'$ and the double cross formula.
10. $t = \ell - \ell' \Rightarrow t^4 = \ell' \circ \ell + \ell \circ \ell'$ is the projection onto $\operatorname{im} t$ spanned by $\operatorname{im} \ell$ and $\operatorname{im} \ell'$ along the line $\ker t = \ker \ell \cap \ker \ell'$, and $1 + t^2 = 1 - t^4$ is the complementary projection.
11. E.g., $\begin{pmatrix} 5 & -25 \\ -2 & -5 \end{pmatrix} - \begin{pmatrix} 0 & -\frac{1}{2} \\ 0 & 0 \end{pmatrix}$
12. (ii) $h = 1 - 2\ell \circ \ell' = \begin{pmatrix} 1 & 5 \\ 0 & -1 \end{pmatrix}$, $k = -\ell - \ell' = \begin{pmatrix} -5 & -12 \\ 2 & 5 \end{pmatrix}$ are involutions satisfying $h \circ k = j$.
13. $t = \frac{1}{2}(j - j^{-1}) = \begin{pmatrix} 1 & -4 & 5 \\ 1 & -4 & 5 \\ 1 & -3 & 3 \end{pmatrix}$
14. $j = \begin{pmatrix} 4 & -7 & 5 \\ 3 & -6 & 5 \\ 2 & -4 & 3 \end{pmatrix}$
15. (i) $(\lambda - 1)(\lambda - 2)(\lambda + 2) = 0$, hyperbolic.
    (ii) $\lambda^3 - 6\lambda^2 - \lambda + 30 = (\lambda + 2)(\lambda - 3)(\lambda - 5) = 0$, hyperbolic.
    (iii) $\lambda^3 + 15\lambda^2 - 25\lambda - 375 = (\lambda - 5)(\lambda + 5)(\lambda + 15) = 0$, hyperbolic.
    (iv) $(\lambda + 1)^2(\lambda - 8) = 0$, dilatation type: eigenspace but no multiple eigenvector [see part (vi)].
    (v) $(\lambda - 1)^2(\lambda - 100) = 0$, dilatation.
    (vi) $(\lambda + 1)^2(\lambda - 8) = 0$, hyperbolic-parabolic: double eigenvector, but no eigenspace [see part (iv)].
    (vii) $(\lambda + 1)^3 = 0$, parabolic: triple eigenvector, no eigenspace.
    (viii) Parabolic (ix) Twist, elliptic. (x) Parabolic.
16. (i) $(\lambda - \lambda_1)(\lambda - \lambda_2) = 0$, $\mathbf{e}_1 = \mathbf{i}$, $\mathbf{e}_2 = \mathbf{j}$; hyperbolic if $\lambda_1 \neq \lambda_2$.
    (ii) $(\lambda - \lambda_1)^2 = 0$, $\mathbf{e}_1 = \mathbf{i}$ double eigenvector, parabolic.
    (iii) $(\lambda - \lambda_1)^2 = 0$, all vectors $\mathbf{e} \neq \mathbf{0}$ are eigenvectors; strain.
    (iv) $\lambda^2 - 2\lambda\rho \cos\phi + \rho^2 = 0$, elliptic if $\rho \neq 0$, $\phi \neq 180°$.

17. *Note*: The answer may depend on the dimension.
    (i) $p \neq 0, 1$; hyperbolic in $\mathbb{R}^2$, dilatation type in $\mathbb{R}^3$.
    (ii) $h \neq 1, -1$; hyperbolic in $\mathbb{R}^2$, dilatation type in $\mathbb{R}^3$.
    (iii) $j^2 \neq 1$, elliptic.
    (iv) $t \neq 0$, elliptic.
    (v) and (vi) $\ell \neq 0$ parabolic in $\mathbb{R}^2$, shear type in $\mathbb{R}^3$.
    (vii) $\phi \neq 0°, 180°$: elliptic. A rotation through 180° is an involution.
    (viii) See under involution.
    (ix) In $\mathbb{R}^2$ hyperbolic, in $\mathbb{R}^3$ dilatation type.
    (x) Strain.
18. (i) $\lambda_1 = 1, \lambda_2 = 3$; $\mathbf{e}_1 = \mathbf{i} - \mathbf{j}$, $\mathbf{e}_2 = \mathbf{i} + \mathbf{j}$
    (ii) $\lambda_1 = 7, \lambda_2 = -2$; $\mathbf{e}_1 = \mathbf{i} + \mathbf{j}$, $\mathbf{e}_2 = 4\mathbf{i} - 5\mathbf{j}$
    (iii) $\lambda = 2$; $\mathbf{e} = -2\mathbf{i} + \mathbf{j}$, $\mathbf{e}' = \mathbf{i} + \mathbf{k}$
    (iv) $\lambda = -1$; $\mathbf{e} = \mathbf{i} + \mathbf{j} - \mathbf{k}$
19. Transition operator, e.g.,

$$g = \begin{pmatrix} 1 & 1 & 0 \\ -1 & 0 & -2 \\ 0 & -1 & 3 \end{pmatrix}$$

Transition equation is

$$\begin{pmatrix} -2 & -3 & -2 \\ 3 & 3 & 2 \\ 1 & 1 & 1 \end{pmatrix} \circ \begin{pmatrix} 1 & 0 & 0 \\ 4 & 5 & 4 \\ -6 & -6 & -5 \end{pmatrix} \circ \begin{pmatrix} 1 & 1 & 0 \\ -1 & 0 & -2 \\ 0 & -1 & 3 \end{pmatrix} = \begin{pmatrix} 1 & 0 & 0 \\ 0 & 1 & 0 \\ 0 & 0 & -1 \end{pmatrix}$$

20. Transition operator, e.g.,

$$g = \begin{pmatrix} 4 & -2 & 3 \\ -3 & 1 & 0 \\ -1 & 0 & 1 \end{pmatrix} \qquad g^{-1} = \begin{pmatrix} 1 & 2 & -3 \\ 3 & 7 & -9 \\ 1 & 2 & -2 \end{pmatrix}$$

21. E.g.:
    (i) $\mathbf{e}_1 = 3\mathbf{i} - \mathbf{j} + 3\mathbf{k}$, $\mathbf{e}_2 = 4\mathbf{i} - \mathbf{j} + 3\mathbf{k}$, $\mathbf{e}_3 = -6\mathbf{i} + 3\mathbf{j} - 6\mathbf{k}$
    Transition equation:

$$g^{-1} \circ f \circ g = \begin{pmatrix} 1 & 0 & 0 \\ 0 & 2 & 0 \\ 0 & 0 & 2 \end{pmatrix}$$

    (ii) $\mathbf{e}_1 = -\mathbf{i} - 2\mathbf{j} + \mathbf{k}$, $\mathbf{e}_2 = \mathbf{i} + \mathbf{k}$, $\mathbf{e}_3 = -2\mathbf{i} + \mathbf{j}$

$$g^{-1} \circ f \circ g = \begin{pmatrix} -1 & 0 & 0 \\ 0 & 5 & 0 \\ 0 & 0 & 5 \end{pmatrix}$$

    (iii) $\mathbf{e}_1 = 12\mathbf{i} - 5\mathbf{j} + 3\mathbf{k}$, $\mathbf{e}_2 = \mathbf{i} - \mathbf{k}$, $\mathbf{e}_3 = 2\mathbf{i} - \mathbf{k}$;

$$g^{-1}\circ g\circ g=\begin{pmatrix}2&0&0\\0&-3&0\\0&0&0\end{pmatrix}$$

(iv) $\mathbf{e}_1 = 3\mathbf{i}+2\mathbf{j}$, $\mathbf{e}_2 = 2\mathbf{i}+\mathbf{j}$;

$$g^{-1}\circ f\circ g=\begin{pmatrix}3&0\\0&1\end{pmatrix}$$

(v) $\mathbf{e}_1 = \mathbf{i}+\mathbf{j}+\mathbf{k}$, $\mathbf{e}_2 = 2\mathbf{i}+3\mathbf{j}+3\mathbf{k}$, $\mathbf{e}_3 = \mathbf{i}+3\mathbf{j}+4\mathbf{k}$;

$$g^{-1}\circ f\circ g=\begin{pmatrix}1&0&0\\0&2&0\\0&0&3\end{pmatrix}$$

22. $$\begin{pmatrix}4&2&1\\-1&1&0\\-2&0&0\end{pmatrix}\circ\begin{pmatrix}4&-8&12\\-1&2&-3\\-2&4&-6\end{pmatrix}\circ\begin{pmatrix}0&0&-\frac12\\0&1&-\frac12\\1&-2&3\end{pmatrix}=\begin{pmatrix}0&0&1\\0&0&0\\0&0&0\end{pmatrix}$$

23. $$\begin{pmatrix}1&-1&0\\0&-1&2\\-1&2&-1\end{pmatrix}\circ\begin{pmatrix}1&-4&5\\1&-4&5\\1&-3&3\end{pmatrix}\circ\begin{pmatrix}3&1&2\\2&1&2\\1&1&1\end{pmatrix}=\begin{pmatrix}0&0&0\\0&0&-1\\0&1&0\end{pmatrix}$$

24. $f(\mathbf{e})=\lambda\mathbf{e}\Rightarrow f^2(\mathbf{e})=f(f(\mathbf{e}))=f(\lambda\mathbf{e})=\lambda f(\mathbf{e})=\lambda^2\mathbf{e}$
25. $f(\mathbf{e})=\lambda\mathbf{e}\Rightarrow f^{-1}(\mathbf{e})=f^{-1}(\lambda^{-1}\lambda\mathbf{e})=\lambda^{-1}f^{-1}(f(\mathbf{e}))=\lambda^{-1}\mathbf{e}$
26. This is immediate from Chap. 9, Exercise 16.
27. $(\lambda-\lambda_1)(\lambda-\lambda_2)(\lambda-\lambda_3)=0$
28. $\det(g\circ f-\lambda)=\det(f^{-1}\circ(f\circ g-\lambda)\circ f)$

$$=(\det f^{-1})\det(f\circ g-\lambda)\det f-\det(f\circ g-\lambda)\qquad \text{if } f \text{ is bijective}$$

For the case when f is not bijective, consult Faddeev and Sominski (1965, problem 937) or Satake (p 157, Example 2).

29. *Hint*: f is volume preserving provided that $\det f=1$. Let $f=1+t^2+t^4\cos\phi+t\sin\phi$ and $t=\ell-\ell'$ be a decomposition of the twist t into the difference of lifts. Set

$$h=1-\ell\circ\ell'-\ell'\circ\ell+\cos\tfrac12\phi\,(\ell\circ\ell'-\ell'\circ\ell)+\sin\tfrac12\phi\,(\ell+\ell')$$
$$k=1-\ell\circ\ell'-\ell'\circ\ell+\cos\tfrac12\phi\,(\ell\circ\ell'-\ell'\circ\ell)+\sin\tfrac12\phi\,(\ell+\ell')$$

Show that $h^2=1=k^2$, $h\circ k=f$, $k\circ h=f^{-1}$.

30. $$\begin{pmatrix}\cos\phi&-\sin\phi\\\sin\phi&\cos\phi\end{pmatrix}=\begin{pmatrix}\cos\frac12\phi&\sin\frac12\phi\\\sin\frac12\phi&-\cos\frac12\phi\end{pmatrix}\circ\begin{pmatrix}\cos\frac12\phi&-\sin\frac12\phi\\-\sin\frac12\phi&-\cos\frac12\phi\end{pmatrix}$$

31. $\operatorname{trace}(1+t+t^2)=\operatorname{trace}(1+t^2)+\operatorname{trace}t=\operatorname{trace}(1+t^2)=1$. See Exercise 10,10 for orientation-preserving invections $1+t+t^2$.

32. False; the twists $\begin{pmatrix} 0 & -1 & 0 \\ 1 & 0 & 0 \\ 1 & 0 & 0 \end{pmatrix}$, $\begin{pmatrix} 5 & 13 & 0 \\ -2 & -5 & 0 \\ 0 & 0 & 0 \end{pmatrix}$ have the same kernel and image.

33. *Hint*: Find the transition operator from the first basis to $\{\mathbf{i}, \mathbf{j}, \mathbf{k}\}$, and the one from $\{\mathbf{i}, \mathbf{j}, \mathbf{k}\}$ to the second basis, and take their product.

36. *Hint*:

$$\operatorname{rank} f = 1 \Rightarrow f = \begin{pmatrix} Aa & Ba & Ca \\ Ab & Bb & Cb \\ Ac & Bc & Cc \end{pmatrix}$$

Consider cases according as $\operatorname{trace} f = 0$ or $\neq 0$.

37. *Hint*: Every lift has a matrix

$$\begin{pmatrix} 0 & 0 & 1 \\ 0 & 0 & 0 \\ 0 & 0 & 0 \end{pmatrix}$$

in a suitably chosen basis.

38. Set $f = 1 + \ell$, $f' = 1 + \ell'$. By Exercise 10.37, $\ell' = g \circ \ell \circ g^{-1}$; Show that $f' = g \circ f \circ g^{-1}$.

42. See Dieudonné p. 52.

43. See Dieudonné p. 113.

Chapter 11

1. (i) Reflection in the bisector plane $x = z$.
(ii) Perpendicular projection onto the equiangular line $x = y = z$.
(iii) Reflection in the plane $x - y + z = 0$.
(iv) Antisymmetric twist $\mathbf{u} \times \mathbf{v}$ with $\mathbf{u} = (1/3)\mathbf{i} + (2/3)\mathbf{j} + (2/3)\mathbf{k}$.
(v) Rotation through 180° around the line $x/3 = y = -z/2$.
(vi) Perpendicular projection onto the plane $3x + y - 2z = 0$.
(vii) Antisymmetric twist $\mathbf{u} \times \mathbf{v}$ with $\mathbf{u} = (1/9)\mathbf{i} - (4/9)\mathbf{j} - (8/9)\mathbf{k}$.
(viii) Perpendicular projection onto the plane $x - 2y + 2z = 0$.
(ix) Reflection in the plane $x - 2y + 2z = 0$.
(x) Perpendicular projection onto the line $x = -y/2 = z/2$.
(xi) Reflection in the plane $3x - 2y + 3z = 0$.
(xii) Reflection in the plane $3x - y = 0$.
(xiii) Perpendicular projection onto the plane $3x - y = 0$.
(xiv) Antisymmetric twist $\mathbf{u} \times \mathbf{v}$ with $\mathbf{u} = 0.8i - 0.6k$.
(xv) Reflection in the plane $7x - z = 0$.
(xvi) Perpendicular projection onto the plane $7x - z = 0$.
(xvii) Rotation through 180° around the line $x/5 = y/4 = z/3$.
(xviii) Perpendicular projection onto the line $x/5 = y/4 = z/3$.
(xix) Perpendicular projection onto the plane $5x + 4y + 3z = 0$.
(xx) Perpendicular projection onto the plane $5x - 8y - 6z = 0$.

(xxi) Reflection in the plane $5x - 8y - 6z = 0$.
(xxii) Antisymmetric twist $\mathbf{u} \times \mathbf{v}$ with $\mathbf{u} = 0.8\mathbf{i} + 0.576\mathbf{j} + 0.168\mathbf{k}$.
(xxiii) Antisymmetric twist $\mathbf{u} \times \mathbf{v}$ with $\mathbf{u} = 0.8\mathbf{i} - 0.48\mathbf{j} + 0.36\mathbf{k}$.
(xxiv) Antisymmetric twist $\mathbf{u} \times \mathbf{v}$ with $\mathbf{u} = 0.64\mathbf{i} - 0.48\mathbf{j} + 0.6\mathbf{k}$.
(xxv) Reflection in the plane $4x + 10y - 3z = 0$.
(xxvi) Reflection in the plane $9x - 5y - 12z = 0$.

2. The transition operators are:

(i) $\begin{pmatrix} -2/3 & 1/3 & 2/3 \\ 2/3 & 2/3 & 1/3 \\ 1/3 & -2/3 & 2/3 \end{pmatrix}$ (ii) $\begin{pmatrix} -6/7 & 2/7 & 3/7 \\ 2/7 & -3/7 & 6/7 \\ 3/7 & 6/7 & 2/7 \end{pmatrix}$

(iii) $\begin{pmatrix} 4/9 & 8/9 & 1/9 \\ -4/9 & 1/9 & 8/9 \\ 7/9 & -4/9 & 4/9 \end{pmatrix}$ (iv) $\begin{pmatrix} 7/11 & 6/11 & 6/11 \\ 6/11 & -2/11 & 9/11 \\ 6/11 & 9/11 & -2/11 \end{pmatrix}$

3. The transition operator is $\begin{pmatrix} 1/\sqrt{3} & 0 & 2/\sqrt{6} \\ 1/\sqrt{3} & 1/\sqrt{2} & -1/\sqrt{6} \\ 1/\sqrt{3} & 1/\sqrt{2} & -1\sqrt{6} \end{pmatrix}$.

4. The transition operator is $\begin{pmatrix} 2/3 & -2/3 & 1/3 \\ 2/3 & 1/3 & -2/3 \\ 1/3 & 2/3 & 2/3 \end{pmatrix}$.

5. The transition operator is $\begin{pmatrix} 5/\sqrt{70} & 0 & 3/\sqrt{14} \\ -3/\sqrt{70} & 2/\sqrt{5} & 1\sqrt{14} \\ 6/\sqrt{70} & 1/\sqrt{5} & -2/\sqrt{14} \end{pmatrix}$.

6. The transition operator is $\begin{pmatrix} 1/3 & -2/3 & -2/3 \\ -2/3 & 1/3 & -2/3 \\ 2/3 & 2/3 & -1/3 \end{pmatrix}$.

7. (i) $g = \begin{pmatrix} 2/3 & 1/3 & -2/3 \\ 1/3 & 2/3 & 2/3 \\ 2/3 & -2/3 & 1/3 \end{pmatrix}$, $g^{-1} \circ f \circ g = \begin{pmatrix} 9 & 0 & 0 \\ 0 & -9 & 0 \\ 0 & 0 & 0 \end{pmatrix}$.

(ii) $g = \begin{pmatrix} 1/3 & -2/3 & 2/3 \\ 2/3 & 2/3 & 1/3 \\ -2/3 & 1/3 & 2/3 \end{pmatrix}$, $g^{-1} \circ f \circ g = \begin{pmatrix} 3 & 0 & 0 \\ 0 & 6 & 0 \\ 0 & 0 & 0 \end{pmatrix}$

(iii) $g = \begin{pmatrix} -2/3 & 2/3 & 1/3 \\ 2/3 & 1/3 & 2/3 \\ 1/3 & 2/3 & -2/3 \end{pmatrix}$, $g^{-1} \circ f \circ g = \begin{pmatrix} 3 & 0 & 0 \\ 0 & 6 & 0 \\ 0 & 0 & 9 \end{pmatrix}$

(iv) $g=\begin{pmatrix}4/9 & -8/9 & 1/9\\ 4/9 & 1/9 & -8/9\\ 7/9 & 4/9 & 4/9\end{pmatrix}$, $g^{-1}\circ f\circ g=\begin{pmatrix}27 & 0 & 0\\ 0 & -27 & 0\\ 0 & 0 & 0\end{pmatrix}$

(v) $g=\begin{pmatrix}2/7 & -3/7 & 6/7\\ 3/7 & 6/7 & 2/7\\ -6/7 & 2/7 & 3/7\end{pmatrix}$, $g^{-1}\circ f\circ g=\begin{pmatrix}-49 & 0 & 0\\ 0 & 0 & 0\\ 0 & 0 & 49\end{pmatrix}$

(vi) $g=\begin{pmatrix}2/11 & -9/11 & 6/11\\ 6/11 & 6/11 & 7/11\\ -9/11 & 2/11 & 6/11\end{pmatrix}$, $g^{-1}\circ f\circ g=\begin{pmatrix}-121 & 0 & 0\\ 0 & 0 & 0\\ 0 & 0 & 121\end{pmatrix}$

(vi) $g=\begin{pmatrix}6/7 & 2/7 & 3/7\\ 2/7 & 3/7 & 6/7\\ -3/7 & -6/7 & 2/7\end{pmatrix}$, $g^{-1}\circ f\circ g=\begin{pmatrix}49 & 0 & 0\\ 0 & 98 & 0\\ 0 & 0 & 147\end{pmatrix}$

8. (i) $\begin{pmatrix}3 & 0 & 0\\ 0 & 3 & 0\\ 0 & 0 & -3\end{pmatrix}$ (ii) $\begin{pmatrix}7 & 0 & 0\\ 0 & 7 & 0\\ 0 & 0 & -7\end{pmatrix}$ (iii) $\begin{pmatrix}-9 & 0 & 0\\ 0 & -9 & 0\\ 0 & 0 & 9\end{pmatrix}$

(iv) $\begin{pmatrix}-3 & 0 & 0\\ 0 & -3 & 0\\ 0 & 0 & 6\end{pmatrix}$ (v) $\begin{pmatrix}1 & 0 & 0\\ 0 & 2 & 0\\ 0 & 0 & -2\end{pmatrix}$ (vi) $\begin{pmatrix}11 & 0 & 0\\ 0 & 11 & 0\\ 0 & 0 & -11\end{pmatrix}$

(vii) $\begin{pmatrix}-17 & 0 & 0\\ 0 & -17 & 0\\ 0 & 0 & 17\end{pmatrix}$ (viii) $\begin{pmatrix}19 & 0 & 0\\ 0 & 19 & 0\\ 0 & 0 & -19\end{pmatrix}$ (ix) $\begin{pmatrix}-21 & 0 & 0\\ 0 & -21 & 0\\ 0 & 0 & 21\end{pmatrix}$

(x) $\begin{pmatrix}27 & 0 & 0\\ 0 & 27 & 0\\ 0 & 0 & -27\end{pmatrix}$ (xi) $\begin{pmatrix}-29 & 0 & 0\\ 0 & -29 & 0\\ 0 & 0 & 29\end{pmatrix}$ (xii) $\begin{pmatrix}-31 & 0 & 0\\ 0 & -31 & 0\\ 0 & 0 & 31\end{pmatrix}$

(xiii) $\begin{pmatrix}33 & 0 & 0\\ 0 & 33 & 0\\ 0 & 0 & -33\end{pmatrix}$ (xiv) $\begin{pmatrix}33 & 0 & 0\\ 0 & 33 & 0\\ 0 & 0 & -33\end{pmatrix}$ (xv) $\begin{pmatrix}-33 & 0 & 0\\ 0 & -33 & 0\\ 0 & 0 & 33\end{pmatrix}$

9. (i) $\begin{pmatrix}-37 & 0 & 0\\ 0 & -37 & 0\\ 0 & 0 & 37\end{pmatrix}$ (ii) $\begin{pmatrix}-41 & 0 & 0\\ 0 & -41 & 0\\ 0 & 0 & 41\end{pmatrix}$ (iii) $\begin{pmatrix}43 & 0 & 0\\ 0 & 43 & 0\\ 0 & 0 & -43\end{pmatrix}$

(iv) $\begin{pmatrix}43 & 0 & 0\\ 0 & 43 & 0\\ 0 & 0 & -43\end{pmatrix}$ (v) $\begin{pmatrix}49 & 0 & 0\\ 0 & 49 & 0\\ 0 & 0 & 49\end{pmatrix}$ (vi) $\begin{pmatrix}51 & 0 & 0\\ 0 & 51 & 0\\ 0 & 0 & 51\end{pmatrix}$

11. There are no symmetric operators, hence no reflections, among the orthogonal operators listed.

12. Permuting the elements of an orthonormal basis, or replacing some element with its negative, will yield another orthonormal basis.
13. (i) $p^2 = p = p^* \Rightarrow (1-2p)^{-1} = 1 = (1-2p)^*$
    (ii) $h^{-1} = h = h^* \Rightarrow [½(1+h)]^2 = ½(1+h) = [½(1+h)]^*$
14. $t^3 = -t = t^* \Rightarrow (1+t^2)^2 = 1 + t^2 = (1+t^2)^*$, etc. $\mathrm{im}(1+t^2) = \ker t = \ker t^4$, $\ker(1+t^2) = \mathrm{im}\, t = \mathrm{im}\, t^4$
15. $\mathbf{w}\cdot f(\mathbf{v}) = (\mathbf{a}\cdot\mathbf{v})(\mathbf{b}\cdot\mathbf{w}) = f^*(\mathbf{w})\cdot\mathbf{v}$ for all $\mathbf{v}$, $\mathbf{w}$ by the star-dot formula. We conclude that $f^*(\mathbf{w}) = (\mathbf{b}\cdot\mathbf{w})\mathbf{a}$.
16. $\ell^* \circ \ell(\mathbf{v}) = (\mathbf{u}_1\cdot\mathbf{v})\mathbf{u}_1$, $\ell \circ \ell^*(\mathbf{v}) = (\mathbf{u}_2\cdot\mathbf{v})\mathbf{u}_2$ are perpendicular principal projections if, and only if, $\mathbf{u}_1^2 = 1 = \mathbf{u}_2^2$, $\mathbf{u}_1\cdot\mathbf{u}_1 = 0$.
    (i) $(\ell - \ell^*)^3 = \ell^* - \ell = (\ell - \ell^*)^*$, $\ker(\ell - \ell^*) = \ker \ell\, n \ker \ell^*$, which is the line spanned by $\mathbf{u}_1 \times \mathbf{u}_2$.
    (ii) $\ker(\ell \circ \ell^*) = \ker \ell^*$, which is the normal plane of $\mathbf{u}_2$; $\mathrm{im}(\ell \circ \ell^*) = \mathrm{im}\, \ell$, which is he line spanned by $\mathbf{u}_1$.
    (iii) $(\ell \circ \ell^* + \ell^* \circ \ell)^2 = \ell \circ \ell^* + \ell^* \circ \ell = (\ell \circ \ell^* + \ell^* \circ \ell)^*$, $\ker(\ell \circ \ell^* + \ell^* \circ \ell) = \ker \ell \cap \ker \ell^*$, which is the line spanned by $\mathbf{u}_1 \times \mathbf{u}_2$; $\mathrm{im}(\ell \circ \ell^* + \ell^* \circ \ell)$ is the plane spanned by $\mathrm{im}\, \ell$ and $\mathrm{im}\, \ell^*$, i.e., by $\mathbf{u}_1$ and $\mathbf{u}_2$.
    (iv) See Chap. 6, Application 4.
17. If $p(\mathbf{v}) = (\mathbf{u}\cdot\mathbf{v})\mathbf{u}$, $q(\mathbf{v}) = (\mathbf{w}\cdot\mathbf{v})\mathbf{w}$ ($\mathbf{u}^2 = 1 = \mathbf{w}^2$); then

$$p \circ q = 0 = q \circ p \Leftrightarrow \mathbf{u}\cdot\mathbf{w} = 0$$

    It follows that $\ell(\mathbf{v}) = (\mathbf{w}\cdot\mathbf{v})\mathbf{u}$, $\ell^*(\mathbf{v}) = (\mathbf{u}\cdot\mathbf{v})\mathbf{w}$ are lifts and $p = \ell \circ \ell^*$, $q = \ell^* \circ \ell$.
18. This is a consequence of Chap. 10, Exercise 9

$$t^3 = -t = t^* \Leftrightarrow t(\mathbf{v}) = (\mathbf{m}\cdot\mathbf{v})\mathbf{n} - (\mathbf{n}\cdot\mathbf{v})\mathbf{m}$$

    where $\mathbf{m}^2 = 1 = \mathbf{n}^2$ and $\mathbf{m}\cdot\mathbf{n} = 0$.
19. $j = \begin{pmatrix} 0 & 0.8 & 0.6 \\ 0 & -0.6 & 0.8 \\ 1 & 0 & 0 \end{pmatrix}$ is an orientation-preserving orthogonal operator because its unique eigenvector $2\mathbf{i} + \mathbf{j} + \mathbf{k}$ belongs to $\lambda = 1$.
20. $a = \begin{pmatrix} 0 & -1/3 & 0 \\ 1/3 & 0 & 0 \\ 0 & 0 & 0 \end{pmatrix}$
21. $r = (1+a) \circ (1-a)^{-1} \Rightarrow (a(\mathbf{e}) = \mathbf{0} \Rightarrow r(\mathbf{e}) = \mathbf{e})$
    $a = (r+1)^{-1} \circ (r-1) \Rightarrow (r(\mathbf{e}) = \mathbf{e} \Rightarrow a(\mathbf{e}) = \mathbf{0})$
22. $\lambda = -1$ cannot be an eigenvalue, because

$$r(\mathbf{e}) = -\mathbf{e} \Rightarrow (1+a) \circ (1-a)^{-1}(\mathbf{e}) = -\mathbf{e} \Rightarrow \mathbf{e} = \mathbf{0}$$

    Therefore, $\lambda = 1$ is the only positive eigenvalue of he orthogonal operator r, which is orientation preserving.
23. $r^* = r^{-1} \Rightarrow a^* = ((r+1)^{-1} \circ (r-1))^* = (r-1)^* \circ ((r+1)^*)^{-1}$

$$= (r^{-1}-1)\circ(r^{-1}+1)^{-1} = (r^{-1}-1)\circ(r^{-1}+1)^{-1}$$
$$= (1-r)\circ(1+r)^{-1} = -a$$

24. It is a one-line proof:

$$\begin{pmatrix} 0 & -\omega \\ \omega & 0 \end{pmatrix} = \omega \begin{pmatrix} 0 & -1 \\ 1 & 0 \end{pmatrix}$$

Indeed, the only antisymmetric twists of $\mathbb{R}^2$ are $\begin{pmatrix} 0 & -1 \\ 1 & 0 \end{pmatrix}$ and its negative.

25. See Exercise 4.11, 4.12, and 10.24.
26. This follows from the principal axis theorem, since the eigenvalues of p are 1, 0, 0.
27. This follows from the principal axis theorem (via the double cross formula), since the eigenvalues of q are 1, 1, 0.
28. Let r be an orientation-preserving orthogonal operator. Recall that r preserves the box product of vectors: For all **a**, **b**, **c**,

$$[r(\mathbf{a}),r(\mathbf{b}),r(\mathbf{c})] = [\mathbf{a},\mathbf{b},\mathbf{c}]$$

Therefore $[r(\mathbf{a}),r(\mathbf{b}),\mathbf{c}] = [\mathbf{a},\mathbf{b},r^{-1}(\mathbf{c})] \Rightarrow (r(\mathbf{a}) \times r(\mathbf{b}))\cdot\mathbf{c} = (\mathbf{a} \times \mathbf{b})\cdot r^{-1}(\mathbf{c})$

$$= (\mathbf{a} \times \mathbf{b})\cdot r^{*}(\mathbf{c})$$
$$= r(\mathbf{a} \times \mathbf{b})\cdot\mathbf{c}$$
$$\Rightarrow r(\mathbf{a} \times \mathbf{b}) = r(\mathbf{a}) \times r(\mathbf{b})$$

Conversely, assume that r preserves the cross product of vectors. Then

$$[r(\mathbf{a}),r(\mathbf{b}),r(\mathbf{c})] = (r(\mathbf{a}) \times r(\mathbf{b}))\cdot r(\mathbf{c})$$
$$= r(\mathbf{a} \times \mathbf{b})\cdot r(\mathbf{c}) = (\mathbf{a} \times \mathbf{b})\cdot\mathbf{c} = [\mathbf{a},\mathbf{b},\mathbf{c}]$$

29. By the double cross formula,

$$s(\mathbf{v}) = \mathbf{v} + (\lambda-1)(\mathbf{e}\cdot\mathbf{v})\mathbf{e} = \lambda(\mathbf{e}\cdot\mathbf{v})\mathbf{e} + (\mathbf{e}'\cdot\mathbf{v})\mathbf{e}' + (\mathbf{e}''\cdot\mathbf{v})\mathbf{e}''$$

where $\{\mathbf{e},\mathbf{e}',\mathbf{e}''\}$ is an orthonormal basis. Hence s is symmetric (see Exercise 11.15) and its eigenvalues are λ, 1, 1. The eigenvector belonging to λ is e; all perpendicular vectors form an eigenspace of dimension 2. It is clear that, for $\lambda \neq 0$,

$$s^{-1}(\mathbf{v}) = \lambda^{-1}(\mathbf{e}\cdot\mathbf{v})\mathbf{e} + (\mathbf{e}'\cdot\mathbf{v})\mathbf{e}' + (\mathbf{e}''\cdot\mathbf{v})\mathbf{e}''$$
$$= (\mathbf{e} \times \mathbf{v}) \times \mathbf{e} + \lambda^{-1}(\mathbf{e}\cdot\mathbf{v})\mathbf{e}$$

30. $\det(s-\lambda) = 0 \Rightarrow (\lambda-1)^2(\lambda-2) = 0$; by the principal axis theorem, s is a perpendicular dilatation in the ratio 2 along the eigenvector $\mathbf{e} = 2\mathbf{i} - 2\mathbf{j} + \mathbf{k}$ belonging to $\lambda = 2$.
31. (i) Elliptic; in fact, an invection with invariant plane $x - y = 0$.
    (ii) Parabolic; invariant plane is $6x - y + 2z = 0$.

32. (i) $z = 0$. (ii) $x + y + z = 0$. (iii) $2x + 2y - z = 0$. (iv) $z = 0$.
33. For a rotation r, $r^* = r \Leftrightarrow r^{-1} = r \Leftrightarrow \phi = 360° - \phi \Leftrightarrow \phi = 180°$.
34. Since f is antisymmetric, by the principal twist theorem it will suffice to show that $(110/111)\mathbf{i} + (11/111)\mathbf{j} + (10/111)\mathbf{k}$ is a unit vector: $110^2 + 11^2 + 10^2 = 111^2$.
35. $f^2 = f \Rightarrow (f^*)^2 = (f^2)^* = f^*$
$f^2 = 1 \Rightarrow (f^*)^2 = (f^2)^* = 1^* = 1$
$f^3 = -f \Rightarrow (f^*)^3 = (f^3)^* = (-f)^* = -f^*$
36. (i) $(2/3)\mathbf{i} + (1/3)\mathbf{j} + (2/3)\mathbf{k}$, $(1/3)\mathbf{i} + (2/3)\mathbf{j} - (2/3)\mathbf{k}$, $-(2/3)\mathbf{i} + (2/3)\mathbf{j} - (2/3)\mathbf{k}$
(ii) $(6/7)\mathbf{i} + (2/7)\mathbf{j} + (3/7)\mathbf{k}$, $-(3/7)\mathbf{i} + (6/7)\mathbf{j} + (2/7)\mathbf{k}$, $(2/7)\mathbf{i} + (3/7)\mathbf{j} - (6/7)\mathbf{k}$
(iii) $(2/11)\mathbf{i} + (6/11)\mathbf{j} - (9/11)\mathbf{k}$, $-(9/11)\mathbf{k}$, $-(9/11)\mathbf{i} + (6/11)\mathbf{j} + (2/11)\mathbf{k}$, $(6/11)\mathbf{i} + (7/11)\mathbf{j} + (6/11)\mathbf{k}$
(iv) $(4/13)\mathbf{i} + (12/13)\mathbf{j} + (3/13)\mathbf{k}$, $-(3/13)\mathbf{i} + (4/13)\mathbf{j} - (12/13)\mathbf{k}$, $-(12/13)\mathbf{i} + (3/13)\mathbf{j} + (4/13)\mathbf{k}$
(v) $(4/9)\mathbf{i} + (4/9)\mathbf{j} + (7/9)\mathbf{k}$, $-(8/9)\mathbf{i} + (1/9)\mathbf{j} + (4/9)\mathbf{k}$, $(1/9)\mathbf{i} - (8/9)\mathbf{j} + (4/9)\mathbf{k}$

37. (i)
$$r = \begin{pmatrix} 1/3 & 1/3 & 1/3 \\ 1/3 & 1/3 & 1/3 \\ 1/3 & 1/3 & 1/3 \end{pmatrix} - \frac{1}{2}\begin{pmatrix} 2/3 & -1/3 & -1/3 \\ -1/3 & 2/3 & -1/3 \\ -1/3 & -1/3 & 2/3 \end{pmatrix}$$
$$+ \frac{\sqrt{3}}{2}\begin{pmatrix} 0 & -1/\sqrt{3} & 1/\sqrt{3} \\ 1/\sqrt{3} & 0 & -1/\sqrt{3} \\ -1/\sqrt{3} & 1/\sqrt{3} & 0 \end{pmatrix};$$
$\alpha = 120°$

(ii)
$$r = \begin{pmatrix} 4/9 & 4/9 & -2/9 \\ 4/9 & 4/9 & -2/9 \\ -2/9 & -2/9 & 1/9 \end{pmatrix} + \begin{pmatrix} 0 & 1/3 & 2/3 \\ -1/3 & 0 & -2/3 \\ -2/3 & 2/3 & 0 \end{pmatrix}; \ \alpha = 90°$$

(iii) and (iv) $\alpha = 180°$, $r = 1 + 2t$

(vi)
$$r = \begin{pmatrix} 1/9 & 2/9 & 2/9 \\ 2/9 & 4/9 & 4/9 \\ 2/9 & 4/9 & 4/9 \end{pmatrix} + \frac{3}{5}\begin{pmatrix} 8/9 & -2/9 & -2/9 \\ -2/9 & 5/9 & -4/9 \\ -2/9 & -4/9 & 5/9 \end{pmatrix} + \frac{4}{5}\begin{pmatrix} 0 & 2/3 & -2/3 \\ -2/3 & 0 & 1/3 \\ 2/3 & -1/3 & 0 \end{pmatrix}$$

38. *Note*: The answer may depend on the dimension.
(i) False; a rotation through 180 is not elliptic. (ii) True in $\mathbb{R}^2$, false in $\mathbb{R}^3$.
(iii) False in $\mathbb{R}^2$; the antisymmetric twist is an orthogonal invection; true in $\mathbb{R}^3$.
(iv) False; $a = 0$ is not elliptic. (v) True. (vi) True in $\mathbb{R}^2$; false in $\mathbb{R}^3$. (vii) True. (viii) True.
39. True; the nonnegative entries must be either 0 or 1. There are six such operators in $\mathbb{R}^3$.
40. True; the integral entries must be either 0, 1, or −1. There are 48 such operators, and they form a group, the group of symmetries of the cube (see Chap. 12).
41. Since $\lambda \neq \lambda'$, we may assume that $\lambda' \neq 0$.

$$\left(1-\frac{\lambda}{\lambda'}\right)\mathbf{e}\cdot\mathbf{e}' = \mathbf{e}\cdot\mathbf{e}' - \frac{1}{\lambda'}f(\mathbf{e})\cdot\mathbf{e}' - \frac{1}{\lambda'}\mathbf{e}\cdot f^*(\mathbf{e}')$$

$$= \mathbf{e}\cdot\mathbf{e}' - \mathbf{e}\cdot\frac{1}{\lambda'}\lambda'\mathbf{e}' = 0 \Rightarrow \mathbf{e}\cdot\mathbf{e}' = 0 \Rightarrow \mathbf{e}\perp\mathbf{e}'$$

42. This follows from the principal axis (twist) theorem. The converse is false, e.g.,

$$t = \begin{pmatrix} 5 & 13 & 0 \\ -2 & -5 & 0 \\ 0 & 0 & 0 \end{pmatrix}$$

is a twist, im t is the horizontal plane, and ker t is the vertical axis; t is neither symmetric, nor antisymmetric.

43.
$$f = \begin{pmatrix} 1 & 0 & 0 \\ 0 & 0 & 0 \\ 0 & 0 & 0 \end{pmatrix}$$

is symmetric; $g^{-1}\circ f\circ g$ with

$$g = \begin{pmatrix} 4 & -2 & 3 \\ -3 & 1 & 0 \\ -1 & 0 & 1 \end{pmatrix}$$

fails to be symmetric (see Exercise 10.20). Note that g is *not* orthogonal.

46. By the principal axis theorem, $s = \lambda_1 p_1 + \lambda_2 p_2 + \lambda_3 p_3$. It is easy to see that $\lambda_\nu \geq 0$, $\nu = 1, 2, 3$. Put $s^{1/2} = \lambda_1^{1/2}p_1 + \lambda_2^{1/2}p_2 + \lambda_3^{1/2}p_3$ and show that $(s^{1/2})^* = s$, $(s^{1/2})^2 = s$, s positive.

49. Put $s = (f\circ f^*)^{1/2}$ (see Exercise 11.46) which is positive definite by Exercise 11.47. Then $r = s^{-1}\circ f$ is orthogonal because $r\circ r^* = s^{-1}\circ f\circ f^*\circ s^{-1} = s^{-1}\circ s^2\circ s^{-1} = 1$. $f = s\circ r$ is obviously satisfied. Uniqueness follows from

$$f\circ f^* = (s\circ r)\circ(s\circ r)^* = s\circ r\circ r^*\circ s^* = s\circ r\circ r^{-1}\circ s = s^2 \Rightarrow s = (f\circ f^*)^{1/2}$$

The other decompositiion $f = r'\circ s'$ can be obtained by starting with $s' = (f^*\circ f)^{1/2}$.

50. $s^* = s$, $r^* = r^{-1} \Rightarrow (r\circ s\circ r^{-1})^* = r^{**}\circ s^*\circ r^* = r\circ s\circ r^{-1}$ $a^* = -a$, $r^* = r^{-1} \Rightarrow (r\circ a\circ r^{-1})^* = r^{**}\circ a^*\circ r^* = -r\circ r^{-1}$. For a counterexample when r is not orthogonal, see Exercise 43.

Chapter 12

1. (i) $\mathbf{i}+\mathbf{j}$, $\cos\alpha = 1/3$
(ii) $2\mathbf{j}+\mathbf{k}$, $\cos\alpha = 2/7$
(iii) $2\mathbf{i}-\mathbf{j}+2\mathbf{k}$, $\alpha = 90°$
(iv) $-\mathbf{i}+3\mathbf{j}+3\mathbf{k}$, $\cos\alpha = 3/22$
(v) $-\mathbf{i}-\mathbf{j}+\mathbf{k}$, $\cos\alpha = -1/26$
(vi) $-\mathbf{i}+\mathbf{j}-2\mathbf{k}$, $\cos\alpha = 1/5$

(vii) $2\mathbf{i} + 2\mathbf{j} + 3\mathbf{k}$, $\alpha = 180°$
(viii) $4\mathbf{i} + 3\mathbf{j} + 3\mathbf{k}$, $\cos\alpha = -4/19$
(ix) $3\mathbf{i} + 4\mathbf{j} - 4$, $\cos\alpha = -5/21$
(x) $3\mathbf{i} + 2\mathbf{j} + 2\mathbf{k}$, $\cos\alpha = -13/21$
(xi) $2\mathbf{i} + 3\mathbf{j} + \mathbf{k}$, $\cos\alpha = -5/23$
(xii) $-\mathbf{i} + 4\mathbf{j} + \mathbf{k}$, $\cos\alpha = 1/3$
(xiii) $2\mathbf{i} + 3\mathbf{j} + 4\mathbf{k}$, $\alpha = 180°$
(xiv) $\mathbf{i} - \mathbf{j} + \mathbf{k}$, $\cos\alpha = 59/62$
(xv) $6\mathbf{i} + 3\mathbf{j} + 4\mathbf{k}$, $\cos\alpha = -30/31$
(xvi) $9\mathbf{i} + \mathbf{j} + 7\mathbf{k}$, $\cos\alpha = -5/6$
(xvii) $4\mathbf{i} + 7\mathbf{j} + \mathbf{k}$, $\alpha = 180°$
(xviii) $4\mathbf{i} - 5\mathbf{j} + 5\mathbf{k}$, $\alpha = 180°$
(xix) $-\mathbf{i} + \mathbf{j} + 2\mathbf{k}$, $\cos\alpha = 29/35$
(xx) $2\mathbf{i} + 4\mathbf{j} - \mathbf{k}$, $\cos\alpha = -5/37$
(xxi) $4\mathbf{i} + 6\mathbf{j} + \mathbf{k}$, $\cos\alpha = -14/39$
(xxii) $\mathbf{i} - \mathbf{j} + \mathbf{k}$, $\cos\alpha = -11/43$
(xxiii) $-6\mathbf{i} + 5\mathbf{j} + 5\mathbf{k}$, $\alpha = 180°$
(xxiv) $2\mathbf{i} + 3\mathbf{j} + 6\mathbf{k}$, $\alpha = 90°$
(xxv) $6\mathbf{i} + 5\mathbf{j} + 5\mathbf{k}$, $\cos\alpha = -35/51$
(xxvi) $\mathbf{i} + 4\mathbf{j} + 3\mathbf{k}$, $\cos\alpha = -1/51$
(xxvii) $-5\mathbf{i} + 4\mathbf{j} + 4\mathbf{k}$, $\alpha = 180°$
(xxviii) $8\mathbf{i} + 5\mathbf{j} + \mathbf{k}$, $\cos\alpha = -55/59$
(xxix) $2\mathbf{i} + 5\mathbf{j} + \mathbf{k}$, $\cos\alpha = 7/13$
(xxx) $-\mathbf{i} + \mathbf{j} + \mathbf{k}$, $\cos\alpha = -71/79$
(xxxi) $5\mathbf{i} + 6\mathbf{j} + 2\mathbf{k}$, $\cos\alpha = -49/81$
(xxxii) $4\mathbf{i} + 4\mathbf{j} + 7\mathbf{k}$, $\alpha = 90°$
(xxxiii) $\mathbf{j} - \mathbf{k}$, $\cos\alpha = -1/99$
(xxxiv) $11\mathbf{i} + \mathbf{j} + 10\mathbf{k}$, $\alpha = 180°$
(xxxv) $5\mathbf{i} + 7\mathbf{j} - \mathbf{k}$, $\alpha = 120°$
(xxxvi) $14\mathbf{i} + 5\mathbf{j} + 5\mathbf{k}$, $\cos\alpha = -0.968$
(xxxvii) $3\mathbf{i} + 4\mathbf{k}$, $\cos\alpha = 0.6$
(xxxviii) $9\mathbf{i} + 5\mathbf{j} + 12\mathbf{k}$, $\alpha = 180°$

2. f is symmetric and can be "orthogonally diagonalized":

$$\begin{pmatrix} \cos \frac{1}{2}\alpha & \sin \frac{1}{2}\alpha \\ -\sin \frac{1}{2}\alpha & \cos \frac{1}{2}\alpha \end{pmatrix} \circ \begin{pmatrix} \cos\alpha & \sin\alpha \\ \sin\alpha & -\cos\alpha \end{pmatrix} \circ \begin{pmatrix} \cos \frac{1}{2}\alpha & -\sin \frac{1}{2}\alpha \\ \sin \frac{1}{2}\alpha & \cos \frac{1}{2}\alpha \end{pmatrix} = \begin{pmatrix} 1 & 0 \\ 0 & -1 \end{pmatrix}$$

Thus f is the reflection in the line that makes an angle $\frac{1}{2}\alpha$ with the x axis.

3. Let $i = \begin{pmatrix} 0 & -1 \\ 1 & 0 \end{pmatrix}$; then $i^* = -i = i^3 = i^{-1}$. We have $i = \ell - \ell^*$, $\ell = \begin{pmatrix} 0 & 0 \\ 1 & 0 \end{pmatrix}$, where $\ell^2 = 0 = \ell^{*2}$ and $\ell \circ \ell^*$, $\ell^* \circ \ell$ are complementary perpendicular projections: $\ell \circ \ell^* + \ell^* \circ \ell = 1$ (see Chap. 10, Exercise 29). Put $h = \ell \circ \ell^* - \ell^* \circ \ell$, $k = \ell + \ell^*$ and show that $h^* = h = h^{-1}$, $k^* = k = k^{-1}$. Moreover, $h \circ k = \ell \circ \ell^* \circ \ell - \ell^* \circ \ell \circ \ell^* = \ell - \ell^* = i$ and $k \circ h = -i$. In fact, h and k are reflections in the lines $\operatorname{im} \frac{1}{2}(1 + h)$ and $\operatorname{im} \frac{1}{2}(1 + k)$, respectively.

Calculate the angle α made by these lines spanned by the vectors $\tfrac{1}{2}(1+h)(\mathbf{v}) = (\ell \circ \ell^*)(\mathbf{v}) = (\mathbf{i}\cdot\mathbf{v})\mathbf{i}$ and

$$\tfrac{1}{2}(1+k)(\mathbf{v}) = \tfrac{1}{2}(1+\ell-\ell^*)(\mathbf{v}) = \tfrac{1}{2}((\mathbf{i}\cdot\mathbf{v})\mathbf{i} + (\mathbf{j}\cdot\mathbf{v})\mathbf{j} + (\mathbf{j}\cdot\mathbf{v})\mathbf{i} + (\mathbf{i}\cdot\mathbf{v})\mathbf{j}) =$$

$$\left(\frac{\mathbf{i}+\mathbf{j}}{\sqrt{2}}\cdot\mathbf{v}\right)\frac{\mathbf{i}+\mathbf{j}}{\sqrt{2}}$$

Then

$$\cos\alpha = \frac{\mathbf{i}+\mathbf{j}}{\sqrt{2}}\cdot\mathbf{i} = \frac{1}{\sqrt{2}} \qquad \alpha = 45°$$

More generally, let $r = \cos\phi + i\sin\phi$ be any rotation. Then we have the decomposition as a product of involutions:

$$r = (h\cos\tfrac{1}{2}\phi + k\sin\tfrac{1}{2}\phi)\circ(h\cos\tfrac{1}{2}\phi - k\sin\tfrac{1}{2}\phi)$$

4.

$1=\begin{pmatrix}1&0&0\\0&1&0\\0&0&1\end{pmatrix}$	$k_1=\begin{pmatrix}1&0&0\\0&-1&0\\0&0&-1\end{pmatrix}$	$k_2=\begin{pmatrix}-1&0&0\\0&1&0\\0&0&-1\end{pmatrix}$	$k_3=\begin{pmatrix}-1&0&0\\0&-1&0\\0&0&1\end{pmatrix}$
$f_1=\begin{pmatrix}0&0&1\\-1&0&0\\0&-1&0\end{pmatrix}$	$f_2=\begin{pmatrix}0&0&-1\\-1&0&0\\0&1&0\end{pmatrix}$	$f_3=\begin{pmatrix}0&0&-1\\1&0&0\\0&-1&0\end{pmatrix}$	$f_4=\begin{pmatrix}0&0&1\\1&0&0\\0&1&0\end{pmatrix}$
$f_1^{-1}=\begin{pmatrix}0&-1&0\\0&0&-1\\1&0&0\end{pmatrix}$	$f_2^{-1}=\begin{pmatrix}0&-1&0\\0&0&1\\-1&0&0\end{pmatrix}$	$f_3^{-1}=\begin{pmatrix}0&1&0\\0&0&-1\\-1&0&0\end{pmatrix}$	$f_4^{-1}=\begin{pmatrix}0&1&0\\0&0&1\\1&0&0\end{pmatrix}$
$g_1=\begin{pmatrix}1&0&0\\0&0&-1\\0&1&0\end{pmatrix}$	$g_1^{-1}=\begin{pmatrix}1&0&0\\0&0&1\\0&-1&0\end{pmatrix}$	$p_1=\begin{pmatrix}-1&0&0\\0&0&-1\\0&-1&0\end{pmatrix}$	$q_1=\begin{pmatrix}-1&0&0\\0&0&1\\0&1&0\end{pmatrix}$
$g_2=\begin{pmatrix}0&0&-1\\0&1&0\\1&0&0\end{pmatrix}$	$g_2^{-1}=\begin{pmatrix}0&0&1\\0&1&0\\-1&0&0\end{pmatrix}$	$p_2=\begin{pmatrix}0&0&1\\0&-1&0\\1&0&0\end{pmatrix}$	$q_2=\begin{pmatrix}0&0&-1\\0&-1&0\\-1&0&0\end{pmatrix}$
$g_3=\begin{pmatrix}0&-1&0\\1&0&0\\0&0&1\end{pmatrix}$	$g_3^{-1}=\begin{pmatrix}0&1&0\\-1&0&0\\0&0&1\end{pmatrix}$	$p_3=\begin{pmatrix}0&-1&0\\-1&0&0\\0&0&-1\end{pmatrix}$	$q_3=\begin{pmatrix}0&1&0\\1&0&0\\0&0&-1\end{pmatrix}$

Multiplication Table for the Group of Rotations of the Cube

	1	k_1	k_2	k_3	f_1	f_2	f_3	f_4	f_1^{-1}	f_2^{-1}	f_3^{-1}	f_4^{-1}	g_1	g_1^{-1}	p_1	q_1	g_2	g_2^{-1}	p_2	q_2	g_3	g_3^{-1}	p_3	q_3
1	1	k_1	k_2	k_3	f_1	f_2	f_3	f_4	f_1^{-1}	f_2^{-1}	f_3^{-1}	f_4^{-1}	g_1	g_1^{-1}	p_1	q_1	g_2	g_2^{-1}	p_2	q_2	g_3	g_3^{-1}	p_3	q_3
k_1	k_1	1	k_3	k_2	f_4	f_3	f_2	f_1	f_2^{-1}	f_1^{-1}	f_4^{-1}	f_3^{-1}	g_1^{-1}	g_1	q_1	p_1	q_2	p_2	g_2^{-1}	g_2	p_3	q_3	g_3	g_3^{-1}
k_2	k_2	k_3	1	k_1	f_2	f_1	f_4	f_3	f_3^{-1}	f_4^{-1}	f_1^{-1}	f_2^{-1}	p_1	q_1	g_1	g_1^{-1}	g_2^{-1}	g_2	q_2	p_2	q_3	p_3	g_3^{-1}	g_3
k_3	k_3	k_2	k_1	1	f_3	f_4	f_1	f_2	f_4^{-1}	f_3^{-1}	f_2^{-1}	f_1^{-1}	q_1	p_1	g_1^{-1}	g_1	p_2	q_2	g_2	g_2^{-1}	g_3^{-1}	g_3	q_3	p_3
f_1	f_1	f_2	f_3	f_4	f_1^{-1}	f_4^{-1}	f_2^{-1}	f_3^{-1}	1	k_2	k_3	k_1	g_3^{-1}	p_3	g_3	q_3	g_1^{-1}	p_1	g_1	q_1	g^{-1}	p_2	g_2	q_2
f_2	f_2	f_1	f_4	f_3	f_3^{-1}	f_2^{-1}	f_4^{-1}	f_1^{-1}	k_2	1	k_1	k_1	p_3	g_3^{-1}	q_3	g_3	q_1	g_1	p_1	g_1^{-1}	g_2	q_2	g_2^{-1}	p_2
f_3	f_3	f_4	f_1	f_2	f_4^{-1}	f_1^{-1}	f_3^{-1}	f_2^{-1}	k_3	k_1	1	k_2	g_3	q_3	g_3^{-1}	p_3	p_1	g_1^{-1}	q_1	g_1	q_2	g_2	p_2	g_2^{-1}
f_4	f_4	f_3	f_2	f_1	f_2^{-1}	f_3^{-1}	f_1^{-1}	f_4^{-1}	k_1	k_3	k_2	1	q_3	g_3	p_3	g_3^{-1}	g_1	q_1	g_1^{-1}	p_1	p_2	g_2^{-1}	q_2	g_2
f_1^{-1}	f_1^{-1}	f_4^{-1}	f_2^{-1}	f_3^{-1}	1	k_1	k_2	k_3	f_1	f_3	f_4	f_2	p_2	g_2	g_2^{-1}	q_2	p_3	g_3	g_3^{-1}	q_3	p_1	g_1	g_1^{-1}	q_1
f_2^{-1}	f_2^{-1}	f_3^{-1}	f_1^{-1}	f_4^{-1}	k_1	1	k_3	k_2	f_4	f_2	f_1	f_3	g_2^{-1}	q_2	p_2	g_2	g_3	p_3	q_3	g_3^{-1}	q_1	g_1^{-1}	g_1	p_1
f_3^{-1}	f_3^{-1}	f_2^{-1}	f_4^{-1}	f_1^{-1}	k_2	k_3	1	k_1	f_2	f_4	f_3	f_1	q_2	g_2^{-1}	g_2	p_2	g_3^{-1}	q_3	p_3	g_3	g_1	p_1	q_1	g_1^{-1}
f_4^{-1}	f_4^{-1}	f_1^{-1}	f_3^{-1}	f_2^{-1}	k_3	k_2	k_1	1	f_3	f_1	f_2	f_4	g_2	p_2	q_2	g_2^{-1}	q_3	g_3^{-1}	g_3	p_3	g_1^{-1}	q_1	p_1	g_1
g_1	g_1	g_1^{-1}	q_1	p_1	g_2^{-1}	q_2	g_2	p_2	p_3	g_3	q_3	g_3^{-1}	k_1	1	k_2	k_3	f_2	f_4	f_1	f_3	f_1^{-1}	f_3^{-1}	f_2^{-1}	f_4^{-1}
g_1^{-1}	g_1^{-1}	g_1	p_1	q_1	p_2	g_2	q_2	g_2^{-1}	g_3	p_3	g_3^{-1}	q_3	1	k_1	k_3	k_2	f_3	f_1	f_4	f_2	f_2^{-1}	f_4^{-1}	f_1^{-1}	f_3^{-1}
p_1	p_1	q_1	g_1^{-1}	g_1	g_2	p_2	g_2^{-1}	q_2	g_3^{-1}	q_3	g_3	p_3	k_3	k_2	1	k_1	f_1	f_3	f_2	f_4	f_3^{-1}	f_1^{-1}	f_4^{-1}	f_2^{-1}
q_1	q_1	p_1	g_1	g_1^{-1}	q_2	g_2^{-1}	p_2	g_2	q_3	g_3^{-1}	p_3	g_3	k_2	k_3	k_1	1	f_4	f_2	f_3	f_1	f_4^{-1}	f_2^{-1}	f_3^{-1}	f_1^{-1}
g_2	g_2	p_2	g_2^{-1}	q_2	g_3^{-1}	p_3	q_3	g_3	p_1	g_1^{-1}	g_1	q_1	f_1^{-1}	f_4^{-1}	f_3^{-1}	f_2^{-1}	k_2	1	k_3	k_1	f_3	f_2	f_1	f_4
g_2^{-1}	g_2^{-1}	q_2	g_2	p_2	p_3	g_3^{-1}	g_3	q_3	g_1	q_1	p_1	g_1^{-1}	f_3^{-1}	f_2^{-1}	f_1^{-1}	f_4^{-1}	1	k_2	k_1	k_3	f_4	f_1	f_2	f_3
p_2	p_2	g_2	q_2	g_2^{-1}	g_3	q_3	p_3	g_3^{-1}	g_1^{-1}	p_1	q_1	g_1	f_4^{-1}	f_1^{-1}	f_2^{-1}	f_3^{-1}	k_1	k_3	1	k_2	f_1	f_4	f_3	f_2
q_2	q_2	g_2^{-1}	p_2	g_2	q_3	g_3	g_3^{-1}	p_3	q_1	g_1	g_1^{-1}	p_1	f_2^{-1}	f_3^{-1}	f_4^{-1}	f_1^{-1}	k_3	k_1	k_2	1	f_2	f_3	f_4	f_1
g_3	g_3	q_3	p_3	g_3^{-1}	g_1^{-1}	g_1	p_1	q_1	p_2	q_2	g_2^{-1}	g_2	f_4	f_3	f_1	f_2	f_1^{-1}	f_2^{-1}	f_4^{-1}	f_3^{-1}	k_3	1	k_1	k_2
g_3^{-1}	g_3^{-1}	p_3	q_3	g_3	p_1	q_1	g_1^{-1}	g_1	g_2	g_1^{-1}	q_2	p_2	f_2	f_1	f_3	f_4	f_4^{-1}	f_3^{-1}	f_1^{-1}	f_2^{-1}	1	k_3	k_2	k_1
p_3	p_3	g_3^{-1}	g_3	q_3	g_1	g_1^{-1}	q_1	p_1	g_2^{-1}	g_2	p_2	q_2	f_1	f_2	f_4	f_3	f_2^{-1}	f_1^{-1}	f_3^{-1}	f_4^{-1}	k_2	k_1	1	k_3
q_3	q_3	g_3	g_3^{-1}	p_3	q_1	p_1	g_1	g_1^{-1}	q_2	p_2	g_2	g_2^{-1}	f_3	f_4	f_2	f_1	f_3^{-1}	f_4^{-1}	f_2^{-1}	f_1^{-1}	k_1	k_2	k_3	1

5. $\frac{1}{2}(r-r^*) = \sin\alpha\, t = \frac{24}{25}\begin{pmatrix} 0 & 2/3 & -1/3 \\ -2/3 & 0 & 2/3 \\ 1/3 & -2/3 & 0 \end{pmatrix}$, $\operatorname{im} t$ is the plane $2x+y+2z=0$.

A pair of perpendicular unit vectors in $\operatorname{im} t$ is $(1/3)\mathbf{i}+(2/3)\mathbf{j}-(2/3)\mathbf{k}, -(2/3)\mathbf{i}+(2/3)\mathbf{j}+(1/3)\mathbf{k}$; they give rise to the perpendicular lifts ℓ and ℓ^* with

$$\ell = \begin{pmatrix} -2/9 & -4/9 & -4/9 \\ 2/9 & 4/9 & -4/9 \\ 2/9 & 2/9 & -2/9 \end{pmatrix}$$

Then

$$h = 1-\ell\circ\ell^*-\ell^*\circ\ell+\cos\tfrac{1}{2}\phi(\ell\circ\ell^*-\ell^*\circ\ell)+\sin\tfrac{1}{2}(\ell+\ell^*)$$

$$= \begin{pmatrix} -20/45 & -20/45 & 35/45 \\ -20/45 & 29/45 & 28/45 \\ 35/45 & 28/45 & -4/45 \end{pmatrix}$$

$$k = 1-\ell\circ\ell^*-\ell^*\circ\ell+\cos\tfrac{1}{2}\phi(\ell\circ\ell^*-\ell^*\circ\ell)-\sin\tfrac{1}{2}\phi(\ell+\ell^*)$$

$$= \begin{pmatrix} 44/45 & -8/45 & 5/45 \\ -8/45 & -19/45 & 40/45 \\ 5/45 & 40/45 & 20/45 \end{pmatrix}$$

Show that h, k are the reflections in the planes $5x+4y-7z=0$, $x+8y-5z=0$, respectively. Indeed, these planes make an angle ϕ such that

$$\cos\phi = \frac{5+32+35}{90} = \frac{4}{5} \qquad \cos 2\phi = \frac{16}{25}-\frac{9}{25} = \frac{7}{25} = \cos\alpha$$

Conclude that $\phi = \frac{1}{2}\alpha$.

6. *Hint*: Substitute $r = 1+t^2+t^4\cos\alpha + t\sin\alpha$ and $r' = -1-t^2+t^4\cos\alpha + t\sin\alpha$, respectively.

7. $r(\mathbf{v}) = (1+t^2)(\mathbf{v})+\cos\alpha t^4(\mathbf{v})+\sin\alpha\, t(\mathbf{v})$
$= (\cos\alpha)\mathbf{v}+(1-\cos\alpha)(\mathbf{u}\cdot\mathbf{v})\mathbf{u}+(\sin\alpha)(\mathbf{u}\times\mathbf{v})$

8. $(1+\cos\alpha)a = (2\cos\tfrac{1}{2}\alpha)(\tan\tfrac{1}{2}\alpha)t \qquad (t^3=-t=t^*)$
$= (2\sin\tfrac{1}{2}\alpha\cos\tfrac{1}{2}\alpha)t = \sin\alpha\, t = \tfrac{1}{2}(r-r^*)$

9. Since $(t\circ t')^* = t'^*\circ t^* = (-t')\circ(-t) = t'\circ t$, $(2)\Rightarrow(3)$ is obvious. Now use Steenrod's formula to show that $[e^{\alpha t}, e^{\beta t'}] = 0 \Rightarrow [t,t'] = 0$.

10. $\frac{1}{6}(1+r+r^2+r^3+r^4+r^5) = \begin{pmatrix} 1/3 & -1/3 & 1/3 \\ -1/3 & 1/3 & -1/3 \\ 1/3 & -1/3 & 1/3 \end{pmatrix}$, which is a perpendicular projection onto the line spanned by $\mathbf{i}-\mathbf{j}+\mathbf{k}$.

Chapter 13

1. $d_1 \circ d_2 = 1 + (\lambda_1\lambda_2 - 1)p$, $d^{-1} = 1 + (\lambda^{-1} - 1)p$ $(\lambda \neq 0)$. Moreover, $\det(d_1 \circ d_2) = \lambda_1\lambda_2 = (\det d_1)(\det d_2)$.
2. $f_v = 1 + \ell_v$ with $\ell_v(v) = (\mathbf{n}\cdot\mathbf{v})\mathbf{e}_v$ $(\mathbf{n}\cdot\mathbf{e}_v = 0)$. Then $f_1 \circ f_2 = 1 + \ell_1 + \ell_2$, $(\ell_1 + \ell_2)^2 = 0$ because $\ell_1 \circ \ell_2 = 0$. Also, $f_v^{-1} = 1 - \ell_v$. Therefore the shears of $\mathbb{R}^3$ with a common eigenspace form a multiplicative group which is isomorphic with the additive group of lifts with the same kernel, which in turn is isomorphic with the additive group of vectors in the plane.
3. Let $f_v = 1 + \ell_v$ with $\ell_v(\mathbf{v}) = (\mathbf{n}_v\cdot\mathbf{v})\mathbf{e}$, $(\mathbf{n}_v\cdot\mathbf{e} = 0)$, and proceed as in Exercise 13.2.
4. Let $\ell^2 = 0 = \ell'^2$. Then

$$(1 + \ell)\circ(1 + \ell') = (1 + \ell')\circ(1 + \ell) \Leftrightarrow \ker \ell = \ker \ell' \text{ or } \operatorname{im} \ell = \operatorname{im} \ell'.$$

It follows from Exercises 13.2 and 13.3 that the condition is sufficient. To show that it is also necessary, we assume that $\ell \circ \ell' = \ell' \circ \ell$ and $\operatorname{im} \ell \neq \operatorname{im} \ell'$. Then $\mathbf{e} \nparallel \mathbf{e}'$ and

$$\begin{aligned}(\mathbf{n}\cdot\mathbf{e}')(\mathbf{n}'\cdot\mathbf{v})\mathbf{e} = (\mathbf{n}'\cdot\mathbf{e})(\mathbf{n}\cdot\mathbf{v})\mathbf{e}' \text{ for all } \mathbf{v} &\Rightarrow (\mathbf{n}\cdot\mathbf{e}')(\mathbf{n}'\cdot\mathbf{v}) = 0 = (\mathbf{n}'\cdot\mathbf{e})(\mathbf{n}\cdot\mathbf{v})\\ &\Rightarrow \mathbf{n}\cdot\mathbf{e}' = 0 = \mathbf{n}'\cdot\mathbf{e} \Rightarrow \mathbf{n} \perp \mathbf{e}, \mathbf{e}' \text{ and } \mathbf{n}' \perp \mathbf{e}, \mathbf{e}'\\ &\Rightarrow \mathbf{n}, \mathbf{n}' \parallel \mathbf{e} \times \mathbf{e}' \Rightarrow \mathbf{n} \parallel \mathbf{n}' \Rightarrow \ker \ell = \ker \ell'\end{aligned}$$

5. $h \in G'$, $g \in G \Rightarrow h' = g \circ h \circ g^{-1} \in G'$, because $\det h' > 0$ by the product rule of determinants.
6. Proceed as in Exercise 13.5.
7. *Hint*: Show that $[f(\mathbf{a}),\mathbf{b},\mathbf{c}][\mathbf{a}',\mathbf{b}',\mathbf{c}'] = [\mathbf{a},\mathbf{b},\mathbf{c}][f^*(\mathbf{a}'),\mathbf{b}',\mathbf{c}']$ by using the double cross formula, the expansion property of determinants, and the star-dot formula.
8. *Hint*: Use the reflection property of determinants.
9. Let $\mathbf{e}_1$, $\mathbf{e}_2$, $\mathbf{e}_3$ be a basis. We have

$$\begin{aligned}\operatorname{trace} f &= \frac{[f(\mathbf{e}_1),\mathbf{e}_2\,\mathbf{e}_3] + [\mathbf{e}_1,f(\mathbf{e}_2),\mathbf{e}_3] + [\mathbf{e}_1,\mathbf{e}_2,f(\mathbf{e}_3)]}{[\mathbf{e}_1,\mathbf{e}_2,\mathbf{e}_3]}\\ &= a_{11} + a_{22} + a_{33}\end{aligned}$$

because

$$f(\mathbf{e}_1) = a_{11}\mathbf{e}_1 + a_{21}\mathbf{e}_2 + a_{31}\mathbf{e}_3 \Rightarrow [f(\mathbf{e}_1),\mathbf{e}_2,\mathbf{e}_3] = a_{11}[\mathbf{e}_1,\mathbf{e}_2,\mathbf{e}_3], \text{ etc.}$$

10. Proceed as in Exercise 13.7.
11. *Hint:* Use the reflection property of determinants.
12. *Hint*: $[f(\mathbf{i}),f(\mathbf{j}),\mathbf{k}] = [f(\mathbf{i}) \times f(\mathbf{j})]\cdot\mathbf{k} = \begin{vmatrix} a_{11} & a_{12} \\ a_{21} & a_{22} \end{vmatrix}$, etc.

14. *Hint*: Calculate inv f from the diagonal form of f.
15. *Hint:* trace $f^{-1} = (\det f)^{-1}(A_{11} + A_{22} + A_{33})$ by the inverse matrix formula (see Chap. 8).
16. *Hint*: Evaluate $[(r + 1)(\mathbf{a}),(r + 1)(\mathbf{b}),(r + 1)(\mathbf{c})]$.
17. *Hint*: Calculate trace r from the standard form of r.
18. *Hint*: $-1 \leq \cos\alpha \leq 1$.
19. *Hint*: Look at the eigenvalues.
20. *Hint*: Calculate rank p from the diagonal form
21. t twist $\Rightarrow -t^2$ projection onto im t; now apply Exercise 13.20.
22. h orientation preserving involution $\Leftrightarrow$ number of eigenvalues $\lambda = -1$ even

$$\Leftrightarrow \deg h \text{ even} \Leftrightarrow \tfrac{1}{2}|n - \operatorname{trace} h| \text{ even}$$
$$\Leftrightarrow |n - \operatorname{trace} h| \text{ divisible by 4}$$
$$\Leftrightarrow\ n - \operatorname{trace} h \equiv 0 \pmod 4 \Leftrightarrow \operatorname{trace} h \equiv n \pmod 4).$$

24. *Hint*: Use some elementary facts about the cubic equation

$$x^3 + ax^2 + bx + c = 0 \qquad a,b,\ c \in \mathbb{R}$$

with discriminant $D = 4a^3c - a^2b^2 - 18abc + 4b^3 + 27c^2$, namely,

$$D > 0 \Leftrightarrow \text{one real root plus a pair of complex conjugate roots}$$
$$D < 0 \Leftrightarrow \text{three distinct real roots}$$
$$D = 0 \Leftrightarrow \text{three real roots, at least two of which coincide}$$

25. *Hint*: Write the characteristic equation of f^{-1}.
27. n invariants, each corresponding to one of the n elementary symmetric functions $\lambda_1 + \lambda_2 + \cdots + \lambda_n$, $\lambda_1\lambda_2 + \lambda_2\lambda_3 + \cdots + \lambda_{n-1}\lambda_n$, $\lambda_1\lambda_2\lambda_3 + \lambda_2\lambda_3\lambda_4 + \cdots + \lambda_{n-2}\lambda_{n-1}\lambda_n, \ldots, \lambda_1\lambda_2 \cdots \lambda_n$.
28. *Hint*: Observe that N normal in $G \Leftrightarrow gN = Ng$ for all $g \in G$.
29. *Hint*: Identify all gN and Ng for $N = \{1, k_1, k_2, k_3\}$. Find another normal subgroup N' via Exercise 13.28.
30. *Hint*: In case of N, the cosets are the six 4×4 blocks in the multiplication table 12.4. The product of blocks is defined as the block containing the product of representative elements. This definition remains undisturbed if the representation of blocks is changed.
31. *Hint:* Set $f: G \to G/N$, $f(g) = gN$, for all $g \in G$ in Exercise 13.30.
32. *Hint*: Show that f has a matrix with all 0 main diagonal in a suitable basis for $\mathbb{R}^2$. Then let g be a suitable involution (see Dieudonné pp. 63, 193.)

Chapter 14

1. (i)

$$e^{\alpha f} = 1 + \alpha f = \begin{pmatrix} 1 & 0 & \alpha \\ 0 & 1 & 0 \\ 0 & 0 & 1 \end{pmatrix}$$

because $f^n = 0$ for $n \geq 2$.

(ii)

$$e^{\alpha f} = 1 + \alpha f + ½\alpha^2 f = \begin{pmatrix} 1 & \alpha & ½\alpha^2 \\ 0 & 1 & \alpha \\ 0 & 0 & 1 \end{pmatrix}$$

because $f^n = 0$ for $n \geq 3$.

(iii) $e^{\alpha f} = \cosh \alpha + f \sinh \alpha$, because $f^2 = 1$.

(iv) $e^{\alpha f} = (1-f) + e^{\alpha} f$, because $f^2 = f$.

(v) $e^{\alpha f} = 1 + t^2 + t^4 \cos \alpha + t \sin \alpha$, because $f^3 = -f$.

(vi) $e^{\alpha f} = h(\alpha) f^3 + h'(\alpha) f^2 + h''(\alpha) f + h'''(\alpha)$ where h, h', etc. are the hyperbolic functions of order 4 (see Exercises 10.4 and 14.8).

(vii) *Hint*: See Exercise 10.22.

(viii) See Satake, pp. 174 and 349.

(ix) See Satake, pp. 43 and 331.

(x) *Hint*: Note that $f = 1 + t$, $t^3 = -t$ (see Exercise 10.14).

(xi) $f^2 = -1$; hence

$$e^{\alpha f} = \cos \alpha + f \sin \alpha = \begin{pmatrix} \cos \alpha + \sin \alpha & \sin \alpha \\ -2 \sin \alpha & \cos \alpha - \sin \alpha \end{pmatrix}$$

(xii) $f^2 = 1$, hence

$$e^{\alpha f} = \cosh \alpha + f \sinh \alpha = \begin{pmatrix} \cosh \alpha + 3 \sinh \alpha & -4 \sinh \alpha \\ 2 \sinh \alpha & \cosh \alpha - 3 \sinh \alpha \end{pmatrix}$$

2. *Hint*: Consider linear operators f by type.
3. *Hint*: Consider linear operators f by type.
4. This follows from the fact that the conjugates of the sum and product are the sum and product of the conjugates.
5. By Exercise 14.4, $e^{\alpha f^g} = (e^{\alpha f})^g$; hence the same transition operator g is used to calculate the matrix for $e^{\alpha f}$ in the new basis as for f. To show that $e^{\alpha f}$ is in fact a linear operator, one has to prove also that each entry of the matrix of $e^{\alpha f}$ is a convergent power series (with respect to any basis). This can be done as, e.g., in Satake, p. 41.
6. $e^{\alpha f} = 1 + \alpha f + ½\alpha^2 f^2$ by Exercise 14.1 (ii). Hence the equation of the streamline passing through the point $\mathbf{v}_0$ is

$$\mathbf{v}_\alpha = \mathbf{v}_0 + \alpha f(\mathbf{v}_0) + ½\alpha^2 f^2(\mathbf{v}_0)$$

Since $\operatorname{im} f$ is a horizontal plane, the streamlines must all lie in horizontal planes.

7. Since f is a twist, the streamlines of f form a family of concentric ellipses. In more detail,

$$e^{\alpha f} = \cos\alpha + f\sin\alpha = \begin{pmatrix} \cos\alpha + 5\sin\alpha & 13\sin\alpha \\ -2\sin\alpha & \cos\alpha - 5\sin\alpha \end{pmatrix}$$

$$\left.\begin{aligned} x &= (\cos\alpha + 5\sin\alpha)x_0 + (13\sin\alpha)y_0 \\ y &= (-2\sin\alpha)x_0 + (\cos\alpha - 5\sin\alpha)y_0 \end{aligned}\right\}$$

$$\Rightarrow \begin{cases} x^2 = (\cos^2\alpha + 10\sin\alpha\cos\alpha + 25\sin^2\alpha)x_0^2 + (26\sin\alpha\cos\alpha + 13\sin^2\alpha) \\ \times x_0y_0 + 169\sin^2\alpha y_0^2 \\ xy = \cdots \\ y^2 = \cdots \end{cases}$$

$$\Rightarrow 2x^2 + 10xy + 13y^2 = 2x_0^2 + 10x_0y_0 + 13y_0^2 = \text{const}$$

$$\Rightarrow \tfrac{1}{2}(4x^2 + 20xy + 26y^2) = \text{const} \Rightarrow (2x + 5y)^2 + y^2 = \text{const}.$$

8. $h(x) = \tfrac{1}{2}(\sin x + \sinh x)$

9. $$h(x) = \frac{x^{n-1}}{(n-1)!} + \frac{x^{2n-1}}{(2n-1)!} + \frac{x^{3n-1}}{(3n-1)!} + \cdots$$

$$h'(x) = \frac{x^{n-2}}{(n-2)!} + \frac{x^{2n-2}}{(2n-2)!} + \frac{x^{3n-2}}{(3n-2)!} + \cdots$$

$$\vdots$$

$$h^{(n-1)}(x) = 1 + \frac{x^n}{n!} + \frac{x^{2n}}{(2n)!} + \frac{x^{3n}}{(3n)!} + \cdots$$

10. $$f = \begin{pmatrix} 0 & 0 & 0 & 0 & \cdots & -1 \\ 1 & 0 & 0 & 0 & \cdots & 0 \\ 0 & 1 & 0 & 0 & \cdots & 0 \\ 0 & 0 & 1 & 0 & \cdots & 0 \\ \cdots & \cdots & \cdots & \cdots & \cdots & \cdots \end{pmatrix}$$

11. $(e^{\alpha f})^{-1} = (e^{\alpha f})^* \Leftrightarrow f^* = -f \Leftrightarrow n = 2$

12. $$e^{\alpha f} = \begin{pmatrix} s''(\alpha) & -s(\alpha) & -s'(\alpha) \\ s'(\alpha) & s''(\alpha) & -s(\alpha) \\ s(\alpha) & s'(\alpha) & s''(\alpha) \end{pmatrix}$$

The addition formulas:

$$s(\alpha + \beta) = s(\alpha)s''(\beta) + s'(\alpha)s'(\beta) + s''(\alpha)s(\beta)$$
$$s'(\alpha + \beta) = s'(\alpha)s''(\beta) + s''(\alpha)s'(\beta) - s(\alpha)s(\beta)$$
$$s''(\alpha + \beta) = s''(\alpha)s''(\beta) - s(\alpha)s'(\beta) - s'(\alpha)s(\beta)$$

In order to express s, s', s'' in terms of the elementary functions, use the third roots of unity.

13. $s(x) = \sin(\tfrac{1}{2}\sqrt{2}x)\sinh(\tfrac{1}{2}\sqrt{2}x)$
14. $e^{\beta f} \circ e^{\beta g} \circ e^{-\beta f} \circ e^{-\beta g}$

$$= \left(1 + \frac{\beta f}{1!} + \frac{\beta^2 f^2}{2!} + \dots\right) \circ \left(1 + \frac{\beta g}{1!} + \frac{\beta^2 g^2}{2!} + \dots\right) \circ \left(1 - \frac{\beta f}{1!} + \frac{\beta^2 f^2}{2!} - \dots\right)$$

$$\circ \left(1 - \frac{\beta g}{1!} + \frac{\beta^2 g^2}{2!} - \dots\right) = 1 + \beta^2[f,g] + \beta^3(\cdots) + \cdots$$

which is asymptotically equal to $e^{\beta^2[f,g]}$.

15. $[\mathbf{v},\mathbf{e}_2,\mathbf{e}_3, \dots ,\mathbf{e}_n] = 0$, $[\mathbf{e}_1,\mathbf{v},\mathbf{e}_3, \dots ,\mathbf{e}_n] = 0, \dots ,[\mathbf{e}_1,\mathbf{e}_2, \dots ,\mathbf{v}] = 0$ are the equations of n hyperplanes. Take the normal vectors $\mathbf{m}_1$, $\mathbf{m}_2$, $\dots$, $\mathbf{m}_n$ of these hyperplanes and show that

$$\mathbf{m}_\nu \cdot \mathbf{e}_\mu = \begin{cases} 0 \Leftrightarrow \nu \neq \mu \\ 1 \Leftrightarrow \nu = \mu \end{cases}$$

17. $[f,g] = 1$ is impossible because trace$[f,g] = 0$ and trace $1 = n$.
18. It is clear that every real number α is a linear transformation of the one-dimensional vector space; conversely,

$$f \in L(\mathbb{R}, \mathbb{R}) \Rightarrow f(x) = f(x1) = xf(1) = \alpha x \qquad \text{for } f(1) = \alpha \in \mathbb{R}$$
$$\Rightarrow f \in \mathbb{R}.$$

19. *Hint*: Show that the restriction of the isomorphism in Exercise 14.18 to $GL^+(\mathbb{R})$ satisfies the requirements.
20. Set $(\exp f)(e^{\alpha x}) = e^{\alpha f(x)}$ for all $x \in A$ and use the exponential property to show that $\exp f$ is a group homomorphism. On the other hand,

$$(\exp g \circ f)(e^{\alpha x}) = e^{\alpha (g \circ f)(x)} = e^{\alpha g(f(x))}$$
$$= (\exp g)(e^{\alpha f(x)}) = ((\exp g) \circ (\exp f))(e^{\alpha x})$$

implies that $\exp(g \circ f) = (\exp g) \circ (\exp f)$.

21. *Hint*: Use the det-trace formula.
22. *Hint*: Show that the Cayley transform of the conjugate is equal to the conjugate of the Cayley transform.
23. See Dieudonné, p. 101.
24. See Dieudonné, p. 62.
25. See Dieudonné, p. 124.
26. $(g \circ f)^* \circ s \circ (g \circ f) = f^* \circ (g^* \circ s \circ g) \circ f = f^* \circ s \circ f = s$
27. $[f,g]^* \circ s + s \circ [f,g] = (f \circ g - g \circ f)^* \circ s + s \circ (f \circ g - g \circ f)$

$$= g^* \circ f^* \circ s - f^* \circ g^* \circ s + s \circ f \circ g - s \circ g \circ f$$

$$= g^* \circ f^* \circ s - f^* \circ g^* \circ s - f^* \circ s \circ g + g^* \circ s \circ f$$
$$= g^* \circ (f^* \circ s + s \circ f) - f^* \circ (g^* \circ s + s \circ g) = 0$$

28. Let a be an operator satisfying $a^* \circ s + s \circ a = 0$ and not having -1 as an eigenvalue. Then if we put $r = (1-a) \circ (1+a)^{-1}$, we have

$$\begin{aligned} r^* \circ s \circ r &= (1+a^*)^{-1} \circ (1-a^*) \circ s \circ (1-a) \circ (1+a)^{-1} \\ &= (1+a^*)^{-1} \circ (s - a^* \circ s) \circ (1-a) \circ (1+a)^{-1} \\ &= (1+a^*)^{-1} \circ (s + s \circ a) \circ (1+a)^{-1} \circ (1-a) \\ &= (1+a^*)^{-1} \circ s \circ (1-a) = (1+a^*)^{-1} \circ (s - s \circ a) \\ &= (1+a^*)^{-1} \circ (s + a^* \circ s) = s \end{aligned}$$

Since $1 + r = 2(1+a)^{-1}$, we have $\det(1+r) \neq 0$, so r does not have -1 as an eigenvalue. From $r \circ (1+a) = r + r \circ a = 1 - a$ we obtain $a = (1-r) \circ (1+r)^{-1}$.

The converse can be proved in a similar fashion.

29. $[f,g] = 0 \Rightarrow f \circ g = g \circ f$

$$\begin{aligned} \Rightarrow e^{\alpha f} \circ e^{\alpha g} &= (1 + \frac{\alpha f}{1!} + \frac{\alpha^2 f^2}{2!} + \cdots) \circ (1 + \frac{\alpha g}{1!} + \frac{\alpha^2 g^2}{2!} + \cdots) \\ &= 1 + \frac{\alpha(f+g)}{1!} + \frac{\alpha^2 (f+g)^2}{2!} + \cdots \end{aligned}$$

which is asymptotically equal to $e^{\alpha(f+g)}$.

30. See Satake, pp. 174 and 349.

31. See Dieudonné, p. 135, and extend by conjugation.

32. *Hint*: Prove the result for the group of orientation-preserving orthogonal operators first (see Dieudonné, p. 135) and extend by conjugation.

33. See Dieudonné, p. 124.

References and Selected Readings[†]

E. Artin, *Geometric Algebra*. New York: Wiley-Interscience, 1975.

T. S. Blyth, *Module Theory—An Approach to Linear Algebra*. New York: Oxford University Press, 1977.

T. S. Blyth, *Set Theory and Abstract Algebra*. New York: Longman, 1975.

L. Chambadal and J. L. Ovaert, *Algèbre linéaire et algèbre tensorielle*. Paris: Dunod, 1968.

Jean Dieudonné, *Linear Algebra and Geometry*. Boston: Houghton Mifflin, 1968.

D. K. Faddeev and I. S. Sominski, *Problems in Higher Algebra*. San Francisco: W. H. Freeman, 1965.

Roger Godement, *Algebra*. Boston: Houghton Mifflin, 1968.

Werner Greub, *Linear Algebra* (4th Edition). New York: Springer-Verlag, 1975.

N. H. Kuiper, *Linear Algebra and Geometry*. Amsterdam: North-Holland, 1963.

H. K. Nickerson, D. C. Spencer, and N. E. Steenrod, *Advanced Calculus*. Princeton, N. J.: D. Van Nostrand, 1959.

C. Pinter, *Set Theory*. Reading, Mass.: Addison-Wesley, 1971.

I. Satake, *Linear Algebra*. New York: Marcel Dekker, 1975.

N. Steenrod, "The Geometric Content of Freshman and Sophomore Mathematics Courses," in the *Report of the Committee on Undergraduate Program in Mathematics*, No. 17. Washington, D.C.: Mathematical Association of America, September 1967

[†] *Note* Another aspect of real linear algebra, real vector spaces with a Lorentz metric, is treated in detail in a forthcoming textbook, A. E. Fekete, *Gateway Geometry*, to be published in 1986 by Marcel Dekker, Inc. A Lorentz metric is a non-positive-definite metric giving rise to *timelike vectors* ($\mathbf{v} \cdot \mathbf{v} > 0$), *spacelike vectors* ($\mathbf{v} \cdot \mathbf{v} < 0$) and *lightlike vectors* ($\mathbf{v} \cdot \mathbf{v} = 0$) such that every plane has at least two linearly independent spacelike vectors. Real vector spaces endowed with a Lorentz metric are of great importance because of the role they play in the theory of relativity, and in noneuclidean (hyperbolic) geometry.

Index